T0302232

Experimental Statistics and Data Analysis for Mechanical and Aerospace Engineers

Advances in Applied Mathematics
Series Editors:
Daniel Zwillinger, H. T. Banks

Advanced Engineering Mathematics with MATLAB, 4th Edition
Dean G. Duffy

Quadratic Programming with Computer Programs
Michael J. Best

Introduction to Radar Analysis
Bassem R. Mahafza

CRC Standard Mathematical Tables and Formulas, 33rd Edition
Edited by Daniel Zwillinger

The Second-Order Adjoint Sensitivity Analysis Methodology
Dan Gabriel Cacuci

Operations Research
A Practical Introduction, 2nd Edition
Michael Carter, Camille C. Price, Ghaith Rabadi

Handbook of Mellin Transforms
Yu. A. Brychkov, O. I. Marichev, N. V. Savischenko

Advanced Mathematical Modeling with Technology
William P. Fox, Robert E. Burks

Introduction to Quantum Control and Dynamics
Domenico D'Alessandro

Handbook of Radar Signal Analysis
Bassem R. Mahafza, Scott C. Winton, Atef Z. Elsherbeni

Separation of Variables and Exact Solutions to Nonlinear PDEs
Andrei D. Polyanin, Alexei I. Zhurov

Boundary Value Problems on Time Scales, Volume I
Svetlin Georgiev, Khaled Zennir

Boundary Value Problems on Time Scales, Volume II
Svetlin Georgiev, Khaled Zennir

Observability and Mathematics
Fluid Mechanics, Solutions of Navier-Stokes Equations, and Modeling
Boris Khots

Handbook of Differential Equations, 4th Edition
Daniel Zwillinger, Vladimir Dobrushkin

Experimental Statistics and Data Analysis for Mechanical and Aerospace Engineers
James Middleton

https://www.routledge.com/Advances-in-Applied-Mathematics/book-series/CRCADVAPPMTH?pd=published,forthcoming&pg=1&pp=12&so=pub&view=list

Experimental Statistics and Data Analysis for Mechanical and Aerospace Engineers

Dr James A. Middleton

CRC Press is an imprint of the
Taylor & Francis Group, an **informa** business
A CHAPMAN & HALL BOOK

First edition published 2022
by CRC Press
6000 Broken Sound Parkway NW, Suite 300, Boca Raton, FL 33487-2742

and by CRC Press
2 Park Square, Milton Park, Abingdon, Oxon, OX14 4RN

CRC Press is an imprint of Taylor & Francis Group, LLC

Library of Congress Cataloging-in-Publication Data

ISBN: 978-0-367-55596-2 (hbk)
ISBN: 978-1-032-06636-3 (pbk)
ISBN: 978-1-003-09422-7 (ebk)

DOI: 10.1201/9781003094227

To my students who taught me
To my teachers who inspired me
And to my family that sustains me

Contents

Preface

Data have always been important for human beings to make informed decisions. Throughout history, people have counted livestock, tallied earnings, compared materials, and developed *models*—ways of thinking about data—that helped them explain how things work, how to improve their lot in life, and how to plan for the future. Until just recently, however, these models were primarily informal, held in the minds of the analyst, or local, shared with a few elites in first, ecumenical and later, scientific circles.

Today, however, as information explodes all around us, human beings need a more democratized access to data and the knowledge that reflecting on data can bring to the business of improving our lot in life. Engineers and scientists are no different. What was once held as knowledge by the few (i.e., statisticians), has now taken on everyday necessity in our fields. To understand new materials, processes, and systems, each of us needs to be able to collect, analyze, and communicate models made from data. But just as importantly, we need to be able to "analyze the analyses" i.e., *critique* the use of data in the world for its veracity, utility, and ethics so that the decisions we make in our careers do good and harm none.

This volume was written as a response to the needs of my students—mechanical and aerospace engineers, primarily, with a few mathematics and physics students thrown in the mixture—who expressed frustration at the density of most statistics texts, and their lack of attention to how the statistics is actually used in "the real world." While they could understand the mathematics well enough, the philosophy of probability and how stochastic events could be modeled and their long-term behavior used to test hypotheses, proved to be confusing. Also, after talking with industry representatives about the data-related skills their workers (my students in their near future) would need, I decided to pare down the mass of statistical concepts, tests, and jargon into those fundamental concepts and skills that could be used to build the rest, and not try to make the book a mile wide and an inch deep in terms of coverage. But I also decided, upon consultation with industry, to take the book further than most introductory texts. They expressed a need for students to understand concepts of factorial design and optimization, as well as computational methods, specifically bootstrap, and data reduction. For that reason, a major portion of this book is dedicated to the General Linear Model as an underlying model that defines nearly all classical statistics students will need to perform at a high level in industry or graduate study with the relatively simple idea of regression as its base concept. But I have also extended the content to include two advanced topics: Bootstrap methods and Principal Components Analysis to enable instructors to tailor the course, using these non-standard topics to the needs of their own industry partners.

I use real data, taken from published articles and reports, extensively throughout the chapters. These data sets are included for students to both practice important skills, but also to extend their understanding, as real data has problems: such as missingness, outliers, skew, and collinearity—the bane of the data analyst. Without real data, one cannot learn how to deal with these problems appropriately.

I hope the reader will forgive some aspects of my style. I tend to be fairly informal, trying to carry on a conversation with the reader. For this reason, there are (bad) jokes included, occasional cartoons and other memes to liven up the topic, and my own ways of thinking about the data as I attempt to solve problems in a manner that reflects the often conflicting ways of thinking I encounter as I try to build the best model of the data I can, given the limitations of the data set and its collection. It is this building of a mindset that is the most important contribution a book on statistics and data analysis can provide. I often say to my students, "Remember Middleton's First Law: There are no rules in statistics, only rules of thumb." I want them to remember that they are the ones who are making sense of the data, and that a thorough analysis requires making decisions about what assumptions hold and what do not. Ultimately, the best answer is not the one that meets all criteria for validity, but one that is good enough to improve the product or the system that the analyst is trying to improve. It is the job of the engineer to make the world a better place, and this "good enoughness" is the means by which statistics can help us do just that.

Many problems include datasets that are just too voluminous to place in the text. Readers using this book have access to course support materials including lecture notes and assessments, data files for use in problems and examples, and Matlab code. These can be accessed at the following book website:

https://docjimbo.wixsite.com/appliedstatistics

List of Figures

List of Tables

Symbols

α	Type I error rate: The probability that a hypothesis will return a false positive decision
β_k	Coefficient (partial slope) in the general linear model
$1-\beta$	Type II error rate: The probability that a test will return a false negative decision
$E(x)$	The expected value of a random variable
μ	The mean of a population
$\overline{x}$	The mean of a sample
$VAR(x)$	The variance of a random variable
σ^2	The variance of a population
s^2	The variance of a sample
σ	The standard deviation of a population
s	The standard deviation of a sample
$cov(X)$	The covariance of a matrix of predictor variables
$\sum$	The sum of a set of values
R^2	The coefficient of determination: A measure of the goodness of fit of a model
$SE_{\overline{x}}$	A non-standard but easy to remember notation for standard error of the sample mean
$\sigma_{\overline{x}}$	The standard notation for standard error of the sample mean
$p(x)$	The probability that a random variable takes on the value x
$p(A \mid B)$	Conditional probability: The probability that event A will occur given event B also occurs
PDF	Probability density function
CDF	Cumulative distribution function
$N\left(\mu, \sigma^2\right)$	The normal distribution function with parameters μ and σ^2
$F_{(df_1, df_2)}$	Value of the F-distribution at degrees of freedom (df_1, df_2)
χ^2_{df}	Value of the χ^2 distribution at *df* degrees of freedom
t_{df}	Value of the t distribution at *df* degrees of freedom
SS	Sum of squares: The sum of the squared deviations of a measure from its expected value
SS_{Total}	Total sum of squares: The total variation expressed by the dependent variable in a regression model
$SS_{Regression}$	Sum of squares regression: variation explained by a regression model
SS_{ε}	Sum of squares error (residual): residual error in a regression model

Part I

1

Introduction

The role of an engineer is to make the world a better place. We do this through the creation of products and systems that, by virtue of their design, are adopted by clients and consumers. The products and systems we envision are intended to improve on the status quo—making work more productive or less tedious, reducing waste and increasing efficiency, or providing end users a better quality of life. If you look around you, no matter where you are, everything you see (and much of what you don't see) is impacted by our work. Much of this impact is good: I am writing this text on my new laptop, comfortable in my air-conditioned home in Arizona. Much, however, stemming from the Law of Unintended Consequences, is bad: Disposal of my old laptop in a landfill will leach toxic materials into the environment and remove valuable rare-earth metals from the supply chain. My air-conditioner is using approximately 45,000 Watt-hours per day, derived from both hydroelectric (renewable, but with serious environmental impact), and fossil fuel (non-renewable), sources.

Understanding the good our products *may* do, while mitigating the bad they *will* do requires more than closed-form calculations. Not every laptop lasts as long as my old one did (6 years!). And, not every air conditioner is as efficient as the one that is currently keeping me cool. Even for the same make and model, there will be variation in use and performance that will impact the overall good, or bad, for any product I may envision. To make sense of this variation, we have to collect, manage, and analyze data, hoping against hope that the data will have enough consistency so that we can make accurate and reliable predictions about the long-term behavior of the products we design.

1.1 Approach of This Book

This volume is about using data as evidence to help engineers make decisions.

Most statistics courses, especially those taught by mathematicians, begin with the theory of probability. This makes sense in that, if we understand the theoretical set that describes all possible behaviors of our product, we can determine the chance that the data we just obtained is due to the impact of the new product design, or if it just due to random variation in environmental conditions.

DOI: 10.1201/9781003094227-1

Our understanding of probability first stems from statistics' roots in the gambling habits of mathematicians. Gerolamo Cardano, a compulsive gambler and mathematician, invented the definition of probability back in 1564 to help him predict the outcomes of his bets. He once bet all his wife's possessions to get stakes at a table. Perhaps to get some of his wife's treasures back, he also wrote the first probabilistic account of effective methods for cheating.

Not surprisingly, his approach used dice as the model for random behavior, and we still use dice, cards, and roulette as canonical examples to help students understand how, though any event may be random, over the long-haul, the fraction of times that event occurs relative to all other events, can be predicted. As a result, Cardano's definition—**If the total number of possible outcomes *n* of an event are all equally likely, and *m* of those *n* actually occur in the event, then the probability of that event occurring is *m/n***—is still the primary definition we use today. We will define probability more rigorously in Chapter 4.

Later, in the mid 17th century, Blaise Pascal formalized the theory of probability because he was interested in understanding the theory behind the gambling habit of his friend, Anton Gombaud, chevalier de Mere. If he could just get the theory down, he would know what the chances would be that a particular hand of cards would turn out a winner, and how much de Mere could be expected to lose in an evening. About 60 years later, Jakob Bernoulli, born into the most prolific family of mathematicians ever, took Pascal's (and Fermat's) formalization and made it theoretical when he proved the **Law of Large Numbers**. To this day, the field of statistics is grounded in the achievements of these reprobate gamblers. But the lesson is this: Statistics was founded upon practical problems that could only be solved modeling real data with mathematical functions.

Cardano's Definition of Probability: If the total number of possible outcomes *n* of an event are all equally likely, and *m* of those *n* occur in the event, then the probability of that event occurring is *m/n*

1.1.1 Data Modeling

This book will take the perspective that engineering statistics is primarily a form of *data modeling*. In other words, we begin with data, we end with models that fit the data (more or less). We make sense of these models using statistical reasoning, including probabilistic reasoning, among other analytic tools available to the engineer such as the calculus, linear algebra, our understanding of physics and economics, and a healthy dose of common sense.

Expect to begin with problems that require an understanding of data, thought experiments and discussions about how the data might best be gathered, as well as considerations of cost and the consequences of errors. These problems and discussions will then require you to gather data, or examine data from real examples where these issues are paramount to the job of an engineer. From the data, and our attempts to solve the problem, we will develop the basic statistical models that are used every day in industry, academia, and research. Probability theory will be used to help us make decisions regarding the reliability of our models, and to predict the future chances of our products and services being successful (or not!). To make this clear and explicit, the following four design principles will be followed in this course:

- **Statistical concepts are learned through use.** Statistics are used to make sense of data. Rather than simply calculating statistics with predetermined data, students will generate their own data through experiments and simulations that model some real or realistic engineering problem. Engaging in analysis of real data makes students more knowledgeable about the uses and limitations of quantitative models, and provides key examples by which they can remember statistical concepts, skills and procedures when they go to industry or graduate education.

- **Probability is a tool for statistical reasoning**. The hardest part of statistics is probabilistic reasoning. The mathematics we will be using is fairly basic: Some essential ideas of calculus and linear algebra; a few matrix operations. The rest is just arithmetic. But the *meanings* behind the data, what they tell us, can be quite complicated to untangle. For this reason, probability models will be developed from data analysis—giving a context for which the degree of uncertainty in a measurement, and in a group of measurements, makes sense.

- **Problems in this book represent a class of problems**. The contexts chosen, and the methods used for making sense of and solving problems in this course are not one-offs. They have been chosen to represent the kinds of problems you will encounter as a professional engineer. While no textbook can cover all the potential issues and dilemmas one might face, we *can* choose problems that utilize common reasoning, common images and representations, and common statistical tools in their solution. These features can be generalized to the broad range of applications you will encounter.

- **We collect data to help us make decisions**. The only reason to go to the expense and trouble of running an experiment rests on the belief that the results of the experiment will help you *make a decision* regarding some course of action your project may take. You may be choosing a material with which to manufacture an automotive part, or you may be comparing the efficiency of two different turbine designs. You may be assessing customer satisfaction, or modeling the degree to which a prosthetic causes unintended muscle fatigue. Whether part, or design, or satisfaction, or fatigue, the data collected is used to *improve the status quo*: making the lives of customers and end-users better. The problems, thought experiments and projects assigned in this course will always have this principle in mind. They are not "exercises." We are not building a mental muscle, the goal is to develop an empirical mindset so that the role of data is put in its proper place in the engineering design process.

1.1.2 Building an Empirical Mindset

Engineering is an empirical profession. The word *empirical* stems from the Greek *εμπειρια*, "emperia," meaning *experience.* What we mean by this is that understanding comes from directly experiencing phenomena. Arcane theories of philosophy aside, the idea that we need to directly observe, and measure things to understand them is a foundation of scientific thinking, and experimentation in particular.

The foundational activity of the experimenter is *observation.*We observe that some materials have certain properties and others do not. We observe that a certain geometrical configuration of steel members yields a truss that is resistant to torsion, while another configuration yields one resistant to compression or tension. We use these observations to make decisions regarding what materials to use when designing a space frame or a chassis, or an airfoil. We use these observations to create products that, by virtue of their design, invoke consumers to use them to solve their everyday problems. But human beings (read "you and me") have limited capacity to discern differences that may be of use in our engineering activity. Aluminum and steel, for example, have similar hardness if we just eyeball them or scratch them with our fingernails. Even for similarly hard alloys, we know that their toughness and resistance to work hardening is different. To observe these differences, we have to invent devices that can sense differing degrees of hardness, let's say, order them by degree, and hopefully, create a scale of measure that allows us to compare measurements arithmetically through subtraction (e.g., a *difference*) or division (e.g., a *proportion*).

Adopting an empirical mindset, the engineer believes that the only way to improve on our current condition is to *be observant*—to collect data either directly or indirectly—to discover and come to understand problems and phenomena in the world. Measurement is the cornerstone of empiricism. Being able to discern how things and events differ, to order those things and events on a scale of measure, and to be able to say that some things at one end of the scale are better than things at the other end is the essence of data collection

Having data available, we must *analyze it*, interrogate it if you will, so that we can build models that will *predict the future behavior* of the problem we are studying. If we were bench scientists, this might be where our investigation ends.—Off to some new problem. But we are engineers. We are not content with understanding a problem. We want to fix it, to *improve* on the status quo. With data collected and analyzed, we have to use the sense we have made to *make decisions* regarding how to optimally utilize what we have learned to create products (tools, vehicles, machines and processes) that, if adopted, would ameliorate the problem, or even eliminate it *as* a problem.

1.2 The Role of Data

But data is not everything, or even the most important thing, in the engineering design process. Creativity, theoretical understanding, knowledge and skill in manufacturing, and ethical decision-making all have their critical roles to play. Data is a means of amplifying these processes, and for connecting them with each other so that a resulting designed product is both innovative *and* practical, and so that it works reliably *and* efficiently. Data and its analysis have four primary roles in the design process, to:

- Help us **understand** phenomena;
- Help us **predict** values of dependent variables given values of independent variables;
- Help us **control** events and processes;
- Help us **make decisions** based on evidence (as opposed to dogma/ideology/tradition); and
- Help us **improve** quality, productivity, applicability, etc.

Understanding is the most important of all these roles. At the beginning of the design process, we are actively researching issues, problems, and resources that might reveal a need we could help fulfill. In discussing this phase with others, I often use the term, "Ass-from-a-hole-in-the-ground" to describe the level of understanding we might initially have when confronting a complex design problem. If we cannot discern what the real problem is—the behavior and interaction of variables that may impact the performance of our product—we have shaky grounds for building it in the first place, or improving it once built.

Prediction is the lynchpin that connects understanding to actionable design decisions. If we can develop a valid and precise model of a phenomenon, we can then use that model to describe the phenomenon under conditions we may not have assessed. It is not enough, for the engineer, to *know that* a material hardens for a time under repeated stress. I are interested in *predicting* the stress when the hardening process reaches a critical point, beyond which a part we are designing may fail.

With good predictive models, we can then design products that optimize the models and **control** the performance of our designed products. The coefficient of performance in a heat exchanger, C_p is a power function of the work used to extract the thermal energy from a closed box, ($C_p = \frac{Q_{cold}}{W} \pm .05$). One can use this knowledge to design optimally efficient refrigeration systems for available power, taking into account the fact that performance will vary as much as 5% higher or lower depending upon random factors in the environment, plus some natural variation in the actual

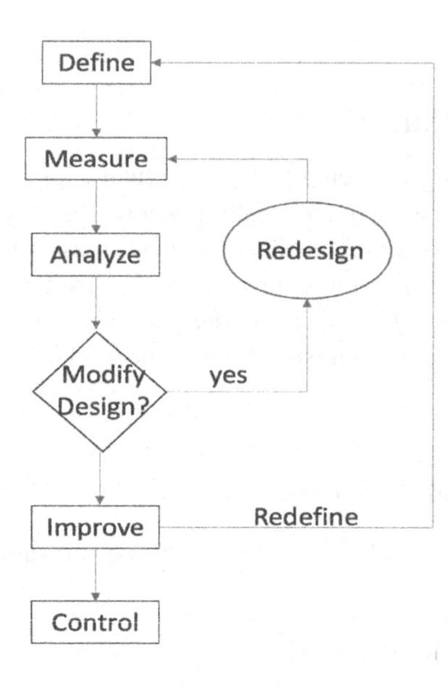

FIGURE 1.1: The DMAIC quality improvement process

manufacturing of the refrigeration units. Further, by varying W, the rate of energy supplied to the unit over time, the efficiency of the unit can be made to vary, thus regulating (*controlling*) the temperature of the unit.

Now that we can control our designs, we can then **make decisions** regarding what parts to use, what input and output parameters make the most sense for the conditions under which our design is expected to perform, how to reduce cost, and other key decisions that will impact the quality and marketability of our products.

Finally, *none* of the roles that data plays is worth a dime (and a dime isn't worth much), if it doesn't yield **improvement** on the current state of affairs in our world. Our heat exchanger is a waste of time and energy, literally, if it doesn't lead to improved efficiency or improved efficacy. In the context of this class, moving from understanding a phenomenon to the creation of predictive models, to process and product control models, and ultimately to key design decisions is only an exercise if it the improvement of the status quo is not kept constantly in mind as a litmus test for where to look for inspiration, what variables to account for in our data collection and analysis, and what decisions to make that balance improvement of performance of a product with improvement of the lives of those who use that product.

While there are many approaches to quality improvement in product design, testing, and implementation, one that provides distinct advantage to the engineer, especially one who is well-versed in data modeling and statistical methods, is the Six Sigma approach (https://goleansixsigma.com). Six Sigma (and other models of improvement such as Kaizen/Lean Manufacturing (Brunet & New, 2003)) utilize data throughout the design process. The basic method that all quality improvement programs use is called DMAIC (Define, Measure, Analyze, Improve, Control, see Figure 1.1.

Many of my students immediately recognize this model just as a variation on the "Engineering Design Process" they learned in their freshman design course. Importantly, everything between Define and Control utilizes data to build, refine, and assess the quality of a product. Even the Definition and Control stages utilize data in important ways: re-Definition is a process that, when a product is developed and refined, it creates new, unexpected problems (the law of unintended consequences) and opportunities (serendipity); Control utilizes data as feedback for the behavior of the designed system, making the system adaptive to new environmental constraints.

We will utilize the DMAIC process throughout this introductory book to situate examples of statistics in use, and to evaluate the ways in which you, the student, respond to thought experiments, to problems and homework assignments, and to exams. Any computer can crank out calculations given a set of data. The engineer must know which statistical methods (and their associated calculations) are appropriate for certain data, under certain circumstances, for certain purposes. Moreover, the engineer must take the outcomes of statistical analyses and use them to make decisions regarding how to best improve the quality of their work.

1.3 References

Ore, O. (1960). Pascal and the Invention of Probability Theory. *The American Mathematical Monthly, 67*(5), 409-419. doi:10.2307/2309286

1.4 Chapter 1 Study Problems

1. Choose an engineering problem that fits your personal interests. Examples might be design of solar desalination units; optimizing airfoil parameters for wind energy generation in a particular location; development of a sensor system for preventing wrong-way driving, or enhancing the performance of a micro-air vehicle, anything you desire to study and improve, really. List the key variables that you will need to measure for each of the four primary roles of data in the design process:

(a) Understand

Answers will vary. But to understand a phenomenon, the data need to be descriptive. Some of the data can even be qualitative even if numbers are involved. It is better to describe something carefully before adding precise measurements, than to measure something carefully without knowing what it is!

(b) Predict

To predict means that we have some kind of mental model of the phenomenon (i.e., that we understand its behavior enough to make some assumptions about future observations. Its scale, limits of its domain (minimum and maximum values), its relationship to other variables, and whether time plays a role in its growth, are all reasonable considerations to take into account when making a predictive model.

(c) Control

To control phenomena means that we have an accurate enough predictive model to account for the situations wherein the product or system we are designing will operate. Control systems require a model of the decision mechanism (generally a set of if-then contingencies) wherein values of the phenomenon are evaluated and appropriate behavioral adjustments to the system we are studying can be made. A good example involves programming a robot to navigate unfamiliar terrain. One must be able to take data from the environment, process the data and from the results make a decision about whether the region the robot is exploring is too hot, too cold, has sounds that resemble cries for help, etc.

(d) Make Decisions

Control involves decision making. There must be feedback from the analysis of data, including the performance of the product or system we are developing. This feedback then allows us to change the settings for the product or system, change the environment in which it operates, or both. Some decisions may be tied to issues of pragmatics, others to theoretical issues of interest.

(e) Improve Quality

Quality is the ultimate goal of design work. Does the system or product we are designing work in such a manner as to be effective, efficient, and reliable? Does it improve the condition of people, or the environment? Does it have any unintended consequences that detract from quality of life and sustainability? These questions are the Big Ideas that should drive our work, and the data we collect should help us address these questions at each stage of the research and development (i.e., *design*) process.

2. How would you go about measuring these variables?

Answers will vary depending on the students' personal interests.

3. What problems do you envision making accurate and precise measurements?

   Think carefully about how little error you need for your measurements to be useful. They don't have to be exact, more accurate measurements can become very costly. They also don't have to be completely consistent. More precise instruments and procedures are also very costly. What is the optimum accuracy and precision you need to be able to make decisions about the performance of the phenomenon of interest?

4. Are the measurements direct or indirect? Is it possible to directly measure the phenomena you are interested in?

   Answers will vary depending on the students' area(s) of interest.

5. Look up an article or website describes an exemplary engineered product. One example is the Leatherman multitool: https://www.leatherman.com/leatherman-difference.html (I bought an original Leatherman PST in the mid-1990s, and I still have it). Examine prototypes of the product. Go look up its patent: https://patents.google.com/patent/US4744272A/en. Put yourself in the mindset of the designer(s):

   (a) What kind of materials would you have chosen to build your product? How would you have decided which materials gave you the best performance?

   Material properties are critical parameters we use to design quality products. But they are only ballpark estimates. How will you take published specifications and determine the specific needs for your product?

   (b) Examine the parts or subsystems of your product. How would you go about testing each to insure that it performs its intended job:

   i. In terms of the performance you want it to accomplish; and

   ii. In comparison to other similar products?

   For the Leatherman, I would want to examine its ability to sharpen and stay sharp through many cycles of use. I would also want to measure the extremes of its function such as the torque required to break the blades, or to loosen the axles and bearings. I would also want to test the ergonomic qualities of the product, and the perceptions of users interacting with it under normal and more extreme conditions.

6. As a novice engineer, how do you envision data helping you as you matriculate through your program?

   (a) How will data factor into the design of your projects, including capstone?

   My capstone students must gather and interpret data at each stage of their product development, from initial conceptualization, to initial prototyping, then through testing and refinement, and ultimately to projecting the project forward to manufacturing in quantity.

   (b) How do you envision data will impact you when you become a professional engineer?

i. As a consumer of reports and standards that present models of data (e.g., materials specs, design requirements, etc).
Every year new materials are produced. New technologies are introduced. New standards are created. All of these are based on data. Understanding how these data were gathered and analyzed helps the engineer spec out the qualities of her/his own designs and model their range of performance prior to actually building and testing them. After one designs a product, extensive testing and refinement is all data driven!

ii. As a producer of reports that tell your boss how well the products you are designing work.
It is not enough to just analyze data and use it to understand, describe, predict, and control. One must also communicate data and its implication to others who may not be sophisticated in engineering science and practice. We must also communicate to our customers, who are not engineers. This is just as important as the analysis itself. Clear description, communication, and translation of fancy math into an argument that people will value and respond to positively makes it possible for good design to actually see the light of day in products that make it to market, are bought by consumers, and are valued by society.

7. Describe each role that data plays in the DMAIC process. How does data help you make critical decisions at each stage of product design and improvement?

- Prior data is most useful in the definition stage. Reports and publications that list testing standards, material properties, and the like help the engineer determine the most important variables to take into account.
- Measurement is critical in that there is no data without measurement. Accuracy and Precision of measurement of the key variables that impact the behavior of a designed product or system insures that error propagation does not become egregiously large.
- Analysis implies creating a model from the raw measurements. Data has no meaning without some kind of model that orders it in importance, delineates its extent and impact, and in the case of quantitative data, describes its shape and pattern. These analyses can then be used to make decisions about the behavior of the phenomenon of interest, and how to modify and improve its behavior.
- Modification uses data to mark out the values of independent variables that, when changed, predict values of a dependent variable, the performance of the product we envision. So the analytic models we create can be used as predictive models in subsequent tests and refinements of our product.
- This refinement is an iterative process in design. As we learn more about our product through testing, we can "tweak" different features to respond to what the data is telling us. In doing so, we successively improve its performance.

- Control is the point where we know enough about our product to produce it en masse, or to use it to impact the world around us to make it better. The culmination of the design process is using our products in a way that changes the lives of people and the environment for the better.

2

Dealing with Variation

Any product you want to engineer will show variation in performance from one instance to the next. For example, you may have noticed that, in a circuits lab, almost none of the resistors that they say are 100 Ohms will show a resistance of exactly 100 Ohms. Resistors, even those from the same batch, can vary as much as +/−10% from their designed value. This is primarily due to tiny differences in the length and/or cross-sectional area of the internal resistive material (usually a wire, wound in a helix around an insulating core). We know that all things being equal, a small change in the length of a wire L will result in a direct, proportional change in resistance R. Change in the cross-sectional area of the wire A is inversely proportional to the resistance as thicker wires show less resistance. These small differences in manufacturing yield batches of resistors that are very close, but not exactly equal, to the advertised value.

To help the engineer manage this potential variation, manufacturers adopt a color coding system.

The gold 4th band on this batch of 10k resistors tells us that any of these resistors has a tolerance of +/−5% (see Figure 2.1). Their actual, measured resistance, can vary from as low as 9.5k to 10.5k Ohms.

You can't escape this fact of nature: Things vary. Even products that you design to be exactly the same as each other will have slight variation in their dimensions, fit, and performance. Luckily, natural variation is not infinite as it exists within natural limits. Fewer than one in one thousand of all 10k +/−5% resistors have a value less than 9.5k and the upper limit is set such that fewer than 1 in one thousand of all resistors with this rating fall above 10.5k. The vast majority fall somewhere in the middle.

2.1 Measurement

Over the millennia, humans have developed means of ordering phenomena, and determining the extent to which one thing has more or less of a certain property than another thing. The great early civilizations like the Egyptians, Chinese, Sumerians, and Indus River civilization pioneered systems of **measurement**: ascertaining the extent, dimension, or quantity of some phenomenon. In essence, when we measure, we assign a numeric value to some property of a phenomenon, comparing it to some

DOI: 10.1201/9781003094227-2

FIGURE 2.1: A sample of 10k Ohm resistors

standard by which all such measurements can be compared. Examples of such standards include *extensive* quantities such as the meter, kilogram, second, Ampere, and Kelvin. Other examples include *intensive* quantities, measurements that combine two or more extensive quantities, multiplicatively, into rates like Torque (kg $\times$ m^2/s^2), or other rates (e.g., heat has units like kpm, kWh, kcal, ft $\times$ lb_{force}, Watt-second, etc.).

Essentially, the ancients discovered that, if we can determine some property of a phenomenon that varies among instances of that phenomenon, we can create a means of comparing instances using that property. This is usually accomplished by choosing an arbitrary standard and calling its extent "1 unit" and then using simple addition: appending 1 unit to another 1 unit to create a scale of measure 2 units in magnitude, and so forth. Our system for measuring length, for example, is merely a process of choosing an arbitrary unit for comparison (say 1 meter), adding multiples of this unit and some fractions of units until the object we want to measure is covered completely and as precisely as necessary.

> **Measurement** is assigning a numeric value to different instances of a phenomenon. Measurements can be *qualitative*, assigning numbers to denote different classes of the phenomenon, or they can be *quantitative*, assigning values by comparing the phenomenon to a common unit.

But some measurements, such as current, prove to be fairly tricky to measure directly. When I fix my electronics at home, the current (Amp) probes I use on my multimeter are imprecise, varying widely in their values for a given circuit. Luckily, thanks to Georg Ohm, I can use known relationships as a proxy for direct measurement. I can measure voltage fairly accurately, and I use the resistors I analyzed at the beginning of this Chapter that have a known resistance. Because $V = IR$, with these values measured well, I have a more accurate measure of current than I could get with a professionally designed current probe. $\frac{V}{R}$ is not current per se, but its values serve as a suitable proxy. We use these kinds of indirect measurements all the time in engineering contexts. For example, we use change in temperature, the average kinetic energy in a system, as a proxy for heat. One thing to keep in mind when using proxy measures is that the error inherent in the measurements, when combined with other measurements, tends to add up. It is good practice to use direct measurements, when their "accuracy and precision" is greater than one would experience with indirect measurements.

> **Indirect Measurement** uses known relationships to create a proxy value for the variable under study. The proxy may be used in place of **direct measurement** when the instruments for direct measurement show a high degree of imprecision.

Physical things vary across length (meters), mass (kilograms), dimension (1,2,3,4... and up!), current (Ampere), and time (second), among others, and combinations of these variables (e.g. 1 Joule of energy = $1\frac{kgm^2}{s^2}$; or 1 Coulomb of charge = $1A \cdot s$). The System Internationale (SI) standard units are generally used as the base upon which the measurement units of all other physical phenomena are built.

Some of the things we measure are not physical. Industrial Engineers, for example, often survey people to determine their opinions about a variety of subjects that might be critical to the development, manufacture, or marketing of products. Opinions are multifaceted, involving beliefs, perceptions, psychological states, and so on that cannot be measured directly. No matter how many questions we ask on a survey, the answers cannot fully explain someone's psychological state. So, any individual's results on a survey are prone to a high degree of inaccuracy.

> **A Survey** is a measurement procedure that attempts to assess a person's opinion at a given moment in time about a given subject, and to compare one person's opinion against those of others.

You may have run across the term "User Experience," or UX for short. Any gamer will recognize that the display of the game they are playing has been carefully designed to provide information about one's performance without being too distracting from the actual gameplay. Applied to engineering, UX is the subjective interaction between a consumer and the product(s) we design. UX design attempts to ascertain the subjective experience of the user as they interface with a product, to determine what features and applications of the product are meaningful and relevant to their needs, goals, and capabilities.

As an example of UX, I had a group of seniors in my mechanical engineering capstone class a few years ago, who designed a prosthetic foot. One of the key questions they had to ask potential users was how comfortable the prototype foot was, relative to prior prosthetics they had used. This involved interviewing each user, using a survey. They also had to assess the gait of the users to determine if the foot improved posture and biomechanics over their prior appliances. To do this, they used video recordings wherein the length of stride, and joint-angle could be assessed. Finally, they used force sensors to determine the amount of energy the foot returned to the user, to compare against biological feet (their muscles and bones). A good combination of physical and psychological measurements enabled them to refine their product so that it could be manufactured to fit many people with only slight after-market modifications.

2.1.1 Natural Variation

All of the examples we have examined in this chapter have addressed **natural variation**. This term means just what you would expect it to. If there are multiple instances of a phenomenon (e.g., multiple resistors, or multiple people needing prosthetics), the measure of each instance will show some difference from the measure of other instances. Resistors vary slightly around some target value. People who need a prosthetic foot will feel the prosthetic differently and so rate its comfort differently. It is pretty intuitive.

Natural Variation is the tendency for instances of a phenomenon to take on different values.

What is not so intuitive is that engineering statistical practice centers around trying to describe this natural variation: its extremes, its typical values, and its shape on the scale of measure we are using to describe our phenomenon of interest.

2.1.1.1 Shape of Data

Let's dive into an example to learn how to describe the natural variation in an engineered product. Chang and Lu (1995) provide measurements of the thickness of 5 steel sheets, each group of 5 taken from batches manufactured at 9:00 am, 11:00 am, 2:00 pm, and 5:00 pm. They did this for 6 consecutive days, resulting in measurements for 120 sheets (5 per batch x 4 times per day x 6 days). The measurements were taken from the right edge of each sheet, the middle of the sheet, and the left edge. These measurements are shown Figure 2.2:

Thought experiment. An engineer might ask any number of questions about this data, trying to figure out the quality of the product he is producing, or if he is a consumer, purchasing. Think through the following questions as a thought experiment. Don't get bogged down calculating anything, just try to make sense of the data in a global way.

1. How consistent are we manufacturing our product? From the data in Figure 2.1, we can see that the measurements DO vary. Without calculating anything, how much, typically, do the measurements seem to differ? We are just looking for an "eyeball" estimate.

	Time											
	9:00 AM			11:00 AM			2:00 PM			5:00 PM		
Date	Right	Middle	Left	Right	Middle	Left	Right	Middle	Left	Right	Middle	Left
6/1	4.4	4.5	4.5	4.4	4.7	4.8	4.7	4.8	4.7	4.3	4.4	4.2
	4.5	4.1	4.0	4.0	4.4	4.5	4.5	4.6	4.4	4.4	4.0	3.7
	4.2	4.2	4.8	4.3	4.4	4.2	4.5	4.4	4.6	4.2	4.0	4.3
	4.1	4.7	4.5	4.5	4.3	4.4	4.5	4.6	4.6	4.0	3.5	4.9
	4.9	5.0	4.1	4.8	4.9	4.0	4.8	4.7	4.9	3.6	4.3	4.3
6/2	4.5	4.4	4.6	3.9	4.0	4.1	4.3	4.4	4.3	4.4	4.4	4.5
	4.4	4.5	4.7	4.4	4.6	4.5	4.7	4.5	4.6	4.0	4.0	3.9
	4.5	4.5	4.6	4.0	3.9	4.0	4.6	4.8	4.7	4.1	4.2	4.3
	4.4	4.5	4.6	4.1	4.3	4.2	4.6	4.5	4.5	4.3	4.3	4.4
	4.6	4.5	4.6	4.0	4.1	4.2	4.7	4.8	4.9	4.2	4.2	4.3
6/3	3.7	4.7	4.6	4.3	4.2	4.3	4.2	4.4	4.3	3.9	3.9	3.9
	4.3	4.7	4.2	4.1	4.3	4.3	4.4	4.1	4.2	4.1	4.1	3.8
	4.1	4.7	4.3	4.4	4.4	4.5	4.7	4.6	4.5	4.0	4.0	4.2
	4.7	4.2	4.8	3.8	3.9	3.9	4.3	4.4	4.2	3.8	3.8	4.0
	4.6	4.1	4.8	4.3	4.1	4.2	4.6	4.6	4.5	3.7	3.7	4.1
6/4	4.2	4.3	4.6	4.3	4.6	4.7	4.1	4.2	4.3	3.8	3.8	4.0
	4.5	4.4	4.6	4.5	4.1	4.2	4.4	4.2	4.6	4.1	4.1	4.1
	4.7	4.8	4.7	4.5	4.6	4.5	4.6	4.5	4.6	4.2	4.2	4.1
	4.5	4.6	4.5	4.4	4.2	4.2	4.5	4.7	4.8	4.3	4.3	4.3
	4.6	4.5	4.4	4.2	4.2	4.1	4.3	4.5	4.3	4.2	4.2	4.3
6/5	4.7	4.8	4.8	4.1	4.2	4.2	4.3	4.4	4.2	4.1	4.1	4.1
	4.7	4.8	4.9	4.2	4.2	4.3	4.3	4.4	4.3	4.0	4.0	4.1
	4.4	4.8	4.4	4.3	4.3	4.3	4.3	4.4	4.4	4.0	4.0	4.2
	4.2	4.3	4.6	4.3	4.3	4.3	4.7	4.6	4.5	4.1	4.1	4.3
	4.7	4.5	4.7	4.4	4.4	4.4	5.0	5.0	4.0	4.3	4.3	4.3
6/6	4.3	4.3	4.4	4.0	4.0	4.0	4.2	4.5	4.4	4.0	4.0	4.1
	4.5	4.5	4.4	4.3	4.0	4.1	4.4	4.4	4.5	4.1	4.1	4.2
	4.6	4.6	4.6	4.3	4.2	4.2	4.5	4.5	4.6	4.1	4.1	4.1
	4.8	4.8	4.9	4.3	3.9	4.2	4.5	4.2	3.8	3.8	3.8	3.8
	5.0	5.0	4.9	4.2	4.2	4.2	4.7	4.7	4.3	3.9	3.9	3.9

FIGURE 2.2: Thickness of steel sheets, measured in three places, for different times across six days of manufacturing

2. If we tell the customer, a typical sheet has these dimensions __________, what would we say? What does "typical" mean to you as a producer? As a purchaser?

3. Are there differences in the characteristics of the sheets depending on what time of day they were manufactured? Why might this be the case?

4. Are there differences in the characteristics of the sheets depending on what position was measured on the sheet? Are sheets of uniform thickness?

5. The manufacturer intends the sheets to be 4.5 mm in thickness with variation of +/− .5 mm. How good did they do in meeting tolerances? Would you depend on them to give you a quality product?

   If you are like me, you can get a general feel for the data just eyeballing the table. It is always good practice, when one has a set of data, to do this eyeball estimate to get to know the data a bit, before jumping in and calculating, or making graphs. For example, we can see that the measurements vary from a low of 3.5mm to a high of 5.0mm. We can also see that there are relatively few very low measurements, and relatively few high measurements. Most seem to be clustered around the intended thickness of ~4.5mm. Examining the data set allows you to set the appropriate viewing window so you don't cut off extreme values, and it alerts you to potential data entry errors (if you saw 45.5mm, for example, you might think someone's fat fingers hit the 5 too many times).

2.2 Distribution

The other questions about quality and tolerance are harder to answer with just an eyeball estimate. To reduce the complexity of 360 pieces of data, it is helpful to make some kind of summary. Expanding on the idea that only a few extreme values occur, while there are typically a bunch of in-between values, we can draw a picture of the data to get a representation of its shape. You have probably drawn bar graphs, boxplots[1], and histograms for data sets back in your middle and high school experiences. But unlike those earlier experiences we are now going to use these graphs as ways to visualize functions that model the long-haul behavior of the variables we measure. To do this requires a little bit of mathematics:

A **Data Distribution** is a function $f(x)$, that assigns a *frequency* for every value of the measured variable.

Frequency is the number of times a value of the measured variable occurs in a data set.

In most textbooks you may have encountered, the word *Distribution* has been defined as a listing of all the values in a data set. This is only partially true. If we look closely at the data for the thickness of steel plates, we can see that some of the steel plates have the same thickness as others, measured in the same places. For example, there is only one measurement at 3.5mm (we say the *frequency* of the value 3.5 mm is 1, or $f(3.5) = 1$), but the frequency of a measurement of 4.5 mm is 45. All I did was counting how many of each value is represented in the data set.

Take a look at Table 2.1, which lists the frequency for each of the 15 values of thickness we measured.

[1]It is assumed that you know how to draw bar graphs and box-and-whisker plots. If this content is fuzzy to you, go to https://www.khanacademy.org/math/statistics-probability/summarizing-quantitative-data/box-whisker-plots/a/box-plot-review

Thickness	Frequency
3.5	1
3.6	1
3.7	3
3.8	9
3.9	12
4.0	28
4.1	33
4.2	48
4.3	53
4.4	41
4.5	45
4.6	31
4.7	25
4.8	16
4.9	8
5.0	7

Table 2.1. Frequency distribution showing $f\,(Thickness)$

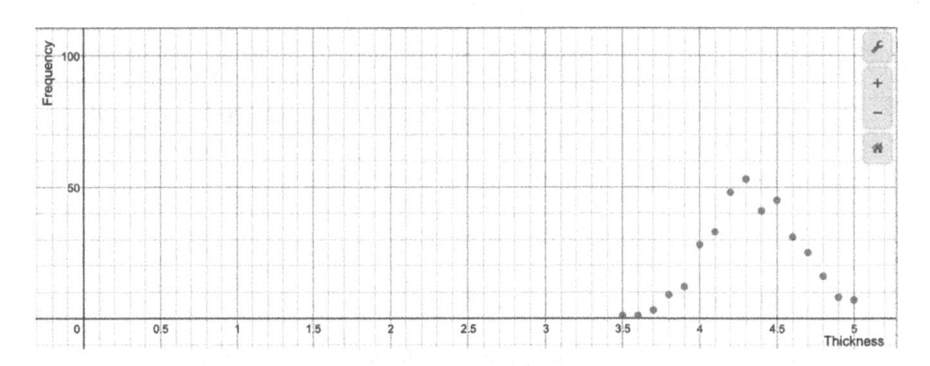

FIGURE 2.3: Frequency plot of steel thickness data

From this table, we can see that there is a one-to-one correspondence pairing each frequency in the right column, to a value of thickness in the left column. This function *f(T)* is plotted in the graph in Figure 2.3:

1. What are the characteristics of *f(T)*?
   (a) Domain
   (b) Range
   (c) Maximum
   (d) Minimum
   (e) Is the domain continuous or discrete? What about the range?
2. If you eyeball *f(T),* describe its shape.
   (a) Is *f(T)* symmetric?
   (b) How many measurements of thickness are out of tolerance?

By treating our data as a function, the usual representations we use for describing functions can also be used to get a handle on much of the complexity of the data. The table simplifies the 360 measurements into only 15 points, so that the variability among measurements can be analyzed effectively. The plot reveals the shape of the data is roughly *bell-shaped* reflecting the greater *mass* of measurements near the center, and lesser mass as one moves from the center of the function towards the minimum and maximum of its domain.

> Many variables you will encounter will show a **bell-shaped distribution**. Such distributions are fairly symmetrical, with the maximum near the midpoint of the independent variable, and thin "tails" as one moves from the midpoint towards the minimum and maximum values of the independent variable.

> **Mass** of a discrete distribution *f(x)* is indicated by the sum of the frequencies $\sum_{i=1}^{n} f(x_i)$, where x_i is a set of n consecutive values of the independent variable. For continuous independent variables, the **density** of the distribution is indicated by the area under the curve $\int_{x_0}^{x_1} f(x)\,dx$, evaluated between some lower bound, x_0, and some upper bound, x_1.

2.2.1 Histogram

The most common way of picturing a distribution you will use in engineering statistics is the **histogram.**

> A **Histogram** is a graph that displays the frequency of data values as bars, across equal intervals of the independent variable. A histogram is used for continuous independent variables. A *bar graph* appears similar to a histogram, but is used for discrete (categorical data) variables.

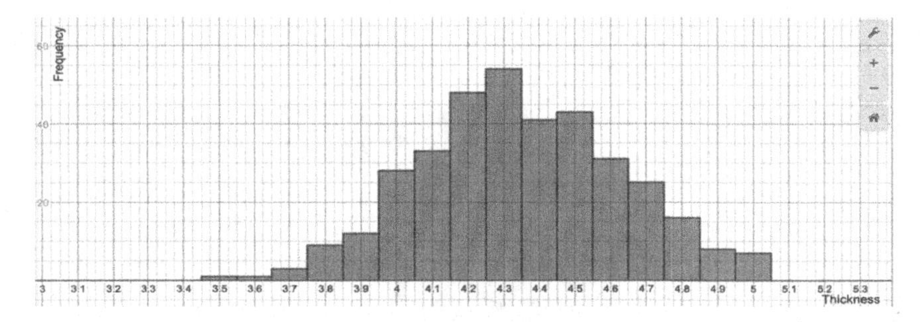

FIGURE 2.4: Histogram of all 360 steel thickness measurements, bin size = .1

Our frequency data for the thickness of the steel sheets is shown in a histogram (Figure 2.4):

From the first glance you can easily interpret this graph: It appears to be just the plot shown earlier with bars representing the frequency, instead of points. The frequency of each value of thickness can be easily read: The most common value, 4.3 mm shows slightly less than 55 of our measurements (53 was the actual frequency). This, however, is an artifact of our data set: We only have 15 values. If we had measured the thickness of our sheets using a more precise caliper, say precise to +/− 0.01mm, it is likely we would have many more values, dispersed in-between the 15 values we could read off with our less precise tool. The values we count as 4.3 mm, might have been read as 4.32, or 4.29 mm. In such cases, the bar with its center at 4.3 consists of not just measurements of exactly 4.3, but *anything in the interval* between 4.20mm and 4.39mm.

Here is another histogram of the same data (Figure 2.5).

Here, the bars are $2\frac{1}{2}$ times the width of our initial bars, resulting in fewer bars, but wider intervals containing our data. The data contained in the bar with center 4.3 in our original histogram is now merged with the data in the bar with center 4.2. This combined "**bin**" is now centered on 4.25 mm, and contains all the data between 4.125 mm and 4.739 mm.

> **Bins** divide a data set into mutually exclusive intervals of equal width. In a histogram, the height of bins are the frequencies of all data values that fall within its interval.

By reducing the number of intervals but widening the values they contain, we lose some of the information in our first graph. This graph tells us that our data isn't perfectly symmetric, and that we have some measurements, in the lower part, which are out of tolerance, but it doesn't give us the count of each value, and it might obscure the fact that none of the measurements are out of tolerance

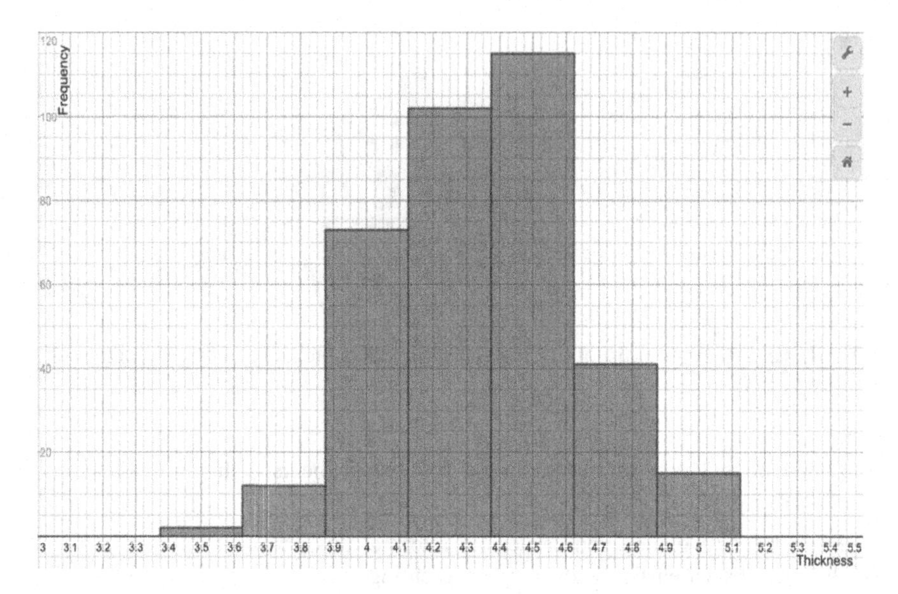

FIGURE 2.5: Histogram of all 360 steel thickness measurements, bin size = .25

in the upper range (>5.0 mm), because the width of the rightmost bar is greater than our maximum measured value.

Of course, you can go the other direction and make bins that are too narrow:

This graph (see Figure 2.6) has bins that are narrower than the precision of our instrument. This results in a graph that is essentially one bar for each value in the data set. If we chose an even smaller bin size, the bars would have the same height, but just be spaced further apart. For situations where there are only a small number of values recorded, such a distribution is appropriate. However, for many distributions, we may have hundreds of values, each with multiple recorded measurements. That is where bin size becomes important to reduce the complexity of the data into a display that retains the important information, but is easier to see the pattern of variation.

The important thing to consider here is that bin size is arbitrary. YOU, the engineer, have to decide what the appropriate bin size is to simplify the data enough to describe its variability. You don't want to oversimplify it or leave it too complex. This is a matter of personal judgement and ethics (one can manipulate histograms in such a way that they mislead the viewer).

2.2.1.1 How to Draw a Histogram

Drawing a histogram is easy. The only tricky part is choosing the right bin size/number of bins to meaningfully display the data.

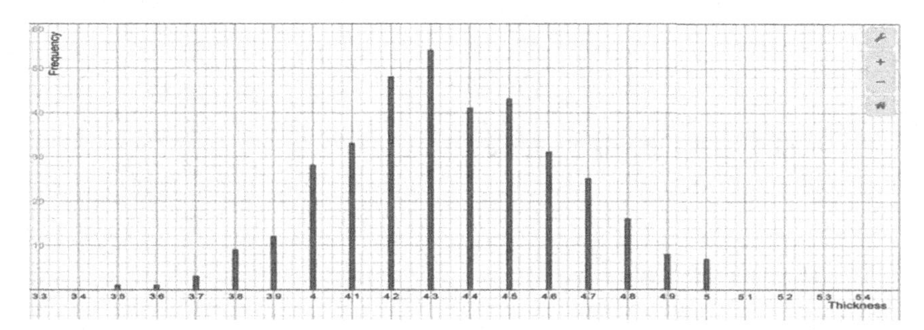

FIGURE 2.6: Histogram of all 360 steel thickness measurements, bin size = .01

(a) *Choose the bin size:* Where the heck do my bins start and end? This is always confusing to students (and many professional engineers). The easy way to think of this is not as actual measurements, but as ranges on a continuous scale (remember, histograms are used for measurements made on a continuous scale). Thickness is a continuous value, theoretically ranging from a low of 0 mm to infinity. So, if you take the range of your data and divide it by the number of bins you want, you will get your bin width.

> **The Range** of a set of data is the difference between the maximum value and the minimum value. It is one representation of the **variability** of a distribution.

> **Variability** is a term that refers to how spread out a set of measurements are. Distributions that show more variability are wider more spread out than distributions that show less variability.

(b) *Choose the number of bins:* In practice, because we have good graphing technology with Matlab, Desmos, Excel, or any number of good statistics packages, we often choose the *number of bins* first, take a look at the graph and then adjust from there. A good rule of thumb to use is to start with 10 bins, divide the range of your data into 10 equal bins, and see what the graph tells you. If it appears to be too complicated, drop down to 6 bins or so. If it doesn't give you much information, try 15 bins. After a few iterations, you will settle upon a graph that has the right resolution for the job. In practice, it is nice to use multiples of 5 or 10 to divide your data into, because most graphing packages use 1, 2, 5, or 10 tick marks to denote the scale of the graph.

Start with your lowest value as the center of your first bin. The interval around it will be:

$$bin\ center \pm \frac{bin\ width}{2}$$

For our example, this would be 3.5mm +/– $\frac{0.1}{2}$ or

$$3.45 \leq x_1 < 3.55$$

The next interval will be

$$3.55 \leq x_2 < 3.65$$

... and so on. The thing to remember is that your bins have to be mutually exclusive. A value of the independent variable cannot be in two bins at once. Using the inequality fully expresses the interval without having to do any deep thinking or calculations involving significant figures.

(c) *Plot the data:* Once you have your bins figured out, and you have chosen your lowest value as the first bin center, you can just draw your histogram as you would a bar chart, labeling the horizontal axis (by convention) with the units of the independent variable, and the vertical axis as frequency.

6. Create a histogram that you feel, accurately reflects the patterns in the data in Table 2.1. Think carefully about the bin size and number of bins that will most accurately reflect the shape of the data.

2.2.1.2 Comparing Histograms

The nice thing about histograms is that, if you place them on the same scale, they are easy to assess against each other. I have provided histograms for the distributions of 120 measurements we took at each position on the sample sheets. The scales are lined up so that the location and height of the bins are all directly comparable.

Comparing these three distributions in Figure 2.7, we can go back to our original questions:

(a) How consistent are we in manufacturing our product?

The width of the distributions gives us a good indication of our overall consistency. We will examine how to measure the width of a distribution in more sophisticated ways than just computing the range in the next chapter. But the intuitive understanding you should take away is that *the distance between measurements is the way we think about variability.* The range is a distance between the minimum and maximum values in a data set.

(b) If we tell the customer, a typical sheet has these dimensions __________, what would we say? Compare it to the midpoint of the distribution. The midpoint of the distribution is an indicator of the "typical" value.

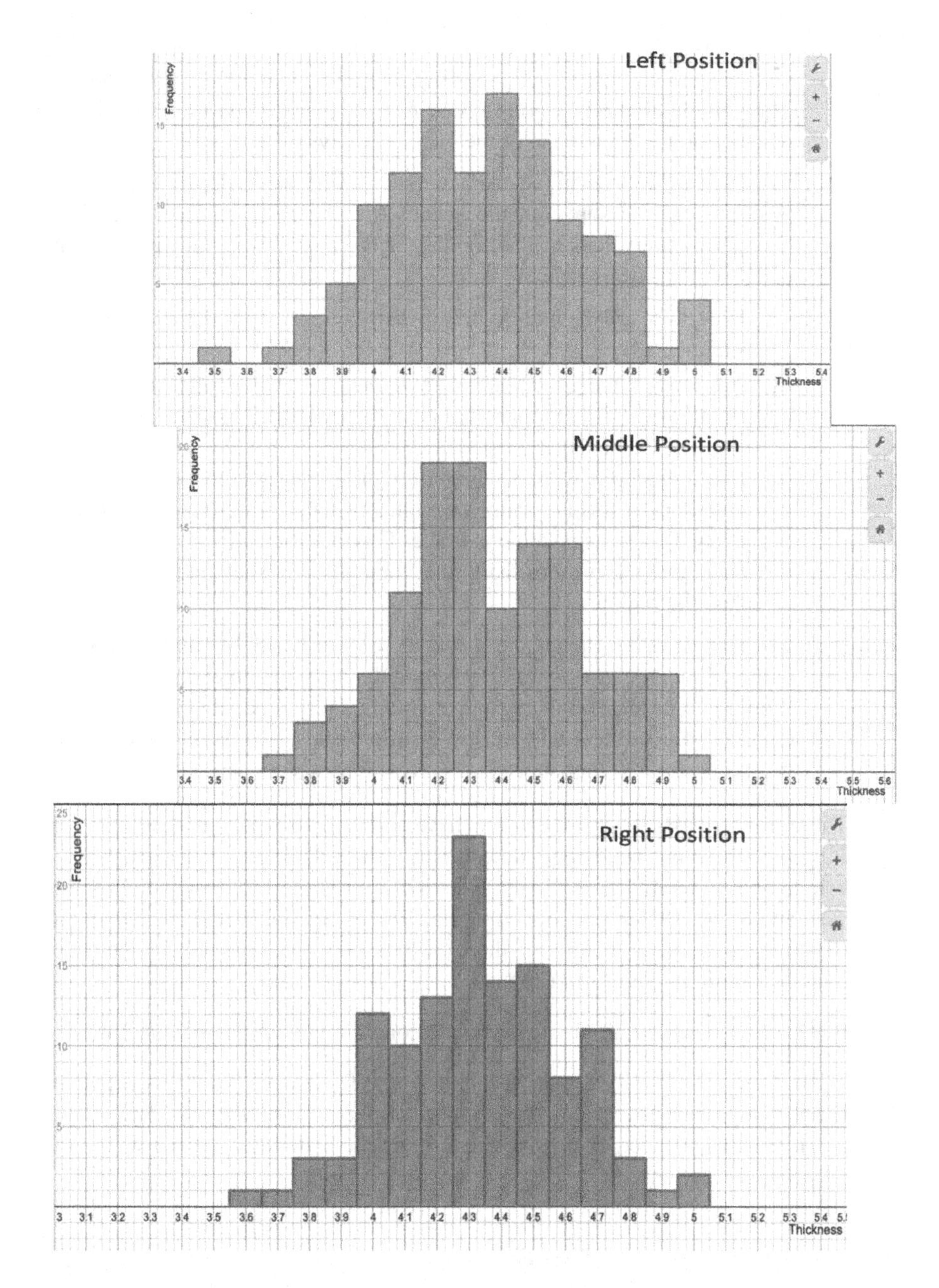

FIGURE 2.7: Histograms of all 360 steel thickness measurements for each position on the sheet

What is meant by "midpoint?" This is not a straightforward question. For now think qualitatively about what the "middle" of a distribution is. Like your professor, most distributions have a middle that is fatter than the rest. This middle clump is a good visual reference when you use a histogram, whether or not the median or the mean is known.

(c) Are there differences in the characteristics of the sheets depending on what position was measured on the sheet? Are sheets of uniform thickness? Comparing the values of the middles of our three histograms gives us the measurements of a "typical" sheet.

i. How close are the middles of the distributions of our three sets of measurements?

ii. Are any of the distributions "skinnier" or "fatter" than the others? What does this say about the consistency of the product at each position of the sheets produced?
In our example, we have three different portions of the steel sheets. If you think of sheets coming through a factory, they are typically cut off a long ribbon using a metal brake. This can stretch the material at both ends, making it more likely that the ends of a sheet are a bit thinner than the middle.

(d) The manufacturer intends the sheets to be of uniform 4.5 mm in thickness with variation of +/− .5 mm. How good did they do in meeting tolerances? Would you depend on them to give you a quality product? Comparing the extreme values (the "tails" of the distributions) gives us an indication of what proportion of our sheets have measurements that are out of tolerance.

There are obviously several sheets that have thinner than expected measurements. If each of these is considered a defect, out of the 120 sheets, we have at least 8 who are out of tolerance on the left side, 8 who are out of tolerance on the right side, and 8 out of tolerance in the middle. Two of our sheets had more than one out of tolerance reading. All told, there are 19 sheets that have at least one out of tolerance reading, out of the 120 total.

7. Is this an acceptable failure rate if you are contracting with the manufacturer?

8. What if there were only 8 *total* specimens out of tolerance? Is this acceptable? If not, you would have to, as the engineer on the project, work to improve the quality of this product, or go elsewhere to get the material you need within the proper specifications.

Summary of Variability (Thus Far).
The important points to remember thus far in the chapter are these:

- Phenomena (objects, forces, relations, beliefs, etc.) vary naturally; This variation is not infinite: It can be described;

- This natural variation can be summarized using a function called a data distribution;
- A distribution pairs each value in your a data set (the independent variable) with its frequency (the dependent variable);
- Tables and histograms are useful tools for making sense out of data distributions;
- When making a histogram, the size and number of bins is the primary consideration;
- Variability is modeled as distances among points in a distribution. The range and other measured of distance are used to describe variability;
- The middle of a distribution describes the location of the typical values in a distribution. The tails of a distribution describe the extreme values;
- For most distributions there is greater mass in the middle of the distribution than in the tails—Mass is just the frequency of data within an upper and lower bound. Comparing the middle part of two or more distributions is a way to assess whether their typical values are different;
- Comparing the tails is a way to assess whether or not the width of the distributions are different.

2.3 Accuracy and Precision of Measurements

Hopefully, in your university career, you have been contemplating one of the most fundamental and astonishing properties of our physical and social world: Its variety. Just take my dog at home. He is a Labrador retriever, distinguishable from the Chihuahua and Great Dane by his relative size, and slobbery demeanor. We say that size (e.g., mass, height), and slobberiness (perhaps ml of saliva) are **variables**, because they hold different values for different phenomena. My lab has many variables, whose values distinguish him from those other breeds. This is true for all of the products you will engineer in your lifetime. As we saw at the beginning of the chapter, even things as uniform and simple as a batch of resistors differ on at least one critical variable: Resistance. Luckily, with some relatively simple tools (like tables of values and histograms), we can reduce all that variability into a neat package that can be described, and decisions made from it.

Variable refers to a measurable attribute of a phenomenon. It is a quantity that can take on different values for different instances of that phenomenon.

Only there is a problem...

One of the things that all measurements have in common is the fact that they are all *wrong*. I measured Baxter, my Labrador the other night and found that his length from snout to tail is about 150 cm. If I used the same procedure and measured him tonight, I would most likely come up with some different measure, say 152 cm or 148 cm.

To tell you the truth, it was difficult lining up the tape along his back, and his tail was especially problematic, as it kept wagging back and forth. But even if I had frozen him in carbonite, and put him in a special jig designed specifically to standardize the measurement of canines, I would still have some small degree of error due to the limits of my jig, the precision of my tape, and other random factors. The graph depicted in Figure 2.8 illustrates how statisticians view these inherent differences in measurements. Despite my best attempts, each time I tried to measure my dog, I came up with a slightly different value. This is similar to the-bell shaped distribution in Figure 2.8. It is a distribution of measurements *all of the same thing.* The center of the distribution represents my average measurement. The degree to which these average measurements differ from the True value is an indicator of the **accuracy** of the measurements. The degree to which all the measurements agree with each other (the spread of the distribution of repeated measurements) is an indicator of the **precision** of the measurements.

Accuracy refers to the difference between a measurement or a set of measurements and the true value of the phenomenon being measured. The center of a distribution is used to assess Accuracy. It is a reflection of the amount of systematic error in my measurements.

Precision refers to the variability of measurements of a phenomenon if repeated. The spread of a distribution is used to assess precision.

2.3.1 Accuracy–Systematic Error

So it bears repeating that *all measurements are wrong*, or more accurately, all measurements have **error** (even if, in some very simple cases, the error may be zero). Errors in accuracy of measurements are typically **systematic**. When they are repeated, they tend to give a consistently high, or consistently low estimate of the true value.

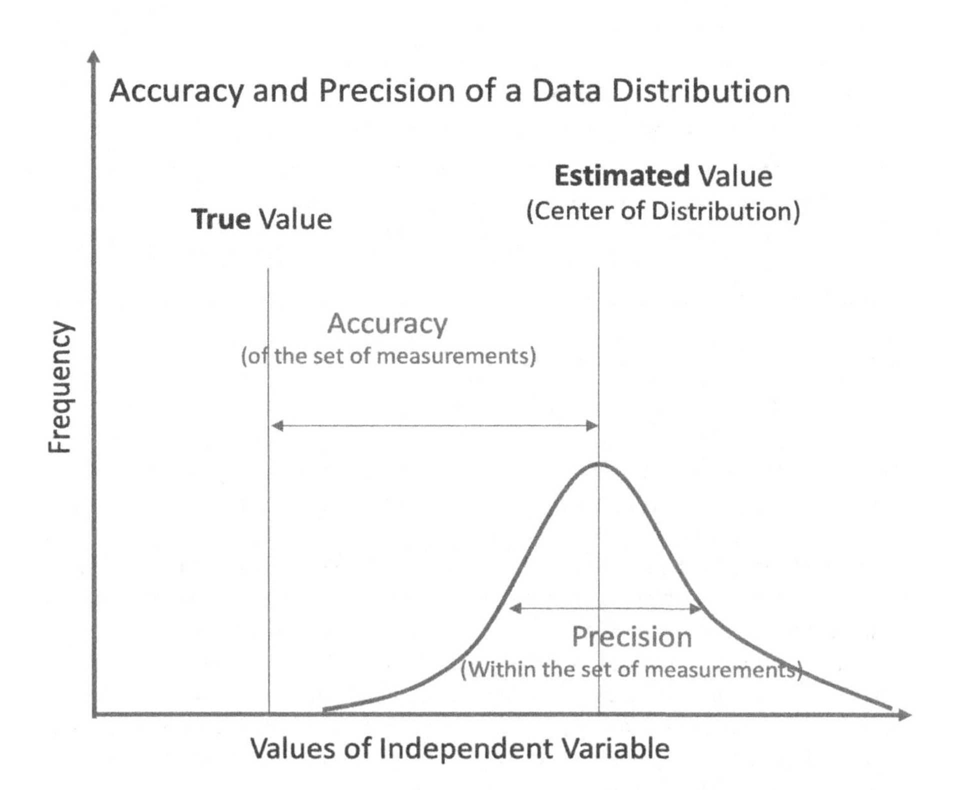

FIGURE 2.8: Accuracy and precision of a sample of measurements

FIGURE 2.9: An object to be measured...

Error is the difference (deviation) between a measured value and a reference value, typically the true value being estimated.

Systematic Error refers to a *consistent* difference (over- or under-estimate of a measurement) between a set of measurements and the true value being measured. Systematic Error is a measure of *Accuracy.*

2.3.1.1 Sources of Systematic Error

Examples of systematic (accuracy) errors include a consistent **improper use of an instrument**. *Parallax,* for example, the tendency that, with an analog gauge, one's line of sight is at such an acute angle, the needle on the gauge appears to be consistently reading a number higher (or lower) than the actual reading. I use this to my advantage when driving with my wife. When she is in the passenger's seat, she looks over at the speedometer and sees that I am going the speed limit of 65 miles per hour. I have accounted for the parallax and have set the cruise control at 74 mph, and she is none the wiser. We both win!

Other common improper usage of instruments includes not accounting for differences in instrument scales. For example, a caliper measuring in mm will show values 25.4 times larger than one using inches. For measuring the volume of liquid in a graduated cylinder, reading the top of the meniscus instead of the bottom will yield values consistently higher than the true value.

Improper use of an instrument must be *consistent*, for its errors to be systematic.

A second common source of systematic error involves improper **calibration** of instruments. A thermometer, for example, may read 1 degree, when in fact it should read zero degrees. To calibrate the instrument, we need to immerse the thermometer in a bath of water at exactly zero degrees (at 1 atm). The thermometer will consistently read 1 degree. We *know* the water is really zero degrees! So, we just subtract 1 from every reading to render an accurate datum.

Calibration is the adjustment of a measuring instrument so that it reads accurately. The most common method for calibration is the 2-point method.

An even better method for calibrating instruments on a linear or exponential scale is a two-point calibration, where we take measurements at two known points (e.g., 0 degrees and 100 degrees), and then use a transformation to equate the instrument readings with the true values they should be registering. This is because many instrument errors are both **offset errors** (just a simple translation of the scale by a few

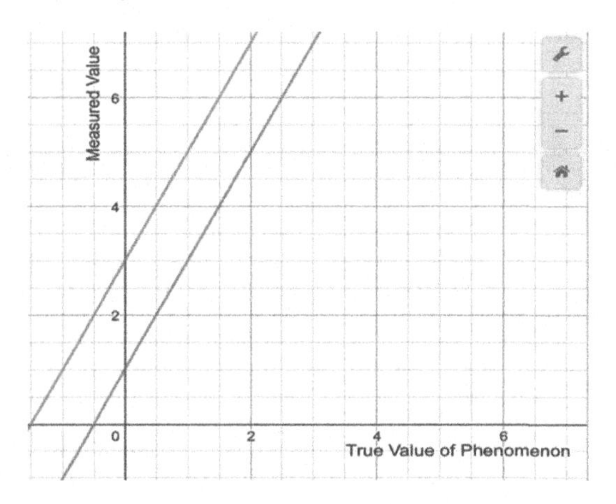

FIGURE 2.10: Offset error is an erroneous translation of the measured values from the true values of the phenomenon being measured

units), and **scale-factor errors** (when the instrument is calibrated to the wrong scale (e.g., imperial vs. SI units) (See Figures 2.10 and 2.11).

> **Offset error** is an erroneous translation of the measured values from the true values of the phenomenon being measured:

> **Scale-factor error** (also called, "Multiplying Error") is a dilation of the measured values from the true values of the phenomenon.

The third common source of systematic error is flawed experimental design. When for example, I measure the temperature of an object without accounting for environmental factors such as convective cooling, my reading may be accurate at the instant I look at the instrument, but it may not be an accurate measure of the object at equilibrium. Measurements, to be accurate, must be performed under **controlled** conditions.

> **Control** (in an experiment) refers to accounting for all known variables that impact the behavior of the dependent variable in the data collection and statistical analysis of an experiment.

2.3.2 Precision–Random Error

If you remember my Labrador, Baxter, I could get a measurement of his length by holding him as still as possible and stretching a tape measure from the tip of his nose

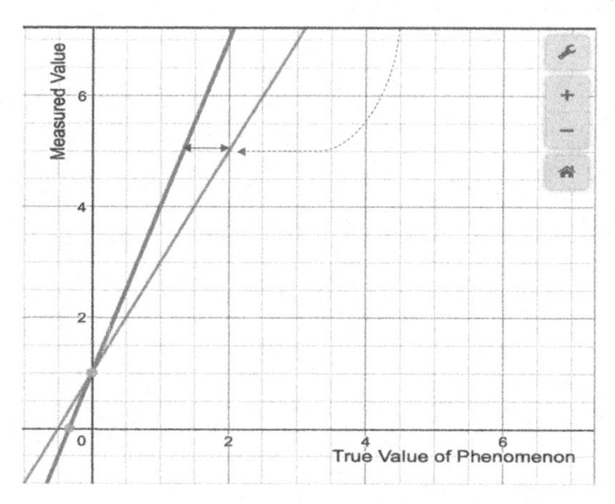

FIGURE 2.11: Scale-factor error (also called, "Multiplying Error") is a dilation of the measured values from the true values of the phenomenon

to the tip of his tail. To account for his wagging tail and general elasticity, I took 10 measurements to see how much they would vary (Yes, I really do these things). Here are the 10 measurements in Table 2.2:

Meas. #	**Value (cm)**
1	150
2	150
3	141
4	145
5	149
6	141
7	145
8	145
9	142
10	143

TABLE 2.2: Ten measurements of Baxter's length

A quick glance at the numbers tells me that my dog is about 1.5 meters in length, but that, any time I measure him *using the same procedure* I could get a value anywhere between about 140 and 150 cm. The histogram in Figure 2.12 shows the distribution of these values:

1. Given I don't really know the exact value of Baxter's length, what value should I use as my *best guess*: 150 cm, 141 cm, 147 or 145 cm?

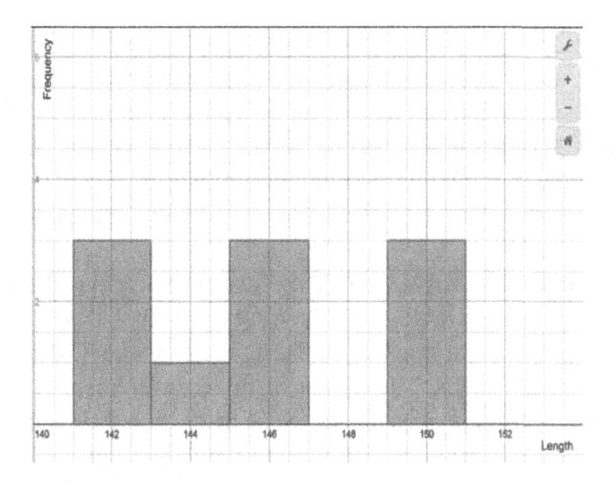

FIGURE 2.12: Histogram of my measurements of Baxter's length. Bin size = 2

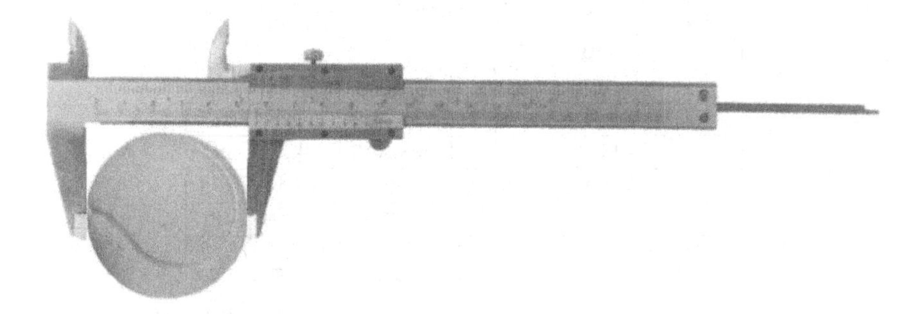

FIGURE 2.13: Measuring the "diameter" of a tennis ball

2. What does "*best*" guess mean when it comes to estimating a measurement?
3. What do you think of the variability of my measurements? Is this a lot of variability or is it what you would expect given the context?

I have used the example of my dog to illustrate the fact that, when we measure *anything* multiple times, the values we obtain will vary. Some things (like Baxter) are very difficult to measure *precisely*.

I often use the example of measuring a tennis ball with my students as a more aerospace example (Figure 2.13). The Coefficient of drag, C_d, of a tennis ball is dependent, in part, on its diameter. But the fuzz on the outside makes precise measurement difficult. The Fuzz is not perfectly uniform. Some strands of felt are sticking out significantly further than the rest, and even those that are not sticking out show uneven texture. Moreover, if you apply a caliper to the ball, the fuzz squishes, so that there is at least 1mm of variation in possible values for the diameter. Even using the word, "diameter" is a misnomer: A tennis ball, though pretty close, is not a perfect sphere. It has flat spaces and seams that make a sphere an inexact model of its true geometry. Unlike Baxter, a tennis ball *will* just lay there for you. But even then, the actual diameter of the ball is not knowable given all these factors. For this reason, every time I take a new measurement of the "diameter" of the ball, I will get potentially different results.

But those results are not *ALL* different. Take a look at the 10 measurements I made for Baxter shown in Table 2.3. They are all pretty close to each other. No matter his true length, I am no more than 9 cm (about 3.5 inches) in error (the range of the data). It is likely, in fact, that most of my measurements (the ones that fall between 141 and 150 cm) are even closer to the true value than 9 cm.

If, let's say, Baxter's "true" length was 145.3697895 cm, I could honestly say that most of each measurement is "true," while just a little bit of each measurement is "error." My measurement of 145 cm, for example has 145 cm that is true, and 0.3697895 cm that is in error. That is only 0.25% error and 99.75% truth!

Meas. #	**Value (cm)**	**True Value**	**Error**
1	150	145.3697895	4.6302105
2	150	145.3697895	4.6302105
3	141	145.3697895	−4.3697895
4	145	145.3697895	−0.3697895
5	149	145.3697895	3.6302105
6	141	145.3697895	−4.3697895
7	145	145.3697895	−0.3697895
8	145	145.3697895	−0.3697895
9	142	145.3697895	−3.3697895
10	143	145.3697895	−2.3697895

TABLE 2.3: Error is the difference between the measured value and the true value of a measurement

Random Error refers to differences in measured values within a set of measurements of the same phenomenon. It is a measure of *Precision*.

We attribute all this *variation in measurement of the same phenomenon* to **random** factors. What we mean by random is that there is no single error that is predictable—They happen by chance. Just take a look at the different error values I computed for Baxter's length. The differences in error for each measurement are not systematic, they are due to random factors. In contrast to Systematic Error, random errors cannot be predicted by consistent misuse of a tool, or consistent poor experimental design. They are caused by the moment-by-moment changes that we see in the real world. Even with the most precise instruments, that measure to the tens of thousandths of decimal places, measures will show varying values depending on small fluctuations in temperature that cause the material of the instrument to expand and contract.

To put Baxter to bed, so to speak, we can see that each measurement I took had some individual error of measurement. Each of those individual errors differ in random ways. But we can also see that, given his "true" length is 145.3697895 cm, I seem to be *consistently underestimating* his true length (the average error is about −2.7 cm). So, the overall set of measurements has some systematic error as well. It may be due to my using a cloth tape stolen from my wife's sewing kit, or it may be that I haven't accounted for the curvature of his tail and back. We could only come to these conclusions about the quality of my measurements, and about estimating the true value of the phenomenon by *taking repeated measurements and looking at their pattern.*

So, we must assume that all measurements have some error, that no single measurement is perfect, and that it is only through taking repeated measurements that we can come to understand the "True" value of the variable we are examining.

Summary of Error of Measurement

Because measurements can take on different values, any measure can be considered a variable quantity. All measurements have error, both systematic and random.

Systematic error is consistent and repeatable; it reduces the accuracy of a measurement;

Random error is not predictable for any individual measurement; it reduces the precision of a set of measurements;

Accuracy is determined by a systematic difference between a measured value or center of a set of measured values, and the true value of the phenomenon being measured is the offset of a distribution from the true value

Precision is determined by the differences in values of a set of measurements: of the spread of a distribution. Taking repeated measurements and looking at their pattern, their distribution—can help us understand the amount of systematic and random error to expect.

2.4 Continuous Versus Discrete Data

From early on, people saw that the accuracy of empirical statistics tends to improve as the number of trials increases. That is to say, when we roll a single die, or when we take the measure of the hardness of a sample of metal, we are not surprised when any particular value appears (say a roll of 3 on the die, or a hardness of HRC 57 for a test piece of very hard steel), but we are not confident that the next sample we take will have the same value as our first. In fact, for both die and steel, we would exhibit surprise if, on the next trial we got exactly the same value. This is because there are multiple values any trial can take on (remember our definition of *variable*). In the simple case of the die, any trial can result in one of only six outcomes, and we can use this set of outcomes to help us determine the fairness of the die. For the steel test piece, there are infinitely many values any test trial can result in, depending on the precision of our testing equipment, but even they will fall within certain established values—you will rarely have a steel test piece with a hardness rating less than HRC 20, or greater than HRC 65.

2.4.1 Discrete Random Variables

For the roll of a die, or for the count of the number of failures of our steel part, we can obtain only integer values. These kinds of events are termed *discrete* because they can only take on a finite number of values. We use the term **Frequency** or **Frequencies** (plural) to denote the *counts* of such variables. These are among the most useful ideas in all of statistics, in that we can just create two or more conditions in an experiment (for example High Temperature, Low Temperature) and count instances of the variable we are interested in (for example Failures of o-rings in a rocket engine). If, under high temperatures, an o-ring we have chosen fails more often proportionally than under low temperatures, we can use this information to make a decision to not employ this particular part under conditions of high temperature. Simple counting!

A Discrete Random Variable refers to phenomena measured on a scale with a countable number of values.

2.4.2 Continuous Random Variables

Other variables can take on (theoretically) infinitely many values in an interval. We might expect our steel sample to have a hardness anywhere between about HRC 55 to HRC 65 depending on random variations in its crystalline structure. For the length of our pencil, its measure can take on any value from zero to (theoretically) infinity, including integer values and all the fractional values between integers.

For continuous random variables, we tend to like to use the arithmetic **mean** to compare different conditions. You all know the mean, it is typically called "the average," but I want you to avoid using this colloquial term. We will reserve the term *mean* to refer to the center of a *distribution* of measurements, all taken under the same conditions. What we mean by "center" is the balancing point of the distribution when all data values are given the same weight. The term "average" is a more general term that may refer to the **median** (the middle value of a distribution) or the **mode** (the most frequent value). Each of these three indices are termed **measures of central tendency**.

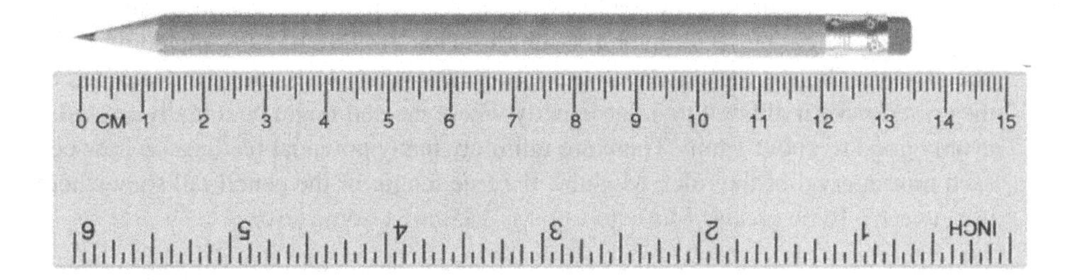

FIGURE 2.14: Measuring the length of a pencil

> **A Continuous Random Variable** refers to phenomena measured on a continuous scale. Theoretically, at least, a CRV can take on infinitely many values.

> **Measures of Central Tendency** describe the center or middle of a distribution of data. The *mean* is the center of mass of the distribution. The *median* is the value of the independent variable at the 50th percentile, and the *mode* is the most commonly occurring value, the one with the highest frequency.

Examples of **Continuous Random Variables** include most of the things we, as engineers are interested in measuring. Hardness of a material, temperature, elongation, mass and density, basically anything that can be expressed in SI units. Though our measuring device may only have a finite number of values it will read (e.g., using a digital caliper to measure the diameter of a flange), the *variable it is measuring* can take on infinitely many values between its most precise units.

2.5 Law of Large Numbers

Why is it that we trust the average of a set of measurements over just a single measurement? Our previous discussion about steel thickness and Labrador retriever length provides some good examples, but we can also show mathematically how the mean represents the true value of a measure, given the assumption that all errors of measurement are random.

Suppose we take a measurement of some simple object (Say a pencil, see Figure 2.14).

We can represent this measure, y_i, as a combination of its actual length x_i and some error in our measurements, e_i. The value we read off our ruler is then

$$y_i = x_i + e_i$$

As I see the Figure 2.14, I would say the pencil is about 135 mm in length. But *is it exactly*? It might be 135 mm or 136 mm. It is hard to tell, because the point of the pencil is not aligned perfectly with the zero on the ruler. Also, the rounded edges of the eraser make it difficult to judge exactly where the end might be. Finally, my ruler is only good to about 1 mm. There are infinitely many potential readings in between each mm interval of the ruler. Mightn't the true length of the pencil fall somewhere in between? To be precise I have to say it is 135mm $\pm$ some error.

$$y_{\mathrm{i}} = x_{\mathrm{pencil}} + e_{\textit{Dr. Middleton and his ruler}}$$

or

$$y_{\mathrm{i}} = 135mm + e_{\textit{Dr. Middleton and his ruler}}$$

If, like measuring my dog, I wanted to get a better estimate for the length of the pencil than my initial attempt, I would take several measurements and "average them" —compute their arithmetic mean. I know this is pretty intuitive, but it is tied to the most basic law of statistics, the **Law of Large Numbers**, which describes the tendency of the total error of measurement of a phenomenon to decrease as the number of measurements increases. This fundamental mathematical law upon which all statistics is founded was invented by Jakob Bernoulli in 1713, though the behavior of this law was understood many years prior by those compulsive gamblers I talked about earlier. Here is the basic logic.

Whenever we take a few measurements of a phenomenon, as I mentioned earlier, we are not surprised if our values differ somewhat. We know that all measurements contain some error, so to have two values that are different just means that one or both have different amounts of error. The problem is, how do we determine which of those values we can trust as the *best* estimate of the true value?

Let's go back to our pencil. We have established that any measurement of its length will be a combination of its true value and the error of measurement.

$$y_i = x_i + e_i$$

The true value of the length of the pencil is y_i. The measured value is x_i, and the error of measurement is e_i.

If I passed this pencil around the class and had each of you measure it, we would not be surprised if we got a few different readings. I have actually done this for a number of years. Table 2.4 shows the results for thirty Mechanical/Aerospace engineering students in one class. Some of them judged the pencil to be closer to 135 mm, some closer to 134 mm, and a brave few judged it to be 136 mm. Which is right?

This used to be an intractable problem for scientists trying to find the distance from the earth to the moon or the planets. Keppler would take his measurements and calculate a value for the distance to Jupiter, for instance, and Tycho Brahe would do the same, but their numbers would be different. Since the true value of the distance from the earth to the moon was unknown, some means for determining the true value, somehow getting beyond the limitations imposed by the error of measurement was needed.

Let's do this with our pencil data. I took the liberty to very carefully, using the magick calipers of Babylon, a fictitious instrument that can measure the true value of an object with no error, and found the true value y_i to be 134.65 mm:

Trial #	**Measured Length (mm)**	**True Length**	**Error**
1	135	134.65	0.35
2	135	134.65	0.35
3	134	134.65	−0.65
4	135	134.65	0.35
5	135	134.65	0.35
6	135	134.65	0.35
7	134	134.65	−0.65
8	134	134.65	−0.65
3	134	134.65	−0.65
10	135	134.65	0.35
11	135	134.65	0.65
12	134	134.65	−0.65
13	136	134.65	1.35
14	134	134.65	−0.65
15	136	134.65	1.35
16	134	134.65	−0.65
17	135	134.65	0/35
18	134	134.65	−0.65
19	134	134.65	−0.65
20	135	134.65	0.35
21	135	134.65	0.35
22	135	134.65	0.35
23	135	134.65	0.35
24	134	134.65	−0.65
25	134	134.65	−0.65
26	136	134.65	1.35
27	134	134.65	−0.65
28	134	134.65	−0.65
29	135	134.65	0.35
30	135	134.65	0.35
Sum	4,040	4,039.50	0.50
Mean	134.667	134.650	0.0167

TABLE 2.4: Measured versus true length of a pencil

If each measurement is a combination of the true length plus some error, you can see that *each measurement* has some error factor e_i. So,

$$134.65mm = x_i + e_i$$

Because we have multiple values of our measurement, we can sum those values.

$$\sum x_{True} = \sum x_i + \sum e_i$$

$$4,039.50 = 4,040 + -0.50$$

In our instance, we have 30 values, so if we divide our equation by the number of observations, we get the arithmetic mean of the true values. This is equivalent to the mean of the measurements plus the mean error.

$$\frac{\sum x_{True}}{30} = \frac{\sum x_i}{30} + \frac{\sum e_i}{30}$$

$$134.65 = 134.667 - .016666667$$

or generally,

$$\overline{x}_{\text{True}} = \overline{x}_{measurements} + \overline{x}_{error}$$

and since the mean of the true values IS the true value,

$$y_i = \overline{x}_{measurements} + \overline{x}_{error}$$

For our 30 trials, notice that the mean of the measurements is very close to the true length of the pencil. In fact, we are within .017mm!

Also notice that the mean of the error terms in our experiment is somewhere between the maximum and minimum error values. Because the error in our experiment is assumed to be **random**, it will tend to vary in equal proportions about a central value: zero. *That* means that the sum of the errors will approach zero as the number of trials approaches infinity. Our mean error -0.017, which is already better than the error of any of our measurements, will get smaller and smaller and ultimately approach zero if we kept adding trials to our table.

The Law of Large Numbers can be stated thusly: As the number of trials of a random process increases, the expected value of the set of trials will approach the true value of the variable.

What this means in English: Size Matters! As the size of a sample drawn randomly from a population gets larger, the mean of the sample gets tends to get closer and closer to the population mean. In other words, we tend to have more confidence in large samples, than small ones, all other things being equal.

Example: Suppose I take the population of all the bicycle frames produced in my factory for one year (N = 10,000), and then take random samples of different sizes from that population (n_1 = 10, n_2 = 100, n_3 = 1,000, and n_4 =

Population				
7	**Sample Size (n=1)**			
6		Sample A	Sample B	Sample C
5	$\bar{x}$	9.00	6.00	4.00
4	error	3.30	0.30	−1.70
7				
4	**Sample Size (n=5)**			
5		Sample D	Sample E	Sample F
6	$\bar{x}$	6.40	5.00	6.40
10	error	−0.70	0.70	−0.70
4				
1	**Sample Size (n=10)**			
9		Sample G	Sample H	Sample I
6	$\bar{x}$	5.90	5.30	6.30
7	error	−0.20	0.40	−0.60
9				
2	**Sample Size (n=20)**			
8		Sample J	Sample K	Sample L
6	$\bar{x}$	6.00	5.30	6.15
6	error	−0.30	0.40	−0.45
7				
2	**Sample Size (n=100)**			
		Sample M	Sample N	Sample O
$\bar{x}$ **5.7**	$\bar{x}$	5.59	5.72	5.71
	error	0.11	−0.02	−0.01

TABLE 2.5: As the size of each random sample gets larger, the average error gets smaller

10,000). After measuring the stiffness of the frames, I compute the means for each sample. As the samples get larger, their means will tend to more and more accurately represent the actual mean stiffness of the population.

The data Shown in Table 2.5 is for a randomly drawn population of 20 measurements. The population size is finite, N is 20. The true value of its mean is 5.7. I have taken three random samples (with replacement) from this population with sample size n=1, n=5, n=10, n=20, and n=100 and calculated the means and the difference in the calculated sample mean from the true population mean. Take a look at the different samples. You can see, as the sample size gets larger, generally the accuracy of the mean as an estimate of the population mean gets better—the error tends to get smaller and smaller as the sample size gets larger. This progression is not uniform, some samples with smaller sample sizes, because of just random chance, have more accurate means than some with larger sample sizes. But overall, looking at this example, I would trust the larger samples as better estimates of the population mean more so than the smaller samples.

Mathematically, we can see that our error in estimating the true mean of the population gets smaller and smaller, so our sample means get closer and closer to each other. This is how the Law of Large Numbers works. The overall tendency is for more accurate estimates of the mean as the size of the sample, randomly drawn, approaches infinity.

2.6 Central Limit Theorem

BUT, we still have some variability in our estimates, even with larger sample sizes. Looking at the example with n=100, the error of our estimates ranges from about −0.01 to about 0.11. Sample O gives us a better estimate than Sample M. But none of the samples n=100 have error even close to the samples where n = 1, 5, or 10.

This finding lead Simon LaPlace in about 1810 (and later, Quetelet) to conjecture that, because of the LLN, samples with larger n will each be, on average, better estimates of the true population mean than samples with smaller n. Because the error of larger samples tends to be smaller than that of smaller samples, if he plotted the distribution of *means* of randomly drawn samples from a given population, it would look Normal (bell-shaped), and the distribution of means would get narrower and narrower as the sample size grew larger and larger. Ultimately if the sample n approached ∞, each sample mean would have zero error, and the mean of the **sampling distribution of the mean** would exactly equal the population mean.

Sampling Distribution is a distribution of a sample statistic, all samples taken randomly from the same population, all samples having the same sample size.

The sampling distribution of the mean is a sampling distribution made up of means, all computed from samples of the same size, all taken randomly from the same population.

There are other sampling distributions, including sampling distribution of proportions, sampling distribution of variances and other statistics. They will become important when we begin to look at inferential statistics, so file this idea away for future use!

What LaPlace and Quetelet developed was nothing less than ***The Most Important Idea in all of Statistics!!!!*** We call it the **Central Limit Theorem**. The Central Limit Theorem (or CLT) states that for any random variable, the sampling distribution of the mean will approach a normal distribution as the size of the samples in the distribution get larger. Further, its mean will approach the population mean and the width of the sampling distribution will tend to grow smaller and smaller as the sample size grows larger and larger.

Central Limit Theorem: Given a sufficiently large sample size, the sampling distribution of the mean for a variable will approximate a normal distribution regardless of the shape of the parent population from which the samples were drawn.

Two important consequences of the CLT are: 1) Given unbiased samples, the mean of the sampling distribution will be the mean of the population (it will be accurate), and 2) its width will grow smaller and smaller as the sample size grows (it will grow more and more precise).

Finally, the most interesting fact about the Central Limit Theorem, is that it holds for *any shape distribution*! What's that? It doesn't matter what the shape of the real distribution of data in a population look like, if I just choose a large enough sample size, the shape of the sampling distribution of the mean will *always* be normal? Yes, Sparky, that *is* what I am saying.

In practical terms, this means:

1. The error of measurement inherent in our work can be divided into two types: Systematic (Accuracy) errors and Random (Precision) errors;
2. The effect of random errors on our estimate of the true value of a measure gets less and less as the number of trials grows larger and larger;
3. Because of this, we get more confident in the mean as our estimate of the true value as the sample size increases;
4. For Continuous Random Variables, the mean of a set of measurements is almost assuredly the best value to use as an estimate of the true value of the variable being measured; and
5. The normal distribution is a good model to use to estimate the mean of a population if we have a large enough sample size no matter what the shape of the parent population is.

2.7 Representativeness

When we take a sample of data from a population of phenomena we are interested in learning about, we try to take it in such a way that the sample represents the population. We would like to base the conclusions we make from the data on the fact that the characteristics we see in the sample reflect the same characteristics in the population. In other words, we are trying to make the case that the location of the sample is at roughly the same spot on the scale of measure as that of the population we drew it from, *and* that the variability we see in the sample is about the same as that of the population we drew it from. To do this, we use **Sample Statistics**. The population characteristics we are trying to estimate with our sample are termed, **Population Parameters**. The word *parameter* is not just jargon. We trust that, through the LLN and CLT, the shape of a sampling distribution of the mean drawn from our population of interest will be shaped fairly normal, that the mean of our sample will approximate the mean of the sampling distribution of all means drawn from the same population, and that the mean of the sampling distribution is equal to the mean of the population. This leads us to two important assumptions about the mean as a measure of the center of a population:

1. For a CRV, the mean of a sample of measurements drawn randomly from a population will almost assuredly be the best estimate of the mean of the population;

2. The sample mean is an *unbiased* estimate of the mean of the population from which it is drawn; and

3. By the CLT, we know that the means of all possible samples of the same size, drawn from our population will have a grand mean equal to the population mean, and that the distribution of that sampling distribution will be approximately normal.

Sample Statistics are characteristics of samples that are used to represent the characteristics of populations. More precisely, sample statistics are numbers used to represent the parameters of the distribution function that describes the population from which the sample was drawn.

Population Parameters are features of populations that correspond to important characteristics of their distribution: Minimum, Maximum, Mode, Median, Interquartile Range, Mean, and Standard Deviation. More precisely, population parameters are variable quantities that, when fixed, fully describe the frequency (or probability) distribution of the population.

If we just had a formula that fully described the distribution function, then the mean would represent the center of that distribution, and be the local maximum (i.e., where the first derivative of the function with respect to x is equal to zero: $\frac{df(x)}{dx} = 0$. The mean would therefore be a *parameter* of this as-yet-unknown function. By changing the mean, we translate the center of the function back and forth on the axis of the independent variable. We will take up this parameter algebraically in the next chapter. For now, understand that when we compute a sample statistic, it doesn't just represent a feature of the parent population from which the sample was drawn, it also represents a variable, a parameter that describes an important feature of the distribution function. There are other parameters describing the variability in the population, the standard deviation being of chief importance. We will describe these parameters more deeply in subsequent chapters.

Practically, representativeness of a sample means that the elements drawn from the population are drawn without *sampling bias*—Each element in the population must have the same chance of being selected as each other element. Note that classes of elements may have different probabilities: For example, if one selected engine blocks from the population of automobiles in the world, there is a higher probability of selecting a cast steel block versus an aluminum block. But for a sample to be unbiased, the probability of selecting any engine block at random (say my 2001 Chevy Suburban) has to be equal to any other block at random (Say my brother's 1997 BMW 530i). That way, 2001 Chevy Suburbans will appear in a sample about as often as they appear in the population; Ditto for BMW 530is. Because no Chevy Suburban has a block made of anything but iron, but some performance models of the BMW 530-i are manufactured with aluminum blocks, the proportion of iron and aluminum blocks in my sample will be approximately equal to the proportions in the population, *but not equal to each other.*

Simply put, an unbiased sample has sample statistics that are exactly the same as its population because it 100% accurately represents that population. Needless to say, this is almost never the case. In practice, we need to reduce *systematic* sampling bias as much as possible by carefully considering our sampling procedure to emphasize random sampling.

Sampling Bias Sampling bias occurs when a sample statistic does not accurately reflect the true value of the parameter in the population from which the sample was taken. A variety of factors influence bias in sampling including taking measures from a limited subset of the population and not insuring that sampling each event is independent of sampling all other events. We try to reduce sampling bias by using simple random sampling.

2.7.1 "Simple" Random Sampling

As we close out this introductory discussion of variation and how we try to get around its complexity to try and describe data, I would like to make a special focus on the sampling technique called, "Simple Random Sampling" or SRS for short. Like its label suggests, SRS *is* simple. All one has to do is assign a unique number to each event in a population, put those numbers in a hat, select an event and record it, put the event back in the hat. Then repeat the process, selecting and replacing, until the desired number of events is chosen.

Simple Random Sampling is a method of selecting a set of events (a sample) from a larger group (a population). Each event is chosen entirely by chance and each event in the population has an equal chance of being included in the sample. *Because of this all possible samples of a given size n are equally likely.*

The reason we like to use SRS is because we want to insure that we don't systematically bias the sample we are collecting. By placing every event into a common pool, then selecting one event at a time from the pool with replacement, we insure that every event has exactly the same probability of being selected as every other event (remember the example of the engine blocks). When we do this, the resulting sample will typically closely resemble the population of events from which it was drawn. The only differences in the sample from the population can be attributed to purely random error, not to any systematic error. By the CLT, if we took infinitely many samples this way, we would see that the means of those samples follow a normal distribution, with very few means way out in the tails of the distribution—most will fall close to the actual value of the population mean.

Sampling with Replacement. So, why replace a sampled event before the next trial? Remember, we are trying to sample, insuring that each event in our population has the same probability of being selected as any other event. This is easy to demonstrate if we use a small population, let's say N=10. When we select our first event at random, its probability of being selected is $\frac{1}{10}$. If we do not replace it back into the population, the next event will have a probability of $\frac{1}{9}$ of being selected, the third will have a probability of $\frac{1}{8}$, and so on. Each successive event sampled has a $\frac{1}{N-n}$ probability of being selected. So, in a sample of 3, it is more likely for the 3rd item to be selected than the 1st. By *replacing* each event prior to randomly selecting the next, we insure that the population always has the same proportion of events prior to each selection. This means that each trial will be unbiased, all will be taken with the same probability, insuring that all samples of size *n* will have the same probability of being selected.

In reality, it is difficult to use SRS for most of the problems we are trying to solve as an engineer. First, some of our populations are huge (think of the number of ball bearings manufactured in a plant in a year). It is impractical to wait a whole year to get the population number, and then assign a number to each ball bearing and select our sample. Knowing what we do about systematic error, if we do not use some form of random sampling, then it is likely that our sample will have some systematic error. Think of only choosing automobiles from GM, instead of all auto manufacturers to determine the proportion of aluminum block engines out there. We call this sampling from a limited subpopulation. If we only selected from GM, we are unlikely to have a sample that is representative of the entire population of auto makes and models. We might have a sample that represents GM cars, but GM does not reflect the larger pool of manufacturers and models in the world. But just engaging in a process that is not random will likely introduce some systematic error. That is where the engineer really comes in. By designing a logical sampling plan, s/he can reduce the systematic error, even if s/he can't completely eliminate it.

There are other types of sampling that we will use later on in this course. But for right now and throughout your career, when someone says, "Random Sampling" think of this case of SRS as the basic idea of obtaining a sample that represents a population without systematic bias.

2.8 References

1. Adeleke, A. A., Odusote, J. K., Ikubanni, P. P., Lasode, O. A., Agboola, O. O., Ammasi, A., & Ajao, K. R. (2018). Dataset on the evaluation of chemical and mechanical properties of steel rods from local steel plants and collapsed building sites. *Data in brief, 21*, 1552–1557. doi:10.1016/j.dib.2018.10.162

2. Chang, P.-L. and K.-H. Lu (1995). "The Construction of the Stratification Procedure for Quality Improvement." Quality Engineering 8(2): 237-247

3. Mehta, R., Alam, F., & Subic, A. (2008). Review of tennis ball aerodynamics. *Sports technology, 1*(1), 7-16.

4. Merchant, M. P. (2005). *Propeller performance measurement for low Reynolds number unmanned aerial vehicle applications*. Masters' Thesis, Wichita State University.

2.9 Chapter 2 Study Problems

1. How do natural variation, and variation due to measurement error each contribute to the variation we find in a sample?

   Any phenomenon we study will vary naturally. As an example, the mechanism for opening my popup RV will vary from one of the same make, model, and year due to random differences in battery charge, windings on the linear actuator that opens the roof, the angle at which the hydraulics are placed by different technicians. I could measure each of these things, and each would show random variation if I take the measurement now, or later. These random issues are due to the limits of the precision of the instruments I use to take those measurements. There will also be some systematic measurement error, whereby the instruments I use are not perfectly calibrated.

2. Choose a phenomenon you might be interested in measuring. It could be anything from an object (its length, mass, volume, etc.), or a process (temperature, flow, coefficient of drag, etc.), or the perceptions of a person (attitudes, comfort, fit, etc.).

   (a) For that phenomenon, list some of the key ways in which it varies naturally. Then list the factors you would want to take into account in taking a representative sample of that phenomenon.

Answers will vary, but suppose I choose something as simple as mass of an apricot (I am eating one right now and I wish it had more mass). Mass of any apricot will differ from any other apricot, even those on the same tree subjected to the same conditions. If I wanted to take a representative sample of the masses of apricots, I might come up with a list of all the different varieties, and find out where they are grown. I would make sure that each variety would have some representation, and I would try to make the proportions of each variety match those in the marketplace. Finally, for each variety I would randomly sample them until I had the proportion I needed to be representative.

(b) **List some of the key considerations you would have to take into account in measuring the phenomenon both accurately and precisely.**

Mass can be measured accurately by calibrating a scale (a spring is generally less accurate overall, than a balance scale, digital scales can be extremely precise) to some known mass (good to an order of magnitude more precise than my instrument). I would want to eliminate too many air currents or vibrations to keep the instrument reading true. I would chop up each apricot into very small pieces and distribute them as evenly as possible in a common receptacle so that I eliminate any differences in center of mass on the scale. Together, these should give me accurate measurements each time. For precision, I need to do the same procedure, the same way, every time.

The table below shows the percentage of carbon measured in samples of steel reinforcing bars (rebar) taken from 11 collapsed building sites in Nigeria (Adelake et al., 2018). The engineers are interested in examining differences in quality in the steel that could explain the collapse of the buildings.

Building Site	Percentage Carbon in Samples of Steel
1	0.339
2	0.311
3	0.345
4	0.324
5	0.351
6	0.315
7	0.259
8	0.329
9	0.330
10	0.169
11	0.291

3. **Does there appear to be much variation in the distribution of carbon percentage—how precise is it?**

This depends on what you mean by "much variation." They are all at about 0.3

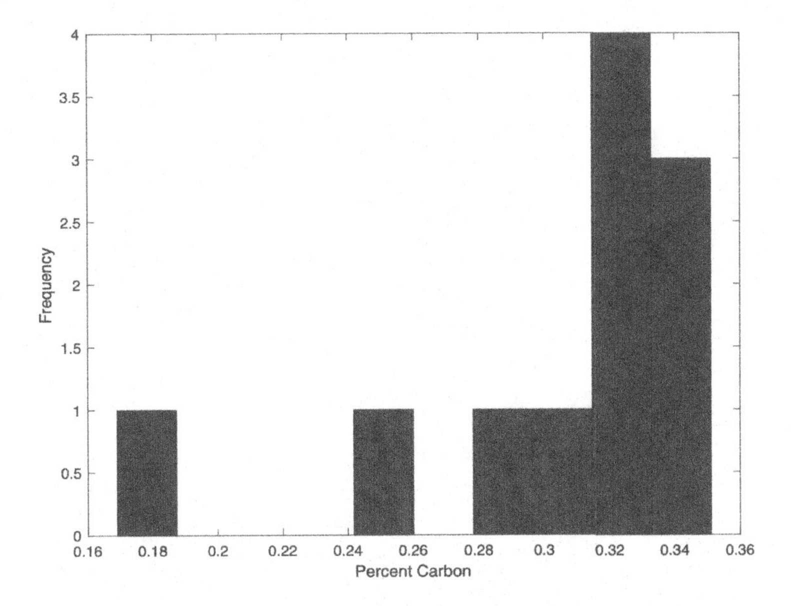

FIGURE 2.15: Histogram of carbon content in construction sites

% +/− 0.05 % or so, with the exception of site 10 which seems to be an outlier. This is pretty precise for specimens of mild steel used for construction.

4. Draw a histogram of the data. What is an appropriate bin size to effectively display the shape of the distribution? Now compare with the histogram in Figure 2.15.

   Using Matlab, I can draw a histogram with 10 bins:

   » hist(data,10);

   10 looks appropriate here. If I use 7 or 8 bins, I don't get much variation in the upper end of the distribution.

5. Is the distribution relatively symmetric, or are there values that appear to be "stretching" the distribution in one direction (i.e. outliers)?

   That one value in site 10 (0.169 %) appears to be stretching the distribution out to the left. It really isn't representative of the bulk of the data, which cluster around 0.3% or slightly above.

6. Suppose the true value of the mean carbon content of **all** steel reinforcing bars in Nigeria is 0.326%. What is the error of estimation for this sample and how accurate is it?

   In Matlab, I can use the function:

   » mean(data);

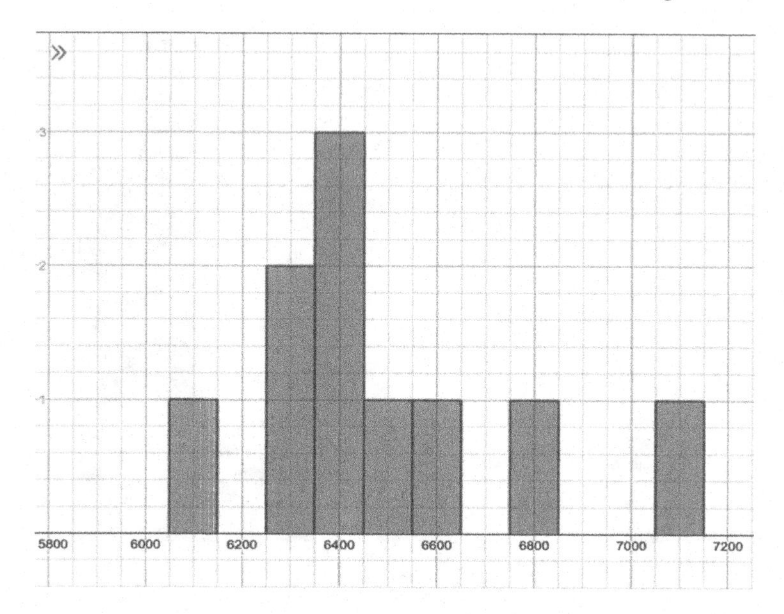

FIGURE 2.16: Histogram of propeller RPM (Merchant, 2005)

This returns the value of the arithmetic mean of our vector of carbon percentages: 0.3057.

So it looks like our sample underestimates the true value by about −0.0203%.

7. **An increase in carbon percentage will increase the yield strength of the steel, but reduce its ductility. If the ANSI standards for reinforcing bar lists maximum carbon content between .22% and .29% depending on the grade of bar, how do you evaluate the quality of this sample against those standards (higher does NOT mean better)?**

   If the standards state that the carbon content should be between .22 and .29%, our sample appears to be quite a bit larger than this interval. The bulk of our values lie above 0.29%, and the outlier 0.169% pulls the mean artificially low. I might conclude that the steel we have tested might be too brittle for use in the intended application.

   For his Masters' thesis at Wichita State University, Monaj Merchant measured the RPM for 10 presumably identical propellers, all under the same conditions (See Figure 2.16).

8. **If you were designing aircraft based on this data, what would be the (approximate) RPM you would expect to find for any given propeller drawn at random from the same population?**

   The middle of the graph is at about 6,400 to 6,500 RPM. Accounting for the more dense section just at and just below 6,400, I would hedge my guess and say about 6,400 RPM. That is about where the mean should lie.

9. **If you randomly drew a propeller from the population Merchant studied, would you be surprised to get a propeller that shows a performance of 7,100 RPM? What about 6,000 RPM?**

   There is only one value of 7,100 RPM or higher out of the 10 propellers. The probability, therefore, is estimated at about 1 in 10 or 0.10. Plus this value is in the skewed tail of the distribution. So I might raise an eyebrow if I randomly drew a value this high. For 6,000, this is much closer in value to 6,400, but there are no values less than 6,100 rpm. So I would also be kind of surprised if I got this measurement (but not as surprised as 7,100). With a small sample size like this, I am kind of uncomfortable making any bold claims.

10. **Knowing that there is always random natural variation in data, what would be the range of RPM that you would expect if you were to take a random sample in the future?**

    I would expect the vast majority of propellers to fall between 6,000 and 7,200 RPM with the bulk clustering about 6,400 RPM.

11. **If, instead of 10 propellers in this sample, this histogram represented 100 propellers, how might you change your answers to question 9 and 10?**

    If I had 100 props and the histogram looked the same, proportionally, I would be very surprised to get either value. If the propellers measured EXACTLY 6,000 or 7,100, I would be shocked, since there are so few in this representative sample.

    Merchant also took measurements of the pitch of his propellers. Here are the 10 measurements.

    3.9, 3.7, 3.67, 3.82, 3.76, 3.76, 3.67, 3.46, 3.24, 2.82

12. **What does this distribution look like? Draw a histogram and describe its center and variation. What number(s) would you design your aircraft for, if this sample is representative of the population of propellers?**

    I first put the data into a column vector, then I used Matlab to draw the histogram:

    » hist(x)

    The center of this distribution is somewhere around 3.6.

    The mean is 3.58 and the median is 3.69. The standard deviation is 0.3271.

    So, if I were designing an aircraft to use the average pitch propeller, I would select something in the range of 3.6 to about 3.7.

13. **Appealing to the Central Limit Theorem, what would be your best guess to be the mean of the sampling distribution of samples of size 10, taken from this population?**

    My best guess is the mean of the sample, 3.58.

14. **What would you expect to happen to the width of the distribution if, instead of 10 props, Merchant sampled 150?**

    The sampling distribution of means taken from multiple samples of size 150 should be narrower than the sampling distribution of means taken from multiple samples of size 10.

15. **Describe how the Law of Large Numbers enables us to estimate a population parameter using sample statistics.**

    As the size of a randomly drawn sample gets larger, the errors of measurement tend to cancel out. So, as the sample gets larger, the errors, on average are less, therefore larger samples are more accurate estimates of their population characteristics, than smaller ones. Because this error gets smaller and smaller as the sample size gets larger, the mean of the sample will approach the expected value of a random draw from the population—the population mean. For the variance, this is just a mean of the squared differences in measurements from the sample mean. That will also approach the population variance as the sample size gets larger by the same logic.

16. **Describe why the sampling distribution of the mean gets narrower as the sample size gets larger via the Central Limit Theorem.**

    Because larger samples are more accurate in predicting the population mean, the distribution of these predictions, a distribution of sample means all taken from the same population and having the same sample size, will tend to be narrower as the sample size gets larger.

17. **Why is simple random sampling an important assumption behind the LLN and the CLT?**

The errors that tend to "cancel out" as the sample size gets larger have to be random. If they aren't random, then some systematic error will be contained in the estimate of the true mean. This introduces inaccuracy in the estimate for each sample that contains the systematic error. For the CLT, each sample is in error by the same amount, so the center of the sampling distribution will be translated by a factor of that systematic error.

3

Types of Data

The reason we collect data, as we discussed in Chapter 1, is to get a handle on all the variation that exists for phenomena in the world. If you will forgive a sports metaphor, the meme below illustrates some of the problems we run into when we collect data in complex engineering applications.

FIGURE 3.1: Head to head comparison of two great athletes

Suppose you had collected data about the two phenomena shown in the meme in Figure 3.1. Each phenomenon has been sampled from the population of professional athletes. They are measured on the same 5 variables: Number of seasons played; Wins; Super Bowl losses; Home Runs hit, and Interceptions thrown.

One of the first things we see when we dig deeper into the data is that there is either a sampling problem, or a problem of **operational definition** of the variables measuring the performance of the phenomena in the population. Our sampling problem concerns the fact that the population from which the two phenomena were sampled is poorly defined. While professional athletes may be a perfectly legitimate population from which to draw data, comparisons of professional athletes on performance has to take into account the features of the sport they play. One cannot just compare electric scooters with racing motorcycles to determine which is "best." This latter consideration gets into the critical idea of how we define "success," what

DOI: 10.1201/9781003094227-3

variables contribute to this outcome, and the extent to which our measurement of these variables is rigorous, i.e., Accurate and precise.

An **Operational Definition** is an explicit, precise description of how the variable under study is measured. It includes the instruments used to do the actual measuring, the scale of measure by which the phenomena is measured, and the procedure by which measurements will be taken.

Ex: *The diameter of Wilson US Open Extra Duty tennis balls was measured using a Mitutoyo 500-196-30 AOS Absolute Digimatic Caliper (precise to 0.01 mm). Each ball was placed in the caliper such that the jaws did not align with a seam. The jaws were then closed to the point where they touched the nap of the ball, and this value was recorded. Then the jaws were closed further, compressing the nap until the experimenter felt significant resistance. This value was recorded. The mean of the two values, outside and inside was used as a measure of the diameter of each ball.*

One of the major tenets of science is that any experiment performed must be replicable. To be able to replicate an experiment, the exact procedures, including the measurement process must be detailed explicitly. Otherwise, a person who wished to validate the results of your study would not know what instruments to use, how they were used, nor potential sources of bias. Many of the problems I find in published reports in engineering research (and in my students' term papers) concern poor or incomplete operational definition. A good rule of thumb to use is: *If your friend can't just take your paper and replicate your data collection procedure precisely, then your operational definition(s) are inadequately articulated.*

Replicability is the capacity of statistical research to be repeated under the same conditions, using the same sampling procedure, instruments, methods, and analyses.

Replication is an attempt to produce the same results as a previous study, under the same conditions. If a study is replicated, the findings are considered robust across the conditions under which the study was performed.

The point is, the measure you choose has to be a valid and fair assessment of the phenomena you are interested in learning about, and the method of measuring the phenomena must be the same for all events in a sample. To ensure that these conditions are met, *it is best to write out a data collection and analysis procedure **prior** to collecting any data.*

But even if our measures are fair and valid (accurate), they can still be imprecise enough to be useless. Again, consider measuring the diameter of a tennis ball. We could measure diameters on scales ranging from 2 values (small, large) to multiple ordered values (small, small-medium, medium, medium-large, large), to something very close to continuous (eg., using a caliper good to +/− 0.01mm). Each of these measures can be useful for different applications. If, for example, we have a threshold diameter wherein by the Magnus Effect, the tennis ball begins to curve with a particular speed and rotational velocity, we may only need to know if the ball meets the threshold, or does not meet it. Hence a 2-point scale is efficient and simple to assess without error. However, for most applications in engineering, we need to measure our phenomena to very precise tolerances. In these cases, the measurements may approach continuity. *Choosing the appropriate scale of measure is critical to being able to adequately describe the phenomenon and to meaningfully compare it to other phenomena to determine which performs at a higher level than the other.*

3.1 Scales of Measure

There are four kinds of measures that a data analyst will be concerned with primarily. Each of these types of measure is fundamentally different in some ways than the others, requiring different mathematical models to make sense of them. One thing that is critical to keep in mind as you begin to design experiments and take measurements is this fact that *the questions you begin with drive the kinds of data you are able to collect.* These kinds of data, in turn, drive the kinds of analyses you can undertake. The analyses constrain the kinds of conclusions you can make about your data, and ultimately, this constrains the kinds of decisions you can make as a product designer or evaluator.

Scale of Measure is the degree to which a measured variable is conceptualized to be categorical, ordered and/or continuous. There are 4 scales of measure:

1. **Nominal,** where all measured values are unordered categories;
2. **Ordinal,** where all measured values have order, but the magnitude of the distance between measurements cannot be determined;
3. **Interval,** where the magnitude of the distance between measured values can be determined by subtraction, but where the origin, or zero of the scale is arbitrary; and
4. **Ratio,** where, because the scale has an absolute zero, the measured values can be compared by division.

These types of measurement are typically called "scales of measure" because they affect the mathematical properties of the variables we are measuring. Some of these scales only involve the use of categorical or unordered data, some use the integers, and some use the rational and real numbers. Just as we know that the mathematical operations that can be used for the integers is different from those we use for the rationals, the kinds of statistics we use for different scales of measure are different to account for these mathematical properties.

I will begin at the lowest level (nominal), and move in order to the highest levels. One thing to keep in mind as you become familiar with these scales is that data collected using the highest scales can be transformed to fit the lower scales, but not vice versa. Data that is collected on a lower scale cannot be transformed to a higher scale.

3.1.1 Nominal Data

Sometimes in industry, we are concerned with comparing how often a particular part succeeds or fails versus another part. We might examine this phenomenon to choose which type of part, from among a variety of options, will provide the most reliable performance in the product we are designing. An example of this might be comparing a brass valve with a stainless steel valve to determine which is best for the intake jet of a personal watercraft engine. In this application, the variable *Type of Valve* is not ordered. That is to say a brass valve is not greater than or less than a stainless steel valve. It is merely different.

In such an analysis, it makes no sense to find a mean value for this variable: here is no arithmetic for determining the average of brass and steel. What we have to do is determine which of the two conditions fails *the most* and which fails *the least.* So, we rig up a few engines that are identical in all respects except for the application of brass valves versus stainless steel valves. We randomly sample valves from our inventory for each of the conditions. And then we run the engines and count the number of failures we find in each.

We call this type of data **Nominal** because we are only concerned with categories. The labels (or names—Nominal means "name-related") for the categories are purely descriptive, they have no quantitative values. In general, the defining characteristics for nominal data include:

1. The categories exist as discrete classes;
2. the categories are non-ordered; and
3. the categories represent a complete description of the phenomenon we are interested in (e.g., we have either stainless steel or brass valves to use).

This latter characteristic is often split into two requirements Our categories must be: 1) mutually exclusive; and 2) collectively exhaustive. **Mutually exclusive** means that any single event can be classified into only one category: our valves can be either stainless steel or brass, but no valve can be both stainless steel and brass at the same time. **Collectively exhaustive** means that there are no other categories of events that we are interested in—all the events we are interested in are included in our analysis. Even if polymer intake valves existed, for some reason, perhaps their durability or thermal properties are sub-optimal, we aren't including them in our analysis as we are *only* interested in stainless steel and brass.

Because these categories aren't ordered, we can't compute mean values for them. Instead, the kind of statistics we compute for nominal data consist primarily of frequencies—counts—of the number of times a particular event occurs. If we collect lots of frequency data, we can then use fractions or percentages to describe *how often* in a sample, particular events occur. Those fractions are related in the long run to the probabilities that those events occur in the population of all possible events.

Nominal Scale Data consists of assignment of phenomena to qualitative categories. Nominal categories have no inherent order. Any number assigned as a value to nominal categories is just a label, it has no magnitude.

One of the most common uses of nominal data in engineering practice occurs in failure analysis. We have to determine if a product we have designed fails at a higher rate than we will tolerate. Reder, Gonzalez, and Melero (2016), for example, studied the failure rates of 420 wind turbines across three nominal categories: Geared drives smaller than 1MW, Geared drives larger than 1MW, and Direct drive turbines. The basic question asked in this analysis was, does any category experience more failures, on average, than the others? Table 3.1 shows their results. Notice that there is a categorical variable being measured: *Type of Wind Turbine Drive*. That variable has three values: Geared < 1MW, Geared ≥ 1 MW, and Direct Drive. Those variables have no inherent order among them. So, the engineers used frequencies (number of failures in a year) and then scaled the counts by the number of turbines to get a failure rate by which they could assess the quality of the three types.

Type of Wind Turbine Drive	**Failures per turbine per year**
Geared < 1MW	0.46
Geared ≥ 1 MW	0.52
Direct Drive	0.19

TABLE 3.1: Failures of wind turbines measured on a nominal scale

3.1.2 Ordinal Data

Some data has order, but we aren't quite sure what the magnitude of the differences between values are. In engineering, we often use decision charts that contain multiple attributes of data to create a *rank order* of design considerations based on a set of criteria we think are important: economic feasibility as well as performance, longevity, aesthetics and so on. It is typically unclear how to mathematically add the data from each of these variables together to create a nice continuous measure, so what we tend to do is choose criteria that rank highest across all the categories.

Suppose we need to design a flywheel for an automobile engine (see Table 3.2). We have a number of candidate materials we could use: 300M steel, 2024-T3 Aluminum, 7050-T73651 Aluminum, Ti–6AL-4 V Titanium Alloy, and a variety of composites. Which one is "best" for our application? A flywheel needs to be able to store kinetic energy, and hold up under continuous use in moderately hot conditions. So, we might look at criteria like fatigue versus density, fracture toughness, fragmentability, and price per mass ratio. We can then rank order the criteria on these variables. The Table 3.2 shows a ranking I did (1 is best, 10 is worst).

	Fatigue/ density	**Fracture toughness**	**Fragment-ability**	**Price/mass**
300 M	5	1	4	4
2024-T3	1	4	4	1
7050-T73651	3	3	4	1
Ti-6AL-4 V	6	8	4	5
E glass-epoxy FRP	2	2	1	3
S glass-epoxy FRP	4	7	1	2
Carbon-epoxy FRP	8	5	2	8
Kevlar 29-epoxy FRP	7	9	2	6
Kevlar 49-epoxy FRP	9	10	2	7
Boron-epoxy FRP	10	6	3	9

TABLE 3.2: Decision chart using rank ordering of different materials across 4 criteria

From the data, we can see that there is considerable variability across each of the criteria. If we look at any single criterion, we can see that the different materials are clearly ordered on each specific variable. For (resistance to) fragmentability, the S glass-epoxy rises to the top, but it ranks fairly low on toughness. Likewise, the Kevlar 49-epoxy FRP ranks high on fragmentability, but is dead last on Fatigue/density. But if we look across our criteria, we can make a good guess at the one(s) that ranks the best overall.

1. From the rankings across variables, what are your top three candidate materials for a flywheel?

Again, like Nominal data, we are primarily concerned with the *frequency* of occurence of a rank. In our data, we see that the Epoxy composites have two materials that rank 1, two materials that rank 2, and one that ranks 3. The metals all rank 4. This fact may be telling regarding how the type of material (composite versus metal) performs. But since the data are ordered, we can combine our criterion a bit differently. For example, we can create an overall rank by taking the median of the ranking for each material across the set of criteria as in Table 3.3.

	Fatigue/ density	**Fracture toughness**	**Fragment-ability**	**Price/ mass**	**Median Rank**	**Overall Rank**
300 M	5	1	4	4	**4**	**4.5**
2024-T3	1	4	4	1	**4**	**4.5**
7050-T73651	3	3	4	1	**3**	**2.5**
Ti-6AL-4 V	6	8	4	5	**6.5**	**7**
E glass-epoxy FRP	2	2	1	3	**2**	**1**
S glass-epoxy FRP	4	7	1	2	**3**	**2.5**
Carbon-epoxy FRP	8	5	2	8	**6.5**	**7**
Kevlar 29-epoxy FRP	7	9	2	6	**6.5**	**7**
Kevlar 49-epoxy FRP	9	10	2	7	**8**	**10**
Boron-epoxy FRP	10	6	3	9	**7.5**	**9**

TABLE 3.3: Decision chart with median rank summary

This tells us that some of our materials have, on average, better overall ranking than others. In the far right column in Table 3.3, you can see how we have translated these ranks into an overall ranking. Handling ties is not as tricky as it seems. All rank ordered data has a sum such that

$$\sum_{i=1}^{n} x_i = \frac{n(n+1)}{2}$$

And, because all ordinal data consists of consecutive integers denoting the ranks, any ordinal variable has a median such that

$$med = \frac{(n+1)}{2}$$

Looking at the equations for the sum and median of a set of ordinal data, you can see that the median is also the mean value.

$$med = \frac{n(n+1)}{2n}$$

$$= \frac{\sum_{i=1}^{n} x_i}{n}$$

Since there are 10 pieces of data, $n = 10$, and the sum of the ranks = 55, the sum of our ranks with ties has to be 55.[1] Beginning at the highest rank (or the lowest in this case), we assign it a 1 because there is no tie. The next two materials, 7050-T73651 and S glass-epoxy FRP have the same median rank across the criteria. They would ordinarily be assigned an overall rank of 2 and 3. But because they are tied, average of the tied ranks $\frac{(2+3)}{2} = 2.5$. The same applies for the next two values. 300 M Steel, and 2024-T3 Aluminum both have the same median rank across criteria. In our overall ranking they would ordinarily be assigned ranks of 4 and 5, but because they are tied we give each $\frac{(4+5)}{2} = 4.5$. The last three ties, Ti-6AL-4 V, Carbon-epoxy FRP, and Kevlar 29-epoxy FRP, all share the same median rank. They would ordinarily be assigned ranks 6, 7, and 8, but because they are tied, we take their mean value $\frac{(6+7+8)}{3} = 7$.

It is important to note that, though we took a mean rank, we are still not treating the mean rank for Carbon-epoxy and Kevlar Epoxy as if they can be subtracted to achieve some kind of absolute distance between the two measures. We are merely using the mean ranks as a way to talk meaningfully about where those materials appear in the distribution of ranked materials.

Ordinal Scale Data consists of *discrete* values that have some meaningful order (e.g., 1st, 2nd, last, etc.). The interval between units may or may not be equal.

Ordinal scale data has the advantage over Nominal data, because the order of the units can be thought of as degrees of quality, or of performance. By ranking data, we can begin to discern which values appear higher than others, more often, and use this information to choose the "best" option from a number of candidates.

1. Sometimes engineers don't want to weight each of the criteria equally. For example, fracture toughness and fragmentability may be considered the most important criteria, because those two predict failure of the part. If you were to weight the criteria, placing special emphasis on toughness and fragmentability, develop a way to compare the different samples.
2. How does the method you developed in question 1 change (or not) the overall rank ordering of the materials?
3. What are other variables that can be described using ordinal scales? Try to come up with 3 different ones, for example: area of home for HVAC installation: $<1{,}000\ \text{ft}^2$, $1{,}000\ \text{ft}^2 \le x < 1{,}500\ \text{ft}^2$, $1{,}500\ \text{ft}^2 \le x < 2{,}000\ \text{ft}^2$, $2{,}000\ \text{ft}^2 \le x$.

Note that in ordinal data, we use the *median* as our primary measure of the center of the distribution. The mean, because the rankings in the data have no magnitude, does not have the same meaning as when data are measured on a continuous scale. This is the advantage of Interval and Ratio Scale data.

[1] This is an example of the now famous story of Karl Friedrich Gauss, one of the most prolific mathematicians ever, when he was 7 years old, proving that the sum of n consecutive integers is the quadratic of the form American Scientist online Volume 94 Number: 3 Page 200 Gauss's Day of Reckoning: A famous story about the boy wonder of mathematics has taken on a life of its own, Brian Hayes

3.1.3 Interval Data

There is quite a bit of inherent imprecision in both nominal scale and ordinal scale data. In our example of the flywheel material, the E-glass Epoxy FRP, has the best rank, but we really don't know how much better it is than the 2024-T aluminum. To be able to compare two or more data points to see how much better (or bigger, or faster, or harder, etc) one is versus another, we need to move up to a scale of measure that allows us to subtract values from each other to determine that difference.

Temperature, for example, is a measure of the average internal energy of a system. I measured the temperature outside today (July 23) in Tempe, AZ. It was a blazing 42 degrees C. Last week, when I was in Johannesburg, South Africa, it was a comfortable 21 degrees C. While I can faithfully state that the temperature in Tempe is 21 degrees warmer than what I experienced in South Africa, I **cannot** conclude that it is twice as warm.

The obvious advantage of using data on an interval scale is the ability to directly compare the magnitude of the measures. Simple subtraction shows the unit difference between the magnitude of any two phenomena. However, to say that one measure is a multiple of another, using proportional reasoning would be in error for interval data. Recall that the internal energy of a system is basically the degree to which the particles in that system are moving, i.e., their kinetic energy (KE). KE for atomic particles ranges from zero for particles that are not moving at all, to about 51 joules for high energy cosmic waves traveling over 99% of the speed of light (this is roughly equivalent to the KE of a baseball thrown at 58 miles per hour (94 kph), Bird, et al., 1995). We know particles at 0 degrees Celsius **are** moving. They **are** moving around, either vibrating in solid form, or careening around in fluid form, just at a lower velocity than at say, 100 degrees C. Regarding my weather measurements, the air particles in Tempe are moving and colliding faster, on average, than those in Johannesburg. But since degrees Celsius (and Fahrenheit) use an *arbitrary* zero, we cannot compare them proportionally—so cannot take the ratio of the two: 42 degrees C is not twice as hot as 21 degrees C. This is why, in calculations that involve proportional relationships, we have to convert temperature to degrees Kelvin, which has a non-arbitrary, exact zero—the temperature at which all particle motion ceases.

The essential properties of interval data are these:

1. The phenomenon is measured on a continuous scale; and
2. There are equal intervals between units on the scale. This means, for example, that the difference between 1 and 2 units is the same as the difference between 7 and 8 units.

Interval Scale Data consists of data drawn from a continuous random variable, with equal intervals between consecutive units.

3.1.4 Ratio Data

If, like degrees Kelvin, phenomena are measured on a continuous scale, with equal intervals among units, and with an non-arbitrary zero, they *can* be compared by their ratio. Engine torque, for example, can be measured using a dynamometer. A standard-equipped 2012 Dodge Charger can produce about 260 pound-feet of torque. The standard 2012 Chevrolet Camaro, can produce about 278 pound-feet. That is a difference of about 18 pound-feet of torque—simple subtraction. But because torque has a non-arbitrary zero, i.e., because zero literally means there is no torque in the direction I am measuring–I can also say that the Camaro has $\frac{278}{260}$, or *106.9%* of the torque displayed by the Charger. Because we can compare data using division we call this highest level of data: **Ratio** data.

Ratio Scale Data consists of equal interval data, drawn from a variable that has a non-arbitrary, or *absolute* zero.

The advantages of interval and ratio data is that we can compare their magnitudes, unlike ordinal data where we can only compare their positions, or nominal data, their frequencies. Table 3.4 shows how each of the levels of data compare with their defining criteria. It is not so important to memorize the criteria for each level, but to recognize that each level constrains the kind of mathematics one can perform on the data, and the kinds of conclusions one can make. With nominal data, one could distinguish the insulative qualities between two types of insulation, say closed cell foam, and Reflectix™. With ordinal data, for example, one might be able to distinguish the insulative qualities of thicker versus thinner closed foam, but not be able to determine the relationship the insulative properties have to the actual cross sectional thickness. But with interval and ratio data, the engineer is able to compare two or more measured samples and model the mean temperature change for each, thus gaining insight regarding the magnitude of the difference by comparing their rate of change.

3.2 Population Parameters and Sample Statistics

When one is comparing one sample to another on various criteria, it is easy to discern differences in the two. This for the drab discovery of the obvious: Any two randomly drawn samples will differ in their center and spread. The questions for engineers is *how much do my samples differ,* and *is this something to be expected given natural variation, or is it something unexpected and therefore due to some systematic factor that influences the difference?* This is where the distinction between a sample and the population from which it is drawn becomes important.

Back in Chapter 2, we defined characteristics of samples as distinct from the populations from which they were drawn. Recall:

	Nominal	**Ordinal**	**Interval**	**Ratio**
Discrete	**X**	**X**		
Continuous			**X**	**X**
Can compare with Frequencies/Proportions	**X**	**X**	**X**	**X**
Uses Mode as Measure of Center		**X**		
Uses Median as Measure of Center		**X**	**X**	**X**
Uses Mean as Measure of Center			**X**	**X**
Can compare the difference between measures using subtraction (i.e., a distance)			**X**	**X**
Can compare the difference between measures using division (i.e., a rate)				**X**
Has arbitrary zero			**X**	
Has true zero				**X**

TABLE 3.4: Features of different scales of data

Sample Statistics are characteristics of samples that are used to represent the characteristics of populations. More precisely, sample statistics are numbers used to represent the parameters of the distribution function that describes the population from which the sample was drawn.

Population Parameters are features of populations that correspond to important characteristics of their distribution: Minimum, Maximum, Mode, Median, Interquartile Range, Mean, and Standard Deviation. More precisely, population parameters are variable quantities that, when fixed, fully describe the frequency (or probability) distribution of the population.

In plain terms, Sample Statistics are just summary numbers that describe corresponding features of the populations from which the sample was drawn. In Algebra, we defined the behavior of certain commonly-encountered continuous functions using their parameters. For example, a quadratic function can be described generally as $f(x) = ax^2 + bx + c$, where a, b, and c are parameters: in this case, real numbers that govern the position of the parabola defined by the function (b determines horizontal shift, c determines vertical displacement), its width (a), and whether it opens up or down (sign of a). Depending on how the function is written, the parameters define the line of symmetry and location of the vertex. Likewise, for exponential functions, $f(x) = ab^x$, a and b are parameters. a corresponds to the intercept of $f(x)$, and b to the growth rate.

3.2.1 Parameters

For these common functions, if we know each of the parameters, the behavior of the function is fully known. I can draw it and compute any value for it across the Real numbers. Also, if the function is differentiable, I only need to compute the derivative to find areas of local maxima and minima, and second derivatives to find points of inflection and regions where it is curve up or curve down.

Because, as we established earlier, distributions are functions with independent variable corresponding to our measurements, and dependent variable corresponding to their respective frequencies (or proportions), knowing the parameters of those functions would fully fix their behavior. This is one of the major pursuits of the field of statistics writ large, and of engineering statistics in particular, figuring out what those parameters might be for a function that models the phenomenon we are trying to understand. The most commonly used distribution functions include the **Binomial** (for nominal scale data):

$$P(X = x|p, n) = \binom{n}{x} p^x (1 - p)^{n-x}$$

Read, "The probability that X is equal to some value x given p and n, is equal to the total combinations of x in n draws, multiplied by the proportion of times x can occur."

where

$$\binom{n}{x} = \frac{n!}{x(n - x)!}$$

The parameters n, the sample size and p, the probability of x occurring in a random draw fully define *P(X)*. We will study the Binomial function in Chapter 5. But for now, just notice that P(X) increases for the first half of its domain, and decreases for the second half, symmetric about a central maximum value.

The **Poisson** distribution also is used to describe nominal data, where we are counting the number of times an event occurs within an interval (say, how many times a particle hits a sensor per hour).

$$P(x|\mu) = \frac{e^{-\mu}\mu^x}{x!}$$

Its parameter, μ the average number of times an event occurs in the interval over the long haul, affects the shape of the distribution. When it is small, $P(x)$ resembles a negative exponential function. But as μ grows larger, the function becomes more symmetric about μ and reaches a maximum at $x = \mu$. We will also take up the Poisson distribution and its uses in Chapter 5

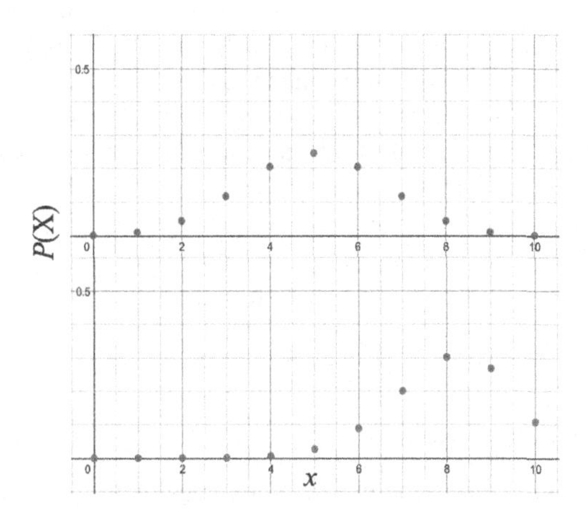

FIGURE 3.2: Shape of binomial distribution as p is varied. n is 10 trials for both illustrated distributions. The distribution on the top is drawn for $p(x) = 0.5$. The bottom distribution is drawn for $p(x) = 0.8$

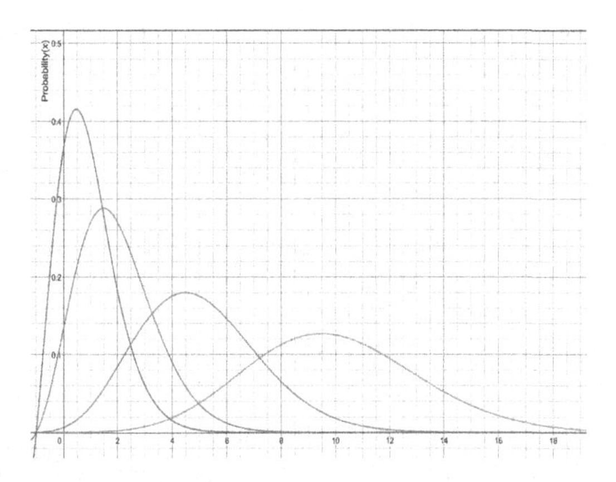

FIGURE 3.3: Shape of the Poisson distribution depends only on the value of μ. After about $n = 10$ or so, the normal distribution can be used as an approximation

The final distribution we will discuss in this introductory chapter is the **Normal** distribution (for interval and ratio data):

$$P(x) = \frac{1}{\sigma\sqrt{2\pi}} e^{\frac{-(x-\mu)^2}{2\sigma^2}}$$

There are a lot of symbols there to account for, but only two *parameters*: the mean of the distribution, and the standard deviation (see Figure 3.4).

The curve defined by this function looks like a bell, with center at μ (where $\frac{dP}{dx} = 0$), and points of inflection ($\mu +/- \sigma$) at (where $\frac{d^2P}{dx^2} = 0$) . When we vary μ, the distribution translates left or right on the independent variable. Varying σ dilates the distribution, making it wider or narrower. Most importantly, *all we have to know are these two values, mu* and σ, *to completely describe the distribution.*

In the 18th century, our old friends de Moivre, Laplace, and Gauss developed and applied this function to the behavior of the binomial distribution as the sample size approached infinity, and to the study of measurement error (Havil 2003). It has since been more associated with Gauss, and all distributions that have this exponentiated quadratic form are termed *Gaussian*, which seems to me like kind of a slap in the face to de Moivre.

3.2.1.1 Population Parameters

We use these functions to describe the characteristics of *populations*. What exactly is a **population**? For our purposes we define a population as all the potential events we are interested in for making inferences from our statistical analyses. For example, one of my capstone design teams designed a portable Computer-Numeric-Controlled Routing table that could be attached to walls and other vertical stock via a vacuum pump (pretty slick!). But one of the questions they had to grapple with was, the table would vibrate as the motors moved the router in the x, y, and z directions. How much did it vibrate on average? How much did the router vary in displacement from the targeted locations? To assess these questions, they examined industry standards for standard CNC machines, found the average displacement (μ) and the standard deviation (σ) and, using the Normal Distribution, created a model of the entire population of standard CNC machines. They then ran their machine many times (over 100), in controlled patterns on standard stock, measuring the actual cut versus the programmed cut. They then compared the mean displacement and its standard deviation with the population they had modeled. They found that the vacuum mounting system did not show significantly greater vibration (using the mean), or greater variation (using the standard deviation) than standard models.

A **Population** is the set of all possible measurements of the phenomenon for which inferences are to be made from the data. Example: When one is testing the efficiency of a wind turbine, we are interested in not just the behavior of a sample, but in the behavior of *all* wind turbines that share the same specifications as the model we are testing.

The lesson here is that, if you have a good model of the population you are interested in describing, and its parameters, you can compare the characteristics of the

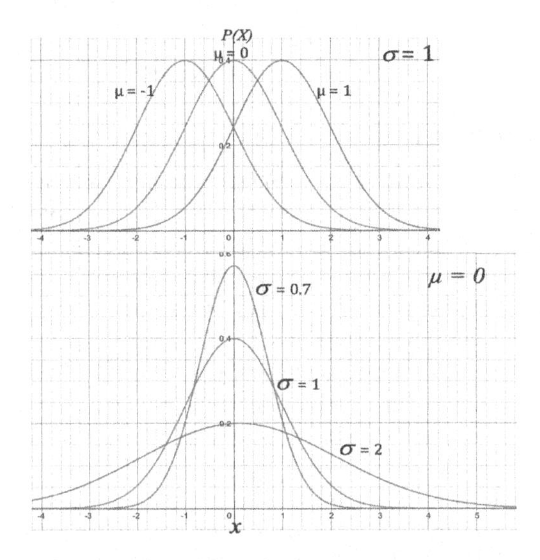

FIGURE 3.4: Effect of changing parameters μ, and σ, on the shape of the normal distribution. Note that the center of each distribution is μ,and the point of inflection is σ

sample to that population to determine if your sample is indicative of that population, or if it shows some significant differences in either its center or its variability.

> A **Sample** is the set of measurements taken from a *subset* of the population for which inferences are to be made. Example: You don't have the money or the time to measure *all* wind turbines of a given design. Instead, you carefully select a subset from the population of wind turbines you are interested in, and use its measurements to infer the characteristics of the population from which it was drawn.

A second lesson is that, when you don't have published population parameters, and when you don't have the resources to measure every single, stinking, phenomenon in a population, you can estimate its parameters fairly accurately by taking sample statistics.

3.2.1.2 What Are the Important Sample Statistics That Model Population Parameters?

Earlier, we found that the different scales of data utilize specific statistics that describe their parameters. Here, we will describe the different sample statistics you will use as an engineer, and how they are computed. We begin with statistics that summarize nominal data, and then move to ordinal, and then interval/ratio, culminating in the Normal Distribution as a general model that can be used to think about the shape of data, and the parameters that describe population distributions.

3.2.1.3 Nominal Data: *n, p*

Because Nominal data just consists of frequencies (counts) within categories, we can only do a few things with the data: Count it for each category, or take its proportion of the total number of observations. The most common population distributions for Nominal data are the Binomial, Poisson, and Hypergeometric probability distributions. Don't worry about the word, *probability*. It just means the proportion of times any value of the independent variable arises. For example, we might be designing an EEG headset to record the brain activity of people involved in Alzheimers rehabilitation. We want to know what proportion of the population of people have large or small head circumference to be able to determine the best number of each size to produce.

Taking a random sample of size 10 from the population of heads from Alzheimers patients, we might come up with the following numbers: 6 people in our sample have large heads, (60%) and 4 people have small heads, (40%). If we were confident that our sample is representative of the population of people who suffer from Alzheimers, we might use these proportions to estimate the proportions in the population. Our best guess would be that about 60% of the population have large heads and 40% have small heads—pretty boring, really.

Once we have this proportion, this estimated probability of large heads figured out, we can use it to estimate the **Expected Value** of the number of people in any randomly drawn sample that should have large heads:

$$E(x_i) = x_i p_i$$

Read, *The expected value of the ith value of x is equal to* x_i *times its probability of occurring.* This is equivalent to saying:

$$E(x_i) = x_i \frac{f(x_i)}{n}$$

Where $f(x_i)$ is the frequency that x_i occurs. Summing the individual $E(x_i)$ provides the Expected Value of the whole, the value with the highest probability of occurrence:

$$E(X) = \sum_{i=1}^{n} E(x_i) = \sum_{i=1}^{n} x_i p_i$$

$$E(X) = \sum_{i=1}^{n} \frac{x_i}{n}$$

Notice the formula for $E(X)$ is equivalent to that of the arithmetic mean. This identity can be confusing when we examine nominal data. It seems akin to finding the average of apples and oranges. But take a look. If we provide the outcome of drawing an apple with the value 0 (a *failure*) and an orange the value of 1 (a *success*) and take a random sample from my fridge (the population of fruit I am interested in), I might get 6 apples (60%) and 4 oranges (40%). Let's find the expected value of an orange $E\left(orange\right)$:

$$E\left(x_i\right) = x_i p_i$$

$$E\left(orange\right) = 1\left(40\%\right) = .4$$

So, over the long haul, I would expect, if I put all my fruit back in the fridge and kept sampling 10 out each time, I would get 4 oranges on average for each sample. This means that I would get 6 apples on average. But zero, the value of x_i for an apple, times 60% is not .6.

For nominal data, we have to define one category with which to compare the other(s). In our example, we defined drawing an orange as a success (perhaps we are making a citrus punch). Because its probability is 0.4, and we know the sum of the probabilities have to be 1, we can express the probability of drawing an apple as:

$$p\left(apple\right) = p\left(\sim orange\right) = 1 - p\left(orange\right)$$

So, the probability of *not drawing an orange* is equivalent to the probability of drawing an apple. Its Expected value can then be computed as the expected value of not drawing an orange.

$$E\left(x_i\right) = x_i\left(1 - p_i\right)$$

$$E\left(apple\right) = x_{orange}\left(1 - p_{orange}\right)$$

$$E\left(apple\right) = 1\left(1 - 40\%\right) = .6$$

3.2.1.4 Symmetry of the Binomial Distribution

We could have defined the draw of an apple as a success (labeling it a 1) and an orange as a failure (labeling it a 0). The result would have been the same, as far as computing the expected value of each. Using the arithmetic mean, we find that with 0 and 1 as codes for drawing an orange (x_i) and drawing an apple ($\sim x_i$), we get the same long-term expectation.

$$E\left(X\right) = \sum_{i=1}^{n} \frac{x_i}{n}$$

$$E\left(X\right) = \frac{\left(0 + 0 + 0 + 0 + 0 + 0 + 1 + 1 + 1 + 1\right)}{10}$$

$$E\left(X\right) = 0.4$$

Expected Value is the long-haul average of a random variable. It is the sum of all possible values, each weighted by its individual probability.

For binomial data like the Alzheimers EEG headset example, or its equivalent—apples and oranges, there are only two categories. We distinguish them by labeling one 0 and the other 1 (Binomial means two numbers. Go figure). Because there are only two, the E(X) for the one we choose as our comparison category (Oranges) is the expected value of the distribution. This expected value is the **center of the binomial distribution:**

$$E(X) = np$$

The variation around the center for binomial data is just those values that don't equal the comparison category (Apples). So the expected value of the variation is equal to the expected value of ~ x or:

$$E(\sim X) = E(X)\,p(\sim x) = np(1-p)$$

Notice that these important statistics, $E(X)$ and $E(\sim X)$, only depend on the number of datum, n, and the probability of a "success" —the probability of an event in the comparison category being selected at random. Since it only has two options, 0 and 1, we will use the binomial distribution as the simplest model to understand how these parameters work. As we gain more options, the binomial distribution becomes **multinomial**, but the same basic sample statistics are used.

For the Poisson distribution, μ is the average number of times an event occurs in an interval (for example, the average number of times a car passes a sensor on a lonely road in Montana each hour). It is the expected value of the Poisson distribution—its long-haul average. Like the binomial, it represents the **center of mass** of the distribution, and as μ grows larger, the shape of the distribution becomes more symmetric about the expected value.

The Summary Statistics for Nominal Data are the proportion, p (also called the *probability*) that a category has relative to the other categories in the data set, and the sample size, n.

3.2.1.5 Ordinal Data: *Med, Interquartile Range*

For ordinal data, though the data are now ordered, we can't just add and subtract numbers to get the expected value. The center of ordinal distributions is best measured using the median. We showed earlier in this chapter that, for rank ordered data, the median is equivalent to the arithmetic mean, and so is the appropriate measure of the expected value. Within a sample, the distribution of ordinal data is always symmetric about the median.

Determination of the appropriate sample statistic to assess variability in ordinal data is more difficult. Because it is kind of in-between nominal and interval scales, people often use methods appropriate for either of these two scales to analyze ordinal data. For the purposes of this introductory course, we will stick to a simple statistic: The **interquartile range** the middle 50% of the distribution.

> **The Summary Statistics for Ordinal Data** are the median, and the interquartile range

3.2.1.6 Interval/Ratio Data: *Mean, Standard Deviation*

You should be noticing that this idea of *Expected Value* keeps coming up when we talk about parameters and the statistics that we use to estimate them. In general, the parameters of population distributions consist of some expected value, and some deviation from the expected value. For the binomial, we had *np* and *np(1–p)*, respectively. For ordinal data, we punted and said that for some cases we treat it as if the data were categorical, and for other cases, as if it were equal-interval. But understand that, in both of these cases, and in more sophisticated renderings of ordinal data, the parameters that model the long-haul distributions of ordinal data utilize the expected value and some average deviation from the expected value as parameters that describe some hypothetical parent population.

> The Parameters of a Population include its *expected value*, and some measure of *deviation* of data from the expected value.
> Sample Statistics that model a population *include these parameters.*

As an example to illustrate how sample statistics are used, refer to Figure 3.5:

In Figure 3.5. We show a histogram of actual data collected. This data has a sample mean of 4,824 psi, and a sample standard deviation of 387 psi. I have overlaid a Normal Curve with the following formula:

$$f(x) = \frac{1}{387\sqrt{2\pi}} e^{\frac{-(x-4,824)^2}{2(387)^2}}$$

Substituting the sample statistics for the population parameters. The resulting curve is not perfect, but it is a very close rendition of what we might envision the data looking like if, using the Law of Large Numbers, we increased the number of measurements to near infinite.

This is what we mean when we say that sample statistics are used to estimate population parameters. If we have taken a good, representative, sample, then the sample mean will be close to the population mean, and the sample standard deviation will be close to the population standard deviation. So, we could fade the background histogram away, and treat the nice differentiatable curve *as if* it were the true population.

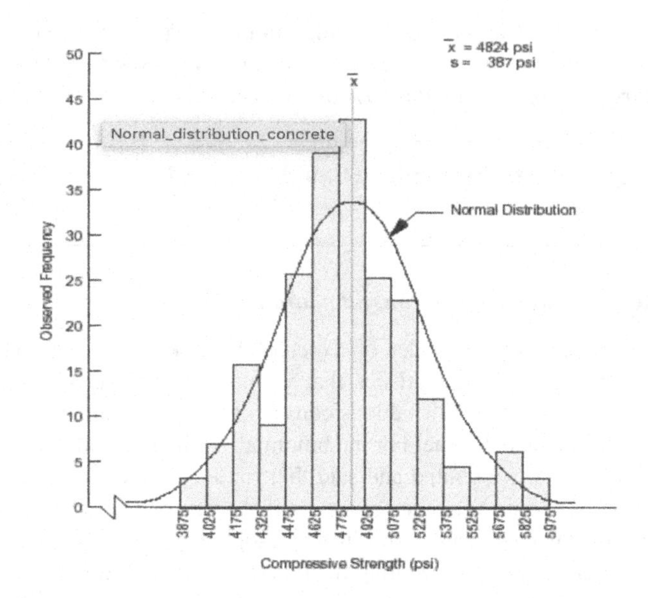

FIGURE 3.5: Compressive strength measurements of Portland cement concrete. https://www.pavementinteractive.org/reference-desk/qc-qa/statistical-acceptance/normal-distribution

There would be some error of estimation, of course, but having the closed form function as our model, allows us to estimate *any value of f(x)* for *any value within the domain of x*. The better our sample, the closer our estimates can be computed.

> **The Summary Statistics for Interval/Ratio Data** are the *mean* and *standard deviation*

3.3 The Sample Mean and Standard Deviation as Robust Estimators

If we are using sample statistics to estimate population parameters, we need to be able to clearly distinguish the two so we know if we are talking about a population or a sample. To do this, we will reserve special symbols for both cases:

- For *Population Parameters*, we use the Greek letters:

  - μ stands for the population **mean**—its expected value. Its location is the center of the Normal Distribution
  - σ^2 stands for the population **variance**—the average squared distance each data point is from the center of the distribution; and
  - σ stands for the population standard deviation. $\mu \pm \sigma$ corresponds to the points of inflection on the graph of the Normal distribution
  - We will also use a capital N to stand for the population size, if it is finite

- For *Sample Statistics*, we use the Indo/Arabic letters:
  - $\overline{x}$ stands for the sample mean, its center of mass. Its location is the center of mass of the sample data
  - s^2 stands for the sample variance—the average squared distance each data point is from the sample mean; and
  - s stands for the sample standard deviation, $\overline{x} \pm s$ is the width of the **mid-spread**, the middle 68% of the data.
  - We will use a lower case n to stand for the sample size.

Calculating these values is pretty straightforward. We all know the formula for the population mean:

$$\mu = \frac{\sum_{i=1}^{N} x_i}{N}$$

Just add up all the numbers and divide by N. For the sample mean, the procedure is exactly the same:

$$\overline{x} = \frac{\sum_{i=1}^{n} x_i}{n}$$

Calculating the population variance is also straightforward: It is the average squared distance each data point is from the mean:

$$\sigma^2 = \frac{\sum_{i=1}^{N} (x_i - \mu)^2}{N}$$

Squaring the distances weights extreme values more than those near the mean, so a distribution with a large variance typically has more extreme values than distributions with smaller variances.

We could have alternatively selected the absolute value of the differences as the numerator for our measure of variation. This is perfectly legitimate, but the mean absolute deviation does not possess the mathematical properties that the squared deviation possesses, namely that a unique local minimum can be established through differentiation. So, to use calculus as a modeling tool, statisticians have historically opted to use the sum of the squared deviations as the numerator of the variance, as our primary measure of variation in a distribution.

The sample variance, s, is not so straightforward. It is computed slightly differently:

$$s^2 = \frac{\sum_{i=1}^{n} (x_i - \overline{x})^2}{n - 1}$$

The only difference is that denominator of the sample variance is 1 less than the sample size. We do this because the sample variance will tend to *underestimate* the population variance without this correction (called *Bessel's correction,* after the mathematician who first proved it). From linear algebra, the sample has reduced degrees of freedom once we fix a parameter. To calculate the variance, we must first calculate the sample mean.[2] Once we fix $\overline{x}$ at some value, we only have freedom to calculate $n-1$ values of $(x_i - \overline{x})^2$ before the final value is fixed. So the average, the mean value of $(x_i - \overline{x})^2$, is also fixed at n−1 iterations. That means that $n - 1$ is the appropriate denominator.

We will see later on in this book, when we compute statistics with two or more parameters that the denominator of the variance is likewise reduced by the number of parameters we are fixing. But notice that, with large sample sizes, the difference subtracting 1 makes becomes negligible, and the sample variance approaches the population variance as *n approaches N.*

The sample standard deviation, s, is just the square root of the variance, s^2.

$$s = \sqrt{\frac{\sum_{i=1}^{n} (x_i - \overline{x})^2}{n - 1}}$$

Think about the standard deviation as a distance—the average distance your data values are from the sample mean. Like the variance, it *is* a distance, in this case similar to the Pythagorean distance you learned in high school Geometry. Unlike the variance, it is scaled *in the same units* as the independent variable, and $\overline{x} \pm s$ corresponds to the points of inflection on the Normal Curve of the population with $\overline{x} = \mu$ and $s = \sigma$.

The sample mean, variance, and standard deviations are **unbiased estimators** of their respective population parameters. We proved this for $\overline{x}$ in Chapter 2 when we showed that the sample mean, over the long haul, will average to the population mean from which samples are drawn (by the Law of Large Numbers).

> **An Unbiased Estimator** is a statistic that will, if the population is sampled and the statistic calculated over and over again, be exactly the same value as the parameter it is estimating on average.

[2]Note that for population parameters, μ and σ, these values are not computed. They are equivalent to their respective formulas, but they define the population distribution. So, the mean is not calculated prior to the variance, hence the variance has a denominator of n (all parameters are free).

The sample variance and standard deviations, s^2 and s, respectively, require a bit more involved proof.

Assumptions:

1. $E(X) = \mu$
2. $E(x_i - \mu) = \sigma^2$
3. $E(x_i - \mu) = s^2$
4. $s^2 = \frac{\sum_{i=1}^{n}(x_i-\overline{x})^2}{n}$ (note that this is from assumptions 2 and 3, and is NOT our actual formula for the sample variance)

Proof:

1. $E\left(s^2\right) = E\left(\frac{\sum_{i=1}^{n}(x_i-\overline{x})^2}{n}\right)$
2. $= \frac{1}{n}E\left(\sum_{i=1}^{n}(x_i - \overline{x})^2\right)$
3. $= \frac{1}{n}E\left(\sum_{i=1}^{n}((x_i - \mu) - (\overline{x} - \mu))^2\right)$
4. $= \frac{1}{n}E\sum_{i=1}^{n}\left((x_i - \mu)^2 - 2\sum_{i=1}^{n}((x_i - \mu)(\overline{x} - \mu))^2 + \sum_{i=1}^{n}(\overline{x} - \mu)^2\right)$
5. $= \frac{1}{n}\sum_{i=1}^{n}\left((x_i - \mu)^2 - nE\left((\overline{x} - \mu)^2\right)\right)$
6. $= \frac{1}{n}\sum_{i=1}^{n}\left(\sigma^2 - n\frac{\sigma^2}{n}\right)$
7. $= \frac{1}{n}\left(n\sigma^2 - \sigma^2\right)$
8. $E\left(s^2\right) = \frac{n-1}{n}\sigma^2$

Step 8 *cannot* be true if our assumptions number 2 and 3 are both true. What *will* make it true is substituting the denominator of $n-1$ in our calculation of s^2. (I know most of you believed me prior to this proof, but there are always some who need to see the math, so here it is!)

The takeaway from this is: *Always remember that the sample variance and its square root, the sample standard deviation have this correction built-in so that they will be unbiased estimators over the long haul.*

3.4 References

Bird, D. J., Corbato, S. C., et al. (1995). Detection of a cosmic ray with measured energy well beyond the expected spectral cutoff due to cosmic microwave radiation. *The Astrophysical Journal*, *441*, 144-150.

Busby, J. T., Hash, M. C., & Was, G. S. (2005). The relationship between hardness and yield stress in irradiated austenitic and ferritic steels. *Journal of Nuclear Materials, 336*(2-3), 267-278.

Reder, M. D., Gonzalez, E., & Melero, J. J. (2016, September). Wind turbine failures-tackling current problems in failure data analysis. In *Journal of Physics: Conference Series* (Vol. 753, No. 7, p. 072027). IOP Publishing.

Sokolsky, P.; Sommers, P.; Tang, J. K. K.; Thomas, S. B. (March 1995). "Detection of a cosmic ray with measured energy well beyond the expected spectral cutoff due to cosmic microwave radiation". *The Astrophysical Journal. 441,* 144

3.5 Study Problems for Chapter 3

Suppose you collected the following data about the failure rates of three random samples containing different rubber formulations of o-rings. Each sample is subjected to the same thermal conditions (0 is coded for failures, 1 is coded for successes):

Trial	Formulation A	Formulation B	Formulation C
1	1	0	0
2	0	0	1
3	0	1	1
4	0	0	0
5	1	1	0
6	1	1	1
7	0	1	0
8	1	0	1
9	0	1	1
10	0	1	0

1. What is your best guess at the probability that each of the formulations fails under the conditions of the test?

   The proportion of failures for each formulation is: A) 4/10 or 0.4; B) 6/10 or 0.6; and C) 5/10 or 0.5. The probability I would expect for each would be estimated by their respective proportions.

2. What is your best guess at the probability that each of the formulations does not fail under the conditions of the test?

   Since $\sim p = (1 - p)$, I estimate the success rate for A to be 0.6, B at 0.4, and C at 0.5

3. Suppose you repeated the experiment. Would it be surprising if Formulation A had a proportion of failures equal to 0.4? What about 0.3?

   Since the expected value of A is $np = 10 \cdot 0.4 = 4\,failures$, I am not too surprised. 3 failures is not too far away from the expected value, so I wouldn't be terribly surprised.

4. What is the overall expected value of these three samples taken altogether?

   The expected value of a failure across all three samples is $4 + 6 + 5 = 15$ out of 30 total trials.

   You and your capstone team have designed a micro air vehicle (MAV) to be used for reconnaissance in major disasters such as earthquakes and typhoons. The copter has to be maneuverable to get in and out of tight spaces in collapsed buildings and debris. So, you compare your vehicle with two other vehicles with different designs to see if your design has better maneuverability than the others.

	Time through obstacle course		
Trial	Your MAV	Alternative MAV A	Alternative MAV B
1	1:43	1:44	1:50
2	1:37	1:35	1:42
3	1:52	1:47	1:38
4	1:29	1:46	1:37
5	1:35	1:38	1:40
6	1:30	1:29	1:35

5. Rank order the data across MAVs for each trial. Does your design appear to be better or worse (or about the same) compared to the alternatives?

	Time through obstacle course		
Trial	Your MAV	Alternative MAV A	Alternative MAV B
1	3	2	1
2	2	3	1
3	1	2	3
4	3	1	2
5	3	2	1
6	2	3	1
Median Rank	2.5	2	1

From this ranking (3 is best), it looks like my MAV performs about the same as Alternative MAV A, though it *may have* performed slightly better than A given its median rank. MAV B looks worse than either mine or A.

6. **Now analyze the data using statistics for ratio scale. Does one of the MAVs appear to be better (or worse) than the others on average? Again, formulate your answer using both an argument about the centers of the distributions and their variability. Is the average distance between the distributions big enough, given their variability, to conclude that one is significantly different from the others?**

   The means and standard deviation for each sample is listed in the table below:

	Your MAV	Alternative MAV A	Alternative MAV B
$\bar{x}$	97.67	99.83	100.33
s	8.66	7.08	5.32

   From this, it looks like all MAVs performed within about 2.5 seconds of each other. My vehicle appears to be better, but it shows more variation in times than either A or B. Vehicle B has the most consistency even though it has the worst mean time.

7. **Draw your best guess of the population distribution for alternative MAV A. Do the same for your MAV. How much do the distributions overlap (no need to calculate, just eyeball the area underneath the curves)?**

   Using the normal distribution, it looks like they overlap quite a bit. The standard deviations are 8.66 seconds and 7.08 seconds respectively. The mean plus or minus the standard deviations represent the points of inflection on the normal distribution, so the means are only about $\frac{1}{4}$ of a standard deviation apart.

8. **Compare the advantages and disadvantages of treating data as ordinal versus interval/ratio scale.**

   The ordinal data obscures the magnitude of the differences in times between the MAVs, whereas the ratio scale allows for more subtle comparisons of the extent to which the MAVs differ in performance. In particular, the variation in samples is highlighted with the ratio data in this example. But the ordinal data is very simple to communicate and interpret.

Type of Wind Turbine Drive	**Failures per turbine per year**
Geared < 1MW	0.46
Geared $\geq$ 1 MW	0.52
Direct Drive	0.19

As discussed earlier in the chapter, failure rates of three different classes of wind turbine were assessed by Reder, Gonzalez, & Melero,(2016).

9. **Given the combined data, which turbine seems to be the most reliable? Which of the Geared Drive turbines presents the best bargain, assuming the initial cost and repair costs of the turbines are approximately equal?**

The direct drive turbine has only about 2 failures, on average, over 10 years, compared to about 5 for each of the geared turbines. So the direct drive seems more reliable, overall.

10. **How would you evaluate the data if the average cost per failure were as follows: Geared < 1MW, $ 1,000; Geared ≥ 1MW, $ 750; Direct Drive, $ 2,000**

    The cost per year for the failures of each turbine looks pretty different. For the Geared < 1MW, the cost of failures is $0.46 \cdot \$1,000 = \$460$. For the Geared ≥ 1MW, the cost is $0.52 \cdot \$750 = \$390$, and the Direct Drive costs $0.19 \cdot \$2,000 = \$380$. So the Direct Drive appears to be the best, using cost versus failure data. Second is the Geared ≥ 1MW, because, though it fails more, it also costs less enough to perhaps justify the failures. Of course my customers might be upset when half of their turbines fail in a given year...

11. **If the company installed 100 turbines in roughly identical sites, about how many of each type would you expect will fail in a year? How much would you expect this to cost in terms of downtime?**

    I would expect about 46 or so of the Geared<1MW to fail, about 52 of the Geared ≥ 1MW to fail, and about 19 or so of the Direct Drive turbines to fail.

12. **What are the advantages of using rates, over frequencies, to describe nominal data?**

    Rates, given enough data, allow for us to use proportions, and the arithmetic mean to estimate expected values. They also allow us to compare across samples of different sizes.

13. **What are other variables that can be described using nominal scales? Try to come up with 5 different ones, at least one example should have more than two categories (for example: Major types of auto transmissions has three levels: Manual, Automatic, and Continuously Variable Transmission).**

    Lots of possibilities can be listed. Success/failure rates of any product. Different manufacturers of similar parts. Different Experimental conditions in an experiment. Geographic locations of solar panels. Economic strata. Etc.

14. **Suppose you were trying to find the effect of using closed cell foam insulation versus Reflectix™ in a revolutionary new beer cooler (hailed by millions in the Southwest as the most important invention since the cordless blender). The sample of coolers with foam insulation shows a mean internal temperature of 8.5 degrees C. The sample of coolers with Reflectix™ insulation shows a mean internal temperature of 4.5 degrees C. What would be your error if you reported that the Reflectix™ sample kept your beer about twice as cold as the foam sample?**

    Because temperature in Celsius is measured on an interval scale, it can only be compared by subtraction. One must convert to Kelvin, a ratio scale, to get the actual temperature ratio.

The table below is a decision matrix of different metals to be potentially used for torque tubes mounted in the leading edge of an airplane wing. I have rank ordered each of the five materials on each of seven criteria related to their strength, ductility, and machinability. The order goes from 5 (best) to 1 (worst). Notice that AL 7075-T6 and AL 2024-T4 have identical rankings of galvanic corrosion and cost/weight ratios.

Material	**Yield Strength (ksi)**	**Elongation (%)**	**Vickers Hardness (HV)**	**Density lb/in3**	**Machinability**	**Galvanic Corrosion**	**Relative Cost per Weight**
Magnesium AZ31B-H24	1	3	1	5	5	5	2
Titanium Ti-6A1-4V	5	1	5	2	1	1	1
Stainless steel 430	2	5	4	1	2	2	5
Aluminum 7075-T6	4	2	3	3	4	3	3
Aluminum 2024-T4	3	4	2	4	3	3	3

15. Which of the alloys is the "best" across all the categories?

    If the median rank across categories is used, and no weighting is applied to each category, then the Magnesium alloy and the two AL alloys rise to the top, tied with a median rank of 3. Titanium falls at the bottom with a median rank of 1, and Stainless Steel is next to last with a median rank of 2. Most of the time, when I do these kinds of comparisons, I consider it a success to *eliminate poorer candidate materials* rather than find the single best example.

16. Suppose not all of the categories are weighted the same. For the intended purpose, stiffening the leading edge of an airfoil, what criteria are the most critical? Develop weights for the criteria based on your current understanding of airfoil design.

    Responses may vary given your experience and understanding.

17. Given your weights in question 8, which alloy appears to be "best?" If it differs from your answer to question 7, explain why.

    Here, some weighted average can be used to distinguish the alloys from each other.

Busby, Hash, & Was, (2005) studied the relationship between Yield Stress and Hardness in Austenitic Stainless Steels. They reviewed several earlier studies and combined the results. These are presented in the table below. The independent variable is the slope of the line of best fit that predicts this relationship. For Study 1, in 304 SS, for example, the authors found that for every unit change in hardness of the material, one would expect a 3.76 change in Yield Stress.

		Yield Stress vs change in hardness
Study	**Type of SS**	**Mpa/kg/mm^ 2**
	304	3.76
1	316	3.43
	347	3.55
	304	3.06
2	316	2.64
	347	2.63
	304	3.83
3	316	3.58
	347	3.79

18. What is the scale of data for yield stress versus change in hardness? How do you know?

Yield stress is measured on a continuous scale with an absolute zero. Hardness is measured on a ratio scale as well. Therefore the ratio of the two is a ratio scale.

19. Treating the data as interval/ratio, what is the mean yield stress versus change in hardness of the SS samples? Does the mean of any of the studies seem to be systematically higher or lower than this overall mean?

Type of SS	$\bar{x}$	s
304	3.55	0.42579338
316	3.21666667	0.50500825
347	3.32333333	0.612318
Total across Alloys	3.36333333	0.47386707

316 SS may have an average yield strength per unit change in hardness less than the average across samples. 304 SS appears to have higher yield strength per unit hardness, and 347 SS appears to be about in the middle.

20. Now take the mean of the data *within* each study. Does any of the types of stainless steel appear to be harder overall than the other types?

    It appears that 316 SS has a lower percent yield stress per unit change in hardness than either 304 SS or 347 SS. The mean for 316 SS is a bit less than 1 standard deviation apart from 304 SS, but is much smaller than 347 SS. I think this may be a systematic difference. Because they used the results of a number of earlier studies, I am hoping that the sample size is large enough to be confident in my conclusion.

21. Take the median of the data *within* each study. Does any of the types of stainless steel appear to be harder overall than the other types using the median?

    The 304 SS appear to have a consistently higher rank across studies, than the 316 SS or the 347 SS.

22. Compare your conclusions for questions 18,19,20, and 21. How is it possible that the mean and median can paint different pictures of the data?

    For 304 SS, there was no difference in my conclusions. But for the 316 and 347 SS, the fact that the actual magnitude of the dependent variable was not taken into account using the rank ordering meant that the differences between these two alloys was not detectable. Using a ratio scale and the mean and standard deviation, I can see that 347 SS appears to have greater yield strength than 316 SS

23. What are the advantages of using the mean as a statistic, if data is measured using an interval or ratio scale?

The arithmetic mean tells us that, all things being equal, a typical measurement will have a value close to the mean, and that values much greater than the mean or much less than the mean are unusual. With ordinal data, the median only tells where the cutoff of 50% of the data lies. It does not have any assumptions about where the data are located (e.g., close or far away, it doesn't matter).

4

Introduction to Probability

To understand how statistical inferences are made, we have to understand the basics of probability. Because in any science we cannot prove most assertions to be true (at least to the level required by mathematicians), we have to rely on the **weight of evidence** that data can provide. Some assertions, like *smoking causes cancer*, or *the deflection of an I-beam, suspended between two points, is the same or less than a solid beam made of the same material and having the same mass,* have so many years of consistent evidence in their favor that we treat them as if they have been proven conclusively, but that is not strictly the case. Not all people who smoke get cancer. Not all I-beams are consistent in manufacture, quality and density of material and so on. Natural variation in these variables causes some small proportion of their populations to violate what are, for the vast majority of instances, true statements.

> **Weight of Evidence** is the accumulated factual support for and against an assertion. The amount of evidence, and the quality of that evidence are used to judge whether an assertion is likely to be true or false, and under what conditions it is true or false.

If we plot the distribution of, say, the deflection of i-beams of a fixed design, and corresponding square-cross section solid beams under a constant load, the distributions might look like this one in Figure 4.1.

The vertical axis on this graph is the proportion (i.e., the **probability**) that a given bending moment occurs in these population of beams. There is very little *overlap* in the two distributions.

1. In your own words, what does it mean when two distributions overlap?

   When distributions overlap, it means that some values are held in common. The more overlap that exists, the more similar the means of the distributions are to each other relative to the variability.

2. What does a small overlap mean? Use Figure 4.1 to contextualize your answer.

   Small overlap means that the two distributions have few common values, and that one may show significantly higher values, on average, than the other.

3. Suppose you choose a beam, at random, from a pile of assorted beams. Estimate how likely it is that your beam is a Solid Beam if its deflection is 30 mm.

   Reading off the graph, 30 mm corresponds to a probability of about 0.4.

DOI: 10.1201/9781003094227-4

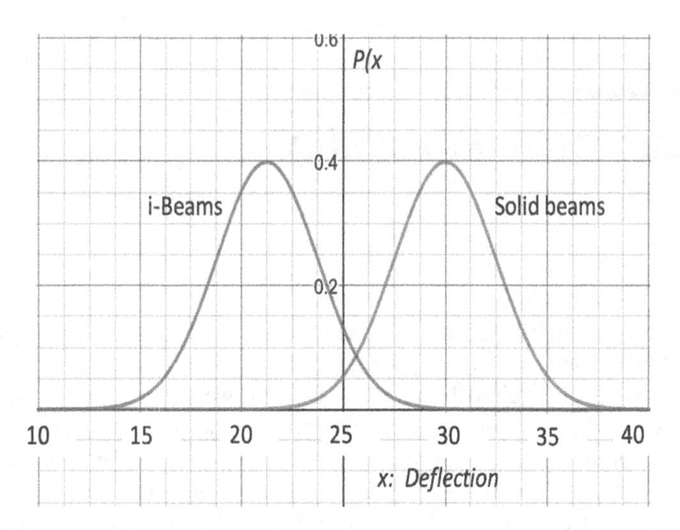

FIGURE 4.1: Illustration of the probability distributions of the bending moments of two types of beams

4. What is a good estimate of the probability the beam is solid, if its deflection is measured at about 25 mm?

   The probability that a solid beam shows a deflection of 25 mm is about 0.05 (reading off the graph). But, the probability that an i-beam shows a deflection of 25 mm is about 0.125. So, the probability there is an overall chance of about 0.175 that any given beam shows a deflection of 25 mm. If we say that each distribution has 100 beams, then our expected value, E(25mm) = 17.5. about 5 of those will be solid. So $\frac{5}{17.5}$ of all the beams that deflect 25 mm should be solid. This reduces to 0.285. $\frac{12.5}{17.5} = 0.715$ will be the probability that any randomly selected beam that shows a 25mm deflection is an i-beam.

   This example illustrates how engineers make conclusions from data regarding the quality of one type of product over another, or the effect that a particular design change has on a distribution of phenomena. We can't say that I-beams are stronger than Solid Beams 100% of the time, but we can say that I-Beams are just as strong or stronger than Solid Beams of the same mass and length the vast majority of the time. The big question is: *How much overlap in the distributions is just enough to say that the two conditions are likely the same versus likely different on their performance on the measured variable.* For our example, the probability is very high that, if I choose an I-beam, my deflection will be below about 26 or 27 mm. If I choose a Solid Beam, it is very likely that its deflection is above about 24 mm. In between 24 and 27 mm, we have a hard time telling the two apart in terms of their performance.

4.1 Simple Probability

So how do we go about calculating the overlap between distributions to determine if they are likely to represent the same or different populations? We will briefly review probability here so that the basis of the calculations we perform can be tied back to good conceptual models.

> **Probability** is the proportion of times an event occurs in a population of events. It is a measure of the likelihood that any event chosen at random will have a given value.

Probability at its most basic definition is a proportion:

$$P(x) = \frac{\text{Number of Events of } x}{\text{Total Number of Events}}$$

Because the denominator is a maximum value, $P(x)$ ranges from 0 when no instances of x exist in the population, to 1, when all **events** in the population are x. Knowing $P(x)$, we also $P(\sim x)$. It is just

$$1 - P(x)$$

An "**event**" is just stats-code for "outcome" or "instance" of a phenomenon. So, if I were to draw an event from the population of solenoids, the probability that the solenoid is a Reynolds Mini Push Pull Solenoid, model number 18-2003, is simply:

$$P(x) = \frac{\text{Number of Reynolds Mini Push Pull Solenoid, model number 18-2003}}{\text{Total Number of Solenoids in Population}}$$

> **An Event** is an outcome of a population for which a probability exists

Sometimes we know the proportions of events in the population prior to collecting data. Probabilities drawn from such **theoretical probabilities** are defined mathematically, and the values of $P(x)$ are known in advance (e.g., the probability of rolling "snake-eyes" with two dice—we don't have to collect data to determine this probability, we just have to buy into the assumption that the dice are fair and identical).

Experimental probabilities, on the other hand, are the results of empirical trials (e.g., the probability that a male, under 25 years old, will be killed in a car crash, or the probability a rivet will break under a certain shear load). Their respective populations are so large, and collecting the data so expensive, that we must make due with estimates taken from samples. Luckily, as we have already seen, there are good theoretical probability distributions we can use in lieu of collecting huge amounts of data: The Binomial, Poisson, and Normal distributions being chiefly useful. Armed with these models, we can compare experimental outcomes to these theoretical distributions to determine their probability of occurrence. In Figure 4.1, for example, we can see that the probability of randomly selecting a beam with a displacement of 30 mm or more (at and above the mean of the solid beam distribution) is 50% for solid beams, and close to (but not equal to) 0 for I-beams.

4.2 Conditional Probability

The kind of experimental question exemplified in Figure 4.1 involves making one assertion contingent upon the truth of another. For example, we might say, *given a beam is an I-beam, what is the probability its deflection under a specific load is about 30 mm?* Such conditional statements restrict the population to a subset. We are not interested in the entire population of beams, only I-beams. Here is another applied example:

Here is data from a random sample of 100 batteries (Table 4.1). The number of hours each battery lasted under a constant load was recorded, along with whether or not the battery leaked fluid.

Hours	**Number of Batteries**	**Number of Leaks**
<58	20	4
58–59	30	3
60–61	40	2
>61	10	1
Total	**100**	**10**

TABLE 4.1: Distribution of the number of hours a sample of batteries lasted, and number of leaks

5. What is your best guess of the probability that any battery, drawn randomly from the same population, will exhibit a life between 58 and 61 hours?

Here, we can just add up the number of batteries that have a life between 58 and 61 hours, 70, and divide by the total number in the sample, 100. This proportion, 0.70 is a good estimate of the probability of randomly selecting a battery with a lifespan between 58 and 61 hours in the population of millions of batteries that exist in the population.

6. What is the expected probability that, given a battery lasts over 61 hours, it will leak?

   Here, the conditional statement, "given... " restricts the sample space from 100 down to 10 batteries. One out of these 10 leak, so the estimated probability is 0.10.

7. What is the estimated probability that, given a battery lasts between 58 and 61 hours, it will leak?

We have 70 batteries in this interval, and 5 of them leak, so $\frac{5}{70} = 0.071$ carriage return.

There are two different relationships among probabilities that these questions force you to contend with. The first is that, the probability of an event falling into at least one of two categories is just the sum of the probabilities of each individual category:

$$P(A \cup B) \quad = \quad P(A) \; + \; P(B) \text{ for independent events}$$

We use the Union symbol, ∪, to denote that we don't care if the event falls in category A or in category B, we count any event from either of the two categories. In question 5, it is easy to see that, since there are $\frac{30}{100}$ between 58 and 59 hours, and $\frac{40}{100}$ between 60 and 61 hours, the probability of randomly drawing a battery that lasts between 58 and 61 hours is just their sum $\frac{30+40}{100} = \frac{70}{100}$ or 70% . These categories are mutually exclusive. Were there to be any overlap in between A and B, we would have to subtract this overlap because it is double-counted.

$$P(A \cup B) \quad = \quad P(A) \; + \; P(B) \; - \; P(A \cap B) \text{ for overlapping events}$$

We use the Intersection symbol ∩ to denote this overlap, where an event counts as *both* A and B, simultaneously. To find the value of this overlap, you find the fraction of one set that is taken up by members of the other set:

$$P(A \cap B) \quad = \quad P(A) \; \times \; P(B)$$

Questions 6 and 7 address what are termed, **conditional probabilities**: the chances an event will occur, given another event has already occurred.

$$P(B|A) = \frac{P(A \cap B)}{P(A)}$$

Here, the probability of B *given A has already occurred* is equal to the proportion of the reference set A, taken up by the overlap between the two. For problem 6, we are only interested in the batteries that last over 61 hours. This restricts our population of interest to only 10 batteries. Since there is only 1 battery in those 10 that leaked, we can estimate the probability of that occurring in the future at about

$$P\left(leak|hours > 61\right) = \frac{P\left(\left(hours > 61\right) \cap leak\right)}{P\left(hours > 61\right)} = \frac{1}{10}$$

For problem 7, we restrict the population to only those batteries lasting between 58 and 61 hours. There are 70 of these.

$$P\left(A \cup B\right) = \frac{30}{100} + \frac{40}{100} = \frac{70}{100}$$

Then, of those 70, we have 5 that leak. So, $\frac{5}{70}$ should be the probability that, given the battery lasts between 58 and 61 hours, it will leak.

$$P\left(B|A\right) = \frac{\frac{5}{100}}{\frac{70}{100}} = \frac{5}{70} = 0.0714$$

From these last two problems you can see that, using a table that lists each of the frequencies for their prospective categories makes it easy to walk through the overlap among the sets to make sense of these conditional probabilities.

Almost all of the probabilities we will be discussing through the rest of the book are conditional probabilities. They will utilize both the relationships for the union of two sets (the addition rule) and intersection (the multiplication rule). Because we will be dealing most often with continuous probability distributions, these sums and multiples will be found using derivatives and integrals, but the basic conceptions introduced here are exactly the same.

Conditional Probability is the chance that an event will occur, given another event with a known probability has already occurred.

Conditional probabilities allow us to test whether or not any two sets of events are **independent:** When the probability that one event occurs doesn't affect the probability that the other event occurs. For example, the chance that my Chevy Suburban fails to start is in no way connected to the probability that the price of sugar will go up today. If this is the case, then

$$P\left(A \cap B\right) = P\left(A\right) \times P\left(B\right)$$

But if B *is* contingent upon A in some way, then the sets overlap, and this rule does not hold. Instead, we find that

$$P\left(A \cap B\right) = P\left(A\right) \times P\left(B|A\right)$$

Setting these two probabilities equal to each other

$$P\left(A\right) \times P\left(B\right) = P\left(A\right) \times P\left(B|A\right)$$

We see that, for the two quantities to be independent:

$$P(B) = P(B|A)$$

If this identity does not hold, then A and B must **not be independent** and **must** have some kind of dependence relation.

Events are **independent** if the occurrence of one does not affect the probability of the other occurring. Mathematically, two events are independent if

$P(B) = P(B|A)$

In other words, because the two probabilities are equivalent, A has no effect on B.

Now the problem is, when we collect data, our probabilities are subsets of A and B, and subject to sampling bias. So when we compute the two quantities experimentally, they won't be exactly the same.

The remainder of this book will be devoted the central question: Just how different does P(B|A) have to be, to be considered impactful? In practical terms we phrase this question like this:

Is the probability of a dependent variable, B occurring significantly different when an independent variable A is taken into account versus when A is not taken into account?

This is the fundamental question of **inferential statistics**—that branch that tries to not just estimate population parameters, but allows us to test hypotheses about the relationships among variables, the degree to which they are independent (or not).

Inferential Statistics is the practice of inferring the properties of a probability distribution that represents a population of measurements. It involves sampling data from a population, and using that data to create models of important characteristics of the population distribution, including its parameters. It is the primary method for experimental hypothesis testing.

4.3 Moments of a Distribution

4.3.1 The Mean as a Moment

It is an important property for any Real function $f(x)$, that we can always find some point that balances out all its values over the independent variable (see Figure 4.2). In statistics, the mean is such a value. Think of a lever as a good model for the mean. When you were young, if you had an older brother, he would challenge you to sit on the seesaw at the playground. You would sit on one end, and he would sit on the

FIGURE 4.2: This distribution of weight is not balanced!

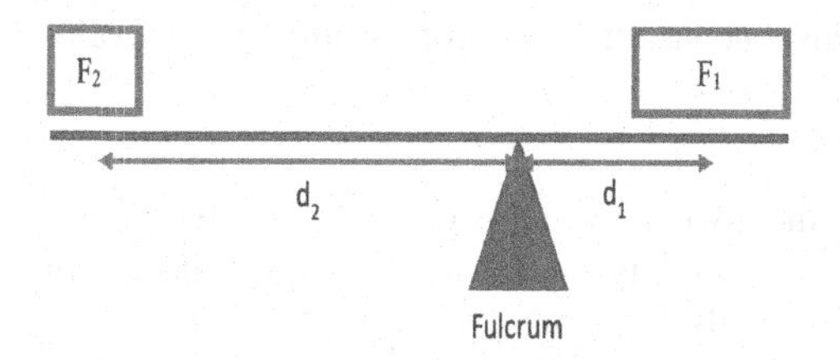

FIGURE 4.3: Archimedes' law of the lever

other. Being so much heavier than you, he shot you into the air and kept you up there, helpless. As you figured out the apparatus, you may have suggested that your brother move closer to the fulcrum of the seesaw so that his and your weight would balance out the beam, allowing both of you to move the other up and down. As a father and a teacher of statistics, I have placed several children at different points on one side of the beam, and asked other kids to predict where they needed to sit on the other side to balance the weight. They get pretty good at it in a few trials.

You may remember from Statics class that a **moment** is just a force applied perpendicular to some lever arm. If two forces are applied at two distances with a fulcrum in the middle, Archimedes' Law of the Lever states that the Moments on each side of the fulcrum must be equal for the lever to balance (see Figure 4.3).

$$F_1 d_1 = F_2 d_2$$

If there are multiple objects placed at different points on the lever, for the lever to balance, the sum of the moments on one side of the fulcrum has to equal the sum of the moments on the other.

$$\sum F_1 d_1 = \sum F_2 d_2$$

In other words, since all $F \times d$ on the left side of the fulcrum equal the $F \times d$ on the right side, when added (accounting for negative values), the sum is zero. The sum of the individual Force-distances must equal zero in a balanced lever. If we have a continuous beam of non-uniform cross-sectional area, we can use calculus to determine the moment:

$$\int_{-\infty}^{\mu} (x-\mu)\,p(x)\,dx = \int_{\mu}^{\infty} (x-\mu)\,p(x)\,dx$$

The mean of the distribution is the value of the independent variable that makes this true. What the mean does is serve as the **center of mass** of the probability distribution.

> The mean is the **center of mass** of a distribution. It is the **first moment** of the distribution about the mean.

4.3.2 The Variance as a Moment

One of the things that drives most statistics students crazy is just why we chose to square the deviations from the mean to get a measure of variation (the variance). One of the reasons for this choice is that the Variance is the second moment about the mean of a probability distribution. The general theory of moments is stated thusly:

$$\mu_n = \int_{-\infty}^{\infty} (x-\mu)^n\,p(x)\,dx$$

Where μ_n is called the *n*th moment of *p(x)*.

As the first moment, the mean is that point where the sum of all the deviations about the mean is equal to zero (the integral of the continuous function *p(x)* from its lower bound to its upper bound is zero). The *variance* is the value where the sum of the *squared deviations about the mean* $[(x-\mu)^2]$ is equal to zero. Like the second moment about the center of a beam, it can be thought of similarly to rotational inertia. In the beam, the moment of inertia is how the mass of the beam is distributed about the center of rotation. In a probability distribution, the variance describes how the accumulated data points are distributed about the center of mass.

> The variance is the **second moment** of the distribution about the mean. It is a measure of how, overall, values of the independent variable differ from the center of mass. It can be thought of as the center of mass of the squared probability distribution. In physical terms, it is the moment of inertia.

4.3.3 Summary: Bringing Probability, Moments, and Sample Statistics Together

- From our work with sample statistics, we found that the mean and variance are the parameters of the Normal Distribution, a particularly useful model for describing the long-haul behavior of random variables;

- From our work with probability, we found that the expected value of a random variable is the mean of its probability distribution;
- From our work with moments of a distribution, we found that the first moment, the center of mass of a distribution, is its mean, and that the second moment, the distribution about the mean, is its variance.

From these discussions we can develop a general theory that merges descriptive statistics, describing samples and populations, with probability, the forecasting of long-haul tendencies:

$$E(X^n) = \int_{-\infty}^{\infty} x^n p(x)\,dx$$

The expected value of a distribution is a moment—the sum of the all the individual moments of each point in the distribution about some central value. If this is true, then the expected value of a random variable is the mean of the probability distribution of that variable, and the variance is the expected value of the squared deviations from the mean of that distribution:

$$E(X) = \mu$$

and

$$E(x-\mu)^2 = \sigma^2$$

$$\left[E(x-\mu)^2\right]^{1/2} = \sigma$$

So, the combined facts that the parameters of the Normal Distribution correspond to its center of mass and its distribution about that center, *and* because those parameters are the expected values of the probability distribution, we can use the Normal Distribution (and other distributions where the parameters are known) to estimate the probability that a sample we have drawn could have come from a population whose parameters we know.

4.4 Probability Density Function and Cumulative Distribution Function

You have probably been sweating a bit, seeing all these integrals, and some pretty arcane functions, like the Normal and Poisson distributions, waiting for some shoe to drop. Well, here it is, but it is not a very big shoe. From your background in calculus and differential equations, you should be expected to understand the process of differentiating and taking the integral of continuous functions and piecewise functions whose intervals are continuous:

- You should understand that the derivative of a function is a function that describes the instantaneous rate of change of the dependent variable (in our case *p(x)*), compared to a unit change in the independent variable (whatever it is we are measuring);
  - A derivative evaluated for some value of the independent variable is the point-slope of the function at that value.
- You should understand that the integral of a function is a function that describes the accumulation of infinitesimally small areas from some lower bound to a value of the independent variable;
  - An integral evaluated across some interval describes the area under the curve between the lower and upper bounds of that interval.
- You should understand the Fundamental Theorem of Calculus as a relationship between a function, and its derivative and integral functions.

$$\int f'(x)\,dx = f(x) + C$$

and

$$\frac{d\left(\int f(x)\,dx\right)}{dx} = f(x)$$

If you truly understand these ideas, this course does not require you to determine the integral or derivatives of the distribution functions we will be dealing with. We have computers and other tech that will do this much faster and with fewer errors. I want you to reserve your time for grappling with what these functions mean, what their values at particular points mean, and how they interact in the process of making inferences from data. So, you *don't* have to figure out the Taylor, or Power series, or other methods for the Normal distribution:

$$\int_{-\infty}^{\mu} \frac{1}{\sigma\sqrt{2\pi}} e^{\frac{-(x-\mu)^2}{2(\sigma)^2}}$$

You just need to know that this value is equal to the area under the curve, summed from negative infinity to the mean. If I told you that this value, evaluated over the integral is equal to 0.50, because the interval is half of a symmetric probability distribution, that should not come as a shock to you. Understanding the features of the distributions, and what the calculus tells us is far more important than doing the algebra.

The functions we will be using in this book will be all lumped together under the term, **Probability Density Functions** or **PDF**s. Briefly, a probability density function is an integrable function, *p(x)*, that satisfies the following conditions:

1. *p* is positive across all values of the independent variable;

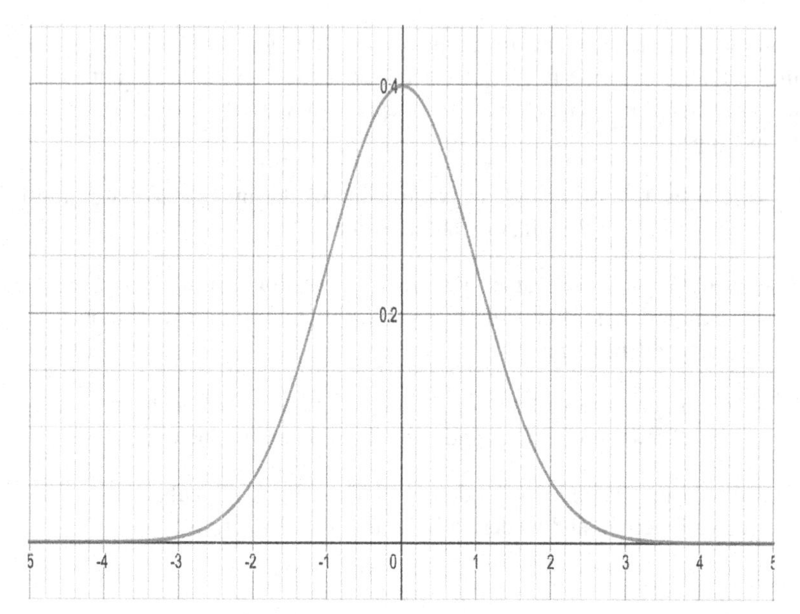

FIGURE 4.4: The normal distribution as a PDF

2. the total area under the curve from its lowest bound to its uppermost bound is equal to 1 (the sum of the probabilities in any sample space must be equal to 1. All possible outcomes are counted); and

3. the probability that x is in some interval of $p(x)$ can found by integrating $p(x)$ over that interval.

The Normal Curve is a PDF, as are a number of other distributions. The binomial, and other nominal distributions, however, are discrete, and therefore not integrable. But, their probabilities *can be summed* across intervals, so they meet criteria 1, and 2, but not 3. So, we call these functions **Probability Mass Functions**. For all intents and purposes, when analyzing data, we treat these functions, conceptually, in exactly the same ways. The only difference is in how the summation across intervals is done. So, for the remainder of this book to make things simpler, we will use **PDF** to refer to both.

1. Using the Normal distribution as shown in Figure 4.4, as our canonical model, think about what its integral function would look like.

   (a) What is its shape?

   The integral function of the Normal distribution is an "S-curve" with lower bound at zero and upper bound at 1, with point of inflection at the mean value.

   (b) What is its minimum and maximum values?

   Lower bound = 0, upper bound = 1.

(c) What is its value at the mean of *p(x)*?

In this graph, the mean is 0, so the integral from negative infinity to zero is half the curve. And because it is symmetric, it should represent a cumulative probability of 0.50.

(d) What is its value at about $-1\ \sigma$ below the mean?

In this graph, $-1\ \sigma$ divides the curve into two regions. The left region looks like it contains an area of about 20% or so of the total area under the curve (in reality, it is closer to 16%).

When we sum across an interval of a PDF, the result is a **cumulative probability**—the probability that any randomly drawn value of the independent variable would fall in that interval. Looking at the symmetry of the Normal Distribution in Figure 4.4, we can see that half of the area under the curve lies at or below the mean, and half lies above the mean. We can conclude that the probability of landing in either region is 50% – 50% . This makes perfect sense, as we know the chance of randomly drawing a value below the mean is $\frac{1}{2}$, and likewise for values above the mean. As a balancing point, the mean affords this nice benchmark with which to evaluate probabilities for symmetric distributions. We could likewise evaluate the integral of this curve to find the cumulative probability of being randomly selected and falling in that interval.

By convention, we evaluate the integrals of PDFs from negative infinity to some upper bound. Even though infinities are typically out of the contextual domain of our data, we use this theoretical distribution, where the probabilities way out in the tails are so infinitesimally close to zero that we can consider them to be negligible. Figure 4.5 shows the evaluated integral of *p(x)* as an "S" curve. As a general rule, all **cumulative distribution functions** (**CDF**s) have this "S" shape, bounded by a lower asymptote of 0, and an upper asymptote at 1. Notice that the point of inflection of the curve is at the mean of *p(x)*, and that the cumulative probability at the mean is 0.5.

2. Using Figure 4.5, what is the approximate cumulative probability that, if I drew a value at random from x, that it would fall below $-1\ \sigma$?

   Reading off the graph, it looks like approximately 0.15 (in reality it is about 0.16)

3. What is the approximate cumulative probability of drawing a value at random that would fall above $-1\ \sigma$?

   $1 - 0.16 = 0.84$

4. What is the cumulative probability that any value at random falls between $-1\ \sigma$ and $+1\ \sigma$?

   $0.84 - 0.16 = 0.68$

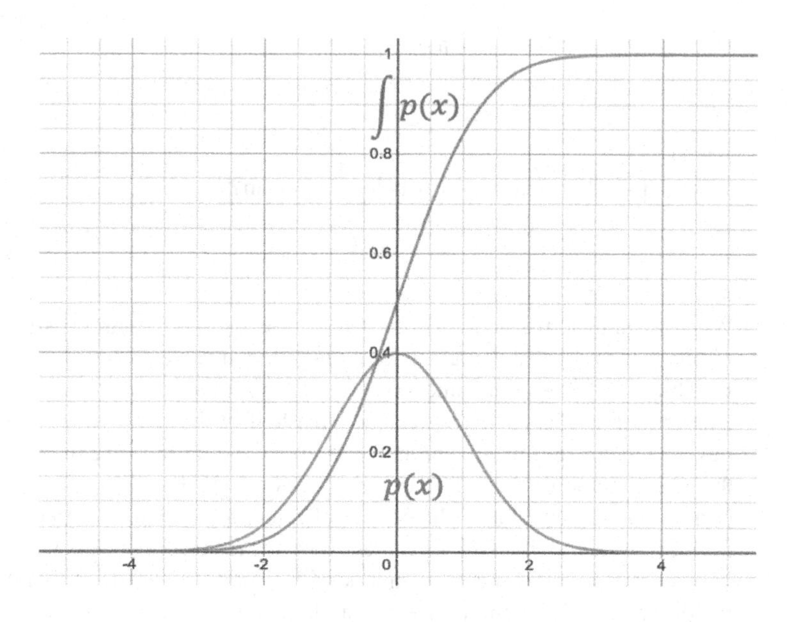

FIGURE 4.5: The relationship between the normal distribution PDF, and its CDF

5. What is the cumulative probability that any value at random falls above $+1\ \sigma$?

   Because of the symmetry, it is the same as the probability that any value at random falls below $-1\ \sigma$

6. What is the cumulative probability that any value at random falls below $-1\ \sigma$ or above $+1\ \sigma$?

This is a disjoint set, so we integrate from negative infinity to $-1\ \sigma$ and then from $+1\ \sigma$ to infinity and add the two sums together. $0.16 + 0.16 = 0.32$

> **Probability Density Function (PDF)** is a function that assigns a probability, *p(x)*, to every value of a random variable.

> **Cumulative Distribution Function (CDF)** is a function that assigns the sum of the probabilities in a PDF, evaluated from a lower bound to every value of a random variable. For continuous distributions, it is the integral of a PDF.

4.5 Summary of Probability

Because we can't, typically, measure every instance in a population, we have to settle with samples. This means that any data we analyze is subject to sampling bias and measurement error, throwing off our estimates of population parameters by some little bit. So, we can't say that our sample statistics ARE the population parameters, but we can say that they fall within some (hopefully small) interval of the actual parameter values. The mean and variance of random variables represent the balancing point of the distribution (μ), and the degree to which the density of the distribution gets less as one moves towards the tails of the distribution (i.e., σ^2 measures how far away, on average, measures are from the center of mass, μ). Moreover, μ and σ^2 are the expected values of the probability distribution over the long-haul.

By appealing to the Normal Distribution and other known Probability Density Functions, if we believe our sample is an adequate representation of its parent population, we can estimate the rest of the population using the closed form equations of those functions. Armed with these powerful models, we can then estimate the probability that a particular would be drawn randomly from that theoretical population.

Cumulative Distribution Functions are just the integral of PDFs, so they give us a function that pairs a cumulative probability (integrated from negative infinity to some bound) with each value of the independent variable.

The rules of probability allow us to add probabilities to compute the union of two or more sets (creating CDFs), and multiply them to estimate their overlap. Together, these simple rules allow us to test conditional probabilities to assess whether or not two variables are independent, or whether there is some kind of relationship between them. This testing of conditional probabilities is the essence of statistical inference.

4.6 Study Problems for Chapter 4

1. The following Venn Diagram shows numbers of smokers interviewed in a recent study, who use some form of nicotine replacement therapy (NRT). The main replacements were nicotine patch, nicotine gum, or a nicotine lozenge. Some smokers in the sample did not use any nicotine replacement.

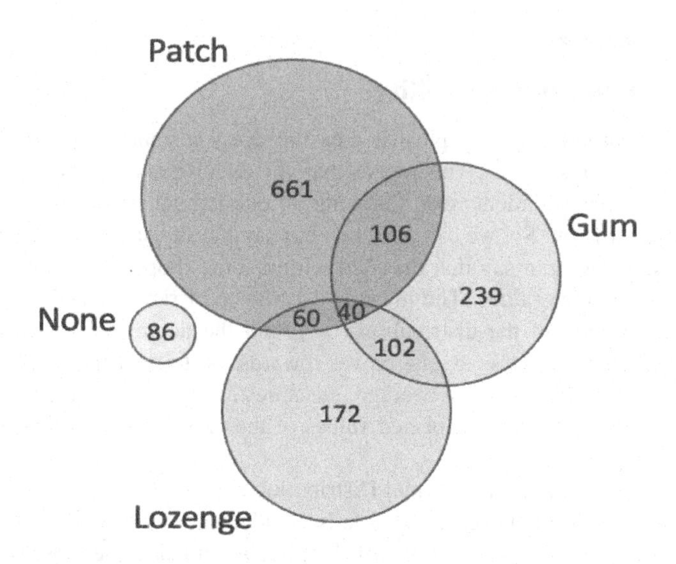

(a) What is your interpretation of this diagram? If you were designing nicotine replacement technology, which type would you choose to best meet a market niche?

From this diagram, it looks like the Patch is most often represented, followed by Gum and Lozenge. There is considerable overlap with Gum and Lozenge users. They seem more likely to, try several different replacement therapies, when compared to the Patch. I would probably target these two niche markets.

(b) What is the proportion of people in this sample who use a nicotine patch?

Adding all the numbers, we get a total in the population of 1,466 respondents. 867 utilize the patch, so $\frac{867}{1,466} = 0.59$ is the proportion using the Patch

(c) What is your best estimate of the probability that any person chosen randomly from the population this sample is drawn from, smokers, uses both the lozenge and nicotine gum?

There are 142 respondents who have used both the lozenge and the gum, as indicated by the overlap in the sets. $p\,(Lozenge \cap Gum) = \frac{142}{1,466} = 0.097$

(d) Given a person uses nicotine gum, what is the estimated probability that they would also use the lozenge?

$$p\,(Gum|Lozenge) = \frac{P\,(Gum \cap Lozenge)}{P\,(Gum)} = \frac{\frac{142}{1,466}}{\frac{487}{1,466}} = 0.292$$

(e) What is your best guess at the probability that a person uses the patch, but NOT gum or lozenge?

867 utilize the patch, 206 use either the gum or the lozenge, or both.

$$p(patch \cap (\sim gum \cup \sim lozenge)) = \frac{867 - 206}{1,466} = 0.45$$

(f) If the proportion of smokers in the US is about 16%, what is the probability that any person drawn at random would be using nicotine gum? What assumptions do you have to make for your estimate to be drawn from this data?

If this sample is representative of the US population of smokers, then the proportions in the sample should be close to the proportions in the populations. I can then use the sample proportions as an estimate of the probabilities of the events happening in the population. Since 59% of the sample use the patch, my best guess is that 59% of the 16% in the general population also use the patch. This represents $0.59 \times 0.16 = 0.094$.

2. An auto parts manufacturer is testing the quality of their brake calipers. They collect data on 100 calipers pulled randomly off the assembly line. The three different types of defects found are listed below:

Defect A:	8
Defect B:	15
Defect C:	9
A & B:	2
A & C:	1
B & C:	5

(a) What is the probability that a randomly drawn caliper has defect A?

$$\frac{8}{100} = 0.08$$

(b) What is the probability that, given a randomly drawn caliper has defect B it also has defect C?

$$p(C|B) = \frac{5}{15 + 2 + 5} = 0.227$$

(c) What is your estimate of the probability that any given caliper drawn off the assembly line at random will have either defect A OR defect C?

$$p(A \cup C) = \frac{8 + 9 + 2 + 1 + 5}{100} = 1 - \frac{15}{100} = 0.85$$

3. The data in the following table are the results of 100 random samples of 2 Solar Cells, tested on their current output. A "Pass" indicates the cell was able to produce 5mA under a controlled light source, a "Fail" indicates that the cell did not produce 5mA current. Assume that the performance of the cells are independent.

Solar Cell 1	Solar Cell 2	Number of Cells
Pass	Pass	56
Fail	Pass	13
Pass	Fail	20
Fail	Fail	11

(a) What is your best guess at the probability that, if you repeated the experiment, at least one of the cells would fail on a randomly selected trial?

$$p(Fail \geq 1) = p(Fail, Pass) + p(Pass, Fail) + p(Fail, Fail)$$

$$= 1 - p(Fail = 0) = \frac{56}{100}$$

(b) What is the probability that, given the first solar cell passed the test, the second cell failed?

For these kinds of conditional probability situations. I always make a table and count from the table: There are 20 that have the first Pass and the second Fail, and there are 76 total that have the first Pass. This is much easier than going through the formulas:

$$p(F|P) = \frac{p(Pass, Fail)}{p(Pass, Pass) + p(Pass, Fail)} = \frac{\frac{20}{100}}{\frac{56}{100} + \frac{20}{100}} = 0.263$$

(c) What is your best estimate of the probability that, repeating the experiment, given Solar Cell 1 failed, Solar Cell 2 would pass on any given trial?

Again, I look to find the situations where the first Cell failed. These total to 24 instances. Then, *of those*, what proportion of the second Cells pass? $\frac{13}{24} = 0.54$

4. A senior capstone design group created a video feed system by which lecture notes written by professors would be captured and displayed on the mobile devices of students who had limited vision. Twenty students participated in the study that tested two different data transfer rates. The slow rate was cheap and easy to implement, but the fast rate seemed to have better resolution. They tested the hypothesis that the fast rate resulted in significantly more intelligible images than the slow rate. Their data is presented in the table below:

	Data Transfer Rate		
Quality of Visual Display	Fast	Slow	**Total**
Intelligible	10	4	16
Not intelligible	0	6	4
Total	10	10	20

(a) If you were this capstone team, would you conclude that one of the data transfer rates is significantly better than the other?

The hidden question here is whether or not quality of visual display is independent of Data transfer rate. So I will compare their respective probabilities:

$$p\,(\text{Intelligible display}|\text{transfer rate} = Fast)$$

$$= p\,(\text{Intelligible display}|\text{transfer rate} = Slow)$$

$$p\,(\text{Intelligible display}|\text{transfer rate} = Fast) = \frac{10}{10}$$

$$p\,(\text{Intelligible display}|\text{transfer rate} = Slow) = \frac{4}{10}$$

Since these two probabilities are not equal, I have to conclude that quality of visual display is NOT independent of data transfer rate, and that a faster rate DOES create a more intelligible display.

This independence could also be tested using the proportions of the poorer intelligibility across transfer rates.

$$p\,(\text{Unintelligible display}|\text{transfer rate} = Fast)$$

$$= p\,(\text{Unintelligible display}|\text{transfer rate} = Slow)$$

$$p\,(\text{Unintelligible display}|\text{transfer rate} = Fast) = \frac{0}{10}$$

$$p\,(\text{Unintelligible display}|\text{transfer rate} = Slow) = \frac{6}{10}$$

(b) In this context, what does "significantly better" mean?

Significantly better means that, with real data, we will always expect some small, insignificant differences between proportions. Extremely large differences, however, are hard to ignore as just random differences. The big issue is, just how large the differences have to be to be considered truly significant.

(c) Suppose the fast data transfer rate costs twice the amount of the slow transfer rate, \$ 1 per minute versus \$ 0.50 per minute. If you were an end user, would you want to pay for this given you will have 45 hours of lecture per class next semester?

In this class, we have 45 contact hours that students would have to stream the data. This would equate to \$ 45 for the fast rate and \$ 22.50 for the slow rate. The magnitude of these dollars is low enough that it seems a good investment to support students with sight limitations with the faster data rate. If the costs were higher and the differential more marked, an administrator might come up with a different conclusion.

5. The growth of the solar industry in the US has been hypothesized by some energy scientists to be more rapid in the Southwest than in other regions in the US. Data from the National Renewal Energy Laboratory (https://openpv.nrel.gov/rankings) is shown below.

	Geographical Region						
Number		**NW**	**SW**	**MW**	**NE**	**SE**	**Total**
of Solar	0 – 99	5	1	8	2	11	27
Installations	100 – 149	1	3	5	3	1	13
	150 +	0	4	0	6	0	10
	Total	6	8	13	11	12	50

(a) What is the probability that, given a state is located in the Southwest, it has over 150 solar installations?

$$p\,(x \geq 150|Southwest) = \frac{4}{8} = 0.50$$

(b) What is the probability that, given a randomly selected state is either in the NE or SE, it has fewer than 100 installations?

$$p\,(x \leq 100|\,(NE \cup SE)) = \frac{\frac{2}{50} + \frac{11}{50}}{\frac{11}{50} + \frac{12}{50}} = 0.565$$

(c) Is there evidence that the number of solar installations is independent of geographical region in the US? Another way to ask this question is: Is the probability of one of the categories of solar installations occurring at random equal to that category appearing given the geographical region it was drawn from?

What I do here is compare the different probabilities for each region.

$$p\,(x \leq 99|NW) = 0.833$$

$$p(x \leq 99|SW) = 0.125$$

$$p(x \leq 99|MW) = 0.615$$

$$p(x \leq 99|NE) = 0.182$$

$$p(x \leq 99|SE) = 0.917$$

I really don't have to compute all of the probabilities. I can see that these are different enough to be pretty confident that different regions of the US have different proportions of solar installations. In particular, the NW and SE appear to have many more small installations than the SW or NE.

Just like problem 4a, we could compute all of the probabilities to gain a clearer picture of the disproportionate distribution of solar installation by region.

6. These are measurements of the length of a center punch (in mm) made by 11 mechanical engineering students in my class:

129	134	130	128
125	128.5	127.5	127
127.6	129	127	

(a) Enter this data into some reasonable statistical package (e.g., Matlab, Minitab, Excel, or use R or Python). What is the mean and standard deviation of this sample of measurements?

$$\bar{x} = 128.4182$$

$$s = 2.2746$$

(b) What is the median and interquartile range of this sample?

$$med = 128$$

$$IQR = 1.875$$

(c) Is the distribution pretty symmetric, or is it skewed in one direction or the other?

There is a bit of a skew in the right direction. The value 134 is somewhat of an outlier

(d) If these sample statistics represent the population from which they were drawn, the universe of possible measurements of the center punch, what is the probability of randomly selecting a measurement of 125mm from this population? What is the probability of randomly selecting 131mm?

If the only values in the population are these, then there is a $\frac{1}{11}$ probability that a measurement would be exactly 125 mm. There is a probability of 0 that a measurement is exactly 131. But, if this is just a sample of values, and we assume that the parent population is continuous, then 131 is close to 1 standard deviation above the mean, so the probability is about 0.16 that we could randomly draw a measurement of 131 or higher, and about 84% that we could draw lower than 131. 125 is a bit below 1 standard deviation from the mean so the probability of drawing a measurement of less than 125 randomly is about 0.16 and 125 or higher about 84% . You will learn to calculate the exact probabilities in the next two chapters.

(e) If the actual mean of the population of measurements is exactly 128.36 mm, and the variance of the population is 4.5, describe the accuracy and precision of the sample. Is it a good representation of the population, given it is only 10 measurements?

Whereas the sample mean is very very close to the population mean (we would call it a good estimate—an *accurate* estimate), the sample variance is

$$s^2 = 2.2746^2 = 5.17$$

This is a bit higher than the population variance, and so the sample doesn't have the best *precision.*

(f) What is the probability that a measurement, randomly drawn from the population of measurements, would fall between $-1\ \sigma$ and $+1\ \sigma$ either side of the mean?

This is taken from the normal distribution. $p(-1\sigma \leq x \leq +1\sigma) = 0.68$

7. The following table is data taken from test flights of an experimental paper airplane (done in my class). All planes have the same design, using the same materials, and were flown from the same height, using the same procedures. The numbers are measured distances each plane flew on its test run (in ft).

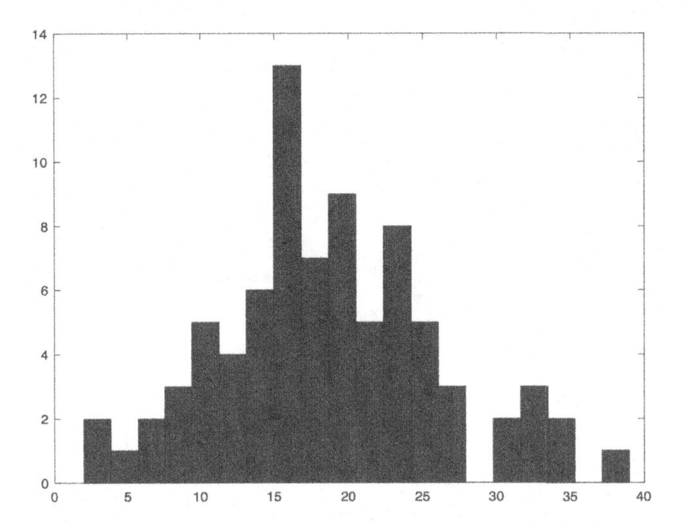

18	18	24	20	16	8	21	32	20
15	23	12	23	32	16	14	32	20
25	14	8	18	31	15	12	23	12
18	10	31	35	16	7	14	23	14
19	10	20	27	16	15	22	27	22
26	16	15	22	11	18	16	6	2
19	34	39	27	20	17	17	10	10
23	16	25	26	13	16	14	2	14
4	19	20	23	8	22	16	25	

8. Draw a histogram of the data. Experiment with the number of “bins.” Start with the default 10 bins, and then draw a few more until you get a histogram that “best” represents the shape of this distribution.

   This uses 20 bins. Anywhere above about 15 bins to about 20 bins gives a nice picture that shows the “bumps” as well as the “valleys” .

9. Draw a boxplot of the data. How does it compare to your “best” histogram in communicating the shape of the distribution?

   This boxplot shows the middle 50% to be between about 15 and about 23. This corresponds closely with the region in which we would expect to find the mean in the histogram. It also shows fairly good symmetry, like the histogram. What it loses is the details, especially at the upper tail of the distribution.

10. What is the mean and standard deviation of this set of data?
    $\bar{x} = 18.5432$

    $s = 7.5813$

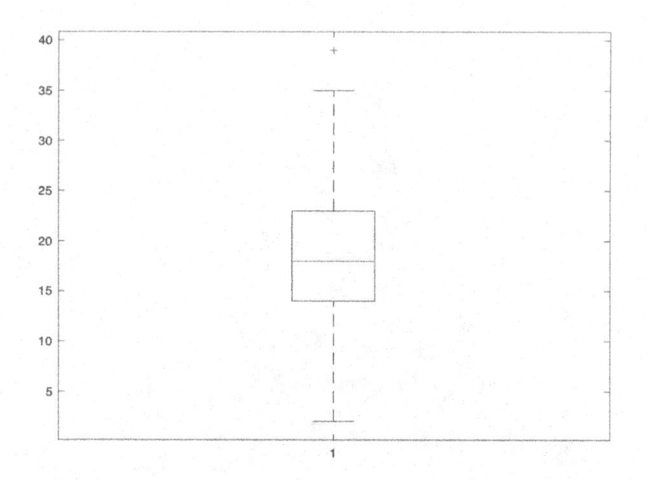

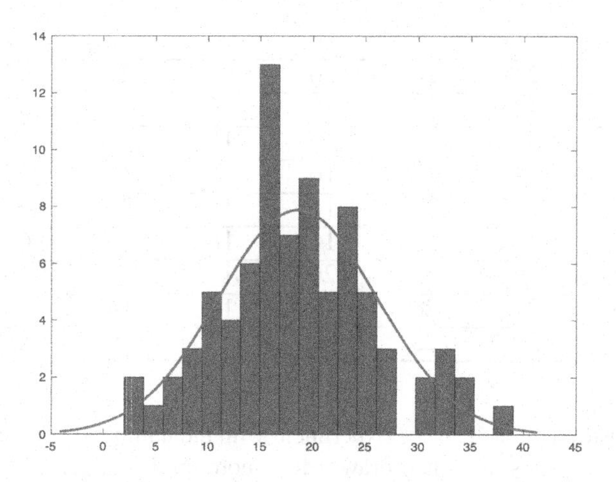

11. Using the sample statistics you have computed, draw the PDF of the parent population from which this data was sampled. What is the probability that any airplane, randomly drawn from this population would have a value *exactly* at the 11. mean?

    From the histogram, it appears that the value at exactly the mean is around 8 out of the 81 total paper airplane measurements. This is approximately $p = 0.099$.

12. What is the probability that any randomly drawn airplane would have a value between $-1\ \sigma$ and $+1\ \sigma$ either side of the mean?

    From the normal distribution, this is $p = 0.68$

13. Compare the value in 6f to your answer to 12. What is the relationship among these cumulative probabilities, and why do you think is this so?

For ANY normal distribution, the probability of any value being randomly selected that falls between +/– 1 *s* from the mean is constant—just about 68%.

5

The Sampling Distribution of the Mean

Thus far, we have examined sample statistics, the population parameters they estimate, and the behavior of certain distributions that are described by these parameters. We have learned to estimate the probability that a given measurement has in a population of measurements estimated by sample statistics. And we know how to estimate the likelihood that a given measurement occurs when it is drawn at random from some interval in a probability distribution.

If we can collect good data in a sample, we can then estimate the parameters of the population distribution and compare values of the independent variables to the dependent variable, $p(x)$. But up to now, we haven't really dealt with sample size very much.

We found out early on, that the larger the sample, the more confident we are that its statistics are good estimates of its population parameters. We trust samples of size 100 more than we do samples of size 10, all other things being equal. That is just an application of the Law of Large Numbers. What we would like to do is take advantage of sample size, so that we can be as accurate as we can in our estimates of population parameters without having to take a sample so large that it becomes unaffordable.

5.1 The General Logic of the Sampling Distribution

Suppose we had a population of measures, and we *didn't* know its parameters. Table 5.1 represents a finite population, wherein every *true value* of the population is recorded (Of course, these are only measurements in real life, but let's pretend they are population values. They are actually real heating loads in BTUs from 120 houses in the Southwest US):

33.13	13.17	14.7	14.71	32.67	14.45
14.1	39.68	40.43	32.73	22.93	11.13

DOI: 10.1201/9781003094227-5

11.45	12.72	11.42	10.47	12.49	28.56
28.66	26.46	33.08	24.77	13.68	14.9
6.01	32.31	36.47	12.33	14.42	18.9
12.43	12.32	11.44	24.26	11.11	14.75
32.31	12.5	33.27	12.59	24.37	32.26
39.81	16.55	11.18	11.8	10.77	36.95
40.03	38.82	11.61	10.8	12.8	36.45
28.42	36.26	12.67	19.36	15.55	8.6
16.69	18.71	39.89	24.29	29.63	36.45
12.28	28.62	16.64	29.91	19.34	14.71
36.81	38.84	29.62	32.21	12.95	11.33
32.48	39.97	15.55	22.89	12.27	28.52
15.16	38.65	11.14	12.73	32.31	29.6
38.98	25.37	14.58	15.36	32.07	12.88
25.36	16.76	24.32	15.16	31.53	11.14
14.66	32.52	25.36	40.19	10.7	10.53
12.45	29.43	16.76	32.53	20.71	28.15
32.49	10.75	12.49	26.47	42.08	12.19

TABLE 5.1: Heating loads in 120 houses in the Southwest US

Let's take one measurement at random from this population: 32.26 BTUs. I really did take this value at random.

1. If this were a *sample*, with size 1, how representative is it of the population?

   From my perspective it is easy to determine how representative this value is of the sample of measurements: I have all the true values on my computer! But for anyone who doesn't have access to all of these values, all they have is 32.26 BTUs. Is it somewhere close to the middle? It has zero variance, so it doesn't model the spread of the population distribution at all. With only one measurement, we can't be sure how it matches up with the population, only that it exists.

   Take a look at a histogram (bin width = 5) for this population in Figure 5.1.

$$\mu = 22.33, \quad \sigma = 10.27$$

   Suppose we took a sample of 5 at random:

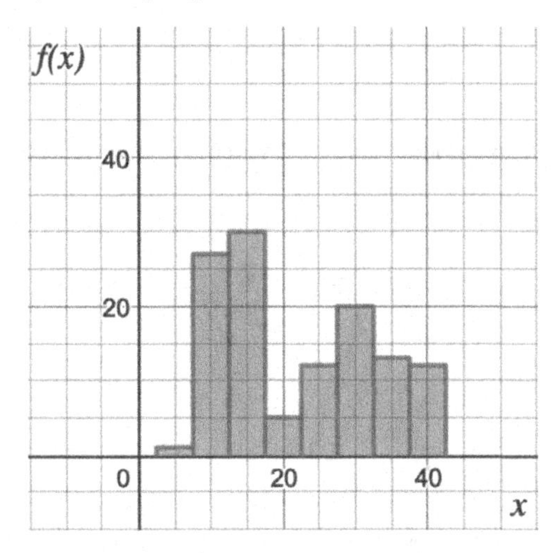

FIGURE 5.1: Histogram of a population of heating loads, 120 houses in the Southwest

40.43
25.36
11.14
39.89
24.32

$$\bar{x} = 28.23, s = 12.25$$

2. Is this sample any more representative? Is it any better than the sample, size 1? How would you tell?

   Without looking at Figure 5.1, I can't say for sure whether this sample is of any more representative of any particular statistic in the population. HOWEVER, because there are multiple data values, there is some variation that can be modeled, and so it does give an indication of the center ($\bar{x} = 28.23$), and spread ($s = 12.25$) of the population from which it was sampled.

   Looking at Figure 5.1, we can see that the mean is pretty close to the actual mean. Closer than the 32.26 we got with our single measure. In this case, the sample is a better estimate of the population location.

3. In other words, would you feel comfortable designing HVAC systems for all homes in the population based on this data? What if we had a sample of 60, half of the population?

   I am much more comfortable with this sample of 5 than a sample of 1, but 5 seems a bit small to me for the Law of Large Numbers to really begin to converge nicely on the population parameters.

We could have taken another sample at random and gotten the following:

10.7
32.31
29.63
31.53
15.55

$\bar{x} = 23.94, s = 10.07$

This seems pretty darn close to the population. But we got this one at random! Our first sample was okay, but not as good as the second. What if, by horrid luck, we got a random sample like this?

12.86
15.09
12.72
12.95
14.10

$\bar{x} = 13.54, s = 1.02$

Wow! That is very precise, but not very accurate at all!

But even this sample, with all values lower than the population mean, has no values that are equal to the population *minimum*. If we think about another bad sample, one with values taken all above the mean, we will find very few samples that incorporate the population maximum and almost ***NONE*** that have the population maximum as its mean value.

This means that, if we were to take repeated, random samples of 5 homes from this population, and plot their means in a histogram, the resulting distribution will have less variability than the population itself!

Since samples of size 10, on average, are better estimates of the population parameters than samples of size 5, we would expect the following: (1) a distribution of *means* all taken from the same population, all of size 10, would have a better overall estimate of the population mean than a distribution of means all with samples of size 5; and (2) the variation of this **sampling distribution of the mean** would be less than the sampling distribution of the mean made up of samples of size 5.

The **Central Limit Theorem**, which we introduced in Chapter 2, is ***the single most important idea in all of statistics***! It states that, if we take repeated, random samples of a given sample size, n, from the same population and calculate their means, the resulting sampling distribution of the mean will approach the Normal distribution as the number of samples increases, the sampling distribution will center on the population mean, μ , and the standard deviation of the sampling distribution will become smaller proportionally by $\frac{1}{\sqrt{n}}$. Now here comes the kicker: ***This is true no matter what the shape of the population!!!***

5.2 Sampling Distribution of the Mean

Take this description of the Sampling Distribution of the Mean and really study it. It will serve as the canonical distribution by which all other distributions can be compared in your statistical career. But to really understand it, we need to generate one using our heating load population.

As stated before, the mean and standard deviation of this *population* are μ = 22.33, σ = 10.27, respectively. We have already taken three random *samples* of 5 homes and found their means to be 28.23, 23.94, and 13.54. Suppose we kept going, taking random samples of 5, computing each sample's means, and then plotting them in a histogram. Here is what we get after 10 samples of n=5 (Table 5.2 and Figure 5.2):

Sample number	$\overline{x}$
1	28.23
2	23.94
3	13.54
4	32.41
5	19.93
6	18.68
7	23.97
8	29.12
9	16.83
10	21.34
Total	**22.78, σ = 5.90**

TABLE 5.2: A Sampling distribution consisting of the means of 10 samples, all with $n = 5$

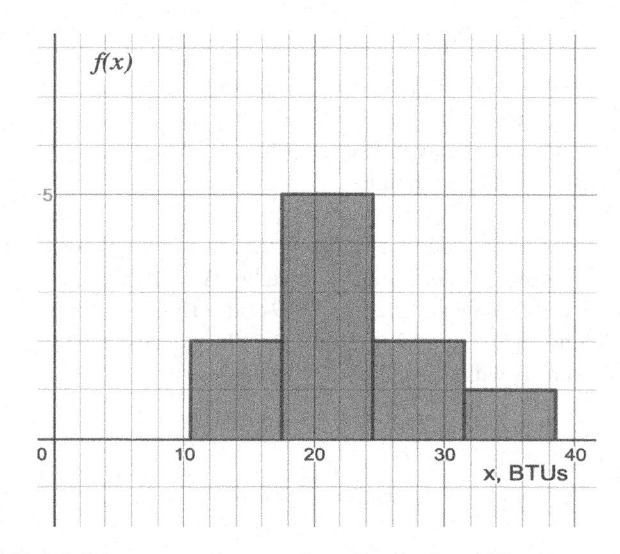

FIGURE 5.2: Histogram of a sampling distribution of $\overline{x}$, 10 samples of 5

The mean of our sampling distribution is pretty darn close to the population mean! Plus, its variation is quite a bit narrower than the population distribution. Now let's take a look at a histogram of 10,000 samples of size 5 all drawn at random from this population (Table 5.3):

Notice that, as the number of samples increases, the histogram becomes more and more symmetric about the population mean, and more and more bell-shaped. Here

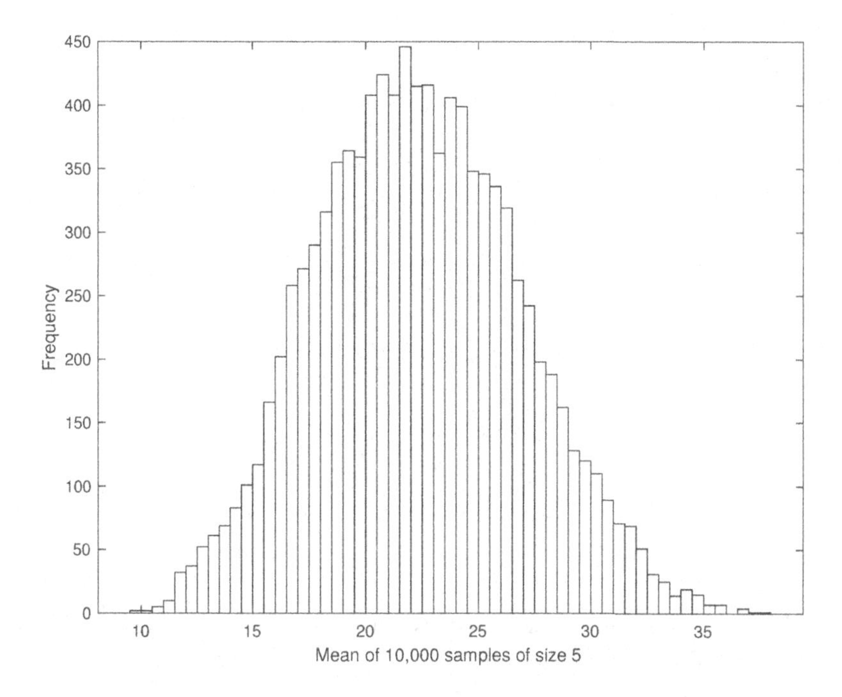

FIGURE 5.3: Histogram of a sampling distribution of the mean using 10,000 samples of size 5 from the heating load data

the mean of these 10,000 sample means is 22.34 BTUs, and the standard deviation is even narrower at 4.50.

Taking this example to the extreme, what do you expect if I played Dr. Evil from Austin Powers, and drew 1 million samples of size 5 (See Figure 5.4)?

Here the mean of the sampling distribution is 22.31, and the standard deviation is 4.51. This is *extremely* close to the population parameters. The mean of our sampling distribution is only off by −.02 BTUs, and the standard deviation is off of the predicted value, $\frac{\sigma}{\sqrt{n}} = \frac{10.27}{\sqrt{5}} = 4.59$, by −.07. Moreover, the curve is becoming decidedly smooth and bell-shaped: Normal.

This is just a crude example of the Central Limit Theorem. Even with small sample size, only 5, the tendency is for the mean of sampling distribution to get closer and closer to the true population mean, and for the width of the sampling distribution to get narrower as a function of the sample size converging upon $\frac{\sigma}{\sqrt{n}}$.

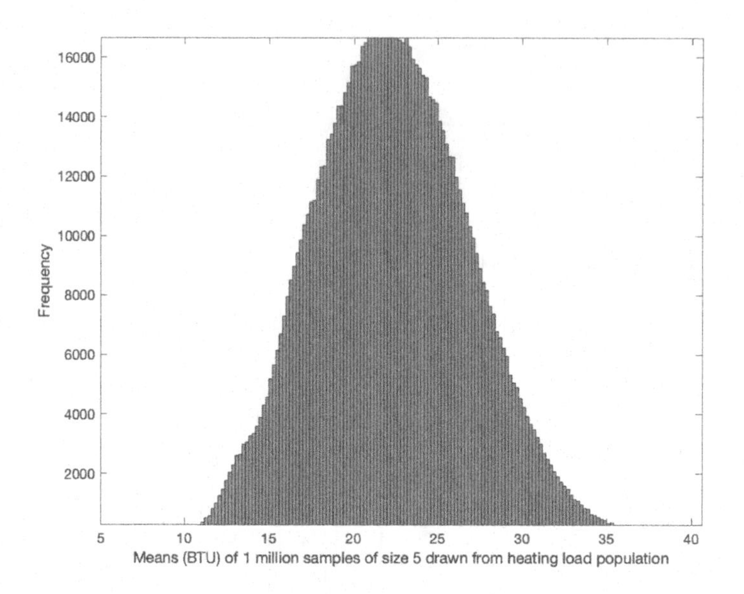

FIGURE 5.4: Histogram of a sampling distribution of the mean using one million samples of size 5 from the heating load data

Sampling distributions made up of smaller samples just take longer to converge to these characteristics.

This quantity $\frac{\sigma}{\sqrt{n}}$, because it is associated with *sampling distributions*, is called by a special name, the **Standard Error of the Mean**. I have given it a special symbol, easy to remember...

$$SE_{\bar{x}} = \frac{\sigma}{\sqrt{n}}$$

You might still refer to it as the "standard deviation of the sampling distribution of the mean," but this is just $SE_{\bar{x}}$ ier.

Just FYI, here is the script I used for generating this sampling distribution in Matlab. You can use it to make sampling distributions of any finite population you can upload and play with. I encourage you to use some data you are interested in, or make up some data that you would like to play with, and examine how the Central Limit Theorem holds *no matter what the shape of the parent population!*

```
% establishing the starting point for iteration and initializing a data

% vector, y

    r

k=1;

y=[];

% iterating a while loop for k iterations

while k < 1000000

%     selecting a random sample of size 5

y = datasample(data,5);

% computing the mean of each sample and placing those means in ybar

ybar(k) = mean(y);

% iterating k to the next step

k = k+1

end

% Transposing the row vector, ybar to a column vector

ybar=ybar'

% Drawing a histogram and computing the mean and standard deviation of
the

% sampling distribution in ybar

histogram(ybar)

mean(ybar)

std(ybar)
```

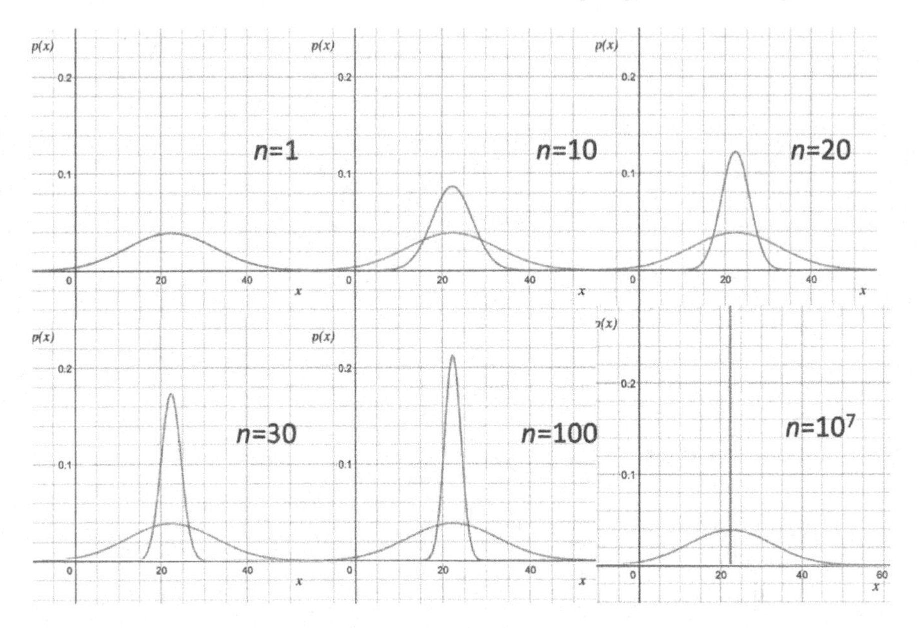

FIGURE 5.5: The sampling distribution of the mean gets narrower as sample size gets larger by a factor of $\frac{\sigma}{\sqrt{n}}$. At $n = 10$ million, the distribution is so narrow, that ostensibly all means in the sampling distribution are virtually indistinguishable from the population mean (i.e., it is almost a perfect estimate)

Now, we don't regularly have the time nor the money to take 1 million samples of any size. Instead, we take advantage of that fact of the Central Limit Theorem that tells us that, *as our samples sizes get larger and larger, the width of their sampling distribution of the mean gets narrower and narrower.*

BEWARE! When we talk about *sample size* affecting the width of the sampling distribution of the mean, we are *not* talking about the number of samples in the sampling distribution, we are talking about the *number of measurements* in each of our 1 million, or theoretically infinite number of samples. This is often a point of confusion for my students, so I want to highlight it here before it throws you off further down the road (and it probably will)... To illustrate this, I have drawn a normally distributed population with parameters equal to our population of house heating loads, $\mu = 22.33$, $\sigma = 10.27$. The narrower curves show how the sampling distribution of the mean changes for different sample sizes (Figure 5.5).

Notice that the distribution gets narrow pretty rapidly as n gets larger, but at about sample size 30, you don't get a lot of change for each additional phenomenon you measure. At about 30 measurements, the sampling distribution of the mean tends to give very good estimates of the population mean, and its variability is small enough that estimates are generally considered pretty precise.

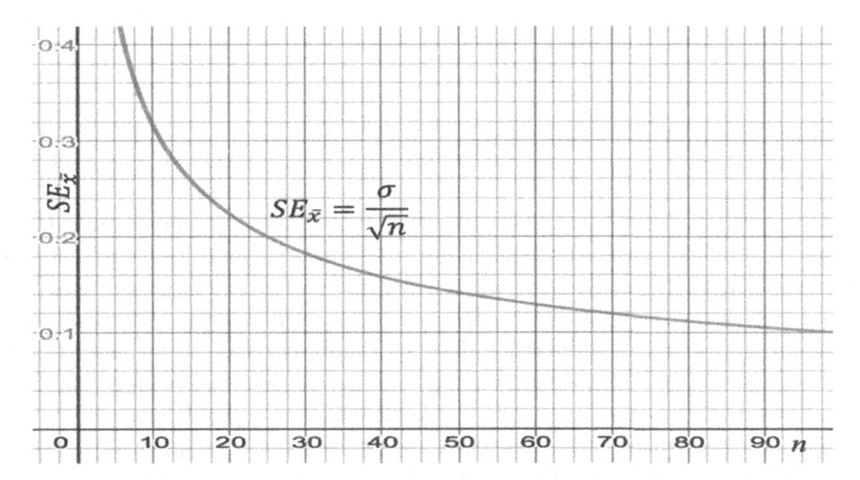

FIGURE 5.6: $SE_{\bar{x}}$ as a function of sample size. This relationship is drawn for the standard normal distribution, where $\sigma = 1$

Some statisticians claim that *30 is a magic number* because of this fact. Don't be fooled! Looking at Figure 5.6, we can see that, at around $n = 30$ the standard error curve gets asymptotically flat, showing smaller and smaller change for each increment of n we add. Increasing sigma (increasing our variability in the population) dilates this curve, increasing the values of n where the curve become more horizontal. The lesson here is, the more variability you have in your measures, the bigger the sample size you will need to get an appropriately small standard error.

Yes, 30 is a good sample size to start with, but if you are dealing with jet engines, you may not have enough money to test 30 engines! Practicality has to play an important role in determining what sample size you choose, as well as your measurement error. If your instruments are extremely precise, and natural variation among phenomena is small, you can get very good estimates of population parameters with samples of 15 or even 10.

5.3 The Standard Normal Distribution

So far, you may think that this book is a love story devoted to the normal distribution. I have to admit that most statisticians do adore this function:

$$p(x) = \frac{1}{\sigma\sqrt{2\pi}} e^{\frac{-(x-\mu)^2}{2\sigma^2}}$$

I, however, have a more mature relationship with it. It is just a model, imperfect, but good enough to help me understand the long-haul behavior of the data I am analyzing. In fact, if I strip some of the unnecessary information away from this function, I can create an even more applicable function that is even easier to understand! The method for this is called, **standardization**. Standardization has two meanings: First, it means to make a sample distribution fit the normal distribution, creating a standard model by which parameters estimated by samples can be compared. Second, it means to take the Normal distribution, translate it so that the mean now equals zero (sliding the distribution to the left or right on the independent variable, so that it is centered at zero), and then dilate it so that its standard deviation is equal to 1. This transforms the distribution to the following form:

$$p(z) = \frac{1}{\sqrt{2\pi}} e^{\frac{-z^2}{2}}$$

You will notice that I have changed our old friend x to a z. To differentiate a *Ztandard Normal Distribution*, statisticians denote the independent variable with a z, sometimes calling them "*z-scores*" and the entire distribution a "*Z-distribution.*"

The thing to remember with the sampling distribution of the mean is, *all Normal distributions can be converted to a Standard Normal Distribution* (Z-distribution) by translating each of the data points by the mean, and then dilating them by the standard deviation (see Figure 5.7):

$z_i = \frac{x_i - \mu}{\sigma}$ for a population, and

$z_i = \frac{x_i - \bar{x}}{s}$ for a sample.

The Standard Normal Distribution is a PDF that converts raw data scores into a distribution that is expressed in standard deviation units. The mean of the Standard Normal Distribution is defined as 0 (zero standard deviation units from the mean), and the standard deviation as 1 (the points of inflection of the Standard Normal Distribution fall on the values $\mu \pm 1$).

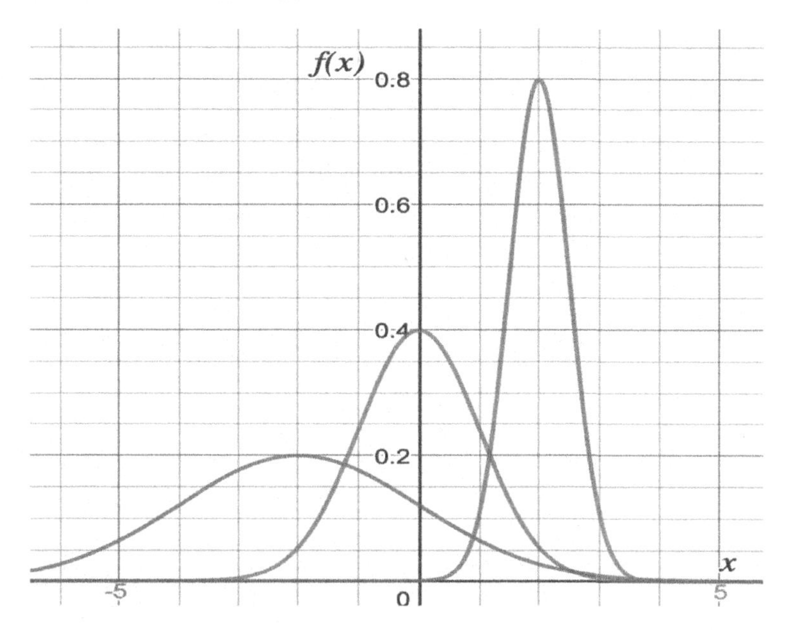

FIGURE 5.7: The standard normal distribution ($\mu = 0, \sigma = 1$), a normal distribution with $\mu = 2, \sigma = 0.5$, and a normal distribution with $\mu = -2$, $\sigma = 2$

Standardization is the process of converting a variable or set of variables to a common scale. The most common form of standardization is to create **Standard Scores** (also called **Z-scores**) by transforming the raw data distribution such that all values of the independent variable are expressed in standard deviation units.

We can do this because there is a one-to-one correspondence between any variable that is normally distributed and the transformed Z-Distribution. Moreover, any value that falls at any multiple of the raw score standard deviation will fall at some integer multiple of the Z-score distribution. Why is this useful you ask? It has to do with our old friend *probability density*.

5.3.1 Probability Density of the Standard Normal Distribution

The general utility of the Standard Normal Distribution is that it provides an anchor by which to assign probabilities (remember, a Normal Distribution is a PDF), to the parameters estimated by samples. Let's take a look at this phenomenon as illustrated in Figure 5.8:

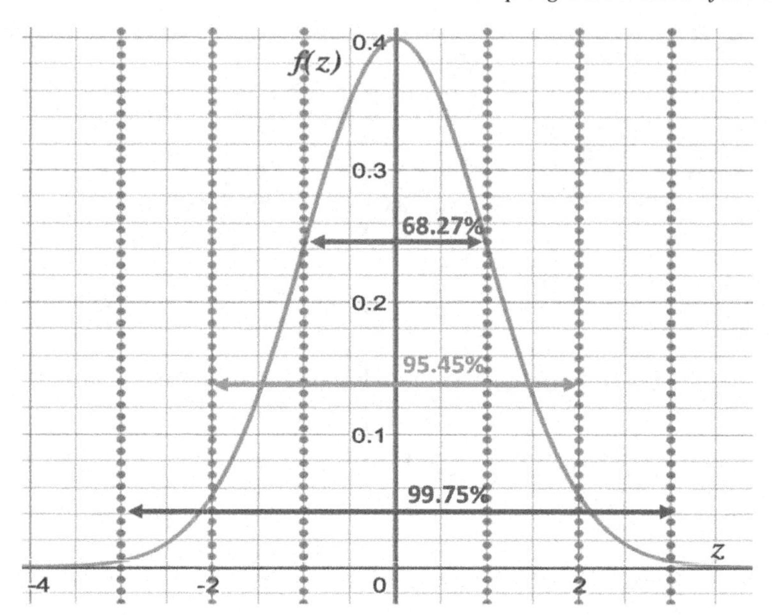

FIGURE 5.8: Probability intervals of the standard normal distribution

Any Normal Distribution has the property that, between -1 and $+1$ standard deviation from the mean, the probability of any value of the independent variable being drawn at random is right around 68% : 68% of all the data in the distribution lies between +/-1 standard deviation. Between +/-2 standard deviations, the probability of a random value being drawn is just over 95% : 95% of all the data in the distribution lies between +/-2 standard deviations from the mean. And lastly, well over 99% of the data fall between +/-3 standard deviations from the mean—the probability of randomly drawing a value between -3 and $+3$ standard deviations either side of the mean is over 99%.

1. Given the Standard Normal Distribution. What is the probability that a value, randomly drawn, will fall at or about 0 on the independent variable?

   From Figure 5.8, we can see that $f(z=0) \approx 0.40$. But if we consider values around z=0, we need to add their probability densities. Because $f(z)$ is a continuous function, this will involve integration. But the probability intervals given in Figure 5.8 can help. From it, we can see that about 68% of the data falls between $\mu \pm 1\sigma$. Most of the data in a normal distribution is close to the mean.

2. What is the probability that a value will fall at +1 standard deviation from the mean?

   $$f(z=0) \approx 0.24$$

3. Given you know the Standard Normal Distribution is symmetric, what is the probability that a value will fall below −2?

   Since 95% of the probability density falls between $\mu - 2\sigma$ and $\mu + 2\sigma$, about 5% falls into the tails (2.5% in each tail). So, the cumulative probability that any randomly drawn value taken from the standard normal distribution falls below $\mu - 2\sigma$ is about 0.025.

4. What is the probability that a value will fall above $\mu + 1\sigma$?

Since about 68% of the probability density falls between $\mu - 1\sigma$ and $\mu + 1\sigma$, about 32% falls in the tails of the distribution outside this interval. Half of that is about 16%

How do we know these things? Well, way back when, somebody's graduate assistant integrated the Z-distribution and evaluated all these cumulative probabilities, and hundreds more! This unlucky person had to put all of these cumulative probabilities, values of the CDF of the Z-distribution, into a table. I have reproduced it for you here in Figure 5.9:

To find, for example, the cumulative probability corresponding to an upper bound of $Z = -1.64$, first find its value in the table. Now move your finger to the right until you find "0.0505." Or we can look it up in the table over to the right, where positive values of Z are recorded (see Figure 5.10).

Luckily, most statistical routines will compute these values for you, but it is handy to have a table with which to look up easy values when you don't want to go to the hassle of writing script. Use this table as a handy time-saver. But *you should commit the 68%, 95%, 99% values for the intervals between +/−1, +/−2, and +/−3 standard deviations from the mean to memory. They will be used continually throughout your statistical career.*

5.3.2 Now Let's Do Some Real Stats with the Normal Distribution!

The total case length of a 7mm Rem. Magnum rifle cartridge is set to SAAMI[1] standards to be 2.500 inches $\pm$ .02 in. Suppose you were manufacturing these cartridges. You pulled a random sample of 30 cartridges off the line and they measured on average 2.51 in, with a standard deviation of .005 in.

1. Is there some systematic problem that is causing your cartridges to be manufactured too long?

[1] Sporting Arms and Ammunition Manufacturers' Institute

Z-value	Cumulative *p*(*Z*)	*Z*-value	Cumulative *p*(*Z*)	*Z*-value	Cumulative *p*(*Z*)	*Z*-value	Cumulative *p*(*Z*)	*Z*-value	Cumulative *p*(*Z*)	*Z*-value	Cumulative *p*(*Z*)
-3	0.0013	-2	0.0228	-1	0.1587	0	0.5000	1	0.8413	2	0.9772
-2.96	0.0015	**-1.96**	**0.0250**	-0.96	0.1685	0.04	0.5160	1.04	0.8508	2.04	0.9793
-2.92	0.0018	-1.92	0.0274	-0.92	0.1788	0.08	0.5319	1.08	0.8599	2.08	0.9812
-2.88	0.0020	-1.88	0.0301	-0.88	0.1894	0.12	0.5478	1.12	0.8686	2.12	0.9830
-2.84	0.0023	-1.84	0.0329	-0.84	0.2005	0.16	0.5636	1.16	0.8770	2.16	0.9846
-2.8	0.0026	-1.8	0.0359	-0.8	0.2119	0.2	0.5793	1.2	0.8849	2.2	0.9861
-2.76	0.0029	-1.76	0.0392	-0.76	0.2236	0.24	0.5948	1.24	0.8925	2.24	0.9875
-2.72	0.0033	-1.72	0.0427	-0.72	0.2358	0.28	0.6103	1.28	0.8997	2.28	0.9887
-2.68	0.0037	-1.68	0.0465	-0.68	0.2483	0.32	0.6255	1.32	0.9066	**2.32**	**0.9898**
-2.64	0.0041	**-1.64**	**0.0505**	-0.64	0.2611	0.36	0.6406	1.36	0.9131	2.36	0.9909
-2.6	0.0047	-1.6	0.0548	-0.6	0.2743	0.4	0.6554	1.4	0.9192	2.4	0.9918
-2.56	0.0052	-1.56	0.0594	-0.56	0.2877	0.44	0.6700	1.44	0.9251	2.44	0.9927
-2.52	0.0059	-1.52	0.0643	-0.52	0.3015	0.48	0.6844	1.48	0.9306	2.48	0.9934
-2.48	0.0066	-1.48	0.0694	-0.48	0.3156	0.52	0.6985	1.52	0.9357	2.52	0.9941
-2.44	0.0073	-1.44	0.0749	-0.44	0.3300	0.56	0.7123	1.56	0.9406	2.56	0.9948
-2.4	0.0082	-1.4	0.0808	-0.4	0.3446	0.6	0.7257	1.6	0.9452	2.6	0.9953
-2.36	0.0091	-1.36	0.0869	-0.36	0.3594	0.64	0.7389	**1.64**	**0.9495**	2.64	0.9959
-2.32	**0.0102**	-1.32	0.0934	-0.32	0.3745	0.68	0.7517	1.68	0.9535	2.68	0.9963
-2.28	0.0113	-1.28	0.1003	-0.28	0.3897	0.72	0.7642	1.72	0.9573	2.72	0.9967
-2.24	0.0125	-1.24	0.1075	-0.24	0.4052	0.76	0.7764	1.76	0.9608	2.76	0.9971
-2.2	0.0139	-1.2	0.1151	-0.2	0.4207	0.8	0.7881	1.8	0.9641	2.8	0.9974
-2.16	0.0154	-1.16	0.1230	-0.16	0.4364	0.84	0.7995	1.84	0.9671	2.84	0.9977
-2.12	0.0170	-1.12	0.1314	-0.12	0.4522	0.88	0.8106	1.88	0.9699	2.88	0.9980
-2.08	0.0188	-1.08	0.1401	-0.08	0.4681	0.92	0.8212	1.92	0.9726	2.92	0.9982
-2.04	0.0207	-1.04	0.1492	-0.04	0.4840	0.96	0.8315	**1.96**	**0.9750**	2.96	0.9985
										3	0.9987

FIGURE 5.9: Z-scores. The standard normal distribution is integrated from left to right that is from $-\infty$ to z. The "Z-table" lists values of the cumulative distribution function—*cumulative probabilities* as a function of z. Common values used in testing hypotheses are listed in bold

Think about this problem a bit. Your typical cartridge is well within tolerances. But what are the consequences of cartridges being .01 greater, on average, than they should be optimally, according to design? Suppose an optimal cartridge uses \$0.20 in brass in its manufacture. This 0.01 inches costs about \$0.0008, 8/100 of a penny. But what happens if you manufacture 1 million of the little buggers? This turns out to represent a cost of \$800 just in brass. This kind of waste cuts into profits, and the profit margin for small arms ammunition is very slim.

Z-value	Cumulative p(Z)	Z-value	Cumulative p(Z)	Z-value	Cumulative p(Z)	Z-value	Cumulative p(Z)	Z-value	Cumulative p(Z)	Z-value	Cumulative p(Z)
-3	0.0013	-2	0.0228	-1	0.1587	0	0.5000	1	0.8413	2	0.9772
-2.96	0.0015	**-1.96**	**0.0250**	-0.96	0.1685	0.04	0.5160	1.04	0.8508	2.04	0.9793
-2.92	0.0018	-1.92	0.0274	-0.92	0.1788	0.08	0.5319	1.08	0.8599	2.08	0.9812
-2.88	0.0020	-1.88	0.0301	-0.88	0.1894	0.12	0.5478	1.12	0.8686	2.12	0.9830
-2.84	0.0023	-1.84	0.0329	-0.84	0.2005	0.16	0.5636	1.16	0.8770	2.16	0.9846
-2.8	0.0026	-1.8	0.0359	-0.8	0.2119	0.2	0.5793	1.2	0.8849	2.2	0.9861
-2.76	0.0029	-1.76	0.0392	-0.76	0.2236	0.24	0.5948	1.24	0.8925	2.24	0.9875
-2.72	0.0033	-1.72	0.0427	-0.72	0.2358	0.28	0.6103	1.28	0.8997	2.28	0.9887
-2.68	0.0037	-1.68	0.0465	-0.68	0.2483	0.32	0.6255	1.32	0.9066	**2.32**	**0.9898**
-2.64	0.0041	**-1.64**	**0.0505**	-0.64	0.2611	0.36	0.6406	1.36	0.9131	2.36	0.9909
-2.6	0.0047	-1.6	0.0548	-0.6	0.2743	0.4	0.6554	1.4	0.9192	2.4	0.9918
-2.56	0.0052	-1.56	0.0594	-0.56	0.2877	0.44	0.6700	1.44	0.9251	2.44	0.9927
-2.52	0.0059	-1.52	0.0643	-0.52	0.3015	0.48	0.6844	1.48	0.9306	2.48	0.9934
-2.48	0.0066	-1.48	0.0694	-0.48	0.3156	0.52	0.6985	1.52	0.9357	2.52	0.9941
-2.44	0.0073	-1.44	0.0749	-0.44	0.3300	0.56	0.7123	1.56	0.9406	2.56	0.9948
-2.4	0.0082	-1.4	0.0808	-0.4	0.3446	0.6	0.7257	1.6	0.9452	2.6	0.9953
-2.36	0.0091	-1.36	0.0869	-0.36	0.3594	0.64	0.7389	**1.64**	**0.9495**	2.64	0.9959
-2.32	**0.0102**	-1.32	0.0934	-0.32	0.3745	0.68	0.7517	1.68	0.9535	2.68	0.9963
-2.28	0.0113	-1.28	0.1003	-0.28	0.3897	0.72	0.7642	1.72	0.9573	2.72	0.9967
-2.24	0.0125	-1.24	0.1075	-0.24	0.4052	0.76	0.7764	1.76	0.9608	2.76	0.9971
-2.2	0.0139	-1.2	0.1151	-0.2	0.4207	0.8	0.7881	1.8	0.9641	2.8	0.9974
-2.16	0.0154	-1.16	0.1230	-0.16	0.4364	0.84	0.7995	1.84	0.9671	2.84	0.9977
-2.12	0.0170	-1.12	0.1314	-0.12	0.4522	0.88	0.8106	1.88	0.9699	2.88	0.9980
-2.08	0.0188	-1.08	0.1401	-0.08	0.4681	0.92	0.8212	1.92	0.9726	2.92	0.9982
-2.04	0.0207	-1.04	0.1492	-0.04	0.4840	0.96	0.8315	**1.96**	**0.9750**	2.96	0.9985
										3	0.9987

FIGURE 5.10: Finding the cumulative probability of a Z-score $p(Z \leq 1.64)$. Working backwards, you can find the Z-value that corresponds to a particular cumulative probability.

What else does this +.01 inches do? It can change the ballistics of the projectiles fired. The pressures in any rifle cartridge are set to precise tolerances. An overage of just 0.01 inches in length, depending on the way in which the extra length is distributed, can increase the volume of the case enough to reduce the internal pressure when the round is fired, subsequently reducing the muzzle velocity of the bullet, and changing the actual trajectory from what is intended. In military and match shooting competitions, this could have serious consequences.

So, we might be losing money and jeopardizing the performance of our product. But we need to establish that our data represents a systematic issue and is not just a random blip that we might expect due to natural variation when we take a sample from the line. We need to restate question 1 to reflect the probability that our sample could have been taken from a population with parameters equal to the designed specifications. The real question we have to ask is:

What is the probability that your sample came from a population with a mean of 2.500 and a standard deviation of .02?

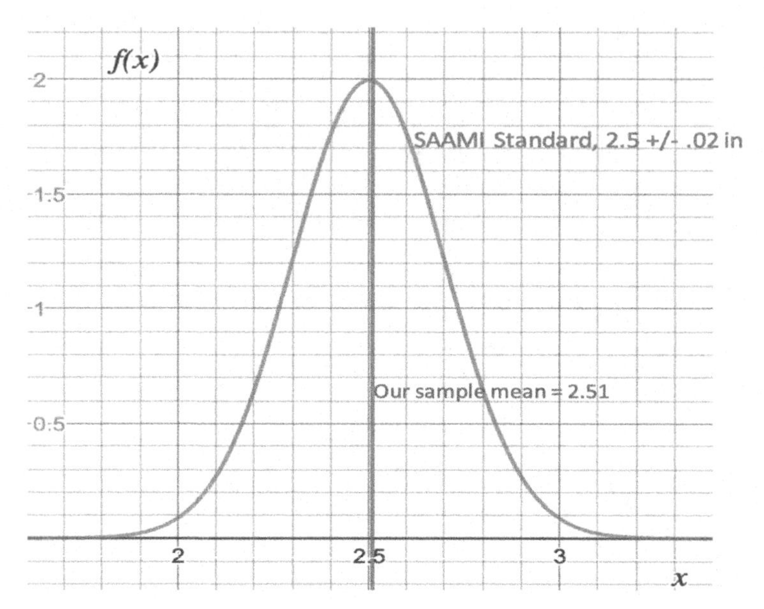

FIGURE 5.11: Intended distribution of 7mm Rem. Mag. Cartridges

Figure 5.11 shows the population represented by these parameters, with our sample mean represented by the vertical line at 2.5, bisecting the distribution. From this graph, it looks like our sample is so close to the mean that there is no significant difference. But there is a problem with this. We didn't just choose a single cartridge. We chose a random sample of 30 cartridges. This sample was *very* precise (s was only 0.005 inches!). In fact, if we look at the proportions of cartridges in this sample, 68% will be between 2.51 +/−.005 (between 2.505 and 2.515 inches). If we had randomly drawn our sample from the population indicated by our SAAMI standards, we would expect to see our data fairly evenly distributed about 2.50, the SAAMI population mean.

Even more interesting, 95% of our data in our sample falls between +/−2 standard deviations from our sample mean (between 2.50 and 2.52). This means that 95% of our sample falls above the SAAMI population mean. This would be expected if our sample was only 1 or two measurements, but 28 out of 30? That seems pretty unlikely...

1. How are we going to get out of this conundrum? Do we trust Figure 5.11, which shows that our sample is all contained well within the desired population, and near the population mean, or do we trust our analysis that shows that 95% of our sample falls above the SAAMI population mean?

5.3.3 The Z-test

The Z-test uses the Standard Normal Distribution as a means of determining the probability that the population estimated by a sample could be the same as some referent population. The basic method is as follows:

Z-test procedure

1. Estimate the sampling distribution for the sample you have collected. This creates a normal probability distribution of *sample means* that are supposed to have been sampled from the reference population, but that use the *sample* mean to estimate their center. Because it is a sampling distribution, its standard deviation, *the Standard Error of the Mean* will be used as an estimate of the population standard deviation, σ.

a. Assume your sample mean is a good estimate of the population mean, μ, from which your sample was collected.

2. Compare your reference population mean, with the sampling distribution

a. Where is the mean of the reference population located, in standard deviation units, in the sampling distribution?

b. Determine the cumulative probability that this could have occurred solely as a function of chance.

c. Make a decision what to do about your results.

Now let's apply this method to our rifle cartridges. First, let's create our model for the sampling distribution of $\overline{x}$. Remember, this is the distribution of all possible *samples* of size 30 that could have been taken from a *population* centered on $\overline{x}$. So we will use $\overline{x} = 2.51$ as our estimated mean of the sampling distribution. The reason we use σ as the numerator of the standard error is that, we begin with an assumption that our sample *was* taken randomly from the reference population of cartridges that fit the SAAMI standard. We are trying to determine how far off that population is from the population of samples that could have given us $\overline{x} = 2.51$.

So, since our assumption is that the sample's referent population is the same as the SAAMI population, we don't have to estimate its variability. We have it, σ. Our sample standard deviation is just an estimate, so it will not be as accurate as the actual population parameter it is trying to estimate. To determine the distance the SAAMI population mean is from our sampling distribution mean, we just subtract:

$$\bar{x} - \mu$$

Then we scale this distance by the number of standard deviations it represents in the sampling distribution. This is just a dilation:

$$Z = \frac{\bar{x} - \mu}{\frac{\sigma}{\sqrt{n}}}$$

Examining this equation, we can see that it is a Z-score, and therefore its probability can be found using the Standard Normal Distribution. This equation is called, the **Z-test of equal means** or the Z-test for short. It tests whether or not $\bar{x}$ and μ are close enough to be considered about the same given random variation in samples, or whether they are far enough apart to conclude that they must represent the centers of different populations, and therefore, there must be something systematic that is affecting the distribution of $\bar{x}$.

In our example, we are asking whether or not the sampling distribution our sample is estimating has the same center as the population it was *supposed* to be drawn from. Let's plug in the numbers:

$$Z = \frac{2.51 - 2.50}{\frac{.02}{\sqrt{30}}}$$

$$Z = 2.74$$

This value, 2.74 standard deviations, or 2.74 *sigmas*, means that, given the variability we expect among samples of size 30, a distance of 0.01 inches is 2.74 standard deviations above the value we expect as the center of the distribution. How likely is it that this could have occurred solely as a function of chance?

I have taken the liberty to draw the sampling distribution of the mean for our problem in Figure 5.12. Remember, it has mean 2.51 and standard error of $\frac{.02}{\sqrt{30}}$. The bold, vertical line is our population mean.

Now, let's take a look at the Z distribution-the Standard Normal Distribution. What is the probability that we could get a difference of *2.74 or more*? We could integrate the PDF, or we could use the handy-dandy table I reproduced for you. I don't want to do the first, so let's look at the table. It is more handy when we just have the statistics and parameters all computed for us. A smart student would go to a statistical package and use its speed to assist her!

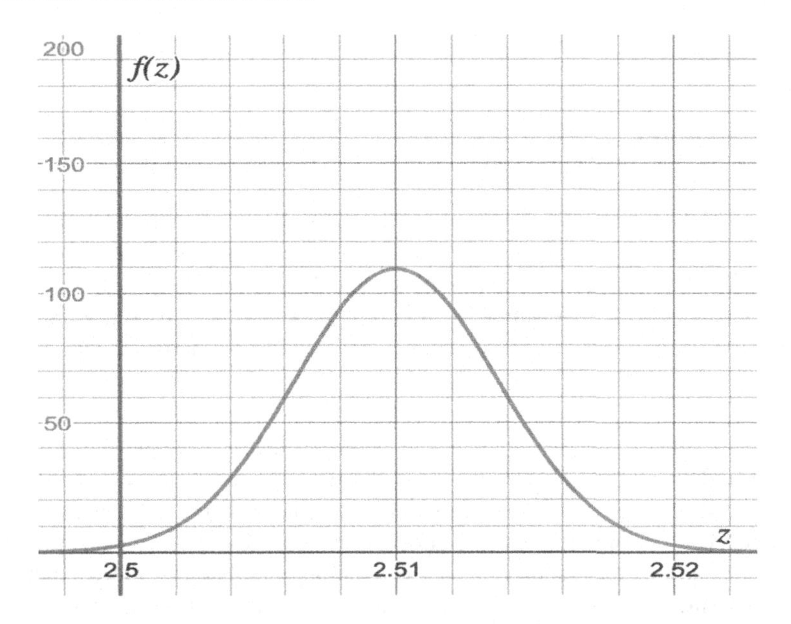

FIGURE 5.12: Sampling distribution of the mean for the population of samples with mean 2.51 and $SE_{\bar{x}} = \frac{.02}{\sqrt{30}}$. The population mean, 2.50, is 2.74 standard errors below the mean of the sampling distribution

Looking at the Z-table, we look up a z-value of 2.74. The table doesn't list 2.74, but it does have values of 2.72 and 2.76. 2.74 will be in the middle of those values. As I read the table, I see p(2.72) = 0.9967, and p(2.76) = 0.9971. So p(2.74) is also between those two values or about 0.9969. That is the cumulative probability that we could have drawn a value from the sampling distribution from $-\infty$ to 2.74. This is any value ≤ 2.74. We want the probability anything could be greater than 2.74. Here is where the rules of probability play in. Because we know that the probabilities in the Standard Normal Distribution sum to 1, and because it is symmetric, we just have to subtract .9969 from 1 to get the area under the curve to the right of .9969 = .0031. Look at Figure 5.12 one more time. If I integrated $f(z)$ from negative infinity to -2.5 standard deviation units, the area under the curve would be 0.0031.

The probability that a sample of size 30 with a mean of 2.51 could have come from a population with a mean of 2.50 and a standard deviation of .02 is about 3 in 1,000: not impossible, but highly *improbable*. As an engineer, you will be placing a bet on this probability. The bet is, if you accept these results, then you would conclude that we could only get these results about 3 times in 1,000 samples of size 30. Because this is so rare, it is more likely that there is something systematic in our machining that is causing the cartridges to be slightly over the specified standard. We now have to compare this result to the consequences of re-tooling machines, changing the alloy of brass we are using, or other fixes to the problem.

5.4 Summary

The Central Limit Theorem finally shows us its usefulness! It assures us that, if we take repeated *samples* all of the same size from a population, the distribution of those samples' *means* will tend to Normal over the long haul. Even more critical, as the size of samples taken from a population increases, the sampling distribution of the means of those samples will tend to get narrower and narrower ($SE_{\overline{x}} = \frac{\sigma}{\sqrt{n}}$).

Because all Normal distributions can be described as a simple transformation of the Standard Normal Distribution, we can describe the distance any sample is from the center of its sampling distribution of the mean in terms of standard deviation units. These standard deviation units are associated with fixed probabilities of samples being drawn at random from the sampling distribution—the population of samples. Using what we know about the Fundamental Theorem of Calculus, we can integrate the PDF of the sampling distribution and find the cumulative probability that sample could have been drawn in any interval we desire. If a sample appears way out in the tail of the sampling distribution of the mean, it is unlikely to have been drawn from the population at random, and we may conclude that there is some systematic factor causing this difference.

5.5 References

Bitra, V. S., Womac, A. R., Chevanan, N., Miu, P. I., Igathinathane, C., Sokhansanj, S., & Smith, D. R. (2009). Direct mechanical energy measures of hammer mill comminution of switchgrass, wheat straw, and corn stover and analysis of their particle size distributions. *Powder Technology*, *193*(1), 32-45.

Herlitzius, H. (1983). Biological decomposition efficiency in different woodland soils. *Oecologia*, *57*(1-2), 78-97.

Kwok, Alison G., "Air movement and thermal comfort in tropical schools," Proceedings of 22nd National Passive Solar Conference, Washington, DC, April 25-30, 1997, p. 25-31.

5.6 Study Problems for Chapter 5

1. Describe how the standard error of the mean is influenced by sample size and variability of measures.

   Because the mean of each sample in a sampling distribution of the mean is a middle value between extremes, the variability in a sampling distribution is less than the variability in the population from which the samples were taken. Given that, on average, samples with larger size, will have better estimates of the population mean than smaller samples, the variability in the sampling distributions containing means of larger samples will be less than sampling distributions containing means of smaller samples. This results in the standard deviation of a sampling distribution to get smaller and smaller as the size of samples increases.

2. The data presented below are the diameters of particles produced in a hammermill processing various species of grass (Bitra et al., 2009). The engineers want to know the distribution of particles to design sieves for air cleaning in farm and industrial applications.

0.76	0.9	0.7
0.75	0.88	0.74
0.69	0.82	0.66
0.64	0.82	0.67
0.49	0.76	0.62
0.76	0.95	0.74
0.72	0.83	0.68
0.63	0.88	0.64
0.65	0.76	0.69
0.56	0.76	0.64

   Examine the sampling distribution of the mean if this were a population. Describe the characteristics of the sampling distribution for the following sample sizes and iterations:

   n=3

   100 iterations

   1,000 iterations

   10,000 iterations

   n=5

   100 iterations

   1,000 iterations

10,000 iterations

n=30

100 iterations

1,000 iterations

10,000 iterations

(a) What is the primary differences you see as the number of iterations increases?

Answers will vary depending on the specific random sampling distributions you produce. BUT the general trend will be that, as the iterations increase, the sampling distribution of the mean approaches the normal distribution.

(b) What is the primary differences you see as the sample size changes?

Answers will vary depending on the specific random sampling distributions you produce. BUT the general trend will be that, as the sample size gets larger, the standard error of the mean gets smaller and get narrower as sample size gets larger.

3. The following data represent a sample of the percent decomposition of different of leaf matter from Alluvial forest floors. The researchers were interested in the different rates in which vegetation could be composted. The population of leaf matter in the study had a mean decomposition of 54.16% with a standard deviation of 10.15% (Herlitzius, 1983).

36.77	60.3
50.88	35.32
21.41	57.85
34.08	96.83
58.74	88.42
85.15	50.72
50.02	

(a) What is the probability that a measurement of leaf matter decomposition is 50% or less in the population?

First, I need to figure out the z-score for $x = 50$:

$$z(x \leq 50) = \frac{50 - 54.16}{\frac{10.15}{\sqrt{13}}} = -1.478$$

Now with this value, I can find $p(z \leq -1.478)$. I look up the value of $p(z = -1.478)$ in a z-table, or have a statistical package calculate it for me. The resulting cumulative probability is:

$1 - p(z = +1.478) = 1 - 0.9292 = 0.0708$.

(b) What is the z-value of a measurement that corresponds to a leaf decomposition of 45% ?

$z(x \leq 45) = \frac{45-54.16}{\frac{10.15}{\sqrt{13}}} = -3.25$. You should be thinking by now, *This is pretty dang far away from a mean of z=0. There should be a probability of less than 1% given the probability intervals Middleton gave me up in Figure 5.8!* but you will look it up or compute it with a handy stats package and find that $p(z \leq -3.25) = 0.000577$.

(c) What is the distance between the population mean and the sample mean, expressed in standard deviation units?

First, I compute the mean of the sample of decomposition data:

$$\overline{x} = 55.88$$

Then I use this to estimate the z-value of the difference between $\overline{x}$ and μ .

$$z(x \leq 55.88) = \frac{55.88 - 54.16}{\frac{10.15}{\sqrt{13}}} = +0.6110$$

4. Is this sample representative of the population, given random variation, or does it seem systematically different? Justify your conclusion.

   +0.6110 is between the mean, 55.88 and one standard deviation above the mean. So, since this is well within $\pm 1SE_{\overline{x}}$, I think the sample is not significantly different from the population. The probability of randomly selecting a sample with a mean of 55.88 from a population with a mean of 54.16 is $p(z \leq +0.6110) = 0.7290$. This, to me seems highly likely that we could have a sample of mean 55.88 (n=13) randomly selected from a population with a mean of 54.16 and standard deviation of 10.15.

5. The mean hydrocarbon exhaust for properly tuned automobiles is around 75 ppm +/−10 ppm. If you take your car into the shop and they measure the hydrocarbon output of your exhaust to be 86 ppm, express this as a Z-score.

   $z(x \leq 86) = \frac{86-75}{10} = +1.1$. This illustrates the difference between a z-score for a particular measurement versus a z-score for a sample. The sample size is 1, so the denominator is just the standard deviation of the population.

6. Alison Kwok did her Ph.D. thesis studying thermal conditions in naturally-ventilated versus air-conditioned schools in a tropical climate. Below is a random sample of the comfort ratings of 40 students in those schools. Comfort was rated on a scale from 1 to 6 where 1 was considered very uncomfortable, and 6 considered very comfortable. The ASHRAE database has a set of thousands of ratings that can be considered a proxy population of comfort ratings of kids in schools around the world. The mean comfort rating across all students in this database is 4.06, with a standard deviation of 1.12.

5	5	4	3
5	5	3	6
3	4	5	6
4	4	5	5
4	4	5	4
5	5	3	5
5	5	3	5
5	3	4	5
4	2	3	5
5	4	5	4

(a) Using a stats package or online calculator to help you, do you think the comfort experienced by this sample, taken from a tropical climate, significantly different from the population of kids around the world?

The mean of this sample is 4.35 "comfort units." We can express this as a z-score relative to the population data:

$$z(x \leq 4.35) = \frac{4.35 - 4.06}{\frac{1.12}{\sqrt{40}}} = +1.528$$

$$p(z \leq +1.528) = 0.9357$$

So, the probability that we could get comfort ratings of 4.35 or less is approximately 94%. That means getting 4.35 or higher is about 6%. This is pretty rare, so I am thinking that the likelihood is low enough that the students in the sample from tropical countries felt more comfortable in their schools, relative to the general population of students in the world.

7. A sample of 20 shafts of the flywheel in a robotic micro-copter motor has been measured to be 0.2508 inches on average by the company who manufactures it. If the diameter has historically been measured to be 0.2500 +/−.0015 inch, how would you find the probability that the manufacturing process systematically produces shafts of 0.2508 inches or more?

Here the population mean and standard deviation are 0.2500 and 0.0015 respectively. Knowing them, and knowing the sample mean of 0.2508, I can use the z-statistic and find its probability.

$$z(x \leq 0.2508) = \frac{0.2508 - 0.2500}{\frac{0.0015}{\sqrt{20}}} = +2.385$$

$$p(z \leq +2.385) = 0.9913$$

Therefore,

$$p(z > +2.385) = 1 - 0.9913 = 0.0087$$

It is highly unlikely that this sample of 20 shafts could have been randomly selected from a population with a mean of 0.2500 and standard deviation of 0.0015.

8. A researcher is studying the characteristics of her newly designed braking system. She tests a random sample of 30 brakes under controlled conditions. She gets a stopping distance of 32 meters, with a standard deviation of +/−5.2 meters. She wants to compare her sample to published reports of typical brake performance under similar testing conditions. The published results show a mean braking distance of 36 meters, with a standard deviation of 6.1 meters.

   (a) If the researcher asked you what you would conclude from her experiment, what would you tell her?

   First, I would have to determine the probability that her sample is likely to have been drawn at random from the population with mean of 36 meter +/−6.1 meters. This this involves finding how far apart the sample mean and population mean are, then scaling it by standard deviation units, accounting for a sample size of 30 braking systems.

$$z(x \leq 32) = \frac{32 - 36}{\frac{6.1}{\sqrt{30}}} = -2.038$$

$$p(z \leq -2.038) = 0.0212$$

   Therefore, there is about a 2% probability that we could have drawn a sample of 30 braking systems from a population with a mean of 36 +/-6.1 at random, and found the sample mean to be 32 meters. If I were the researcher, I would be very happy and conclude that my braking system results in significant reduction of the braking distance for the conditions under which the sample was tested.

9. A researcher is characterizing the performance of high performance skis that utilize actuators to stiffen the skis as stress increases (e.g., in high speed turns). She tests a random sample of 25 skis under controlled conditions. She gets a mean turning radius of 3.2 meters, with a standard deviation of +/−2.1 meters. She wants to compare her sample to published reports of typical ski performance under similar testing conditions. The published results show a mean turning radius of 4.3 meters, with a standard deviation of 2.5 meters.

   (a) How good a job has she done designing the skis, in terms of their turning performance?

   This problem is identical in nature to problem 8. The researcher wants to know if her new actuator-based stiffening system impacted the turning radius of the skis. hard return.

   First, I determine the probability that her sample is likely to have been drawn at random from the population with mean of 4.3 meter +/− 2.5 meters. This this involves finding how far apart the sample mean, 3.2 and population mean, 4.3 are, then scaling it by standard deviation units, accounting for a sample size of 25 skis.

$$z(x \leq 32). = \frac{3.2 - 4.3}{\frac{2.5}{\sqrt{25}}} = -2.200$$

$$p(z \leq -2.200) = 0.014$$

This z-value is way in the left tail of the normal distribution, and the probability is very small, about 1%, so I would conclude that, if I repeated the experiment infinitely many times, only about one sample in 100 would show a mean turning radius of 3.2 or lower.

10. **Why do we like to use the normal distribution to make statistical claims?**

    Aside from "That is what everybody else does... " I would say that, 1) because many natural phenomena exhibit this bell-shaped behavior in their distribution of measurements, 2) because the CLT proves that, over the long haul, the sampling distributions of the mean becomes normally distributed *regardless of the shape of the parent population*, and 3) because its PDF can be expressed as a closed-form equation allowing us to use calculus and the mathematics of continuous, real numbers, it is fairly easy to understand, and the mathematics is a close model to many applied situations.

11. **Why is it logical to use the standard deviation of the population for the Z-test, as opposed to the standard deviation of our sample?**

    This trips students up a lot. We will see, when we really get into hypothesis testing, that the assumptions we have to make are that the sample *does* come from the population, and if the z-test statistic falls way out in the tail of the distribution, that it is *unlikely* that our sample came from the population. Because of this assumption, we need to use our best information regarding the probability distribution of the population. For that reason we use the population standard deviation in our z-test to compute the standard error of the mean.

12. **What are the assumptions you have to make for the Z-test to hold?**

    The assumptions of the one-sample z-test are:

1. The data are sampled from a population that is a continuous, random variable.
2. The sample is a simple random sample from the population.
3. The population standard deviation, σ, is known.

Technically, there is another assumption: That the population from which the sample is drawn is normally distributed. BUT, because of the CLT, the sampling distribution of the mean WILL be normally distributed. So for most applications, the normal distribution is a reasonable model to use. It just may take a large sample for it to converge quickly enough for the z-test to be a good model.

Part II

Testing Hypotheses

6

The Ten Building Blocks of Experimental Design

As engineers, we design new things to solve problems of import to society: New vehicles, new devices, new media, and new systems and processes. We trust that our products are *better* in some way that current products, or that they improve the lives of human beings because prior to their introduction certain jobs or activities, or ways of thinking were impossible.

Additionally, the consequences of our actions can have serious consequences: both good and bad. We want to improve the world through our work, but we also want to be sure that we don't harm anything or anyone in the process. Mistakes in, and unintended consequences of, mechanical and aerospace engineering have led to disasters, lost lives, pain and sorrow, as well as improved health, more efficient use of resources, and more convenient and faster transportation, to name a few.

To provide assurance that our work DOES improve on current products and practices, we engage in **experimentation**. Briefly, experimentation is just the collection of data where we introduce one or more **treatments** to determine their impact on some phenomena we want to improve. If the data show that we run faster, live longer, or experience more satisfaction, without causing any unintended harm, we can conclude that our engineering work has been a success, and we can move our products or systems to market.

Experimentation is a process of testing hypotheses, predictions of the relationship among two or more conditions, with theoretical rationale for why the relationship exists. Typically, hypotheses are devised to explain the behavior of a phenomenon, or predict the results of some treatment on the behavior of the phenomenon.

A **Hypothesis** is a logical statement that proposes a set of conditions that result in a predicted (i.e., unknown) outcome. Additionally, a hypothesis proposes a theoretical reason the outcome is a logical predicate to the antecedent conditions. For a hypothesis to be scientific, it must be testable, that is, capable of being falsified.

DOI: 10.1201/9781003094227-6

A **Treatment** is means of manipulating phenomena, to induce some change in their behavior. When a treatment is applied to a group of phenomena and these compared to non-treated phenomena, any difference in their behavior can be attributed to the effect of the treatment.

6.0.1 Notation

The following symbols will be used to describe the different steps in the experimental designs described in this chapter (see Table 6.1):

Step	Symbol
Selection of the sample(s) or experimental unit(s)	X
Random assignment to an experimental condition	R
Blocking units, or other variables, into sub-samples	BK
Administering a treatment to a group	T
Observing (measuring) results	O

TABLE 6.1: Notation for describing the defining features in Experimental Designs

6.1 Basic Experimental Designs

Ten commonly used experimental designs will be described. They include the following:

1. One-Shot Case Study
2. One-Sample, Pre-Post
3. Static Sample Comparison
4. Random Sample
5. Pre-Post Randomized Sample
6. Factorial, including Randomized Block
7. One-Shot Repeated Measures

8. Randomized Factors Repeated Measures
9. Ex-post-facto (Quasi-experiment)
10. Time-series

6.1.1 One-shot Case Study

The One-shot case study is a design in which one sample is selected, a treatment is administered, and results are then measured or observed. In terms of the experimental procedure, we can symbolize it as follows:

X–>T–>O

Where X in this case is a sample typically of a single, or of a small number of cases. T is the treatment—the innovation you wish to introduce, and O is the observation/measurements you take. The arrows show steps you go through in effecting this design, the first steps appear left-most in the diagram, and the consequent steps following each successive arrow.

This is NOT a true experiment, in that there is no **control condition** by which to gauge any natural changes in the conditions of the case not attributed to the treatment. An example of this would be to take a car engine, put on an experimental PVC valve, and measuring crankcase emissions following installation. Because there is no baseline measurement, there can be no clear understanding of the changes the PVC valve has caused over the previous emissions control devices. Also notice that there is no random assignment of engines to conditions (we are only using 1 engine or a small number of engines, and one treatment condition). All you can do is *describe* the behavior of your engine with the PVC.

The One-Shot Case Study Design is useful to gather baseline data, or to determine tolerances for a designed treatment. It also is a way to inexpensively record any unintended consequences of the treatment (e.g., excessive heat, loss in power) that can be factored into the design in later stages.

Simplicity, ease, and low cost represent strong potential advantages in this design.

6.1.2 One-sample, Pre-post Design

In this design, one sample is collected, and its behavior is measured prior to treatment. Then the treatment is added, and a measurement is taken after. The difference between measurements pre-observation versus post-observation is considered to be due to the addition of the treatment.

This design is diagrammed as follows:

X–>O_1–>T–>O_2

Notice that in this design, like the one-shot case study, we cannot account for natural variations that may occur between observations that are NOT a function of the treatment. But, we can say that the treatment **caused**, at least *in part*, the differences in observed measurements, because we have taken measurements of the baseline behavior of our sample. For our PVC valve example, we would select a sample of candidate engines, measure the crankcase emissions to establish a baseline, then install our new valve and measure the emissions a second time. The difference between emissions AFTER the valve installation versus BEFORE, is considered the *effect* of the treatment. We cannot say that our valve caused ALL of the measured differences in the observations, but we can say it caused some of those differences. We cannot generalize the results of our experiment to all engines of the type we tested, because we have not taken a random sample from the population of engines, as have no assurances that our sample is representative of the population. This is the one major drawback of this design.

If our independent variable is time-varying, meaning that we expect its behavior to not be static, but to change over time, we might want to see how our treatment impacts the rate at which the variable changes. We can extend the pre-post model indefinitely. For example:

$$\mathbf{X\text{–>}O_1\text{–>}O_2\text{–>}T\text{–>}O_3\text{–>}O_4\text{–>}O_5}$$

With equal intervals of time in between observations, we have what is called a **time-series** study (described later in this chapter). We can then see what patterns exist in the behavior of our sample prior to treatment, and then see how the sample is changed by the treatment and observe if these changes remain the same, accelerate, or decay over time. Often we can fit a curve to such data. Change in the slope or intercept of the curve at or closely following each treatment event is considered evidence of its effect.

This design is frequently used in design research to see if there are changes in your product's performance as you tweak the performance of different design elements. It is also used in market research to determine change in customer attitudes or preferences towards a product you are developing.

The major advantages of this design are that it is comparatively cheap, requiring no random sample, effective in that it does directly measure the effects of the treatment, and because it can be modified to accommodate time-varying independent factors. Its major disadvantage, again, is that, because there is no random sampling, the results are more difficult to generalize to the entire population of interest.

6.1.3 Static Sample Comparison

The static sample comparison design (Also called Static Group Comparison in social sciences) essentially takes two samples that differ on some important factor, and compares them to determine the effect of that factor. An example of such a study would be to select two engines, one that has a PVC valve installed and one that does not, and compare the emissions of the two. The difference in emissions would be considered to be a function of the PVC valve being installed versus not installed.

X–>T–>O

X——> O

Notice how only one of the samples gets the treatment. This design requires that we assume our samples are *functionally equivalent* for all of their characteristics related to the treatment and observed changes in their behavior. If the two groups are not comparable—if they differ on some important, but unknown, variable, the results of your study may be confounded, because it was impossible to tell if the differences measured were solely a function of the treatment itself.

In the case of engines, two from the same batch may establish functional equivalence. In the case of customers, two samples, each taken from different cities, may result in differences not attributable to your treatment. In that case, you really want to try to use a TRUE experimental design if at all possible (described below in section 6.1.4).

Another way to build the case for your hypothesis is to collect many samples under this model, and look at the differences across all of these samples. The more samples you take with similar results, the more credence your overall hypothesis will have.

The major advantages of the Static Sample Comparison design is its ability to compare two "functionally equivalent" conditions to examine the effect of the treatment. Its major disadvantage, like all designs that do not use random sampling, is the potential presence of confounding variables and sampling bias that make the comparison samples non-equivalent at the start of the study. This is a form of errors of accuracy.

6.1.4 Random Sample Design

The single most powerful and conceptually simple experimental design you will encounter is the Random Sample Design. This one establishes the functional equivalence of the sample you are interested in—the one in which you build in the treatment, with a control sample—one that differs from your treatment group only by random chance, *not by any systematic factors.*

The procedure for the Random Sample Design (see Figure 6.1) is to identify the population from which you will take your samples (say the most common automobile engines in the United States), and randomly choose two samples from that population. It is best to use **sampling with replacement**, due to its statistical simplicity, but **sampling without replacement** may be done if you build in a large enough sample, and/or use a correction factor to account for the different probabilities of selection of your individual cases.

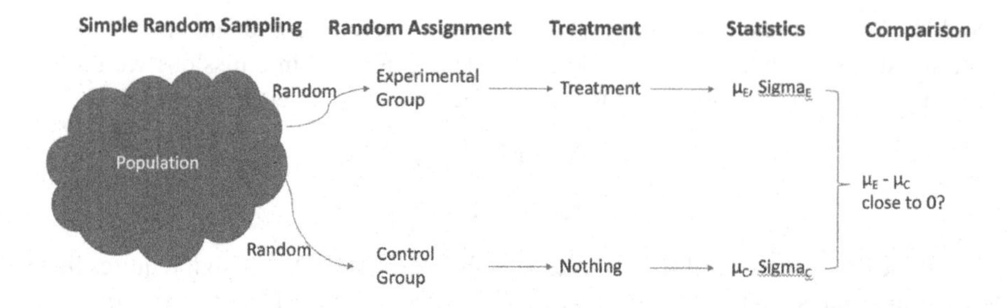

FIGURE 6.1: Random sample experimental design

A **True Experimental Design** is a method of establishing causal inferences about the effect of a treatment on the expression of one or more dependent variables. True experiments utilize random sampling of phenomena, random assignment of phenomena to treatment conditions, and researcher manipulation of the treatment.

This design can be diagrammed as:

R->X->T->O

R->X——->O

Because both treatment and control conditions were randomly assigned, the two samples can ONLY differ as a function of random chance prior to treatment. Therefore the researcher can assume any differences in behavior following treatment is due solely to the effect of their treatment.

This simple design was developed by Sir Ronald Fisher to help determine the effects of different agricultural practices in the early 20th Century. The advent of new fertilizers, pesticides, crop rotation systems, plant hybridization and the like revolutionized agriculture because Fisher and others developed statistical methods for determining the degree to which one treatment is better than another. And THIS work revolutionized medicine, social sciences, empirical sciences, and engineering because of their universal applicability.

The most commonly used statistical tests for this design include the two-sample *t*-test, and One-Way Analysis of Variance (ANOVA). But it must be noted that any test that can generate a sampling distribution for the measure you are interested in analyzing can be applied to this universal design (e.g., X^2, Wilcoxon, etc.)

Its advantages are obvious. You have a simple, logical design that provides an accurate probability estimate that, aside from random variation, your two samples are the same, and that any post-treatment effects measured in your experimental condition are due *solely to the effect of your treatment.* Disadvantages for this design

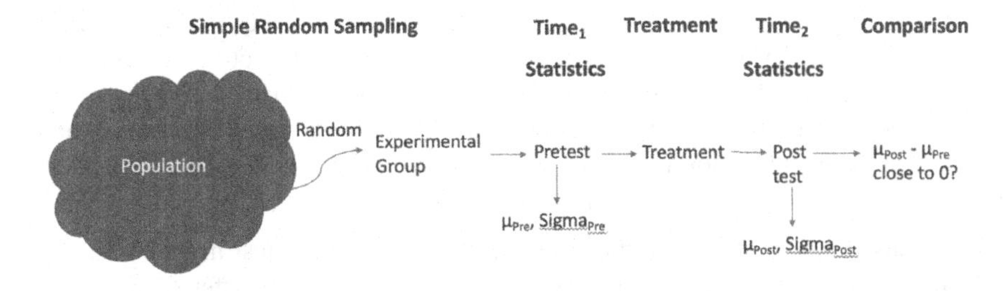

FIGURE 6.2: Pre-post randomized design

include impracticality of collecting random samples for expensive, or rare, or ethically tricky populations. Human subjects, in particular are difficult to randomly sample, so we often take to using **convenience samples** and using other kinds of controls to establish the equivalence of control versus experimental groups.

A **Convenience Sample** is a sample that is not randomly selected from the population it is supposed to represent. Instead, it typically consists of the best phenomena we have close to hand. Because there is no random sampling, convenience samples are prone to bias.

6.1.5 Pre-post Randomized Sample

Sometimes, you want to know about **growth** over an interval of time, or where you want to check on the comparability of the control and treatment groups. Such designs are called "Pre-Post Randomized Sample" designs. We diagram them like this:

$$\mathbf{R\text{->}X\text{->}O_1\text{->}T\text{->}O_2}$$

$$\mathbf{R\text{->}X\text{->}O_1\text{———->}O_2}$$

Like the Random Sample Design, the experimenter has assurance that the treatment group and control group do not differ across any dimensions due to systematic sampling bias, but they now have a way to check on some important differences that may be due to chance. We are all familiar with taking two samples from the same population (say, playing Yahtzee or some board game), and having them differ significantly from each other. This happens due to random chance. But if our two samples in an experiment differ on levels of our dependent variable, we have an elevated probability of making an error of prediction. So, by *measuring the dependent variable prior to the treatment*, and then after the treatment, we can subtract the measures to determine the degree to which the treatment group has grown or *changed* as opposed to the control group (which we hope hasn't changed at all See Figure 6.2).

A matched pairs *t*-test or OneWay ANOVA on the differences in pretest versus post test is a good way to test hypotheses for this design, but other tests may be used depending on the type of data and its shape. More sophisticated methods for accounting for pre-test differences can be analyzed using Repeated Measures Analysis of Variance and its non-parametric analogs (see Repeated Measures Designs in section 6.1.7), or Analysis of Covariance (ANCOVA).

It is also important to note that one may take several pre-test measures across which the experimenter wants assurances that the groups do not differ significantly. For example, measuring the pre-treatment bend modulus of two samples of trusses in automobile chassis might be important information to determine the extent to which one sample's modulus changes as a result of altering the skin of the car from steel to polymer or composite materials. But it might also be important to determine the degree to which the torsion for the two samples differs. If the torsion differs across the samples the specific effect of the new skin material on bend may be difficult to determine, and likewise, the bend may impact the torsion. In this case, both outcome variables are considered pre- and post-measures.

Advantages of this design are powerful with respect to the kind of claims you can make. Because you took one or more pre-tests, you can algebraically account for these random differences when analyzing your post-treatment measurements. Because you took a random sample, you know that these variations pre-treatment are random, and therefore you can just subtract them from the pre-test mean to equilibrate the two samples!

Disadvantages of the pre-post randomized sample design are relatively few. Sometimes you cannot get a pre-test measure. Sometimes the test itself may alter your sample (say taking a bend test for the chassis strut we discussed earlier. Deformation, particularly out of the elastic zone for the material, can cause metals to exceed their yield strength, and just by performing a pre-test, you are making the strut more susceptible to bend and more susceptible to failure).

6.1.6 Factorial Designs

Many of the experiments we conduct in engineering require us to examine the interactions among treatment conditions. We may wish to discover whether or not our Jacuzzi jet improves as a function of its housing material *and* its nozzle taper, for example. In other words, we now have three questions of interest when we introduce two experimental variables: 1) Is the performance of our jet improved by changing the Housing Material, 2) by changing the Nozzle Taper, or 3) by some combination of the two? Obviously, the more we introduce experimental **factors** into our experiment, the more combinations of interactions and the more complicated the interactions become that we have to account for.

Factorial designs are typically described in terms of the number of categories in each variable like this: Categories of Factor A x Categories of Factor B. If A has two levels, and B has two levels, we can diagram this 2x2 factorial design like this:

		Treatment A			
		A_1	A_2	Marginal Effects of A across B	Main Effect B
	B_1	$\mu_{A1\,B1}$	$\mu_{A2\,B1}$	Effect A across B_1 $\mu_{A1\,B1} - \mu_{A2\,B1}$	$\mu_{B1} - \mu_{B2}$
Treatment B	B_2	$\mu_{A1\,B2}$	$\mu_{A2\,B2}$	Effect A across B_2 $\mu_{A1\,B2} - \mu_{A2\,B2}$	
	Marginal Effects of B across A	Effect B across A_1 $\mu_{A1\,B1} - \mu_{A1\,B2}$	Effect B across A_2 $\mu_{A2\,B1} - \mu_{A2\,B2}$		
	Main Effect A	$\mu_{A1} - \mu_{A2}$			

FIGURE 6.3: 2x2 factorial design

A_1 B_1: R->X->T->O

A_2 B_1: R->X->T->O

A_1 B_2: R->X->T->O

A_2 B_2: R->X->T->O

Notice that A has two levels and B has two levels. Each level of variable A is paired with a level of Variable B so that **all levels of each factor are paired with all levels of the other factor(s)**. Once these experimental conditions are identified, the population is sampled randomly and phenomena assigned to the conditions.

The following table (Figure 6.3) shows this factorial design a bit more explicitly. Notice how the marginal values for each row or column allow us to isolate the overall effects of levels of the two treatment factors. The internal cells show that each of the combinations A_iB_i must be tested against each of the other to determine whether or not the value in each cell differs significantly from the values in the other cells.

A **Factor** in an experimental design is an independent variable measured on a nominal scale. In other words, each treatment, or pre-existing variable hypothesized to impact the measured outcome.

Once we choose factorial designs, the number of combinations of levels of potential treatment variables can become staggering. They can be combined with Blocking (see discussion of Randomized Block Factorial Designs in section 6.1.7) to isolate covariates and their differential effects on different combinations of the treatment levels, and we occasionally have empty cells, where there are no instances of a particular combination of levels. These are all problems the researcher must think about ahead of time when designing her or his experiment so that the appropriate number of observations can be collected, and the analytic framework mapped out *ahead of time.*

Also, it is important to note that, as the number of sub-samples increases (for Randomized Block designs) or as the number of treatment combinations increases (for full Factorial Designs), the number of observations you need to have enough power to detect a significant difference, if it exists, grows proportionately. A good rule of thumb if you are using methods that reference the Normal distribution, is to create a table like the one above, and plan to include *at least* 30 observations per cell. This will generally allow the researcher to reference a sampling distribution that closely approximates the Normal distribution, and generally is economical for most studies. Of course, the cost and level of precision you wish to use are both affected by *n*, so as you require more precision in your estimates, you will need to up your sample size, greatly increasing the cost of your study.

Advantages of factorial designs include the ability to account for interactions among treatment variables or treatments and pre-existing variables in the sample. By examining interactions, we can see if our designs work optimally under particular sets of constraints, but not for others. Disadvantages include increased complexity of analysis, underestimating the Standard Error (and thus increasing the probability of making a false-positive claim) and of course, cost.

6.1.7 Randomized Block Factorial Designs

In cases where some pre-existing characteristic (**covariate**) of the phenomenon we are trying to test may interact with the treatment, we use blocking to equilibrate our experimental and control samples by dividing them into sub-samples according to levels of the covariate. This is often called "stratification," and the sampling procedure, "**Stratified Random Sampling**."

> **Stratified Random Sampling** in a Randomized Block design is selecting multiple random samples from pre-existing levels of one Factor, referred to as a "Block." These multiple samples are then used to account for the effect of the Block, reducing the overall standard error of the analysis (increasing its power to detect a significant difference in levels of the Treatment).

This design can be diagrammed as:

Level 1:	**R–>X–>T–>O**
	R–>X——->O
Level 2:	**R–>X–>T–>O**
	R–>X——->O
Level n:	**R–>X–>T–>O**
	R–>X——->O

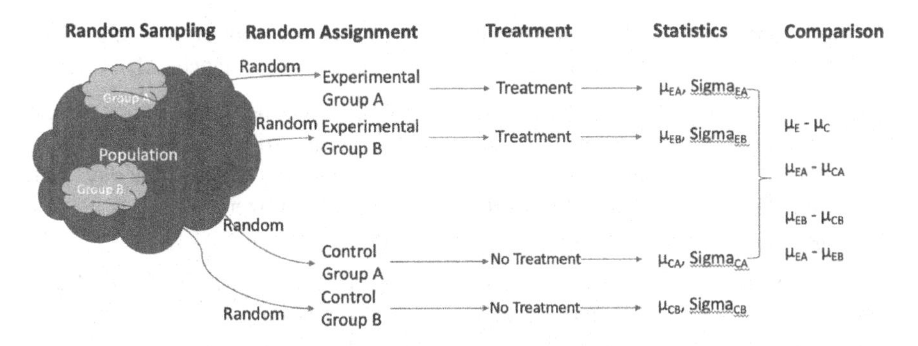

FIGURE 6.4: Randomized block experimental design. Note: A stratified design should sample from *all* significant strata in the population or be subject to bias for undersampling particular strata

Where Level_n is some important value of the covariate. In market research, these levels often correspond to demographic characteristics of consumers such as Race/ethnicity, age cohort, level of income, etc (See Figure 6.4).

A **Covariate** in an experimental design is an independent variable that is hypothesized to have an effect on the expression of the independent variable, but is not the treatment(s). Covariates may be discrete (nominal categories), or continuous. In Randomized Block Designs, the covariates are Factors.

In mechanical contexts, such levels might be different materials for the housing of a hot-tub jet and their impact on the time to failure under a pressure of 200 PSI: ABS Plastic, Stainless Steel, Brass, or Urea Resin. The experimenter would divide the potential sample into four sub-samples corresponding to housing material. Then she would randomly assign jets within each sub-sample to experimental and control conditions.

The important thing to remember about Blocking is that, when some critical variable that might interact with the treatment condition exists in a population, it is often beneficial to account for the variability in outcome that this co-variate might cause. By sampling within levels of the covariate and performing essentially n different Random Sample experiments, one can determine the effect of the treatment overall, and also determine the effect differentially for each level of the covariate.

Analysis of Randomized Block experiments is a bit more tricky than for Random Sample experiments. Because Levels of the covariate exist as pre-existing characteristics of our sample, we cannot randomly assign our phenomena to them (you can't assign a steel Jacuzzi jet to the "plastic" condition). As such, we can't compare across levels by just doing $\binom{n}{x}$ 2 sample t-tests. We have to determine the variability associated with levels of the covariate and either 1) subtract it out of the overall variability using Algebra, or 2) factor it in using Analysis of Co-Variance (ANCOVA).

While blocking according to characteristics of the sampling frame is most typical of this design, blocking could be based on other relevant attributes. For example, if, in a marketing study, people are to be treated during different times of the day, such as morning and afternoon to study preferences for hand sanitizers, we might block a morning and an afternoon group within each treatment condition.

The advantages of a Randomized Block Design generally outweigh the disadvantages of additional cost and difficulty of analysis. Extraneous (noise) variables plague the mechanical engineer, because so many of our designed products depend on the simultaneous proper functioning of other elements. If we can block our sample across differential performance of these other elements, we can isolate the effect of our variable we are calling the treatment. Randomized Block designs are also easy to use as a building block for more complex designs. They are the perfect starting point to begin to design any experiment you might want to conduct.

6.1.8 One-shot Repeated Measures

Repeated Measures designs use each phenomenon in a sample as its own control. If, for example one were to examine, over time, the oxidation of AL-6XN stainless steel in our Jacuzzi jet given repeated applications of a corrosion-inhibiter on a regular schedule, one might diagram the study as follows:

$$\mathbf{X\text{–>}T\text{–>}O_1\text{–>}T\text{–>}O_2\text{–>}T\text{–>}O_3\ldots}$$

Here, as you can see, we have repeated treatments and observations. The degree of corrosion over time is our dependent measure, and we can easily measure build-up in fixed intervals after treatment. If this build-up grows slowly, or not at all, we can claim that the corrosion inhibiter application is effective, because we can be reasonably confident that our sample of AL-6XN stainless steel does not differ in substantial ways from the general population of samples of this grade of steel.

What we lose, is the ability to measure the corrosion in a sample that receives the treatment, from another control sample, that does NOT receive the treatment, and therefore the effect of the treatment cannot be accurately quantified.

This design is really just a series of simple one-shot case studies in succession. It only contributes the additional information about repeated or continued treatment. And like other one-shot studies, we generally use one-shot repeated measures in the early stages of design so we can model the behavior of our designed part and use this knowledge to improve the design to a point where we can do a true randomized experimental design.

Advantages are cost and ease of use. Disadvantages of one-shot repeated measures designs fall neatly into the fact that, because we have not randomized our sample, nor provided a control condition, we do not know how much of the change we are measuring is "natural" —i.e., NOT caused by our treatment. That can only be done with randomization and control.

6.1.9 Randomized Factors Repeated Measures

The Randomized Factors Repeated Measures design is a true experiment that involves both the ability to compare our preferred treatment against some other standard. This other standard (symbolized T_2 here) may be a control condition (i.e., T_2 may be no treatment), but we still take repeated measurements to determine the "natural" effect of time on the sample.

$$\mathbf{R\text{->}X\text{->}T_1\text{->}O_1\text{->}T_1\text{->}O_2\text{->}T_1\text{->}O_3\ldots}$$

$$\mathbf{R\text{->}X\text{->}T_2\text{->}O_1\text{->}T_2\text{->}O_2\text{->}T_2\text{->}O_3\ldots}$$

Here the diagram shows two different treatments, but we may include an indefinite number of conditions depending on the nature of the experimental question we have contrived to study.

There are special problems associated with analyzing repeated measures, particularly because in each **panel** or repetition of the treatment/observation phase, the outcome measure in subsequent phases is dependent on the outcome measure in previous phases. In nearly all of our statistical tests, we assume independence of observations. While each unit in our sample may be treated and measured separately and independently, within a single unit, time confounds independence. You will recall our discussion of the bend in automobile struts. Each time we measured the struts, we bent them, causing a bit of work-hardening. The same kind of problem may occur when we administer drugs on a repeated schedule. Drugs may build up in the patient's system to an extent to where subsequent treatments are not JUST the current administration, but a build-up of chemicals that are all working in concert. Likewise, for an airplane engine, repeated stalls may impact the integrity of engine parts due to stress caused by the increase in torque.

Repeated measures analysis will allow us to model some of these trends, but it may reduce our ability to determine the unique contribution of treatments in each subsequent analysis. We will study how to overcome these problems when we get to our use of the General Linear Model for higher order experimental designs (in a few chapters).

> A **Panel** in a Repeated Measures design is a set of samples, typically Treatment and Control, that are measured multiple times to determine the change over time, if any, caused by a hypothesized treatment.

6.1.10 Ex-post-facto

Sometimes, particularly when using human subjects, or when it is too expensive or impractical to randomly assign phenomena to different experimental conditions, we have to make due with existing groups or existing samples:

X–>T–>O

X——->O

Our diagram of the design is basically the same as the Random Sample design with the exception that here we fail to randomly assign phenomena to the experimental versus control conditions.

Such designs are often called "**Quasi-experimental**" because they retain the basic structure of two comparison groups of observations, only one of which gets the treatment. They differ from true experiments in that we cannot be sure that the treatment sample and the control sample only differ as a function of random chance. In the case of human subjects, we can be pretty sure they DO have some meaningful differences that we are unable to account for statistically.

> A **Quasi-Experiment** is a design that uses two existing groups for comparison, rather than random sampling from a population. Groups are randomly assigned to treatment and control conditions in the same manner as a Random Sample Design, and analyses are conducted exactly the same way. Generalization is limited, however, because the two existing groups will differ across the dependent variable, making it hard to determine what, of this difference, is attributable to the impact of the treatment.

In such cases, we must try to equilibrate the two contrasting conditions as much as possible prior to administering the treatment. This may take the form of giving pre-tests, or taking pre-measurements of the phenomena, and then creating from those pre-tested cases, two or more sets that have equivalent values. This is often called **matching**. The degree to which the experimenter can logically convince his or her audience that the experimental and contrast conditions are equivalent for all intents and purposes, will determine the extent to which the results of such studies are considered rigorous or believable.

Blocking, or *stratification* as it is typically called in these types of quasi-experiments, is also a common method for generating similarity between conditions for Ex-post-facto designs, as it can reduce known variability that exists within groups, and relegate that to separate groups that can be analyzed separately. This is a type of statistical control, as opposed to experimental control wherein phenomena are assigned randomly to a control condition.

> **Matching** is selecting from a pre-existing sample, cases that are similar across different attributes in an attempt to equilibrate treatment and control samples prior to application of the treatment.

Another means for overcoming the lack of randomization is to test the contrast group *after* performing the experiment:

$$X \text{->} T \text{->} O_1 \text{———>} O_2$$

$$X \text{———>} O_1 \text{->} T \text{->} O_2$$

Such **Counterbalanced** designs can demonstrate that the treatment, which worked for the experimental condition over the contrast condition, NOW works for the contrast group as well. If the nature and extent of the recorded effect are the same for the contrast condition as they were for the experimental condition, then the case that the treatment caused the effect is supported.

Like its true experiment counterpart, the Random Sample Design, this simple quasi-experiment can be augmented to include more than one treatment, Blocking, repeated measures and other useful designs to help understand the nature of a set of treatment effects, the impact of covariates, and other factors. Unlike randomized designs, the more factors one must account for, the more difficult it is to say with confidence that any difference found is solely a function of the treatment conditions.

The lesson here is to use randomization if at all possible, but do not hesitate to use Quasi-experimental methods if your situation makes randomization too costly or impractical. You will have to do much more up-front work to equilibrate your samples, and your argument about the probability of making a Type I or Type II error will most likely be under-inflated.

6.1.11 Time Series

Time series studies use periodic measurement of a phenomenon to establish a baseline behavior pre-treatment that then can be compared to consequent behavior post-treatment. Here the researcher is interested in behavior that is time-varying in some systematic way. If the function between the measurement and time can be established through regression or some form of curve fitting or modeling, then any deviation from the baseline function can be seen as having been caused by the treatment:

$$X \text{->} O_1 \text{->} O_2 \text{->} O_3 \text{->} O_4 T \text{->} O_5 \text{->} O_6 \text{->} O_7 \text{->} O_8$$

Figure 6.5 illustrates for a fictitious example, that following treatment by a certain chemical, a sample of metal exhibited accelerated corrosion (i.e., its slope is greater following treatment), AND that the treatment appeared to cause a displacement discontinuity—a—near instantaneous jump in the amount of corrosion measured.

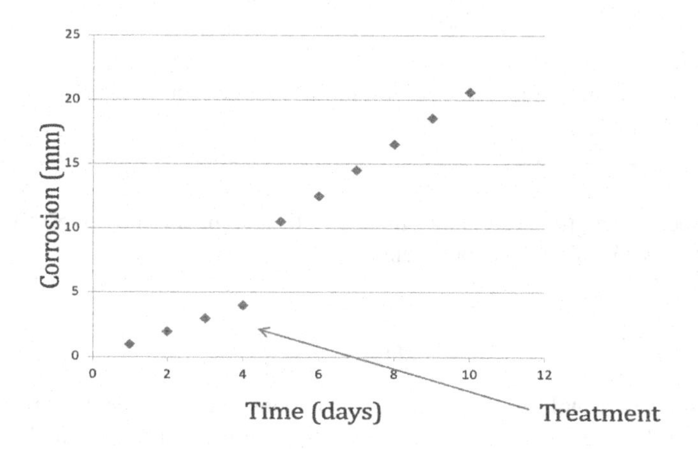

FIGURE 6.5: A discontinuity or change in slope (or both) indicates an effect of a treatment in a time series

Because nothing else has changed in the study, we attribute this discontinuity and the accelerated rate of deposition to the introduction of the treatment chemical.

Notice here that this design is similar to the repeated measures design, with the exception that the treatment is generally not repeated, but the measurements are. Also here we typically do not randomly assign phenomena to the experiment, though random assignment, when practical, is always a good choice.

A clear disadvantage to time-series designs is the fact that they may take time! Drug interactions, effects of deposition of soils and other outcomes often take a lot of time, and this has to be factored into the design. Sometimes we can "fudge" a bit and use heat to substitute for time in chemical reactions, or we can try to model simpler systems and then project a longer time scale.

6.2 Summary

This list of common designs is by no degree exhaustive. As we have shown, many researchers combine these designs in a modular fashion to craft a study that helps them answer the question that they need to find out the answer to. You will encounter such arcane (but useful) methods as *nested* or *mixed* designs later on in this book. You may also see a kind of time-series called *regression discontinuity.* No manual can ever list all the possible or even all the useful designs one might encounter. BUT, *these ten building blocks are the basic modules from which all other designs can be crafted.* If you understand these designs, their logic and the kinds of analyses and statistical tests they utilize, you will be able to answer nearly any experimental question that industry will dream up for you.

You must also recognize that not all engineer's statistical repertoire is tied up in experiments. Much of your work will focus on describing and predicting the behavior of certain phenomena, or discovering the relationships that exist, if any, among two or more variables without the existence of a treatment. These methods are not experimental, but are just as valuable to us. We will examine them in depth later on in the book.

6.3 References

Brownlee, K. A. *Statistical theory and methodology in science and engineering.* New York: Wiley, 1960.

Campbell, D. and Stanley J. *Experimental and quasi-experimental designs for research and teaching.* In Gage (Ed.), *Handbook on research on teaching.* Chicago: Rand McNally & Co., 1963.

Cornfield, J. and Tukey, J. W. Average values of mean squares in factorials. *Annals of Mathematical Statistics*, 1956, 27, 907-949.

Cox, D. R. *Planning of experiments.* New York: Wiley, 1958.

Fisher, R. A. *The design of experiments.* (1st ed.) London: Oliver & Boyd, 1935.

Winer, B. J. *Statistical principles in experimental design.* New York: McGraw-Hill, 1962.

6.4 Study Problems for Chapter 6

1. Describe how a treatment is different, in terms of experimental design, than a pre-existing variable. A good context to use is varying the wing design as a treatment for improving the coefficient of lift of an airplane and then comparing that treatment to airplanes that have a standard wing design; versus recording the coefficient of lift from a random sample of airplanes that employ different wing designs.

   While treatments and pre-existing variables are often treated the same mathematically, experimentally, a treatment is an experimental condition that is manipulated by the experimenter. One therefore assigns experimental units (ball bearings, engines, people, etc) to different treatments. Experimenters cannot assign units to pre-existing variables. Animals cannot be assigned to species, for example. The species exist prior to any treatment.

2. What are the features that distinguish a True Experiment from a Quasi-Experiment? Discuss how these features change:

   (a) The nature of sampling;

   A true experiment ideally samples from a population at random. A Quasi-Experiment utilizes pre-existing samples, where random assignment is not used. In addition, for true experiments, samples are randomly assigned to treatment conditions. Quasi Experiments typically do not randomly assign to treatment conditions.

   (b) The way in which the samples are analyzed; and

   The analysis, comparing group statistics, is exactly the same, mathematically.

   (c) The claims that can be made from each method.

   Because of this random assignment, in true experiments, we can attribute the post-treatment differences in samples solely to the effect of the treatment. This is not true for Quasi-experiments. We have to make much more effort to justify the functional equivalence of samples before we can make a claim about the treatment effect.

3. Suppose you were studying the impact that different working fluids have on the performance of a heat exchanger. You are primarily interested in the fluids' coefficient of expansion and its impact. What other variables related to working fluid might impact a heat exchanger's performance? How might you account for these extraneous variables in the design of your experiment?

   The ambient temperature around the heat exchanger matters a lot, as does the flow rate at which the fluid moves through the exchanger. Certain design elements, such as fins, or baffles in the heat exchanger also impact the performance. To account for these and potentially other variables, I might use blocking, creating random samples from different levels of each of these factors, or I might account for them using a factorial design, wherein I could model their interactions.

4. Under what conditions is convenience sampling a better option than random sampling?

   When the ethical or economic/practical conditions make it impossible or impractical to use random sampling. I often say that it is impossible to randomly assign humans to surgeries. To study the effects of different surgical procedures we have to study them after the fact.

5. What restrictions are placed on the kinds of claims that can be made using convenience samples versus random samples?

Because convenience samples will differ on some variables other than the treatment, we cannot attribute any pre-treatment differences to random chance. We have to assume there are some systematic biases present. Hopefully those biases are negligible, but we are never completely sure.

6. What are the necessary features for an "educated guess" to be a true scientific hypothesis?

   An hypothesis must have a prediction of the outcome of an event, *and* a theoretical rationale for why that predicted outcome is the most likely.

7. You are studying the comfort of cockpit seats in a military helicopter. You have identified one key factor that you want to manipulate: The height of the lumbar support. What covariates exist that might impact this Factor's expression on the dependent variable, and how can you address this in the sampling process?

   Human body size is a key one. Also the shape and distribution of weight will matter. Lastly, different people have different posture and existing back problems that will affect their feelings of comfort using a lumbar support. To account for these factors, I would Block from different body sizes and weights, perhaps eliminating anyone with back problems from the study, or at least accounting for the fact that some subjects may have back problems of different types. I would biometric analysis of posture to account for any differences in comfort that may be caused by this factor.

8. How does matching samples assist the engineer in making claims about the effect of a treatment?

   Matching helps make the case that samples are functionally equivalent prior to treatment. Coupled with random sampling, this is a uniquely powerful method of control in an experiment. Without random sampling, this is a key strategy to make the argument that a treatment effect did or did not make a difference.

9. Describe the basic logic of repeated measures. What is the role of a panel in this design?

   Repeated measures designs assume that the dependent variable is time-varying, pre and post treatment. By recording behavior prior to treatment and then post treatment, one can determine the extent that the treatment impacted each experimental unit, and then one can analyze the cross-units average effect. Panels serve as either multiple replications of the experiment, and/or as blocks that examine the treatment effect across different covariates in the population.

10. What are the advantages and disadvantages of the 10 different research designs discussed? Discuss the following comparisons:

    (a) Ease of implementation;

    (b) Cost (e.g., money, time, personnel);

(c) **Power for making claims about the effect of a treatment.**

In general, the easiest designs to implement are those that do not involve random sampling or random assignment. Quasi-experiments and one-shot case studies are classic examples of this. As more blocks or covariates are added, one must begin to account for their interactions, increasing sample size dramatically. This increases the cost of the experiment, its logistics, and the complexity of analyses. In general, as the complexity of designs grows, the sample size must grow to insure enough power to detect non-random effects or non-random relationships among variables.

11. **What are the key features that distinguish the Randomized Block Factorial Design from the Factorial Design in general?**

    The major difference in the Randomized Block design over the Full Factorial design is: 1) whether or not the experimenter is interested in the additional factors in the analysis or whether the factors are used to account for potential confounding. In Randomized Block designs, the additional factors are used to reduce the error of prediction of a model. In Full Factorial designs all of the independent variables (factors) are of interest as potential effects, including interactions among factors.

12. **What are the major advantage(s) of the Pre-post Randomized Sample design over the Random Sample Design?**

    The Pre-post Randomized Sample design enables modeling the natural change in a variable for the control group, and the additional change caused by the treatment in the experimental group.

13. **What two features of a time-series curve inform the researcher about the effect of a treatment? Be able to discuss how these two features indicate an effect.**

    A discontinuity indicates that the treatment made a change in the level of expression of a dependent variable, while a change in slope indicates that the dependent variable has impacted the rate at which a variable is expressed over time.

14. **How might repeated testing of a panel introduce error into the measurement process in a repeated measures design?**

    Fatigue, test effects and other indications of the non-independence of events in subsequent observations in a panel may cause some change in the expression of the dependent variable *not* due to the treatment.

15. **How is the Static Sample Comparison distinguishable from the Ex-Post Facto design?**

    For many cases they are so similar as to be considered the same because both use pre-existing groups to form the comparison. BUT, an Ex-Post Facto design will use these pre-existing groups as a factor in the analyses. In other words The Ex-Post Facto design is interested in the expression of the dependent variable both by treatment and by existing strata in the population from which the samples were drawn.

7

Sampling Distribution of the Proportion

Suppose you were working as a quality control engineer, in a company manufacturing, among other things, bearings to be used in high speed rail applications. The bearings have to be within precise tolerances due to the extreme RPMs at which the drivetrain of the engines perform. In layperson's terms, the bearings need to be as round as possible. But in addition to the roundness, the bearings must also be resistant to thermal deformation due to the high temperatures at which they are expected to perform. In layperson's terms, the dimensions of the bearings should not change dramatically at top-end speed. You randomly sample 200 bearings off the line, taking measurements of the radial displacement of the bearings under maximum expected operating temperatures.

Of the 200 bearings you find the following:

- 16 bearings fail, showing radial displacement greater than 5% of the original diameter;

Is this level of performance something you would consider typical, out of 200 bearings, or atypical, given the overall failure rate you are designing for is 5%? Think through this problem as we progress through this chapter. We will return to it following our discussion of the binomial distribution.

These types of situations, assessing the failure rates of engineered products, are central to Quality Control. We know that some products will fail due to natural variation. Some will fail due to mis-application such as running our trains at too high RPM. But the vast majority should work well within the environment for which we have designed them. In our example, we have 16 bearings (total) failing, for an overall failure rate of 8%. We only want 5%, but we know we are unlikely to get exactly 5% —a bit more or fewer are likely to fail than expected. The probabilistic question we have to ask is, *given the size of our sample, how likely is it that the results we achieved would occur in a distribution of sample statistics all taken randomly from random of size 200, all taken randomly from the same population of bearings?*

Recall in Chapter 2, that a sampling distribution is a distribution of a sample statistics, all samples taken randomly from the same population, all samples having the same sample size. In essence it is a distribution that shows the probability of drawing a sample with given statistics, in a population of samples. In Chapter 5, we note that *population of samples* is a theoretical idea. We don't actually take repeated

DOI: 10.1201/9781003094227-7

samples infinitely many times and plot the probabilities of their sample statistics. We rely on the distributions that describe these populations: The Binomial, Poisson, Normal, etc., to provide long-haul probabilities of our results occurring. Let's see how this works with our quality control example.

> Our overall success rate is 92%. If our sample is a good, representative sample, taken from our population of manufactured bearings, we would expect about 8% to fail in the population overall. But we are skeptical, because we know there might be some sampling error (e.g., George, our least reliable machinist may have worked two shifts that day).
>
> So, we take a look at the variability in all the possible samples of 200 we could have taken, using a theoretical probability of success at 0.95—our intended success rate, and a theoretical probability of failure at 0.05—our intended failure rate.

Recall back in Chapter 4, the expected value of a proportion is the value of the referent category (in our case, successes), times its proportion.

$$E(x_i) = x_i p_i$$

$$\mu = E(x_i) = x_i \frac{f(x_i)}{n}$$

For a binomial distribution, $x_{success} = 1$, so

$$\mu = E(1) = 1\frac{(1)}{n} = np(1)$$

$$\sigma^2 = E(1-\mu)^2 = np(1)(1-p(1))$$

A **Binomial Variable** is a discrete random variable composed of trials that only have two outcomes (0 and 1). The summary statistics estimating the parameters of the Binomial Distribution are p and n, such that:

$$\mu = np, \text{and}, \sigma^2 = np(1-p)$$

1. Reason about how you might go about estimating the probability of getting 16 or more successes in a random draw of 200 bearings from this population.

   There are lots of ways of thinking about this. If you think that the reference distribution is normal, the mean of the population should be about $200(0.95) = 190$ successes and $200 - 190 = 10$ failures. The distribution should vary around 190 by $200(0.95)(0.05) = 9.5$. So about 68% of our data should fall somewhere between 180.5 and 199.5 successes. 16 successes is way low in the distribution. Again, if I assumed a normal distribution, I would compute the z-test and see how probable getting only 16 successes out of 200 trials would be.

But I am not sure that these success and failure proportions are distributed normally. For one thing, they aren't a continuous random variable, this violates one of the assumptions of the z-test! If only I could estimate the sampling distribution of the proportion of successes vs. failures, I could estimate the probability more closely.

7.1 Sampling Distribution of a Proportion: Binomial Distribution

The question we are asking in this example is about the likelihood of us of obtaining a sample drawn from a population with particular parameters. This is the same logic as we explored in Chapter 5, examining the likelihood that a given sample has a particular mean value when its referent population has known parameters.

For parameters we have $p(X = success) = 0.95$, and $n = 200$, and our parameter estimates from our sample are $p(X = success) = 0.92$, and $n = 200$. So now we have parameter estimates and a sample with which to draw a sampling distribution. But what shape is the sampling distribution of the proportion when we only have 0 and 1 as an outcome of our trials and not a continuous random variable? Lets go back in history a bit to the drunken gamblers that gave rise to statistics as a field. We will return to this problem in a few pages.

7.1.1 Bernoulli Process

In the latter part of the 17th century and beginning of the 18th century, Jakob Bernoulli worked out much of what we now call probability theory (Bernoulli, 1713). In his most famous (and final) work, *Ars Conjectandi* (The Art of Conjecturing), Bernoulli formally proved the Law of Large Numbers, in essence, working from a simple process of randomly selecting outcomes that only have two (e.g., binary) values, "successes" and "failures," —"this," or "that." He had to make sure that every trial had the same probability of drawing each outcome, and that there were only two possible outcomes (1 = the outcome in which Bernoulli was interested and 0 = everything else). This was the beginning of our requirement that outcomes in nominal data be both mutually exclusive and collectively exhaustive (Chapter 3). This process has become so important to the field of statistics that any trial that fits these requirements is called a **Bernoulli Trial**, and the whole procedure is termed a **Bernoulli Process**. The reason I remember this now is that I didn't remember this in a qualifying exam in my statistics program, and was ridiculed by my professors for forgetting something so basic.

A **Bernoulli Process** is selection of events outcomes, through repeated trials, from a discrete random variable with only two possible outcomes. The outcomes must be collectively exhaustive and mutually exclusive. Any single trial using such a process is called a **Bernoulli Trial.**

One misconception I had (briefly) when studying probability as an undergrad was that, like a coin, both outcomes in a Bernoulli Trial had equal probabilities of occurring. This is not the case. Suppose you were to toss a cylindrical coffee can, defining a success as when the can landed on its circular face, and a failure when the can landed on its side. I don't know what the probability of each of those outcomes is, but I don't think they are equal. I suspect that my coffee can will land more on its side than on one of the two faces. This is still a legit Bernoulli Trial. We can test whether or not a coin is fair by examining the number of heads versus tails in a finite number of trials by comparing the actual distribution of heads to the theoretical distribution. Suppose I tossed a coin 100 times and it turned up heads 65 times. Do you think the coin is *fair* (i.e., that the long-haul probability of tossing a head on any random trial is equal to 0.50)? We know the expected value of a fair coin given 100 trials is:

$$\mu_{heads} = 100\,(0.5) = 50 \text{ heads}$$

But we also know it will vary somewhat and that getting *exactly* 50 heads in 100 trials is unlikely to happen. Again, like Chapter 7, we can appeal to the sampling distribution of the proportion, in this case the proportion of heads to tails, to help us. The expected variance around μ_{heads} is estimated at:

$$\sigma^2 = 100 \cdot 0.50\,(1 - 0.50)$$

$$\sigma^2 = 25$$

$$\sigma = 5$$

If this were the normal distribution and the coin fair, we should expect about 68% of the outcomes of our trials to fall in the interval $\mu \pm \sigma = (45, 55)$, and 95% to fall in the interval $\mu \pm 2\sigma = (40, 60)$. In other words, it is only 5% probable that we could get a value either below 40 or above 60. This is pretty low. If it were me, I would say that the coin is *probably* not fair given these low chances. But I still don't know if the Normal Distribution is a good model for the sampling distribution of the proportion (It is, you know that, and I know that, but we have to pretend like we don't know it so that we can explore *why* and how this is so).

7.1.2 Binomial Distribution

For a Bernoulli process with only two outcomes, the distribution of the proportion is easy to generate. You have done it before in 6th or 7th grade and again in 8th or 9th grade when you looked at Pascal's triangle and the binomial theorem. I know it is a lot, but I am asking you to think back to the dusty recesses of your memory to recall that the product of two monomials (two linear expressions) results in a binomial expression:

$$(x + y)^n = \sum_{k=1}^{n} \binom{n}{k} x^k y^{n-k}$$

Where $\binom{n}{k}$ is called the *binomial coefficient*, and is essentially, the number of combinations of x and y given n trials.

$$P(X = x|n, p) = \binom{n}{x} p^x (1 - p)^{n-x}$$

Read, *The probability that the number of successes equals x, given n trials and a probability of success in each trial of p, equals the number of combinations we get for x, given n trials times the probability of a success accumulated across those trials, times the probability of not being successful accumulated across n trials.*

Where the binomial coefficient (number of combinations of x, given n) is calculated by:

$$\binom{n}{x} = \frac{n!}{x!\,(n - x)!}$$

$\binom{n}{x}$ is often read, *n choose x.*

Now let's see how this progresses using a binary tree. Since there are only two outcomes: 0 and 1, in the first trial we would expect either one 0 or one 1. In trial 2 we expect to see 4 combinations: 00, 01, 10, and 11. In Trial 3, we accumulate from the previous trials, and now have 8 combinations: 000, 001, 010, 011, 100, 101, 110, 111. The progression is exponential, with 16 combinations after 4 trials and 32 after 5, (2^n) after n trials (See Figure 7.1).

If we rewrite these combinations, using the binomial coefficient to help us compute, we can see a nice progression known as Pascal's triangle. Figure 7.2 shows the number of combinations of x successes in n trials. The total number of combinations across is the sum of each row, and the number of successes is the kth element in each row counted from the left, or the n–kth element counted from the

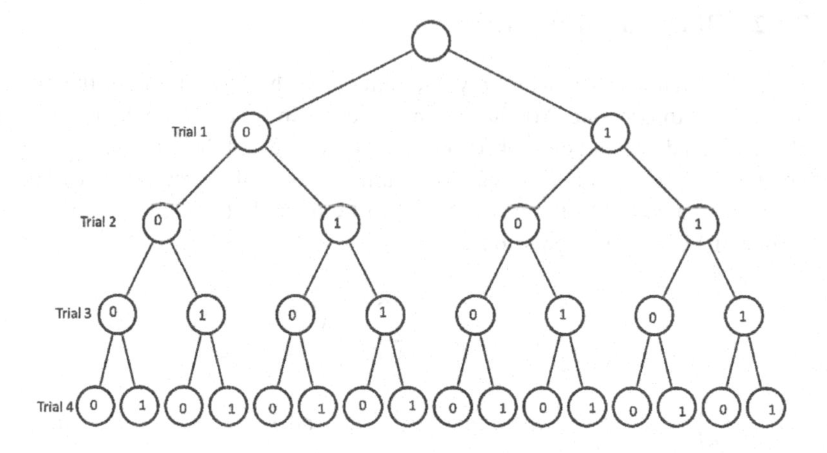

FIGURE 7.1: Combinations of successes (1) and failures (0) for 4 Bernoulli trials

0th trial 1
1st trial 1 1
2nd trial 1 2 1
3rd trial 1 3 3 1
4th trial 1 4 6 4 1
5th trial 1 5 10 10 5 1
6th trial 1 6 15 20 15 6 1
7th trial 1 7 21 35 35 21 7 1
8th trial 1 8 28 56 70 56 28 8 1
9th trial 1 9 36 84 126 126 84 36 9 1
10th trial 1 10 45 120 210 252 210 120 45 10 1

FIGURE 7.2: Pascal's triangle. Shows the combinations of x successes in n trials. x is the n-k^{th} element in each row, n is the cumulative number of trials

right. If we want to know the combinations of successes we might find given 6 trials we go down to the 6th row and see there is 1 way of getting 6 successes (all successes), 6 ways of getting exactly 1 success, 15 ways of getting 2 successes, 20 ways of getting exactly 3 successes, and so on. Notice that the distribution for any sample size is symmetric.

Armed with this model, we can find p and ~ p easily. Let's ask the question: what is the probability of getting 3 successes in 4 trials? Using Pascal's triangle (see Figure 7.2) we go down to the 4th row, find x=3 successes by counting over two spaces and we find that there are 4 combinations that give us 3 successes. The total number of combinations is 1+4+6+4+1 = 16, so

$$p(X = 3|4, 0.50) = \frac{4}{16} = 0.25$$

We read this as *the probability that the number of successes is equal to 3, given 4 trials, each with a probability of success being* $\frac{1}{2}$, *is equal to 0.25.* Just to show that this value equals the binomial formula we can substitute x, n, and p and evaluate for $p(X = 3)$:

$$\begin{aligned} P(X = 3|4, 0.50) &= \binom{4}{3} 0.50^3 (1 - 0.50)^{4-3} \\ &= \frac{4!}{3!(4-3)!} \left[0.50^3 (1 - 0.50)^{4-3}\right] \\ &= \frac{4 \cdot 3 \cdot 2 \cdot 1}{3 \cdot 2 \cdot 1 (1)} \left[0.50^3 (0.50)^1\right] \\ &= \frac{4}{1} [0.125 (0.50)] \\ &= \quad 0.25 \text{ Ha!} \end{aligned}$$

Because of the iterative nature of the distribution, the binomial coefficients grow exponentially as the number of trials increase. Pascal figured this all out algebraically, and first applied it to his gambling habit, with the restriction that $p(x) = \frac{1}{2}$. Bernoulli's great contribution, and the one that caused us to use his name for the process of generating this distribution, was to prove that *the binomial formula worked for all values of* $p(x)$ and that, for samples with a large number of trials, *the binomial distribution approximated the Normal Distribution very closely.*

7.1.3 Binomial Probabilities in an Interval

If we take repeated samples of a given size using a Bernoulli process, we can graph the sampling distribution of all the possible outcomes (a sampling distribution of the proportion). Just like the sampling distribution of the mean, we can add up the probabilities across intervals of the to determine cumulative probabilities. In Figure 7.3, we can see that the probability of randomly drawing any value from n=0 heads to n=7 heads in a sample of 15 throws of a coin is 50% by adding up the probabilities of each of the bars. This is obtaining a cumulative probability mass, same as the normal distribution, except the function is not continuous, and therefore we can't use integration. We just have to use summation:

For any interval $k = x$

$$P(k \leq X \leq x|n, p) = \sum_{i=k}^{x} \binom{n}{x_i} p^{x_i} (1 - p)^{n-x_i}$$

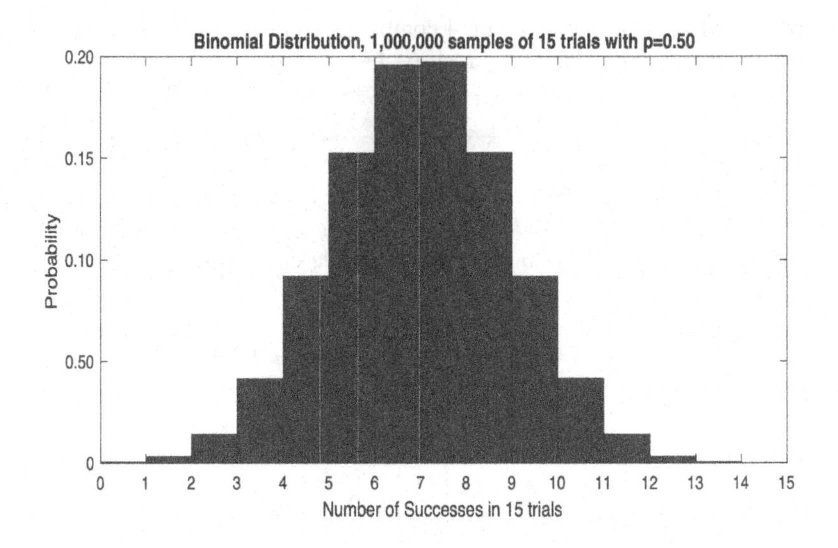

FIGURE 7.3: Binomial distribution of samples of size 15 with p = 0.50

2. Suppose you are a quality control engineer studying the extent to which batteries manufactured for your cochlear implants meet specifications. In general, 97% of the batteries from the company you have contracted from meet specifications.

   If a sample of 5 batteries is selected, what is the probability that at least 4 of them meet specifications?

   Here we are appealing to the binomial distribution with n=5 and p=0.97, yielding the following sum:

$$P(0 \leq X \leq 4|5, 0.97) = \sum_{i=0}^{4} \binom{5}{x_i} p^{x_i} (1-p)^{5-x_i}$$

   We start by evaluating $P(X)$ at X=0:

$$P(X=0|5,0.97) = \binom{5}{0} p^0 (1-p)^{5-0}$$

$$P(X=0|5,0.97) = \frac{5!}{0!\,(5-0)!}\left[0.97^0 (1-0.97)^5\right]$$

$$P(X=0|5,0.97) = \frac{5\cdot 4\cdot 3\cdot 2\cdot 1\cdot 0}{1\,(5\cdot 4\cdot 3\cdot 2\cdot 1\cdot 0)}\left[1\,(1-0.97)^5\right]$$

$$P(X=0|5,0.97) = (0.03)^5$$

$$P(X=0|5,0.97) = 2.43x10^{-8}$$

Then, using the same procedure, we find

$P(X=0) = 2.43 \times 10^{-8}$

$P(X=1) = 3.93 \times 10^{-6}$

$P(X=2) = 0.00025$

$P(X=3) = 0.0082$

$P(X=4) = 0.13279$

$P(X=5) = 0.8587$

The sum of the probabilities of success across that interval is 0.1413 or about 14%. So we can conclude that it is only 14% probable that we could have gotten 4 *or fewer* successes in a sample of 5. This means that it is about 86% probable that we would get 5 successes out of 5 trials. Of course we could have just looked at $P(X=5) = 0.8587$!

7.1.4 Using the Symmetry of the Binomial Distribution

Now I don't know about you, but I am pretty lazy when it comes to plugging in numbers. I would rather have the computer do it for me, but sometimes I just have to plug and chug. For this problem, you may have noticed that $P(X \leq 4)$ is the same as $1 - P(X > 4)$. There is only one value of the distribution where $X > 4$, and that value is when $X = 5$. I would much rather solve this problem only making one calculation instead of making 5 and then summing them all together. Let's try it:

$$P(0 \leq X \leq 4|5, 0.97) = \sum_{i=0}^{4} \binom{5}{x_i} p^{x_i} (1-p)^{5-x_i}$$

$$(X > 4\,|5, 0.97) = \binom{5}{0} p^5 (1-p)^{5-5}$$

$$(X > 4\,|5, 0.97) = 0.8587$$

$$P(0 \leq X \leq 4|5, 0.97) = 1 - 0.8587$$

$$P(0 \leq X \leq 4|5, 0.97) = 0.1413$$

In situations where you need to make a whole lot of calculations to find the cumulative probability, it may be more efficient to find $1 - P(\sim x)$. $\sim x$ is called the **logical complement** of x, and is defined as the set of outcomes in the sample space that are mutually exclusive of x.

Try this one on your own, then take a look at my reasoning:

3. The antilock brakes you have just designed display the following behavior: 5 displayed severe hydroplaning, 15 displayed slight hydroplaning, and 80 displayed no significant hydroplaning in a standard wet road skid test. Find the probability that out of eight tests you run that 1 or more will be exhibit severe hydroplaning.

Here is my reasoning on it:

I have to take the data and make it binary. So, I have 5/100 displaying severe hydroplaning.

Using the binomial distribution, I have n=8, x=1, p=0.05. Also, I want to find the probability that 1 or more will exhibit severe hydroplaning. That is the logical complement of finding 0 that exhibit severe hydroplaning (1-p(0)):

So, $P(x \geq 1) = 1 - P(x < 1) = 1 - P(x = 0)$.

$$P(X = 0|8, 0.05) = 1 - \frac{8!}{0!\,(8!)}.05^{0}.0.95^{8}$$

= 1-.6634

= 0.3366

Assumptions of the Binomial Distribution:

(a) *Binomial*: Events must be classified as a 0 (failure; not an apple) or 1 (success; apple).

(b) The probability of success is the same for each trial.

(c) *Independence*: Events must be independent. One event occurring should not impact the probability of another event occurring. This is another way to say it uses **Simple Random Sampling**.

7.1.5 The Normal Approximation to the Binomial Distribution

4. Using the same scenario as in problem 3, What is the probability that, if I ran 30 tests, 8 or more exhibit severe hydroplaning?

"What the heck, Middleton," I hear you saying, "I am not going to sum 32 probabilities. I don't even want to do the logical complement of 8 calculations... " There has got to be an easier way!

Leave it to the gamblers to prove that the Normal distribution is a good approximation of the binomial for relatively large n. In a book published just after his

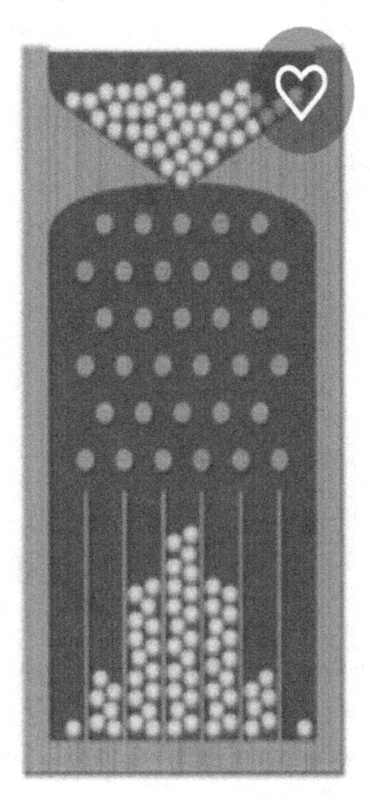

FIGURE 7.4: Galton's bean machine

death, Jakob Bernoulli showed that, as the number of trials of a Bernoulli Process increases, the resulting binomial distribution closely approximated the Normal Distribution. Sir Francis Galton, about 80 years later, wasn't satisfied. He created what, in essence was a plinko machine (see Figure 7.4) by drawing out (very precisely), a symmetric binary tree and placing pins at every node in the tree. Dropping little beans (now we use marbles or little steel ball bearings) down the center of the machine, he was able to reproduce a bell-shaped curve, closely resembling the Normal Distribution using a physical model (See Figure 7.4). Galton's bean board has been used ever since as THE model of the binomial distribution and its relationship to the Normal Distribution in every statistics class ever taught (which is, of course, a gross hyperbole).

But hold on, you say! The first few rows of the binomial distribution don't look at all like the normal distribution. They aren't even close to continuous, and there are so few combinations that, any number of curves could fit the outlines of their graphs. . .

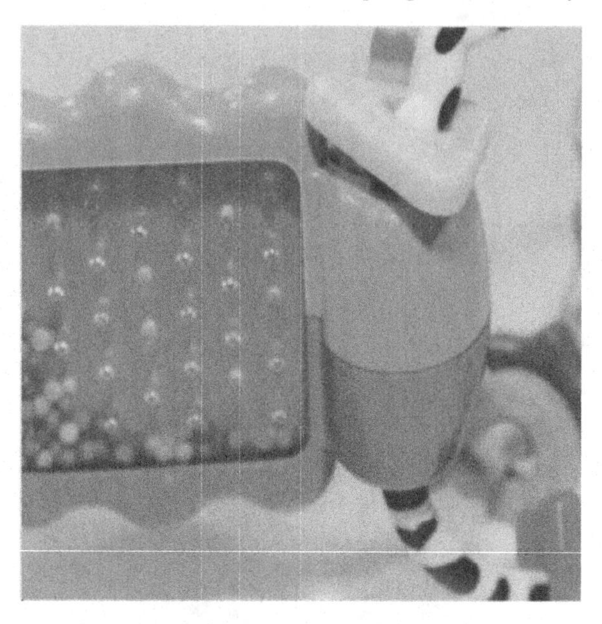

FIGURE 7.5: The normal distribution is so simple, even a child can understand...

Yes. This is true. So, we have to figure out when is a good time to use the normal approximation to the binomial versus using the binomial formula.

A good rule of thumb to follow is when $np(1\text{-}p)$ is greater than 9, the normal approximation to the binomial will generally be close enough, and you don't have to calculate all the factorials! You can still use the binomial formula, up until the point where your computer can't handle the factorials, at which point it may even switch automatically to the normal approximation without telling you.

When the variance is less than 9, you really should use the binomial formula, because it will produce the exact value of $p(x)$. Figure 7.6 illustrates the relationship between sample size and probability.

When to use the Normal approximation to the Binomial Distribution: Because when sample size gets larger, the binomial distribution approximates the Normal distribution, and this happens even more quickly as p approaches 0.50 (i.e. as the distribution becomes more symmetric), we can use the Normal Distribution as a nice proxy for the binomial. Generally $np(1\text{-}np)$ should be greater than 9 at a minimum before applying this approximation (Schader & Schmidt, 1989).

So now we know! At the very beginning of this chapter we tried to figure out whether or not the failure rate of the ball bearings we were producing in our

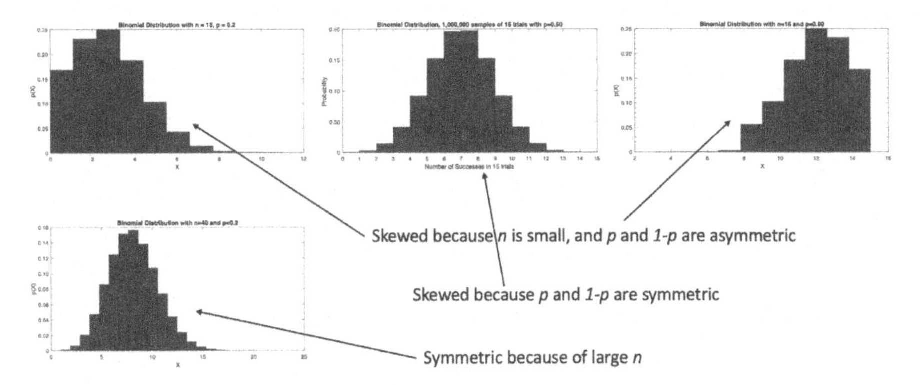

FIGURE 7.6: Binomial distribution as a function of both n and p

factory was acceptable or not. The question we were asking can now be framed in terms of the binomial distribution as a sum of the individual probabilities of obtaining a success given we have 184 successes, a sample size of 200, and a desired probability of a success being 95%.

$$P(X \leq 93|100, 0.95) = P(X = 0) + P(X = 1) + P(X = 2) + \ldots + P(X = 93)$$

Man! I wish there was some approximation that we could use that was close enough to make a decision, but didn't require me to do 94 different calculations and then sum them up! Thanks to Bernoulli, there is: *The Normal Approximation to the Binomial Distribution.*

The normal approximation uses exactly the same logic as the Z-test we learned in Chapter 5. We just find the distance between the mean value we got in our sample (93 successes), and the mean we would expect under our design constraint μ, and then scale it by the standard error of the sampling distribution. Because the binomial has only two values, the standard error is the same as σ for the population. Had our problem had 3 or more categories, we would have to divide by n just like the standard error of the mean.

$$Z = \frac{x-\mu}{\sqrt{np(1-p)}}, \quad \mu = np, \quad \sigma = \sqrt{np(1-p)}$$

For our problem, $\mu = 200 \cdot .95 = 190$, and $\sigma = \sqrt{190(0.05)} = 3.082$. $\sigma^2 > 9$, so we are pretty sure we can use the normal approximation to the binomial to solve the problem.

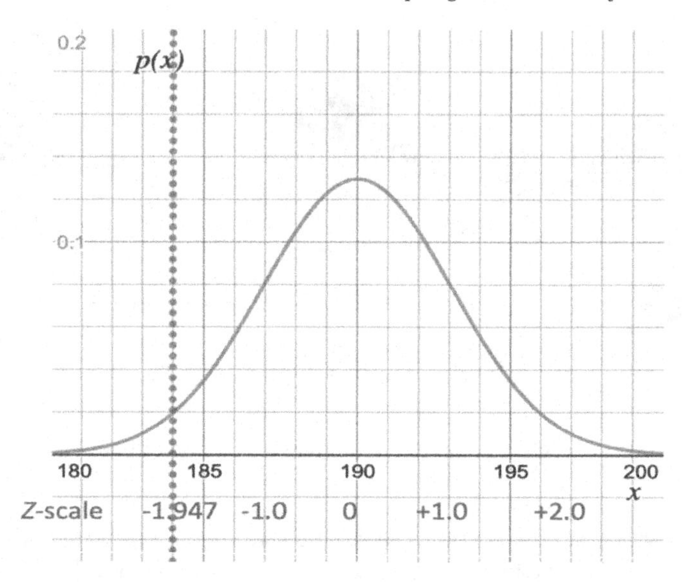

FIGURE 7.7: X≤93 successes corresponds to a Z-value of −1.917 standard deviations from the mean

Figure 7.7 shows the normal distribution with these values as parameters

$$Z = \frac{184 - 190}{3.082}$$

$$Z = \frac{-6}{3.082}$$

$$Z = -1.947$$

Looking this value up in the table, or computing it with a statistical package, we get a cumulative probability of a Z-value less than or equal to −1.947 equal to 0.0258.

Examine the area under the curve to the left of our Z-value. Only about 2.6% of the area of the normal curve falls in this region. Also, to hammer home the point, Z=−1.947 corresponds to a value of 184 successes in the original scale of the problem. Since there is a one-to-one correspondence between Z-values and **any** normal distribution, we can use the Z-distribution to express the PDF for any problem that utilizes a normal distribution as a probability model. But just how good is the Normal Approximation to the Binomial?

Using the binomial formula for this, we get an answer that is pretty close:

$$P(X \leq 184|200, 0.95) = P(X = 0) + P(X = 1) + P(X = 2) + \ldots + P(X = 93)$$

$$= 0.0238$$

I would say a result within two tenths of 1%, in exchange for not having to compute all those sums, is worth it in this case.

7.1.6 Sampling with and without Replacement

One of the major assumptions of the binomial distribution, and most distributions we deal with is that the samples are drawn using Simple Random Sampling (SRS). If you recall, SRS involves sampling by taking one item from the population, recording its measure, then replacing it back into the population before drawing another item. This insures that the probability that any item is drawn at random is the same for all draws, and that all samples of size n have the same probability of being drawn.

When this assumption is violated, each successive item drawn has a larger probability of being drawn. If we take this to the extreme, the last item drawn has a 100% probability of being selected. If the sample size is large compared to the size of the population, these unequal chances wreak havoc with the behavior of our sampling distributions. For this reason, many finite populations with small N have to have some correction applied so that the probabilities of selecting items without replacement can be accounted for.

Bernoulli trials without replacement involve selecting one item from the population, then selecting another, and another and so on until the sample is complete. If the sample size is small relative to the population size, the differences in probabilities among phenomena in the sample are negligible. But if the sample size is large relative to its population, we need to account for it by appealing to the **Hypergeometric Distribution**.

7.1.7 The Hypergeometric Distribution

The Hypergeometric Distribution is identical, conceptually, in every way to the binomial with the exception that it accounts for sampling *without replacement* from a finite population! All trials are Bernoulli Trials (0 and 1 are the only outcomes). Like the binomial distribution, the number of successful events (1s) is just the sum of successes across trials:

$$\sum_{i=1}^{n} x_n$$

The probability of x successes in n trials, drawn without replacement from a population of size N, with the number of successes in the population being S is pretty straightforward, conceptually:

$$p(X = x|n, N, S) = \frac{\binom{S}{x}\binom{N-S}{n-x}}{\binom{N}{n}}$$

$$\binom{S}{x}$$

is the number of ways x can be drawn from a sample of size S.

$$\binom{N-S}{n-x}$$

accounts for the fact that we are sampling without replacement. $N - S$ is the size of the population minus the number of successes in the population. $n-x$ accounts for the size of the sample and the number of successes found in the sample.

$$\binom{N}{n}$$

Accounts for the number of ways a sample of size n can be taken from a finite population of size N.

The parameters of the Hypergeometric Distribution can be written as sums of probabilities:

$$\mu = \left(\sum_{i=0}^{N} \frac{\binom{S}{x}\binom{N-S}{n-x}}{\binom{N}{n}}\right), \text{ and } \sigma = \sqrt{\frac{nS(N-S)}{N^2}} \cdot \sqrt{\frac{N-n}{N-1}}$$

These expressions can be simplified to a closed form:

$$\mu = \frac{nS}{N}, \text{ and } \sigma = \sqrt{\frac{nS(N-S)}{N^2}} \cdot \sqrt{\frac{N-n}{N-1}}$$

And further simplified to show us that the Hypergeometric distribution is just a simple extension of the binomial.

$$\mu = np, \textbf{ and } \sigma = \sqrt{\frac{np(1-p)(N-n)}{(N-1)}}$$

$\frac{N-n}{N-1}$ is of special importance here. Sampling without replacement from a finite population doesn't change the expected value of the number of successes, but it does change the expected value of the variance. Examining σ we see that it is exactly the same as that of the binomial distribution with the exception that $\frac{N-n}{N-1}$ is a correction factor accounting for the proportion of observations that the sample takes up given N, adjusting the distribution to accurately reflect the population parameter.

Assumptions of the Hypergeometric Distribution:

(a) *Binomial*: Events must be classified as a 0 (failure; not an apple) or 1 (success; apple).

(b) *Known number of successes in finite population.*

(c) *Sample Size n*, must be greater than or equal to 5% of the population N. This assumes that N is finite.

(d) *Sampling without Replacement.* The Hypergeometric distribution applies a correction factor for finiteness to the Binomial distribution.

Applying the Hypergeometric Distribution, given you understand the Binomial, is again, pretty straightforward. I am providing the exact same scenario as problem 2 in this Chapter, with the only difference being that we will draw our sample of bearings from a finite population without replacement.

5. Suppose you are a quality control engineer studying the extent to which batteries manufactured for your cochlear implants meet specifications. In general, 97% of the batteries from the company you have contracted from meet specifications.

   If a sample of 5 batteries is selected without replacement, what is the probability that at least 4 of them meet specifications?

$$\mu = 5 \cdot 0.97 = 4.85$$

$$\sigma = \sqrt{\frac{4.85\,(0.03)\,(50-5)}{(50-1)}} = 0.1336$$

From these estimates, we would expect a little fewer than 5 successes in every draw of 5 batteries. That is good! The variation around that expected value is only 0.1336. That is also very good! So, what is the probability of getting 4 or 5 successes in a draw?

$$p\,(X \geq 4|5, 50, 45) = p\,(X = 4) + p\,(X = 5)$$

$$p\,(X = 4|5, 50, 45) = \frac{\binom{45}{4}\binom{50-45}{5-4}}{\binom{50}{5}} = 0.3516$$

$$p(X = 5|5, 50, 45) = \frac{\binom{45}{5}\binom{50-45}{5-5}}{\binom{50}{5}} = 0.5766$$

$$p(X \geq 4|5, 50, 45) = 0.3516 + 0.5766 = 0.9282$$

So, we expect about 93% of the time to get 4 or more successes in a random sample of 5 measurements. Is this good enough? Do we want to stay with the company who is supplying our batteries? Should we go with another company? Those are not statistical questions, but practical ones that our analysis can help us make, given our tolerance for failed batteries, and given the consequences of failed batteries (remember, these are for cochlear implants. People won't be able to hear if the batteries fail!).

Like the binomial distribution, we can use an approximation if the population size is very large, relative to the sample size. In such cases, $\frac{N-n}{N-1}$ becomes close to 1, making the differences in probabilities due to sampling without replacement, negligible. What is the approximation to use? The binomial! If $\frac{N-n}{N-1}$ becomes 1, then the binomial and hypergeometric distributions become identical, with the same parameters and the same PDF and CDF.

6. Compare our results here in Problem 5, with the results we achieved in Problem 2. How does the finite correction factor for the Hypergeometric Distribution change as N and n change?

The ratio of n and N is very important here. As n gets smaller and smaller, $N - n$ becomes closer and closer to $N - 1$, making the correction factor ratio closer to 1. If n is one, then the correction factor becomes 1 and the variance of the hypergeometric distribution is the same as the variance of the binomial.

Poisson Distribution

I hope you have noticed that, for finite distributions like the Binomial and Hypergeometric distributions, the number of observations matters in determining what model is most appropriate and most practical to use. In addition, the way in which phenomena are sampled matters, sometimes changing the distribution to account for finite versus infinite populations, and for sampling performed with replacement (SRS) or without replacement.

Quite a number of applications of proportions as statistics are bounded within some continuous interval (for example, dust particles are deposited on light detectors within time intervals, making the detectors less efficient over time; or within an area of land, as animals may migrate at infrequent times, but end up at a particular destination). If we know the long-haul probability of observing one success in a particular interval, and if we know that probability is stable within that interval, *and* if there is a zero probability that we will find more than 1 success in that interval, we can estimate the probability of any number of successes appearing using the **Poisson Distribution**.

Because determining whether or not a discrete variable with binary outcomes can be reasonably modeled with a Poisson distribution, you have to think very carefully about its assumptions.

Assumptions of the Poisson Distribution:

(a) *Binomial*: Events must be classified as a 0 or 1. Only one of these outcomes is considered a legitimate event (e.g., success = 1).

(b) *Independence*: Events must be independent. One event occurring should not impact the probability of another event occurring.

(c) *Homogeneity*: The mean number of events is assumed to be the same for all intervals measured.

(d) *Equal areas of opportunity for an event to occur*: All intervals in which the event is to be measured must be equal in size. Because only 1 event can happen in an interval, we have to sum across the number of intervals. To do this, we need to have them all equivalent.

What we do is partition the continuous interval into equal sub-intervals where the probability of more than one event occurring in a sub-interval is at or near zero. Then we can count the number of sub-intervals we would expect to find an event occurring across the entire interval of sub-events. The following problem will illustrate such a situation:

7. The number of micro cracks appearing in the alloy skin of a commercial aircraft with over 1,000 hours of service is approximately 0.005 cracks per square meter. If an aircraft has 520 square meters of skin:

   (a) What is the probability that there are zero micro cracks in an aircraft?

   (b) What is the probability that there are more than 48 microcracks in an aircraft?

Each "interval" here is 1 m^2 of an aircraft skin. We have no reason to assume that any part of the aircraft has more or less micro-cracks (this may in fact be wrong, but we assume homogeneity until we have evidence we can test otherwise). Further, a micro-crack appearing in one portion of the aircraft has no bearing on a crack potentially appearing in another portion. And, finally aside from some probable leftover in the last bit of the skin we cover, all intervals are equal at 1 m^2. From this analysis of the situation, we can conclude that a Poisson process is the most appropriate model to apply.

If we take the Binomial distribution and follow it as $n \to \infty$, we can see that it converges on an exponential distribution with one parameter:

$$\lim_{n \to \infty} \binom{n}{x} p^x (1-p)^{n-x} = \frac{e^{-\mu}\mu^x}{x!}$$

$$P(X = x|\mu) = \frac{e^{-\mu}\mu^x}{x!}$$

Note, μ is often symbolized alternatively as λ . I use μ here, because it ties the $E(x)$ of the Poisson Distribution to the $E(x)$ of the Binomial and Normal Distributions. They all have the same meaning—as the mean value of x in the distribution.

$$E(x) = np = \mu$$

$$\sigma_x = \sqrt{\mu}$$

Looking at Problem 7, we can first evaluate μ and σ_x

$$\mu = np = 520 \cdot 0.005 = 2.6$$

We expect about 2.6 microcracks on a given airplane

$$\sigma_x = \sqrt{2.6} = 1.6125$$

What is the probability that there are zero micro – cracks in an aircraft?

We expect a deviation of about +/−1.6 cracks for any airplane. Now to find the probability of getting a perfect plane—the probability that we will find zero cracks!

$$P(X = 0|2.6) = \frac{e^{-2.6}2.6^0}{0!}$$

$$= .07427$$

What is the probability that there are more than 48 micro-cracks in an aircraft?

The expected value of the number of cracks per plane remains the same:

$$\mu = np = 520 \cdot 0.005 = 2.6$$

But now, like we have seen before, since we are evaluating the probability over an interval, we need to sum the probabilities in each subinterval:

$$P(X > 48|2.6) = P(X = 49) + P(X = 50)$$

$$= \frac{e^{-2.6}2.6^{49}}{49!} + \frac{e^{-2.6}2.6^{50}}{50!}$$

$$= 0.6799$$

This probability tells us that, even though there is a small chance of a crack happening in any given square meter of an aircraft's skin, it is pretty likely that any randomly chosen aircraft will have may micro-cracks, potentially causing a safety problem.

1. The average number of robocalls detected at my house each hour during election season is 4. What is the probability that:
   (a) I get 2 robocalls in 5 minutes?
   (b) There are fewer than 2 robocalls in 3 hours?

Since we have hours as an initial unit, we can break it up into subareas of 5 minutes. There are 4 robocalls in 12 5-minute intervals, yielding an average of 1/3 of a robocall per 5 minutes.

$$\mu = 1/3$$

What is the probability that I get 2 robocalls in 5 minutes?

$$P(X = 2|0.3333) = \frac{e^{-0.3333}0.3333^2}{2!}$$

$$= 0.0398$$

What is the probability that are 3 or fewer robocalls in 3 hours?

Now we are looking at an interval of 3 hours. Given we have 4 robocalls per hour, we expect 12 robocalls in 3 hours.

$$\mu = 12$$

$$P(X \leq 3|0.3333) = P(X = 0) + P(X = 1) + P(X = 2) + P(X = 3)$$

$$P(X = 3|12) = \frac{e^{-12}12^3}{3!} = 0.0018$$

$$P(X = 2|12) = \frac{e^{-12}12^2}{2!} = 0.00044$$

$$P(X = 1|12) = \frac{e^{-12}12^1}{1!} = \textit{very close to } 0$$

$$P(X = 0|12) = \frac{e^{-12}12^0}{0!} \text{even closer to } 0$$

$$P(X \leq 3|12) = 0.0023$$

Unfortunately, it looks like I will be having a lot of robocalls this campaign season! The probability of getting 1 per hour is just too low...

Like the binomial distribution, at higher μ ($\mu \geq about 10$) the Normal distribution can be used as a close approximation. Figure 7.8 shows how the Poisson distribution changes shape and becomes more symmetric as μ, the mean number of events in an interval, grows.

Because the Poisson distribution is the limit, as x approaches infinity, of the Binomial distribution, when p is very small, some authors advocate that it is easier to use

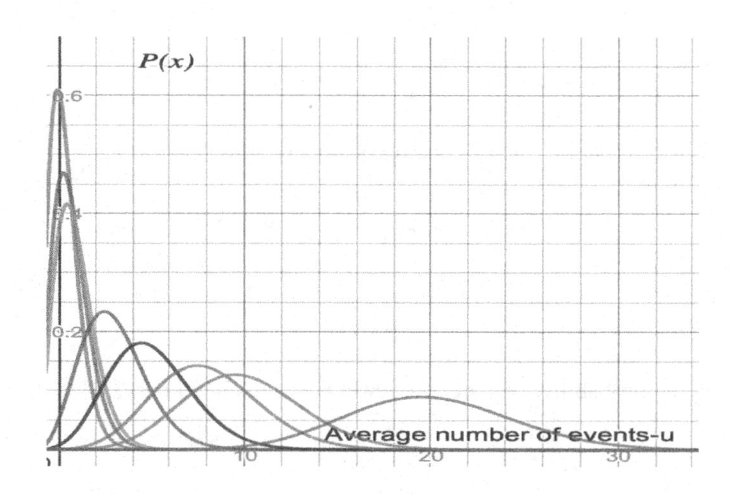

FIGURE 7.8: The Poisson distribution becomes more normal as μ grows

the binomial formula as an approximation of the Poisson. In general, however, it is a good rule of thumb to use the proper reference distribution for its application. Computers have made most approximations obsolete, being able to handle large means, small *p*s and other difficult computations with ease.

7.2 Summary

The sampling distribution of the proportion is manifest in a number of ways depending on the number of categories in which one is computing proportions (probabilities), and the size of the parameters, n and p that fix its geometry. In this Chapter, we have focused only on categorical events that have binomial outcomes those that can only be classified as a 0 or a 1. When we can use Simple Random Sampling, such events are best modeled with the Binomial Distribution. When we use Sampling Without Replacement from a finite population, we are best advised to use the Hypergeometric Distribution. When the sample size gets very large, the Binomial converges to a Poisson distribution, and when the mean of the Poisson distribution gets large, it converges to the Normal Distribution.

All of these models of the proportion of successful events occurring in n trials, can be thought of as cases of a larger model—the Normal Distribution. The logic of hypothesis testing for each of these distributions, therefore, is exactly the same as the logic of finding a cumulative probability (CDF) across some region of the independent variable. For these discrete distributions, however, because they are not integrate-able, we compute each separate value of $P(X)$ in the interval and sum them up to get the cumulative probability. For situations where $P(X)$ requires many computations, using the logical complement of $P(X)$, $1 - P(X)$, can be a more efficient method.

7.3 References

Pearson, K. (1925). James Bernoulli's Theorem. *Biometrika,17*(3/4), 201-210. doi:10.2307/2332077

Poisson, S. D. (1837). *Recherches sur la probabilité des jugements en matière criminelle et en matière civile precédées des règles générales du calcul des probabilités par sd poisson*. Bachelier.

Ramsey, P. H., & Ramsey, P. P. (1988). Evaluating the normal approximation to the binomial test. *Journal of Educational Statistics, 13*(2), 173-182.

Schader, M., & Schmid, F. (1989). Two rules of thumb for the approximation of the binomial distribution by the normal distribution. *The American Statistician, 43*(1), 23-24.

7.4 Study Problems for Chapter 7

1. **A hip joint replacement part you designed is being stress-tested in your laboratory. The probability of any given part successfully completing the test in the past has been shown in the past to be 0.80. You sample 10 parts at random and test each independently. You find that 2 of your parts fail. Are you satisfied that your sample's performance isn't too far off what you might have predicted?**

This is an opportunity to use the binomial distribution! I assume there are lots of parts in the population, and that 10 is a small sample relative to the population size. I don't know if it is sampled with or without replacement, but without knowing the population size, I have to use the binomial as my best available model. I have 2 parts fail, so I am going to convert that to successes (10 total – 2 failures = 8 successes):

$$P(8 \le X|10, 0.80) = \sum_{i=8}^{10} \binom{10}{x} 0.8^{x_i} (1 - 0.8)^{10-x}$$

We start by evaluating $P(X)$ at X=8:

$$P(X = 8|10, 0.80) = \binom{10}{8} 0.80^8 (0.20)^{10-8}$$

$$= 0.302$$

We then follow the same procedure for $P(X = 9|10, 0.80)$ and $P(X = 10|10, 0.80)$ and add the three sums together to find a cumulative probability:

$$P(8 \le X|10, 0.80) = 0.678$$

We could have alternatively found the probability of 2 or fewer success and subtracted from 1.

With only about a 2/3 chance of randomly getting 8 or more success given a sample size of 10, I would say this is pretty poor performance.

2. **Flaws occur in parachute canopy material (ripstop nylon) on average about .1 per in every 1 square meter of fabric. If 25 square meters were inspected, what is the probability that there are 1 or fewer flaws?**

   Because I am assuming there is an infinite (or very large) number of square meters of parachute canopy material in the population, and that my average rate of flaws is about 0.1 in every square meter, I see this as an application of the Poisson distribution. Because we have 25 square meters, the mean number of flaws in 25 square meters is expected to be $0.1 \times 25 = 2.5$

$$P(X \leq 1|.2.5) = \sum \frac{e^{-\mu}2.5^{n}}{n!}$$

$$\frac{2.5 \times 2.5^{0}}{0!} + \frac{2.5 \times 2.5^{1}}{1!} = 0.287$$

   This means I have about a 30% probability that in every 25 square meters of canopy material, I will have 1 or fewer flaws. THAT means that I have about a 70% probability that I will have greater than one flaw in each of my 25 square meters of fabric! Too much for me given a single hemispherical parachute takes about 38 m^2 of material.

3. **You are developing a new broccoli-flavored tooth polish. Suppose you randomly select 20 customers without replacement from a group of 100 product testers. You have read good research that shows about 25% of the American population hates broccoli. What is the probability that 2 or fewer will like your new product? What do you conclude about the marketability of your product?**

   Here, we have sampling without replacement from a relatively small population. We have a known population N and probability. This is where the hypergeometric distribution can help us. But first we have to figure out what a "success" is. 25% of the population hates broccoli, that means 75% like it. So S is 75, not 25.

$$p(X = x|n, N, S) = \frac{\binom{S}{x}\binom{N-S}{n-x}}{\binom{N}{n}}$$

$$p(X \leq 2|20, 100, 75) = p(X = 0) + p(X = 1) + p(X = 2)$$

$$p(X = 2|20, 100, 75) = \frac{\binom{75}{2}\binom{100-75}{20-2}}{\binom{100}{20}} = \text{very}$$

   close to zero (my stat package can't compute this low)

$$p(X = 1|20, 100, 75) = \frac{\binom{75}{1}\binom{100-75}{20-1}}{\binom{100}{20}} = \text{very close to zero}$$

$$p(X = 0|20, 100, 75) = \frac{\binom{75}{0}\binom{100-75}{20-0}}{\binom{100}{20}} = \text{very close to zero}$$

= *about zero* that 2 or fewer will like it.

This is pretty low! That means that it is almost certain that more than 2 will like it! Maybe I have a niche market! But seriously, I would want to choose another number of people who will like it, like 20 or more, to really test this product! Try this on your own.

4. **You are designing a wrong-way driving warning system. Collecting data, you record the number of cars driving the wrong way on the freeway each month. You find that the average for Arizona is 2.5 wrong-way drivers discovered every month. What is the probability that 10 or more cars will be caught wrong-way driving in a year?**

   Because this scenario has an average number of "successes" (if wrong-way drivers are a "success"), within an "area of opportunity" (each month—a time interval), and because there are infinitely many time intervals, we can use the Poisson distribution as a useful model with which to compute probabilities. Again, like problem 2, because we are modeling the number of wrong-way drivers in a year, and our mean is 2.5 per month, we have to multiply 2.5 per month by 12 months to get the mean for 1 year.

$$2.5 \times 12 = 30$$

$$P(X \geq 10|30) = \sum \frac{e^{-\mu}30^{n}}{n!}$$

   I really don't want to compute 21 individual probabilities and sum them, so I will use a stat package or an online calculator to do it for me. So I input a vector, called x, that has the numbers 10 through 30. Then I find the individual probabilities of each of these entries (Here I am using Matlab):

```
»y = poisspdf(x,30);
```

   Then I sum the probabilities:

```
pois=sum(y);
```

pois = 0.5483

It is very probable that I will have 10 or more wrong-way drivers detected in a year on Arizona Highways.

5. The probability of a mobile phone switch being defective is 0.02. Which of the following statements is true?

   (a) In a shipment of 100 chips, two will be defective;

   (b) The expected number of defective chips in a shipment of 500 is ten.

   (c) In a shipment of 1000 chips, it is certain that at least one will be defective.

   These are all binomial probabilities:

   i. $P(X = 2|100, 0.02) = \binom{100}{2} 0.02^2 (1 - 0.02)^{100-2} = 0.273$

   You could also use the normal approximation of the binomial for this. But most computers can handle it. However you do it, having a 1 in 4 chance of 2 being defective is pretty high! I might want to check on our manufacturing process. So, while two defects is probable, it isn't necessarily destined to happen.

   ii. $P(X = 10|500, 0.02) = \binom{500}{10} 0.02^{10} (1 - 0.02)^{500-10} = 0.126$. This is more like it. But I still think this is high, given there could be 8 or 9 or 10, etc., defects somewhere around the expected value of

$$\mu = 0.02 \times 500 = 10 \textit{ defects}$$

$$\sigma = \sqrt{0.02 \times 500 (0.98)} = \pm 9.8 \textit{ defects}$$

   iii. It is not certain that, in a shipment of 1,000 chips, that at least 1 will be defective. The probability is very high (>0.99999), but it still isn't certain.

   (d) For Problem 5, what is the probability that, in a simple random sample of 50 phones, the number of defective phones is greater than 1?

$$P(X > 1 |50, 0.02) = \binom{50}{1} 0.02^1 (1 - 0.02)^{50-1} = 0.264$$

   Remember to keep the inequalities straight. This interval is not inclusive of 1, so I computed

$$P(X > 1 |50, 0.02) = 1 - p(X = 0|50, 0.02) - p(X = 1|50, 0.02)$$

   (e) Suppose for Problem 5, the random sample of 50 phones was taken from a population of 1,000 phones returned for service. What is the probability that the number of defective phones is greater than 1?

Now, because we have a finite population with sampling without replacement. We need to correct for the ratio of sample size to population size:

$$p(X = x|n, N, S) = \frac{\binom{S}{x}\binom{N-S}{n-x}}{\binom{N}{n}}$$

$$p(X > 1|50, 1000, 20) = p(X = 2) + p(X = 3) + \ldots + p(X = 1000)$$

OR

$$p(X > 1|50, 1000, 20) = 1 - [p(X = 0) + p(X = 1)]$$

By now you should be using a statistics package or online calculator (Here is the Matlab code)!

```
» h = hygepdf(x,1000,50,20)
»p=sum(h);
»p=1−p
```

I get the cumulative hypergeometric probability to be 0.264. Now is this acceptable that of any 50 phones sampled, there is a 1/4 chance that at least one is defective? That seems a bit high to me. It all depends on the consequences of your decision. Will customers complain? Will returns cut into profit margin? The numbers give you the probability, but your decision is to interpret it given the consequences that your company (and you) face if you make the wrong decision.

(f) **A manufacturer of halogen bulbs knows that 3% of the production of their 100 W bulbs will be defective. What is the probability that exactly 5 or more bulbs in a carton of 144 bulbs will be defective?**

This is a common issue faced by statisticians. While it may seem like our population is finite, N=144, and we have the success rate in the population 3%, so our expected success rate is $\mu = 0.03 \times 144 = 4.32$, we don't have the sample size. Alternatively, if we have a sample of 144, we don't know if the carton is a simple random sample (it probably isn't). So we are in a quandary whether or not the hypergeometric or binomial probability distribution is the most appropriate for the problem. In this case, we really don't have all the information to use the hypergeometric, so we have to fall back on the binomial and hope that our sample of 144 (a sample of convenience) is representative enough of the population for our probability to be considered an accurate estimate. Here we could use the normal approximation of the binomial, or compute it using a statistics package (Matlab code here):

```
y = binopdf(x,144,0.03)
```

Where x is a vector of integers from 5 to 144.

If we sum the vector of individual probabilities, My statistics package returns a cumulative probability of 0.434. That is a pretty high probability that 5 or more bulbs will be defective! We may need to see what manufacturing machines are causing such poor performance.

(g) The rate of radiation measured (using a Geiger counter) in a particle physics lab is known to be about 3000 counts per hour.

i. If I run the Geiger counter in the room for a five second interval, how many counts do I expect to see?

ii. What is my uncertainty for the number of expected counts in five seconds?

iii. What is the probability of measuring 2 or more particles in 5 seconds?

These are all pointing to the Poisson distribution

1. There are 12 five second intervals in a minute and 60 minutes in an hour making an expected value of $\mu = 3000 \times \frac{1}{60} \times \frac{1}{12} = 4.17$ counts in each 5 second interval.

2. $\sigma = \sqrt{\mu} = \sqrt{4.17} = 2 \pm .04$

3. $P(X = 2|4.17) = \frac{e^{-4.17} 4.17^2}{2!} = 0.134$

I then repeat this calculation for n=3, n=4, all the way up to infinity, or I find $1 - [p(x = 0) + p(x = 1)] = 0.920$

8

Hypothesis Testing Using 1-Sample Statistics

In Chapter 4 we talked about how we are playing a game: Trying to establish the probability that a given sample we draw comes from a population we are interested in. In Chapter 5, we focused closely on the sampling distribution of the mean as a general model for probability density functions and how, by integrating across some interval in a PDF, we could determine the likelihood that a sample drawn at random could have displayed a mean value significantly different from the mean of a referent population. In Chapter 6 we examined ways to design experiments so that we can determine the cause that one or more samples might come from different populations primarily due to our manipulation of the independent variables through the application of some treatment. To bring us up to date, in Chapter 7 we saw how to compute probabilities of samples taken from distributions that use proportions as their main statistics, using the same logic as the sampling distribution of the mean. Now it is time to bring all of this together so that we can formally test hypotheses experimentally in such a manner that we have good weight of evidence for or against some innovation we want to introduce. As engineers, we are trying to improve upon the status quo—the current state of affairs in the world. We may dedicate our lives to providing safe drinking water in areas where waterborne illness is rampant. We may try to extract energy from the sun, wind and waves. We may devote our energies to entertainment, building amusement park rides, or fast cars, or apps or sports equipment. The point is, we look at the world and feel dissatisfied. We want to improve on what is by designing what will be.

> **Example Problem:** As wind-turbine designers, we know that turbine blades bend significantly under load. Published data for a 10 m blade shows an average deflection of 1 m +/– .24m behind the hub, when subjected to wind speed of 13.5 m/s, causing inefficiencies in transfer of energy.

Our Solution: We have read a new article by Momeni et al (2019) on how putting a pre-bend in a turbine blade can assist with its efficiency under an expected load (See Figure 8.1). Using this as inspiration, we design a new turbine blade for high-load application with exactly the same design as our earlier blades that have an average of 1 +/– 0.24 m bend under load, only our design now has a pre-bend of 1m at the tips of the blades.

DOI: 10.1201/9781003094227-8

FIGURE 8.1: Putting a pre-bend in a wind turbine blade (theoretically) increases its efficiency under a typical load

Using a Random Sample design with the published data representing the control condition, we draw a random sample of 36 turbine blades from our inventory and find that, under a wind velocity of 13.5 m/s the blade deflection averages 0.9 m. Does our sample indicate that our new turbine blades have, on average, less bend than the population?

1. Of the different sampling distributions we could use, which one is most appropriate for this experiment?

   Because deflection is measured on a continuous random variable (displacement), the most appropriate sampling distribution is the sampling distribution of the mean. We still have to assume that the displacement is randomly distributed about some central value, but this is a pretty safe assumption.

2. What is the expected value of the sampling distribution?

   The sample mean, $\overline{x}$ is the $E(x)$ of a continuous random variable.

3. What is the expected variation about the mean of the sampling distribution?

   The standard error of the mean, $\frac{\sigma}{\sqrt{n}}$ describes the variation of the sampling distribution, assuming a normal distribution. We know the intended population parameters, so we don't have to use the sample standard deviation, s, to estimate the $SE_{\overline{x}}$.

To determine if our empirical data indicates that our design significantly changes the bend in the turbine blades, we have to examine the logic of hypothesis testing.

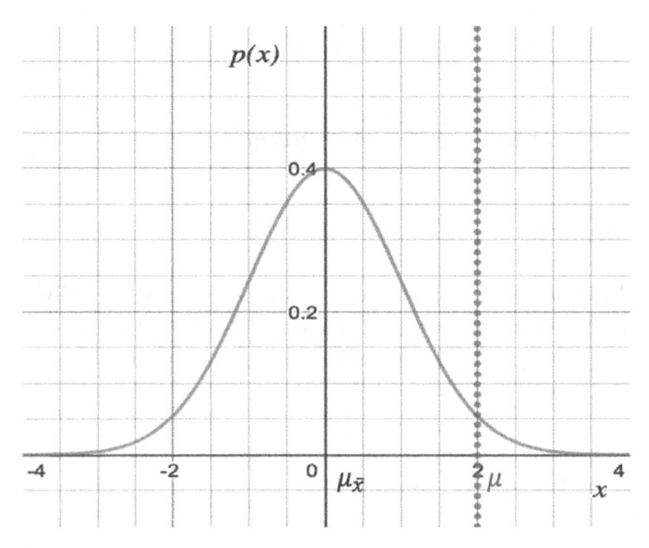

FIGURE 8.2: The value of a population of measures (dotted vertical line) 2 standard deviations above the mean of the sampling distribution (red)

8.1 Philosophy

Hypothesis testing is making a claim about the likely performance of some data, given some assumptions about its parent distribution. As in the Z-test, if we know the population parameters and we assume our sample mean represents a population of sample means in a sampling distribution, we can make claims regarding how probable it is that the value of our population mean could represent the center of the sampling distribution.

In Figure 8.2, the mean of the population of measures from which the sampling distribution was *supposed* to have been drawn is 2 standard errors above the mean of the sampling distribution. Since the sampling distribution, by the CLT, will tend to the mean of the population from which the samples were theoretically drawn, this almost looks like an error of accuracy. How could it be so far off from the target population? The answer lies in experimental design: if we actually *drew the sample randomly from the target population in the first place*, then subjected it to some treatment, we *want* its performance to be different from the population performance. If it was drawn at random, we can only attribute this large difference to random chance, and because we are using the Z-distribution as our model for the sampling distribution of the mean, we estimate that the probability we could get a mean difference of 2 standard errors is about 1 – 97.5% (integrating from negative infinity to Z=+2), about 2.5%.

To reiterate, the probability that our sample mean, given standard deviation and sample size, could be 2 standard errors from the population mean from which it was originally drawn is only 2.5 in 100! If I obtained this kind of result, I would assume that there was some systematic effect of my treatment.

8.1.1 Falsification

The key notion in this kind of process is what was termed, **Falsification**, by the philosopher Karl Popper. What is meant by falsification is that, since it is very difficult to prove any empirical hypothesis *true*—To do so would mean one would have to do an exhaustive search for *all instances* of a phenomena. Instead, we can prove the logical converse of a hypothesis *false* instead. Whaaaaa????!!!! This sounds like a semantic game that a philosopher *would* play. Let's take a look at our turbine blade example to illustrate the problem:

1. We believe that our pre-bend treatment to our wind turbines *does have an effect* causing the blades to not bend beyond the hub of the tower under the designed-for load. We wouldn't have designed and built them if we didn't.
2. If we want to prove that our treatment does cause the improved performance, we would have to take all wind turbines and find, in every instance, that our treated blades perform in this manner. Just one blade that doesn't, means that our proof is shot down in flames;
3. BUT, if we assume that the turbine blades originally came from the population of status quo blades, and try to show that their performance *is unlikely to have occurred in this population*, we can falsify the assumption that the population of treated blades is the same as the reference population and conclude that the treatment *likely made a difference.*

Falsification is a means of proving a hypothesis incorrect in an attempt to build evidence that an alternative hypothesis is a better explanation for the phenomenon. It is a method of proof by negation.

In essence, this method of falsification is saying to yourself, "if I can prove that my data did not come from the population that represents the status quo, it must represent a population that is different from the status quo." That is the purpose of engineering design, to design and produce products and systems that improve upon the status quo (Read: "improve" to mean "does not come from the status quo population"). The longer you look, finding disconfirmations, the more you support your original assumption, that the treatment does improve performance of your designed product. In engineering and other scientific fields, we tend to be conservative when we test hypotheses: We like to say that there has been no effect, even if there might be, if the evidence is sketchy. That way, our customers don't go throwing out old products or retooling their manufacturing systems to account for this new, sketchy, evidence of improved performance when our conclusions may be incorrect. In other words, if you don't have good evidence that your product is better than some other product in some demonstrable way, you should redesign your product to do a better job. Passing it off as superior without good evidence of efficacy costs money, and in cases like medicine or public safety where the performance really does mean life or death, consequences of people purchasing a mis-attributed product can be severe. In our example, if our turbines don't actually improve performance, it would cost the customer somewhere in the six-figures to replace them, for no demonstrable increase in energy produced.

8.1.2 The Double-Negative: The Null Hypothesis

If we can disprove a hypothesis, it gives weight to alternative explanations of the data. For our turbines, if we disprove that there is no difference between the performance our new pre-bent blades and the population norm, it gives weight to our belief that there really is a difference due to our clever design. Did you catch this double negative? By refuting the hypothesis that our design made no improvement, we are supporting the case that our design really was effective. This method of verification of an argument by negation is more than just semantics.

Remember in the Random Sample experimental design, we draw two samples from a population and assign one sample to an Experimental condition where we administer some treatment, and we assign the second sample to a Control condition, where we do nothing. We have to assume that, aside from random variation, the two samples are functionally equivalent, each equally representing the population from which they were both drawn. After administering our treatment, if the performance of our treatment group is, on average, different from that in the control group, we can conclude that the two groups do not, now, represent the same population. We can attribute this difference to the effect of the treatment. That is how Falsification works in hypothesis testing.

The process of hypothesis testing is as follows:

Hypothesis Testing General Procedure

1. Create a mathematical statement that specifies the effect you expect from the application of your treatment. We call this statement the **Working Hypothesis, or Alternative Hypothesis.** We can symbolize this statement as H_1.

   $H_1: \overline{x} \neq \mu$

   In English, this means "my sample of exquisitely designed widgets is so good, it no longer represent the population of just ordinary widgets from which it was originally drawn. I know this because my sample mean value is so far away from the population mean as to be unlikely to have just happened as a function of chance."

2. AFTER you have specified your H_1, you then specify the actual hypothesis you are going to try to falsify. This is a mathematical statement that indicates that your sample came from the reference population, and its expected value is no different than that of the population:

   $H_0: \overline{x} = \mu$

   We call this statement the **Null Hypothesis**. It is symbolized as H_0.

3. Determine the distance the expected value of your sampling distribution must have from the mean to be considered a probable "real" difference.
4. Once these hypotheses are specified, take your samples from the population, using random sampling, or some other method of equilibrating the samples to the parameters of the population.
5. Assign one sample to the Experimental condition, and if you have a second sample, assign it to the Control condition, representing the untreated population.
6. Apply your experimental treatment.
7. Measure the phenomena for each sample taken.
8. Compare the expected value of the Experimental sample, with the population expected value represented by the Control sample. Use the appropriate sampling distribution that accounts for the scale of measure, sample size, and statistic estimating the population parameter.
9. Assess the cumulative probability that the population expected value has in the sampling distribution of the statistic, given a sample size *n*.
10. Make a decision whether or not the cumulative probability of the difference in your sample statistic and population parameter is likely to have happened as a function of chance, or whether it is likely enough to be a *significant difference*.

A **Null Hypothesis:** is a statement, put in mathematical form, that there is no difference between the expected value of measured phenomenon, and that of a reference population from which the phenomenon is supposed to have been drawn. It is usually specified as an equivalence relation between the two expected values:

$H_0: \mu_{\overline{x}} = \mu$

but it can also be directional, depending on if the expected value of the phenomenon, being measured under just random chance is expected to be less than or equal to the expected value of the reference population,

$H_0: \mu_{\overline{x}} \leq \mu$

or if the phenomenon being measured is expected to be greater than or equal to that of the population due to random chance.

$H_0: \mu_{\overline{x}} \geq \mu$

An **Alternative Hypothesis**, also called a **Working Hypothesis**, is a statement of the logical converse of the Null Hypothesis. It is, colloquially assumed that the Alternative Hypothesis is what is *really* expected by the researcher, being the most viable explanation of any systematic difference found between the expected value of the sample from that of the referent population. Because it is the logical converse of the Null Hypothesis, it can also be expressed as either bi-directional,

$H_1: \mu_{\overline{x}} \neq \mu$

or directional if the treatment is expected to *systematically* increase the value of x compared to the population,

$H_1: \mu_{\overline{x}} > \mu$

or if the treatment being measured is expected to *systematically* reduce the value of x compared to the population

$H_1: \mu_{\overline{x}} < \mu$

Following this procedure, we are looking to reject H_0 by looking at the distance, and seeing how big the difference is relative to the standard error of the sampling distribution. The Z-test is the simplest form of this hypothesis test. Since we don't know, we use as its estimate:

$$Z = \frac{\overline{x} - \mu}{\frac{\sigma}{\sqrt{n}}}$$

See how Z is just a distance, scaled in standard deviation units on the Normal distribution? If the magnitude of Z is a relatively large value, then the probability we could have gotten as a function of random chance is very low, so we can reject H0, and accept H1 as the best alternative explanation. Applying this to the wind turbine problem we started the chapter with, we have the following:

$$\mathbf{H_0:} \mu_{\overline{x}} \geq \mu$$

this is the logical complement of our Alternative Hypothesis. If our experiment didn't come out the way we anticipated, then this is what we would expect to see. We could alternatively specify $\mathbf{H_0}$ as

$$\mathbf{H_0:} \mu_{\overline{x}} - \mu \geq 0$$

$$\mathbf{H_1:} \mu_{\overline{x}} < \mu$$

because we are trying to create a turbine that bends *less* than the population average. Likewise, we could rewrite $\mathbf{H_1}$ as

$$\mathbf{H_1:}\ \mu_{\overline{x}} - \mu < 0$$

We collected our data, and came up with the following as estimates:

$$\mu = 1m, \sigma = 0.24m$$

$$\overline{x} = 0.9m, n = 36$$

With these, we can compute our Z-test:

$$Z = \frac{0.9 - 1}{\frac{0.24}{\sqrt{36}}}$$

$$Z = \frac{-0.1}{0.04}$$

$$Z = -2.5$$

So, $\overline{x} - \mu$ is -2.5 standard deviations away from 0, the value we would expect for $\mu_{\overline{x}} - \mu$ if the null hypothesis were true. The probability of this obtaining a Z of -2.5 σ away from the mean, or lower is:

$$P(Z < -2.5) \quad = \quad 1 \; - \; P(Z \geq -2.5) \quad = \quad 0.0062\,.$$

This is very small, only 6 in 1000 samples of size 36 would be this far away from the population mean just by random chance. The best alternative to thinking that our sample behaves the same as the population is to assume that $\mu_{\overline{x}} < \mu$, reject H_0, and accept H_1 as the best alternative.

Hooray, we are pretty sure our design worked and that we are going to make lots of money with it! *But isn't there a chance we might be wrong?* Oh yes, yes there is...

8.2 The Consequences of Being Wrong: α, β, and Type I, and Type II errors

Now, because we use probability distributions to model all this variation in data, even if we get a huge magnitude of Z, meaning that $\overline{x} - \mu$ is a big difference relative to the variability in the system, there is still some small probability that we *could* have gotten this big difference solely as a function of random chance. Our turbine data shows that this probability for $Z < -2.5$ is equal to about 6 in 1000, so it is possible that we may have made an error because our sample was somehow biased, or because of some systematic measurement error. How comfortable are we with our judgment? That is a question every engineer has to make whenever s/he analyzes data to make a decision that might cost money, hours, labor, or reputation. Table 8.1 illustrates the results of a hypothesis test. You can see that there are two major errors that can be just due random chance.

		Your Decision	
		Reject H_0	Fail to Reject H_0
In the Real World	H_0 is False	*Correct Decision* **Confidence** $(1-\alpha)$	**Type II error** (β)
	H_0 is True	**Type I error** (α)	*Correct Decision* **Power** $(1-\beta)$

TABLE 8.1: Consequences of hypothesis tests. Incorrectly rejecting a null hypothesis is a Type I error. Its probability is symbolized α. Failing to reject the null hypothesis is a Type II error. Its probability is symbolized β

In our wind turbine example, we have a small, but possible chance that our design didn't in fact improve the performance of the turbines, but that, just do to random chance we got 36 turbines blades that didn't really reflect the characteristics of the population of blades we manufactured. If this were the case in the real world, H_0: $\mu_{\overline{x}} - \mu \geq 0$ would be true, and H_1: $\mu_{\overline{x}} - \mu < 0$ would be false. So, the magnitude of our Z-value is large due to random chance, and not our design—an error. This is an example of a **Type I error**. We also call Type I errors, *false positives*, because we get a positive result (rejecting the null hypothesis in favor of our alternative hypothesis), but the positive result is not true.

If, however, our design is excellent, our blades *would* actually improve the efficiency of the turbines, but we had obtained a Z-value at or near zero, that is a **Type II error**. These are often called *false negatives* because we didn't detect a true effect of our design (See Figure 8.3).

The problem with errors is that there are real consequences associated with them. Saying a new blade improves the efficiency of a wind turbine when it doesn't

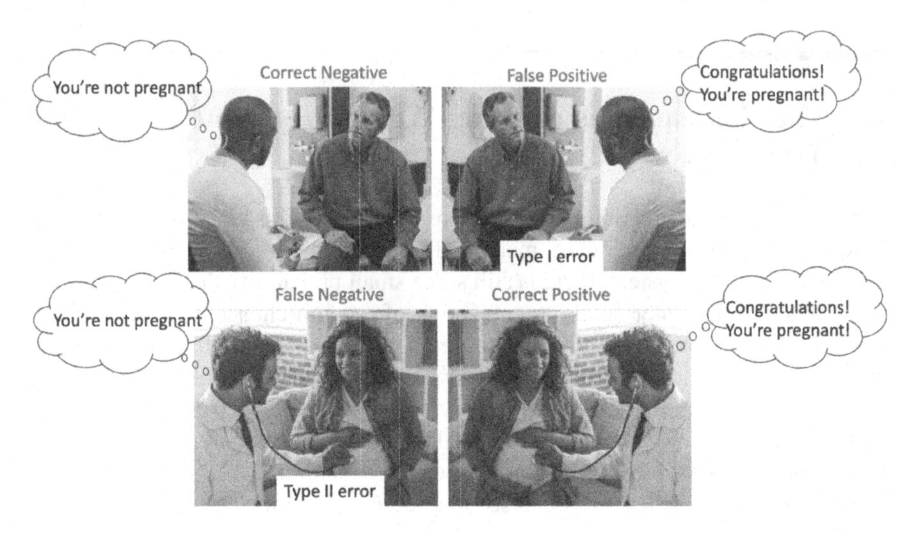

FIGURE 8.3: Types I and II errors illustrated in a ridiculous context. adapted from https://towardsdatascience.com/understanding-confusion-matrix-a9ad42dcfd62

(Type I error), or saying the new blade doesn't improve efficiency when it does (Type II error) each can impact your business, and the subsequent use, customer satisfaction, and repeat business you may have. In many cases, we work on projects that have potentially life-threatening, or life-enhancing consequences. These errors, therefore, are to be avoided if at all possible.

1. What are the potential consequences of making a Type I error in our wind turbine case?

   If we falsely conclude that our turbines perform better than the population in general, we may begin manufacturing them at scale, replacing them in wind farms, *with no demonstrable improvement in performance!* This may cost in the millions of dollars, and have associated reduction in customer satisfaction, lost revenue, litigation, etc.

2. What are the potential consequences of making a Type II error?

   A false negative—when our turbine blades really are superior, but our test didn't detect a big enough improvement to clearly distinguish the effect from random chance—results in our failure to manufacture new blades, resulting in no extra revenue, and in no improvement of our ability to harvest energy from wind. On the other hand, it does no harm! We aren't making the situation worse, we are just failing to capitalize on potential opportunity.

   The point here is, we have to deal with these possibilities, even when we plan and execute data collection and hypothesis testing well. As they say, sometimes "Shit happens."

Type I error is falsely rejecting a Null Hypothesis. It is often referred to as a "False Positive" result. The probability of making a Type I error is symbolized as α. It can be thought of as the probability that the difference observed *is* a function of random chance.

Type II error is failing to reject a Null Hypothesis when it is in fact false. A Type II error is often termed "False Negative" result. The probability of making a Type II error is symbolized by β. β can be thought of as the probability that the difference observed is too small to detect a difference that actually exists.

8.2.1 Type I Error Rate: α

Theoretically, the probability of making a False Positive decision in a hypothesis test ranges, like all probabilities, from 0, meaning there is no chance whatsoever, that you would falsely reject the null hypothesis, to 1, meaning that no matter what you do, you are doomed. In reality, we have the ability to hedge our bet, so to speak, by setting the largest α we will tolerate, and in doing so, we can then mess around with sample size to get the right n that gives us enough **power** to detect a significant difference in our sample mean from the status quo population mean. The trick is choosing this α so that it doesn't restrict us too tightly, because, in general, the smaller the α , the larger the difference in our sampling distribution and the population mean has to be able to reject H_0.

In Figure 8.4, you can see that if we had a Z-value of +/− 1.96, the cumulative probability between these values is 0.95. What this means is, 95% of all the samples taken with a given size n, taken from a population with $\mu = \mu_{\bar{x}}$ and $\sigma = \frac{\sigma}{\sqrt{n}}$ would have a difference of less than +/− 1.96 σ. The Type I error rate for this sampling distribution is $1 - 0.95 = .05$. Notice how α is the cumulative probabilities in the tails of the distribution $\alpha = p(Z < 1.96) + p(Z > 1.96)$. If we reduce our α, we make these tails smaller, thus increasing the distance our sample mean must be from the population mean to be considered small enough that we would tolerate. If we went to the extreme, where α gets very close to zero, the difference our sample statistic would have to be would be infinite. The only way to do this is to increase n astronomically, thus making our hypothesis test completely impractical due to cost (time, human resources, and money). So, we have to choose an α that is reasonable, given the severity of making a Type I error, and an n that will give us enough **power** to detect that difference in means, at a minimum.

How do we reduce our chances for making such decision errors and increase the probability that our decisions will be correct? We mess around with sample size!

It turns out that the **power** to detect a significant difference in our sample mean

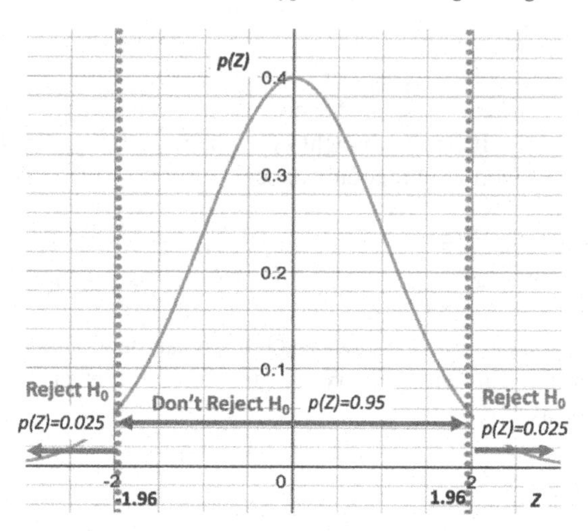

FIGURE 8.4: Probability intervals for $\alpha = 0.05$. If the Z-test value falls in the region where $p(Z) = 1 - \alpha$, the mean difference is considered too small to be significant. If the Z-test value falls in either of the tails, the mean difference is large enough to be considered highly improbable if H_0 is true, and therefore, we choose to reject H_0 in favor of H_1

from the status quo population mean, all else being equal, is a function of sample size. Recall that, for the Z-test, the standard error of the mean, our index of the width of the sampling distribution of X is inversely proportional to $\sqrt{n}$. The larger the sample size, the more precise the sampling distribution is. This narrowing down of the sampling distribution, relative to the distance between its mean and the population mean, is an index of power—the probability that the experiment will produce a true rejection of the null hypothesis. Looking at the z-test algebraically we can see this:

$$Power = 1 - \beta = p\left(\overline{x} \neq \mu | \mu\right)$$

$$\sim P\left(Z_{1-\alpha=0.95} \neq \frac{\overline{x} - \mu}{\frac{\sigma}{\sqrt{n}}}\right)$$

The rule of thumb you should take from this relationship is that, by increasing n you can improve your chances of detecting a real difference–a real *effect*–of your treatment. The Z in this relationship is the **critical value**, the value of the Standard Normal Distribution that demarcates the middle $(1 - \alpha)\,\%$ from the tails $(\alpha)\,\%$. In Figure 8.4, we used +/– 1.96, the Z where $\frac{\alpha}{2}$ is 0.025, but **you** need to choose the appropriate α for the situation you are studying. There are no hard and fast rules for this. If you are working on kidney dialyses equipment, you might want to be conservative and choose an α of 0.01 (Z = +/−*2.33)* or even 0.005 (Z = +/−2.58), because making a false positive decision could critically harm a patient using your equipment. If you are just exploring a relationship and you aren't going to build some life-altering technology with the information you gather, you might use an α as high as 0.1 (Z = +/−1.28). The point is, ***you have to make this ethical decision, weighing the potential consequences, both positive and negative of your experimental design.***

> The **Critical Value** is the value of the test statistic that divides a sampling distribution into two regions with probability α, and $1-\alpha$. As an example for a Z-test, dividing Z-distribution into two regions $\alpha = 0.05$ in the left tail, and $1 - \alpha = 0.95$ in the remainder of the distribution, is −1.645. As the Type I error rate gets smaller, its critical value grows smaller in magnitude moving further towards the tails of a distribution.

> The **Rejection Region** is interval of a distribution wherein the cumulative probability is equal to α. Typically this interval is placed in the tails of distributions indicating a large relative distance between a population mean and the mean of a sampling distribution.

One of the most commonly employed Type I error rates is $\alpha = 0.05$, or a one in 20 chance that a rejection of the Null Hypothesis is in error, and is in fact a False Positive.

In Figure 8.5, we see that any mean difference that is greater than 1.645 standard errors below the mean has a probability that is less than 0.05, and is considered improbable enough to be considered likely a real difference. As I emphasize, α depends on the reference population, reflects the null hypothesis, and is set *a priori*. Once α is fixed, β is also fixed for any given difference in means and given sample size. Knowing the difference in means—the effect we want to make—we can alter the sample size to change our power using just a little Algebra (see figure 8.6).

Because you will know μ and σ from the population, you have to decide two things: 1) What is the *smallest* detectable difference, $\mu_{\bar{x}} - \mu$, I will accept as being practically significant?, and 2) what is the *largest* likelihood of making a False Positive (α) will I accept? With these parameters, n becomes the only variable, narrowing the width of the sampling distributions as n increases to the point where the probability of making a correct decision is optimal. The following example will illustrate this for the Z test.

3. You are designing new chassis for the SAE baja car. You know that 1060 steel will provide the strength and rigidity you need, but you want to cut weight, and therefore are considering 7000 series Aluminum. You feel that the minimum

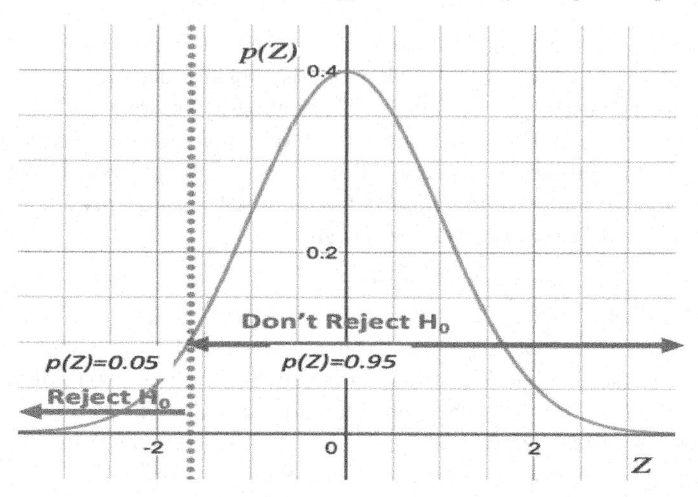

FIGURE 8.5: Rejection region for $\alpha = 0.05$

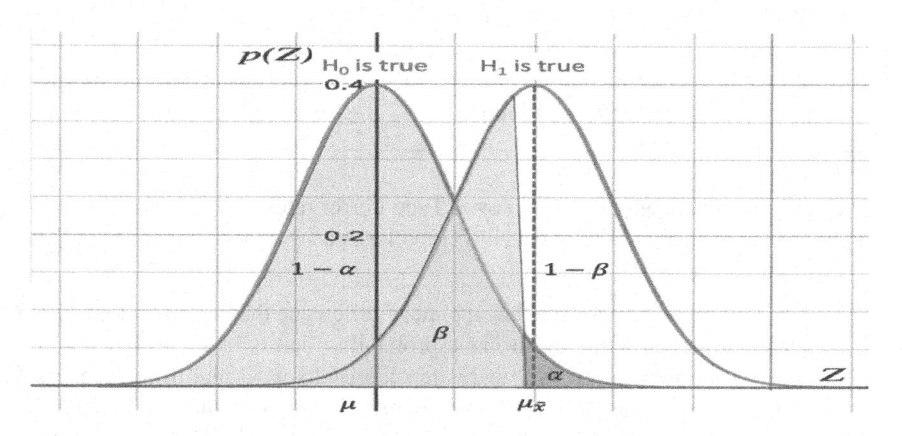

FIGURE 8.6: Null and alternative distributions illustrating Type I error rate (α), type II error rate (β), confidence $1 - \alpha$, and power α

yield strength you will need to still have a rigid enough frame, but with enough toughness to not break apart on the track, is 469 Mpa. You look up the yield strength of 1060 steel and find it to be 485 Mpa +/− 40 Mpa.

(a) State the null hypothesis and alternative hypothesis for your question.

(b) What is an appropriate α for this study, given consequences you make a False Positive Decision?

(c) What is the probability of making a correct rejection of the null hypothesis?

The working hypothesis I would use for this experiment is that the samples of Al are somehow different in yield strength than the 1060 steel. Since I want to know if the AL is significantly less strong than the steel, I have a *uni-directional* or **one-tailed** hypothesis.

$$H_1: \mu_{\overline{x}} \quad < \quad \mu$$

Given this is the real question I am asking, the Null Hypothesis, the hypothesis we are really going to test, is its logical converse:

$$H_0: \mu_{\overline{x}} \quad \geq \quad \mu$$

α for this study is, again, your choice. A good rule of thumb is to begin thinking about α at about 0.05, or 5%. This gives you a 1 in 20 chance of making a Type I error, and usually doesn't sacrifice power too much (or alternatively, require a costly, large sample). Let's use this value for our power estimate:

To figure out my power, I just use the Z-test. I am trying to figure out the sample size needed to obtain a mean difference of −16, Mpa, given a σ of 40. If I compare this to the Z-value at $\alpha = 0.05$, we can see that this value is −1.645 standard errors from the mean of the population if H_0 is true. If H_1 is true, −1.645 is at a different location in the sampling distribution of $\overline{x}$. If we can find its value in the H_1 distribution, then we can integrate from $-\infty$ to that value to get β.

$$1 - \beta = p\,(RejectH_0|H_1)$$

$$1 - \beta = p\left(Z_{1-B} \leq \frac{\mu_{\overline{x}} - \mu}{\frac{\sigma}{\sqrt{n}}} +/- Z_\alpha\right)$$

You can see that this probability is just the difference between the computed Z-test value and the critical value of Z_α.

In our example, we have the following:

$$1 - \beta = p\left(Z_{1-B=} \leq \frac{469 - 485}{\frac{40}{\sqrt{n}}} - (-1.645)\right)$$

Now, if we fix n, our sample size, we can estimate our probability of correctly rejecting the null hypothesis—our power. Let's say we have a sample size of 25.

$$1 - \beta = 1 - p\left(Z_{1-B} \leq \frac{469 - 485}{\frac{40}{\sqrt{25}}} + (1.645)\right)$$

$$1 - \beta = 1 - p\left(\frac{-16}{8} + 1.645\right)$$

$$1 - \beta = 1 - p(-2 + 1.645)$$

$$1 - \beta = 1 - p(Z_{1-B} \leq -0.365)$$

Looking up this value, or integrating the Z-distribution from $-\infty$ to -0.365, we get a probability of

$$1 - \beta = 0.635$$

So, with only 25 measurements of Al, our power to detect a difference of 16 MPa, given our Type I error rate is set at $\alpha = 0.05$, is only 65%. This is way too low! I want a power of at least 80% or so to make my expenditure of time and money worthwhile. So, my only alternative is to increase my sample size. If I increase n to 36, I get

$$1 - \beta = 0.775$$

Much better! So, just increasing by 11 measurements gives me a fighting chance at detecting a real difference in my Al sample versus Steel. If I further increased the sample to 49, my power is 0.876. This will cost more, then I will have to purchase and process more Al, but it will give me good power so that I don't commit a Type II error.

The key here is to think of these concepts, Type I error rate, Type II error rate, Confidence and Power, as an optimization problem. There is a sweet spot with sample size that gives you enough Power to detect a real difference, is cost-effective, yet still keeps the Type I error rate at a low enough percentage to feel comfortable with the results of rejecting the Null Hypothesis.

8.3 How Many Tails? Or Knowing Your Ass From the Hole in the Ground

You will have noticed that some of the hypotheses we have explored in our examples have been kind of vague: I want to find out if my sample has a *different* mean than the population. These kinds of hypotheses are called, **Two-tailed**, because we accept any mean value that falls into either tail of the sampling distribution (beyond the critical value) as being significantly different from the mean of the sampling distribution. Other hypotheses are called, **One-tailed**, because we have good evidence from

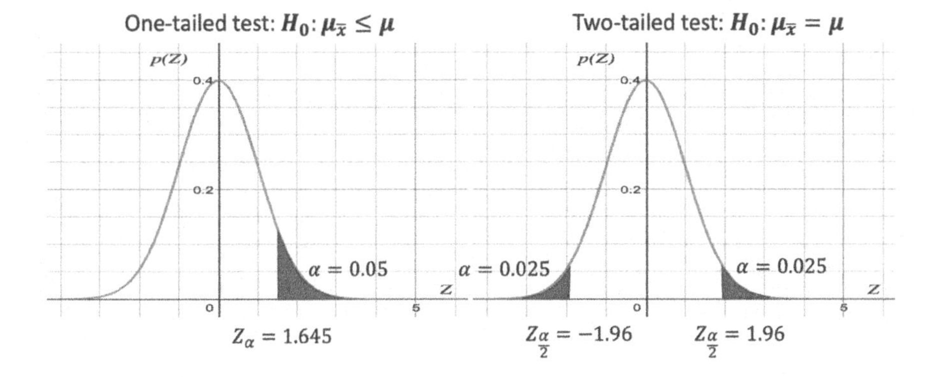

FIGURE 8.7: The region of rejection for a one-tailed test has a smaller critical value, Z_α , than those for a two-tailed test $-Z_{\frac{\alpha}{2}}$ and $+Z_{\frac{\alpha}{2}}$

the physics or materials science, or the psychology, that the proposed treatment will *improve* the performance of our sample. Improvement can be focused on the *upper tail* of the distribution as in increasing the horsepower of an automobile. Improvement can also be focused on the *lower tail* of the distribution, as in reducing the fuel consumption of an automobile.

I often call two-tailed hypotheses, "Ass from a hole in the ground" tests. If you don't know your ass from a hole in the ground, it is a good idea to figure out the two if you are designing a product to apply to one or the other. Similarly, if you don't know if a treatment will improve or inhibit the performance of your variable, you should perform a two-tailed test.

If, however, you have reason to believe that your treatment will almost assuredly influence a phenomenon in a positive or negative direction, you should perform a 1-tailed test. Figure 8.7 will illustrate why.

The critical value for a one-tailed test will be closer to the mean of the sampling distribution, than critical values for a two-tailed test, given α for the test is fixed. This gives more power to detect a significant difference if there is little to no possibility that the treatment condition would have the opposite effect of that expected under H_1. The tricky thing for the analyst to keep in mind, is for a 1-tailed test, to specify H_0 and H_1 properly. A good rule of thumb, in my experience, is to sketch the distribution and draw the region of rejection before writing out the formal hypotheses. That way, if you have a left-tailed hypothesis, you will remember that your critical value will be negative, not positive.

A Two-tailed Test is a hypothesis test that splits the Type I error probability into two regions: One in either tail of the distribution. Generally speaking we split α into two equal regions, each with cumulative probability $\frac{\alpha}{2}$. This test is bi-directional:

$H_0 : \mu_{\overline{x}} = \mu$

$H_1 : \mu_{\overline{x}} \neq \mu$

A One-Tailed Test is a hypothesis test that locates all of the Type I error probability in either the left *or* the right tail of the distribution. This hypothesis is directional:

One-tailed test, right-tailed. Critical value is positive:

$H_0 : \mu_{\overline{x}} \leq \mu$

$H_1 : \mu_{\overline{x}} > \mu$

One-tailed test, left-tailed. Critical value is negative:

$H_0 : \mu_{\overline{x}} \leq \mu$

$H_1 : \mu_{\overline{x}} > \mu$

8.3.1 Confidence Intervals for the One-sample Z-test

There are two approaches to testing hypotheses which are mathematically equivalent, they will get you the same results, every time but that have slightly different procedures. We have already been introduced to the first, the Test Statistic method. This is where we calculate a value of Z, and then compare it to the critical value of Z_{α}. If $Z_{compute}$ falls out further in the tails of the distribution than Z_{α}, we say it represents a *significant difference* between the mean of the population our sample represents, and the referent population:

$$Z = \frac{\overline{x} - \mu}{\frac{\sigma}{\sqrt{n}}}$$

But as we saw in Chapter 5, we can transform any value of x into a Z-score. We can also transform a Z-score into the units of the independent variable (eg., Mpa). This is simply multiplying Z by the standard error, dilating the distribution back to the original width, then adding it to the mean of the sample to re-center the distribution about $\overline{x}$. In doing so, we demarcate a region that represents the **confidence interval**, the region where the mean of $1 - \alpha$ % of all samples drawn from the sampling distribution would likely fall, if they were drawn at random. Any values outside this confidence interval are considered highly unlikely. So, if μ is not found in the confidence interval around the mean, it is unlikely that the sampling distribution was

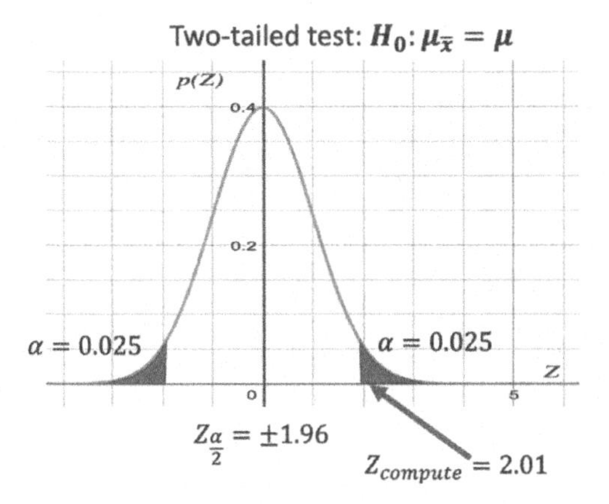

FIGURE 8.8: When $Z_{compute}$ falls in the rejection region of the sampling distribution, because its probability is less than 0.05 of occurring as a function of chance, considering the null is true, we reject H_0 in favor of H_1

taken from a population with μ as its center. The equations below show these intervals for the Z-distribution (See Also Figure 8.8).

$$\textbf{Two-tailed: } CI = \overline{x} \pm Z_{\frac{\alpha}{2}} \frac{\sigma}{\sqrt{n}}$$

$$\textbf{One-tailed (right): } CI \leq \overline{x} + Z_{\alpha} \frac{\sigma}{\sqrt{n}}$$

$$\textbf{One-tailed (left): } CI \geq \overline{x} - Z_{\alpha} \frac{\sigma}{\sqrt{n}}$$

1. You are studying the extent to which the frozen smoothie dispenser you have designed accurately fills the machine's cups. Past research on your company's dispensers show that they fill an average of 12.6 fluid oz +/− 1.2 fluid oz. You have taken a random sample of these machines and outfitted them with a new nozzle. Taking a sample of 16, you find the mean number of ounces each cup is filled is 12.0 Does your dispenser systematically pour less fluid, compared to the old machines?

First, we need to know if we have a 1-tailed or 2-tailed test. Since we are only interested in whether or not the new dispenser is underfilling cups, this is a one-tailed (left-, or lower-tail) test.

$$1 - \alpha\%CI : \mu \geq \overline{x} - Z_{\alpha} \frac{\sigma}{\sqrt{n}}$$

Just to make sure I understand the test, I draw a picture (See Figure 8.9):

We can see, in this picture, for the Z-distribution, the right tail is the region of

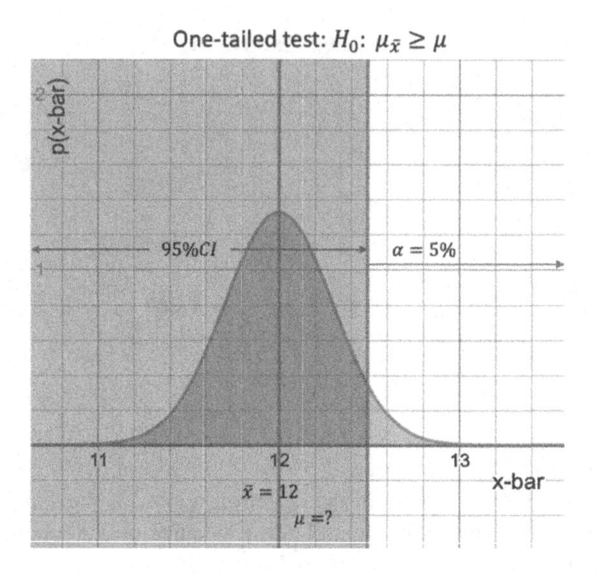

FIGURE 8.9: Drawing representing the confidence interval and rejection region for smoothie dispenser problem

rejection. That means that the confidence interval is in the $1 - \alpha$ direction (to the left of the rejection region). Again, I would choose $\alpha = 0.05$ as my Type I error rate, because the consequences of making a false positive decision are not major. Given I know the critical value $Z_a = -1.645$, I can transform this distribution into units of the independent variable, fluid ounces.

$$1 - \alpha\%CI : \mu \leq \bar{x} - Z_\alpha \frac{\sigma}{\sqrt{n}}$$

$$1 - \alpha\%CI : \mu \leq 12.0 - \left(-1.645 \frac{1.2}{\sqrt{16}}\right)$$

$$1 - \alpha\%CI : \mu \leq 12.4935$$

This distribution is the *same* as that showing our rejection region in Z-units. This, however is just rescaled and translated to the same scale, fluid ounces, as our original problem (see Figure 8.10). Our population $\mu = 12.6$ fluid oz. This value falls outside the 95% confidence interval and therefore can be considered less than 5% probable that we could have drawn this sample from a population with a mean of 12 and standard deviation of 1.2 fluid ounces. This indicates that our machine is systematically underfilling cups relative to our intended design.

Just to prove that this method is equivalent to the Test Statistic method for the Z-test, here are the calculations:

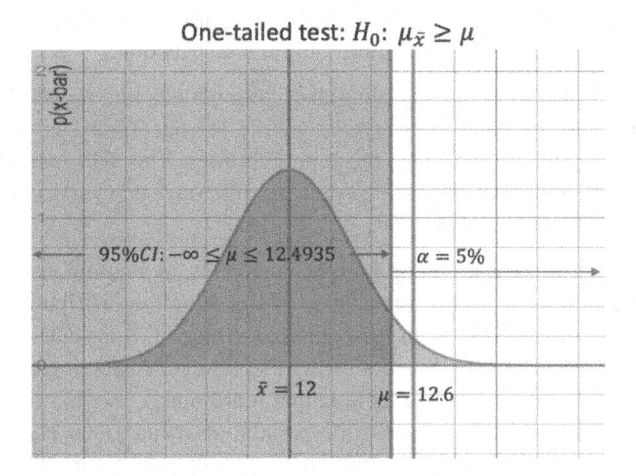

FIGURE 8.10: The population mean, μ, lies above the $1 - \alpha\%CI$, indicating that it is unlikely that our sample could have been randomly drawn from a population with a center of 12.6 and a standard deviation of 1.2 fluid ounces

$$Z = \frac{\overline{x} - \mu}{\frac{\sigma}{\sqrt{n}}}$$

$$Z = \frac{12.0 - 12.6}{\frac{1.2}{\sqrt{16}}}$$

$$Z = -2$$

Since 2 is further in the tail than the critical value $Z_a = -1.645$, we can reject the null hypothesis and assume your dispenser is systematically dispensing less than the old machines.

8.4 Summary of Z-test

The Z-test is the model that I want you to have in mind when thinking of any basic hypothesis test. The logic of other, more complicated tests is the same. All we have to do is figure out what the appropriate reference distribution is, create a sampling distribution of our statistic, and compare a population parameter with a critical value that represents our Type I error rate, in that sampling distribution.

Once we know what our sampling distribution looks like (generally a Normal distribution, or some close variant), we should set our Type I error rate and estimate our power *a priori*, based on those probabilities. This will insure that our test is designed well, with enough power to detect a difference that we deem practical and useful, but also that is cost effective.

We can test hypotheses using both test statistics, and confidence intervals. The critical value of the test statistic is a value of the sampling distribution that demarcates the region of rejection from the confidence interval. It is scaled in Z-units (for the Normal distribution), or in units of the theoretical distribution that models our PDF. The confidence interval is that area that represents $1 - \alpha$ %, the region of the distribution where we expect to find the population mean given H_0 is correct. Confidence intervals are scaled in the units of the independent variable, allowing for direct comparison of population parameters. If the population mean is not found within the confidence interval, then we can reject H_0 in favor of H_1.

8.5 One Sample *t*-test

I must admit at this time, I have pulled a fast one on you... you will probably never use a Z-test in your lifetime (or rarely at best). We just almost never know the actual population parameters of what we are testing. This is particularly true of the population variance. We can theoretically set a population mean by assuming, under the null hypothesis, that the difference in the average measurements of our treatment group and our control group will be zero. So, under the null hypothesis, the mean of the *population of differences between groups*, is theoretically zero. But even with this, we have a hard time creating a justifiable estimate of the population variance without some data to help us estimate it. That is where beer comes into the picture!

8.5.1 Guinness and the Invention of *t*

October 1899 was an auspicious beginning at the Arthur Guinness, Son & Co. brewery in Dublin, Ireland. It was at that time that a newly-graduated chemist from Oxford, William Seeley Gosset became a brewer. The company had taken note of the recent advances in chemistry and biology, and decided to introduce scientific methods into brewing. In a fit of sanity, they hired a slew of young, bright minds into junior management to effect this vision. Gossett was concerned with quality improvement, and so he had an engineer's heart. In particular, he wanted to be able to figure out how to use small samples to make claims about big phenomena. At the time, most

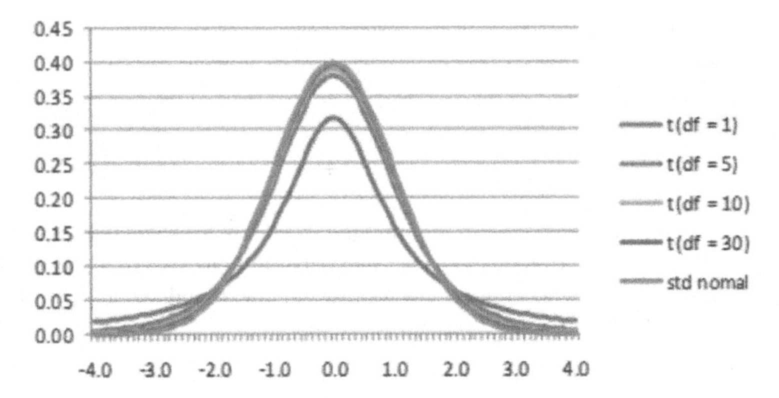

FIGURE 8.11: *t*-distributions for different degrees of freedom. Notice how at about *df*=30, the distribution is very close to the normal distribution. In fact, like the binomial and Poisson distributions, as $df \to \infty$, *t* becomes closer and closer to Z.

statistical thinkers worked in the world of economics, and had access to large groups of people, plants or animals to experiment on. Gossett worked on the problem of small samples, and in particular, the problem that he almost never had a population variance with which to compute the value of z! This work led to the discovery of a new distribution, one that is based on the Normal, but that uses a sample variance to estimate that of the population. In doing so, Gossett had to deal with the issue of degrees of freedom, because sample size became an important new parameter in this distribution. Since his work in the early 1900s, Gosset's *t* distribution—*Student's t,* it was called because Guinness forbade him to use his surname in any publications, afraid that their trade secrets would be stolen—would become the most cited and utilized statistical test in the history of the field. Is there no problem beer cannot solve?

8.5.2 The One-sample *t* Distribution

The *t* distribution can be conceptualized with exactly the same logic as the z-distribution, with just some minor differences. As a PDF, its graph is bell-shaped, centered on μ . But its width is a bit narrower than the normal distribution, and its tails are a bit fatter (See Figure 8.11). As a PDF, it serves the same function as the Z, if we can describe the distance between a the mean of a sampling distribution and that of the population from which it is supposed to have been sampled, given its variability, we can integrate from $-\infty$ *to t* to find the probability that the two means could be that far apart or greater, just due to random chance (i.e., if H_0 is true).

The reason these subtle differences exist is in the relative variability of samples used to estimate a population standard deviation, σ. For the Z-test, because the population μ and σ are known ahead of time, we do not need to estimate their values.

We only need to compare μ to $\overline{x}$ and see how many standard errors apart they are. For the ***t*-test**, we don't know σ so we have to estimate it using *s*.

$$t_{(\alpha,df=n-1)} = \frac{\overline{x} - \mu}{\frac{s}{\sqrt{n}}}$$

Notice that the one sample *t*-test is computed in exactly the same way as the one-sample z-test. The ONLY difference in the *z* and *t* distributions, in terms of procedure, is that we have to use our best guess, *s* as a substitute for σ, and account for the size of the sample by looking up the value of *t* for *n-1* degrees of freedom.

Likewise, the procedure for determining the confidence interval for *t* is exactly the same as that for Z:

$$1 - \alpha\%CI = \overline{x} \pm t_{(\alpha,df=n-1)}\frac{s}{\sqrt{n}}$$

It is the interval in the *t*-distribution that, under the null hypothesis, should contain the population mean. Again, like the confidence interval for *Z*, it is scaled in the original units of the independent variable.

> **Student's *t*-test** is a means of hypothesis testing, assessing the difference in two mean values: The mean of a sampling distribution (estimated by the sample mean), and the mean of a population. It differs from the Z-test, in that it utilizes the sample standard deviation, *s* as an estimate of the population σ. The *t*-distribution has the following parameters:
>
> μ estimated by the sample mean
>
> σ estimated by a function of *df*
>
> Since there are two parameters needed to fix *p*(*t*), for any given degree of freedom, and one of those parameters is estimated (*s*), the degrees of freedom are reduced to *n*-1.

Let's look at an example that will illustrate these similarities in *t* vs *z*, but highlight the important differences that make *t* more useful, in general.

1. A researcher is studying the braking distance of newly designed brake calipers. She tests a random sample of 30 brakes under controlled conditions. She gets a stopping distance of 32 meters, with a standard deviation of +/– 5.2 meters. She wants to compare her sample to published reports of typical brake performance under similar testing conditions. The published results show a mean braking distance of 34 meters.

(a) What is the appropriate test the researcher should conduct to determine if her brakes perform differently than the general population of brakes? Justify your answer

(b) Perform the appropriate test you selected in a. Use confidence interval approach, and interpret your results. Assume alpha is .05.

Here is how we go about testing the working hypothesis that the new brake calipers shorten the average stopping distance significantly, compared to the status quo calipers. The hypothesis is two-tailed:

$$H_1 : \mu_{\text{new calipers}} \neq \mu$$

$$H_0 : \mu_{\text{new calipers}} = \mu$$

What is the appropriate test the researcher should conduct to determine if her brakes perform differently than the general population of brakes? Justify your answer.

We know the population mean, but not the population standard deviation, so we need to use a one-sample t-test. Stopping distance is a continuous random variable, so the normal distribution is a good assumption, given our reasonably large sample size, so t should be a good model to use.

Perform the appropriate test you selected in a. Use the 2-tailed confidence interval approach, and interpret your results. Assume alpha is .05.

$$1 - \alpha\%CI = \overline{x} \pm t_{(\alpha, df=n-1)} \frac{s}{\sqrt{n}}$$

$$1 - \alpha\%CI = 32 \pm t_{(\alpha, df=n-1)} \frac{5.2}{\sqrt{30}}$$

We now have to find the critical value of $t_{(0.05, df=29)}$. Figure 8.12 shows values of t for commonly used Type I error rates and degrees of freedom. In the t-table, we first select α a priori, then we split it into $\frac{\alpha}{2}$ =0.025 for each tail of the two-tailed hypothesis. Then, knowing our degrees of freedom ($n-1 = 29$), we find the critical value of t is +/−2.043 (Figure 8.13).

Substituting our critical value into the Confidence Interval we find

$$1 - \alpha\%CI = 32 \pm 2.043 \frac{5.2}{\sqrt{30}}$$

$$1 - \alpha\%CI = 32 \pm 1.939$$

$$1 - \alpha\%CI = [30.060, 33.939]$$

Degrees of Freedom	Type I Error Rate				
	$\alpha = 0.001$	$\alpha = 0.01$	$\alpha = 0.025$	$\alpha = 0.05$	$\alpha = 0.10$
1	318.3088	31.8205	12.7062	6.3138	3.0777
2	22.3271	6.9646	4.3027	2.9200	1.8856
3	10.2145	4.5407	3.1824	2.3534	1.6377
4	7.1732	3.7469	2.7764	2.1318	1.5332
5	5.8934	3.3649	2.5706	2.0150	1.4759
6	5.2076	3.1427	2.4469	1.9432	1.4398
7	4.7853	2.9980	2.3646	1.8946	1.4149
8	4.5008	2.8965	2.3060	1.8595	1.3968
9	4.2968	2.8214	2.2622	1.8331	1.3830
10	4.1437	2.7638	2.2281	1.8125	1.3722
11	4.0247	2.7181	2.2010	1.7959	1.3634
12	3.9296	2.6810	2.1788	1.7823	1.3562
13	3.8520	2.6503	2.1604	1.7709	1.3502
14	3.7874	2.6245	2.1448	1.7613	1.3450
15	3.7328	2.6025	2.1314	1.7531	1.3406
16	3.6862	2.5835	2.1199	1.7459	1.3368
17	3.6458	2.5669	2.1098	1.7396	1.3334
18	3.6105	2.5524	2.1009	1.7341	1.3304
19	3.5794	2.5395	2.0930	1.7291	1.3277
20	3.5518	2.5280	2.0860	1.7247	1.3253
21	3.5272	2.5176	2.0796	1.7207	1.3232
22	3.5050	2.5083	2.0739	1.7171	1.3212
23	3.4850	2.4999	2.0687	1.7139	1.3195
24	3.4668	2.4922	2.0639	1.7109	1.3178
25	3.4502	2.4851	2.0595	1.7081	1.3163
26	3.4350	2.4786	2.0555	1.7056	1.3150
27	3.4210	2.4727	2.0518	1.7033	1.3137
28	3.4082	2.4671	2.0484	1.7011	1.3125
29	3.3962	2.4620	2.0452	1.6991	1.3114
30	3.3852	2.4573	2.0423	1.6973	1.3104
40	3.3069	2.4233	2.0211	1.6839	1.3031
60	3.2317	2.3901	2.0003	1.6706	1.2958
100	3.1737	2.3642	1.9840	1.6602	1.2901
∞	3.0902	2.3263	1.9600	1.6449	1.2816

FIGURE 8.12: Values of student *t* distribution for common degrees of freedom and commonly selected Type I error rates

Since $\mu = 34$ is not in the confidence interval, we can conclude that our brakes performed significantly differently than the status quo population. In our case, they reduced the mean braking distance an average of about 2m. This mean difference could only happen about 5% of the time, if the null hypothesis were true.

Degrees of Freedom	Type I Error Rate $\alpha = 0.001$	$\alpha = 0.01$	$\alpha = 0.025$	$\alpha = 0.05$	$\alpha = 0.10$
1	318.3088	31.8205	12.7062	6.3138	3.0777
2	22.3271	6.9646	4.3027	2.9200	1.8856
3	10.2145	4.5407	3.1824	2.3534	1.6377
4	7.1732	3.7469	2.7764	2.1318	1.5332
5	5.8934	3.3649	2.5706	2.0150	1.4759
6	5.2076	3.1427	2.4469	1.9432	1.4398
7	4.7853	2.9980	2.3646	1.8946	1.4149
8	4.5008	2.8965	2.3060	1.8595	1.3968
9	4.2968	2.8214	2.2622	1.8331	1.3830
10	4.1437	2.7638	2.2281	1.8125	1.3722
11	4.0247	2.7181	2.2010	1.7959	1.3634
12	3.9296	2.6810	2.1788	1.7823	1.3562
13	3.8520	2.6503	2.1604	1.7709	1.3502
14	3.7874	2.6245	2.1448	1.7613	1.3450
15	3.7328	2.6025	2.1314	1.7531	1.3406
16	3.6862	2.5835	2.1199	1.7459	1.3368
17	3.6458	2.5669	2.1098	1.7396	1.3334
18	3.6105	2.5524	2.1009	1.7341	1.3304
19	3.5794	2.5395	2.0930	1.7291	1.3277
20	3.5518	2.5280	2.0860	1.7247	1.3253
21	3.5272	2.5176	2.0796	1.7207	1.3232
22	3.5050	2.5083	2.0739	1.7171	1.3212
23	3.4850	2.4999	2.0687	1.7139	1.3195
24	3.4668	2.4922	2.0639	1.7109	1.3178
25	3.4502	2.4851	2.0595	1.7081	1.3163
26	3.4350	2.4786	2.0555	1.7056	1.3150
27	3.4210	2.4727	2.0518	1.7033	1.3137
28	3.4082	2.4671	2.0484	1.7011	1.3125
29	3.3962	2.4620	2.0452	1.6991	1.3114
30	3.3852	2.4573	2.0423	1.6973	1.3104
40	3.3069	2.4233	2.0211	1.6839	1.3031
60	3.2317	2.3901	2.0003	1.6706	1.2958
100	3.1737	2.3642	1.9840	1.6602	1.2901
∞	3.0902	2.3263	1.9600	1.6449	1.2816

FIGURE 8.13: Finding the critical value of t at 29 df, and $\frac{\alpha}{2} = 0.025$

We could have also used the t-test statistic to test this hypothesis, even though it was not asked for in the problem.

$$t_{(0.05,df=29)} = \frac{32 - 34}{\frac{5.2}{\sqrt{30}}}$$

$$t = -2.1067$$

Comparing our computed value of t to the critical value from the table, −2.043 (the negative is for the left tail of the distribution), we can see that our mean difference is in the rejection region, and we can reject H_0 and accept H_1 that our newly designed brake calipers made a difference.

In general, it is much more common to NOT have the population standard deviation at hand. We may have a performance standard expressed as a mean value but typically these are not associated with any variability measure. So, the t-test enables us to get on with it and estimate our sampling distribution. For most examples in engineering sample sizes of 30 give us adequate power, but if the parent distribution is clearly normal, and the distribution of the sample is fairly normal, sample sizes as small as 15 may be adequate. Just to be sure, power estimation for the t-test follows the same procedure as the Z. The power for our example is only 0.5306. If we desire more power, we can increase our sample size.

8.6 Summary of Basic Hypothesis Testing

Testing a hypothesis is a process of determining whether or not the characteristics of some sample, drawn from a population, match those of the population after some treatment has been applied.

The key characteristic we use to compare samples to normally distributed populations is the mean. We know the sampling distribution of the mean will tend, over the long haul, to the normal distribution. With this, we use the normal distribution as a model to estimate the probability that we could have drawn a sample with our measured mean just by chance. If the likelihood that the sample could have been drawn from the population just by chance is very low, then it can be concluded that it is more likely that the treatment could have cause this difference.

The hypothesis we actually test with our data is called the Null Hypothesis. This is the logical converse of the Alternative Hypothesis. It is a statement of the status quo—that is, a statement that there is no real difference between the population our treatment sample is measuring, and the population from which it was drawn prior to treatment.

The probability of making a false positive decision, an incorrect rejection of the Null Hypothesis, is called a Type I error. The consequences of these errors are generally more egregious than false negatives (Type II errors), so we try to be conservative fixing a Type I error rate at or lower than 10%, and generally lower than 5%. The actual Type I error rate one selects for making a decision in a hypothesis test should be done a priori, taking into account the severity of the consequences of making such an error.

The power of a hypothesis test is the probability that it will return a true rejection of the Null Hypothesis, when the null is, in fact, false. This means that the power of your experiment is the probability that your hypothesized treatment really did make an effect. Power is directly related to sample size: The larger the sample, the greater the power you have to detect a real difference. To determine power, one must first fix the Type I error rate, the minimum difference in means one will count as a real difference, and the sample size. One can optimize the power and sensitivity of a hypothesis test by adjusting the Type I error, power, and sample size to fit cost, time, and impact constraints.

The *Z*-test is the basic model for all hypothesis tests. This test of differences in means assumes that the difference between the population parameters and sample statistics are purely random prior to treatment. The Z-test differs from the *t*-test procedurally in that, for the Z-test, we know the population mean and standard deviation, while for the *t*-test, we do not know the population standard deviation.

The one-sample *t*-test is a test of mean differences when the population mean is known, but the population standard deviation is not. We then have to resort to using the sample standard deviation as an estimate. We pay a small price in degrees of freedom because of this. As the sample size gets beyond about 30, the *t* and *Z*-tests become so close as to be practically the same.

The actual hypothesis test compares the distance the population mean is from the center of the sampling distribution. The sampling distribution is centered on the sample mean, and gets narrower as a function of sample size. In other words, the standard error of the mean gets smaller and smaller as our sample size gets larger and larger.

There are two methods for testing hypotheses: 1) Making a $1-\alpha$ % Confidence Interval; and 2) computing a test statistic. A confidence interval is the region of the sampling distribution, scaled in units of the independent variable, where we expect to draw samples at random, given the Null Hypothesis is true. If the mean of the population is not found in the confidence interval, it is improbable (at $1-\alpha\%$) that our sample could have been drawn from a population with parameters μ and σ. A test statistic is a value of the PDF (*Z* or *t*, depending on the known parameters) that demarcates the region of rejection from the region where the difference in means would be expected to fall $1-\alpha\%$ of the time given the Null Hypothesis were true. If the calculated value of the PDF falls in the rejection region, this is less than $\alpha\%$ likely to occur as a function of random chance, and we will reject the Null Hypothesis in favor of the Alternative.

8.7 References

Biau, D. J., Jolles, B. M., & Porcher, R. (2010). P value and the theory of hypothesis testing: an explanation for new researchers. *Clinical orthopaedics and related research, 468*(3), 885–892. doi:10.1007/s11999-009-1164-4

Thornton, Stephen, “Karl Popper", *The Stanford Encyclopedia of Philosophy* (Fall 2018 Edition), Edward N. Zalta (ed.), URL = <https://plato.stanford.edu/archives/fall2018/entries/popper/>.

Momeni, F., Sabzpoushan, S., Valizadeh, R., Morad, M. R., Liu, X., & Ni, J. (2019). Plant leaf-mimetic smart wind turbine blades by 4D printing. *Renewable energy, 130*, 329-351.

Pearson, E.S., Plackett, R. L., & Barnard, G. A. (1990). Student: a statistical biography of William Sealy Gosset. Oxford, UK: Clarendon Press.

Pearson, K., Fisher, R., & Inman, H. (1994). Karl Pearson and R. A. Fisher on Statistical Tests: A 1935 Exchange from Nature. *The American Statistician, 48*(1), 2-11. doi:10.2307/2685077

Sanborn, B., DiLeonardi, A.M. & Weerasooriya, T. J. dynamic behavior mater. (2015) 1: 4. https://doi.org/10.1007/s40870-014-0001-3

8.8 Study Problems for Chapter 8

1. What is the logic of falsification in hypothesis testing?

   By assuming there is no effect of a treatment, we are, in essence stating that the only difference between a sample and its population are due solely to random chance. That means, that the further a measured effect is from the center of a population, the more unlikely it is to just be a function of random chance. Falsifying the null hypothesis, therefore, lends support to the alternative hypothesis, which is that the treatment does, in fact, cause some difference in performance on the dependent variable.

2. What is the role of a Probability Density Function in hypothesis testing?

   A PDF enables us to model the likelihood that a measured difference in a sample statistic and the population parameter is due to random chance.

3. Why is the sampling distribution of the mean so important for testing hypotheses?

Partly this is because so many of the variables we measure are continuous random variables, and that the normal distribution serves as a good model for naturally varying phenomena. But also, we fully know the shape of the Normal distribution, and how it varies as its parameters change. Moreover, the standard error of the mean gets smaller as a function of sample size, so we are able to effectively optimize the power of statistical tests. Lastly, as we saw in Chapter 7, even binomial and other categorical relationships can be modeled using the Normal distribution if the number of trials is large enough.

4. Describe the relationships between Type I error, Type II error, Power and Confidence.

   Type I error is a false positive conclusion—falsely rejecting the null hypothesis.

   Type II error is a false negative conclusion—falsely failing to reject the null hypothesis.

   Power is the probability that one will reject the null hypothesis when in fact it is false.

   Confidence is the probability that we do not falsely reject the null hypothesis.

   Each of these conclusions of an experiment, with the exception of Type I error rate, which is chosen a priori, is impacted by sample size. The larger the sample, the narrower the probability distribution (sampling distribution) we are using as a model for the long-haul behavior of our population of samples. Narrowing the sampling distribution results in greater power, and a smaller confidence interval. If we fix both Type I error rate and Power a priori, we can optimize for the smallest sample size needed to provide adequate assurance (Type I error rate) and enough power, to make an experiment practical to run.

5. What are the primary differences between the one-sample Z-test and the one-sample *t*-test?

   These two tests are identical in terms of formula, with the exception that the *t*-test uses the sample standard deviation to estimate the population sigma. This results in a reduction in the degrees of freedom of the model. To fix the t-distribution, one must know the mean, standard deviation, and degrees of freedom.

6. Suppose you were a design engineer for a custom bicycle manufacturer. You want to determine if your ultralight frames are lighter but stronger than the industry standard frames so you collect some data. Each frame in the random sample below was tested for up to 200,000 compressive cycles in a pneumatic load tester, with an applied load of 1,200 N. The table shows their mass as well as the number of cycles each frame lasted until failure. The published value of the mean ultralight road racing bike is 1.59 Kg +/− 0.23 Kg, and lasts 140,000 cycles +/−37,263 cycles.

Frame	Mass (Kg)	Cycles
1	1.86	119,316
2	1.52	200,000
3	1.895	120,358
4	1.63	108,762
5	1.415	131,907
6	1.525	100,595
7	1.64	108,623
8	1.46	200,000
9	1.3	160,356
10	1.515	166,294
11	1.485	181,966
12	1.2	200,000

(a) Given your question, should you run a two-tailed test, a right-tailed test, or a left-tailed test? Justify your answer.

Because I am trying to improve bike frames, I would say that I am trying to make lighter, more robust frames, than the industry standard. Those are one-tailed tests because the directionality of the effect on mass is negative, and the effect on cycles is positive.

(b) State the null and alternative hypotheses for your study.

$$H_{0,\text{ mass}} : \mu_{\overline{x}} \geq \mu$$

$$H_{1,\text{ mass}} : \mu_{\overline{x}} < \mu$$

$$H_{0,\text{ cycles}} : \mu_{\overline{x}} \leq \mu$$

$$H_{1,\text{ cycles}} : \mu_{\overline{x}} > \mu$$

(c) Given each frame in the sample costs \$3,500 to produce and sells for \$8,000, what are the potential consequences for your company if you commit a Type I error. A Type II error?

A false positive (Type I error) will mean that my company will spend lots of money developing the manufacturing processes for the frames, and that customers will spend lots of money for frames that don't improve upon the industry standard.

A false negative (Type II error) will mean that my frames really are an improvement, but I don't realize that. Therefore, I am missing out on a great opportunity to make money. My riders are missing out on the opportunity to use advanced equipment.

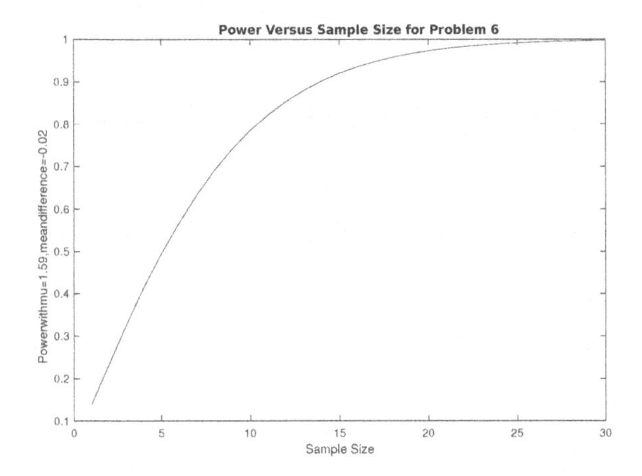

FIGURE 8.14: Power vs sample size curve for Problem 6e

(d) **Given the cost and potential profit in part c, what is an appropriate Type I error rate for your study? What is an appropriate Power? Justify your answer as if you were the engineer in charge of the project.**

I think that somewhere between 0.05 and 0.01 is a good area to select a Type I error rate. The bicycles cost a lot to manufacture, and destroying frames means I can't reuse the parts. So, I would probably have to use a smaller sample size, necessitating a Type I error rate that is in the higher end of this range. For Power, I think somewhere between 0.70 and 0.80 is appropriate to hedge my bet to be able to detect a real improvement on both of these variables.

(e) **For the Power and Type I error rate you selected, what is the sample size you would need to find a mean difference between your sample and the population of 0.2 Kg? Compare that to the current sample size. Is it adequate?**

Here, I will use Matlab, a useful package that most Mechanical and Aerospace students have free access to, to help me. You may use any statistical package you wish or write your own code. It is not a good practice to try to calculate these things by hand, as too many errors may occur without a way to trace them. The procedure I am showing here graphs the power for different sample sizes. You can see that I have used the population parameters in the brackets, and that the mean difference of 0.02 is negative (I want the sample to show less mass than the population—one tailed—resulting in a minimum mean of my sample needing to be 1.39 kg):

```
» nn = 1:30;
» pwrout = sampsizepwr('z',[1.59 0.23],1.39,[],nn);
» plot(nn,pwrout);
```

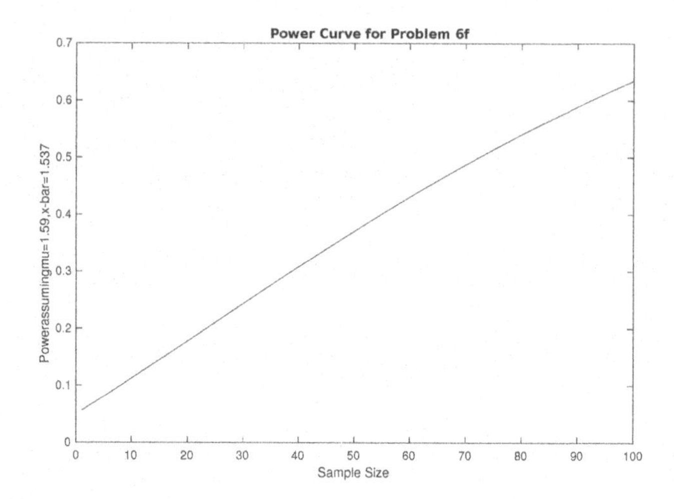

FIGURE 8.15: Power vs sample size curve for Problem 6f

From the graph in Figure 8.14 you can see that the power I have to detect this difference of –0.2 kg is about 0.8 with 12 bike frames tested.

(f) **What is the actual power of your study, given the data provided?**

Here, I need to look at the real mean difference, so I calculate $\overline{x} = 1.537kg$ this is a difference of $\overline{x} - \mu = 1.537 - 1.59 = -0.053kg$.

I can then use the same Matlab code:

```
» nn = 1:100;
» pwrout = sampsizepwr('z',[1.59 0.23],1.537,[],nn);
» plot(nn,pwrout);
```

I had to change the range for Sample Size to get any kind of curvature to the graph (see Figure 8.15). But as you can see, our sample with size 12 only has a power of about 0.12 or 0.13. We would have to greatly raise the sample size if we wanted to detect a mean difference as small as -0.053. In fact, the curve shows that we would need well over 100 frames to have any kind of shot at a power of 0.75 to 0.80.

(g) **Run the appropriate hypothesis test at your selected Type I error rate with the given data. (be able to do this using the confidence interval method and the test statistic method). What do you conclude?**

For the test statistic method, we can use a statistical package like Matlab. The most important thing you can do for any statistical package is to arrange your data as a matrix, with column vectors that consist of measurements of a single variable. If you have multiple variables, you need to have a matrix with one column for each.

```
»[h,p,ci,zval] = ztest(x, 1.590 , 0.23 , Alpha, 0.05 , 'Tail', 'right')
```

where x is a column vector consisting of the mass data. The calculation being done in the background is the following:

$$Z_{calc} = \frac{\bar{x} - \mu}{\frac{\sigma}{\sqrt{n}}} = \frac{1.537 - 1.590}{\frac{0.23}{\sqrt{12}}} = -0.798$$

The critical value for a one-tailed hypothesis at $\alpha = 0.05$, $Z_{critical} = -1.645$

Z_{calc} is definitely closer to center of the distribution than the critical value, so we fail to reject the null hypothesis. Matlab returns the value of the test statistic, and the actual probability we could have gotten $Z_{calc} = -0.798$ as a function of random chance, 0.7873. This is clearly a very high probability, so this confirms that the null hypothesis should not be rejected. Our frames don't appear to be significantly lighter than the industry standard.

For the confidence interval method, Matlab performs the following computation:

$$95\%CI : \mu \geq \bar{x} - Z_{\alpha}\frac{\sigma}{\sqrt{n}}$$

$$95\%CI : \mu \geq 1.537 - \left(1.645\frac{0.23}{\sqrt{12}}\right)$$

$$95\%CI : \mu \geq 1.4279$$

Since μ is clearly outside this interval, this is what we would expect under the null hypothesis about 95% of the time if we repeated this experiment. Of course your answers will vary depending on the value of Type I error rate you felt was appropriate for this problem.

7. **The tensile strength of 14.5–16.9 μm diameter Dyneema® (Ultra High Molecular Weight Polyethelene) fibers has been published as 4.24 ± 0.39 GPa. You are testing fibers to be used in lightweight composite body armor. Dyneema fibers are fairly inexpensive.**

   (a) **What assumptions about the sample and population do you need to have to trust the results of your test? If you want a sample of fibers to have a mean tensile strength of at least 4.65 GPa, with a Type I error rate of 0.01, what is an appropriate power and sample size for your test?**

   As this is a z-test, I need to assume that the population from which my sample is drawn is normally distributed, and that the sample is a random sample. Since Dyneema is relatively expensive, I will choose a power of 0.85. If I go much higher, the sample size will be too impractical for time constraints, given the use of my strain gauge equipment. I can use Matlab to help me with the power:

   ```
   « nn = 1:30;
   ```

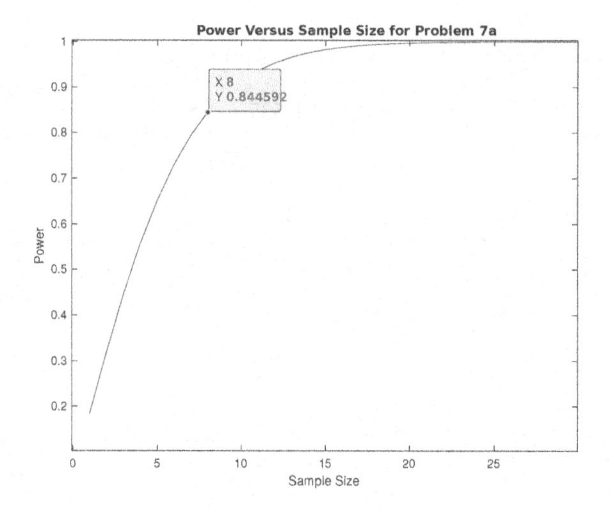

FIGURE 8.16: Power versus sample size curve for Problem 7a

« pwrout = sampsizepwr('z',[4.24 0.39],4.65,[],nn);

« plot(nn,pwrout);

Here, in Figure 8.16, you can see that a sample size of about 8 gives us a power of 0.84

(b) **Using the sample size you estimated in part a, you compute a mean tensile strength of 4.64. Is your sample significantly greater in tensile strength, than the population?**

$$Z_{calc} = \frac{\bar{x} - \mu}{\frac{\sigma}{\sqrt{n}}} = \frac{4.64 - 4.24}{\frac{0.39}{\sqrt{8}}} = 2.900$$

The critical value of Z at $\alpha = 0.01$ is $Z_{critical} = +2.33$

Z_{calc} is further in the right tail of the distribution than $Z_{critical}$, so we can conclude that our fibers are significantly stronger, on average, than the published specs for Dyneema.

(c) **Are you happy with your results, given your goals of the study? Explain why or why not.**

I am happy! My fibers are very likely to be stronger in tensile strength, so I can manipulate this into stronger, lighter body armor. This will, in turn save lives. I am confident in this because my Type I error rate is nicely conservative.

8. **A machining company claims that their $\frac{1}{4}$" length, 0.12" outside diameter, 0.08" inside diameter, 0.02 wire diameter, steel coil springs have a spring constant of 46 pounds per inch (this is from a real catalog). You take a random sample of these**

springs, and anneal them to make them a bit softer so that they will elongate beyond their 0.16" maximum. You take a random sample and test them, finding the following spring constants:

45.7488	45.7257
44.5823	45.6763
46.2937	45.6269
45.8245	45.1175
46.5571	44.5013
45.7223	46.2642
44.6894	45.4632
45.7148	45.8512
45.8558	46.1446
45.773	45.2638
45.7194	46.7111
45.3903	45.7751

(a) Given your question, should you run a two-tailed test, a right-tailed test, or a left-tailed test? Justify your answer.

This is a one-tailed test, as I am only interested in whether or not the annealing caused the spring constant to decrease.

(b) State the null and alternative hypotheses for your study.

$$H_0 : \mu_{\overline{X}} \geq \mu$$

$$H_1 : \mu_{\overline{X}} < \mu$$

(c) Given each spring is about \$5, what are the potential consequences for your company if you commit a Type I error, if you use 250,000 springs on average, each year.

At \$ 1.25 million per year, that is a lot of money to waste on a spring that doesn't measure up to the design constraint that I have placed on it.

(d) What are the potential consequences of making a Type II error if you estimate the profit you get from each product the spring is in use, is \$20?

My company may lose out on up to \$5 million in additional revenue each year. Because I didn't detect a true difference that the annealing made.

(e) Given the cost and potential profit in parts b and c, what is an appropriate Type I error rate for your study? What is an appropriate Power? Justify your answer as if you were the engineer in charge of the project.

For this, I may lose a lot of money with a Type I error (big consequence), and I may forego up to \$5 million per year with a Type II error (Missed opportunity). So, I think the chance of making a Type I error should be rather small, say 0.01, and my power should be relatively high, say 0.80.

But I could run a power analysis to optimize these constraints against real costs and benefits.

(f) **What is the actual power of your study, given the data provided.**

This is a one-sample t-test, since I don't know the population standard deviation. So I need to compute the sample mean and standard deviation. I first enter the data into a column vector, x. Then I use Matlab to compute the mean and standard deviation:

» mean(x);

45.6663

»std(x);

0.5545

With these, I can take a look at the overall power, substituting the sample mean and standard deviation, the population mean and sample size:

»power = sampsizepwr('t',[45.6663 0.5545], 46 ,[],24,'Alpha', 0.01 ,'Tail','right');

0.6726

So, this particular study has low power. Ideally, I would have figured out the minimum mean difference I would tolerate, optimize to get the sample size I needed for a power of 0.85, and then collect my sample.

(g) **Run the appropriate hypothesis test at your selected Type I error rate with the given data. (be able to do this using the confidence interval method and the test statistic method). What do you conclude?**

» [h,p,ci,stat]=ttest(x,46,0.01, 'left'); Returns the following:

h=1, therefore the null hypothesis is rejected

p=0.0036, the computed probability of obtaining a mean difference of this much is about 4 in 1000

ci = [-inf,45.9493], $\mu = 46$ is not in this interval, therefore we must reject the null hypothesis and conclude that our annealing did, in fact, significantly reduce the spring constant.

stat reveals a t-statistic at $n - 1 = 23df$ of -2.9480. This is way beyond the critical value of $t_{23} = -2.500$. This is consistent with the confidence interval method.

9. **Two engineers utilize the same data to test the same hypothesis. The first uses a one-sample z-test. The second uses a 1-sample t-test. The z-test returns a test statistic whose probability is greater than the alpha level assigned by the researcher (.05). The t-test returns a test statistic whose probability is less than the alpha level assigned by the researcher.**

   (a) **Show how this is possible. What is the reason the t-test came up with its results when the z-test did not?**

$$Z = \frac{\overline{x} - \mu}{\frac{\sigma}{\sqrt{n}}} \quad t = \frac{\overline{x} - \mu}{\frac{s}{\sqrt{n}}}$$

If the sample variation, estimated by s is a poor estimate of the population variation, σ, then Z and t will return very different results. When s underestimates σ , the t-test may be over-inflated, and therefore may return a significant result when Z does not. If this happens, the researcher has made a Type I error.

(b) If you were the supervisor of these two engineers, which of the two tests would you include in your technical report, and why?

While the two look the same, the t-test has fatter tails than the Z test, and therefore is not quite as powerful as Z. This is because the Z test utilizes the actual population standard deviation, while the t-test has to use the sample standard deviation as an estimate. For this reason, if you have the population standard deviation, it doesn't pay to use just an estimate. I would use the Z-test!

10. Going beyond this chapter. Experiment at Home: Take a box of spaghetti to represent your population (spaghetti is cheap. I don't want you retrofitting your fuel injectors in your car). Now take a random sample of, say 10 strands. Make sure none are broken. Assign that random sample to a control group. Now take a second random sample of 10 strands (no broken ones!). Assign that random sample to your experimental group. Immerse your experimental group in boiling water for 10 seconds. Take them out and let them dry. Now bend the strands from each sample, one by one, until they break, measuring the displacement of the end of the strands at the point of breaking.

(a) Did the treatment change the material properties of the spaghetti? How do you know, one way or the other?

We are looking to see if the pre-boiled spaghetti has a different bending moment (either significantly more or significantly less) than the non-boiled spaghetti. If the experimental treatment, boiling, has an effect, the mean of its sample should be different than the mean of the non-boiled spaghetti.

(b) We don't have any recorded population parameters for spaghetti bending, so the one-sample Z-test and one-sample t-test don't apply to this experiment. How can you use the data you have to answer your question?

Because the two samples came from the same population, their mean difference should be zero under the null hypothesis for a Random Sample design. So, you can subtract the means and use a mean of zero as your population mean. Estimating the standard error has to account for the fact that you have two samples, each with a different standard deviation.

(c) Make a guess regarding the Alternative Hypothesis. What do you really predict will happen from the treatment?

Because some of the strands of the amylose (starch) will have been broken, I think the pre-boiled spaghetti should bend, on average, more under the same load as the non-boiled.

(d) Make a guess regarding the Null Hypothesis. What are the mean values we are comparing?

The Null Hypothesis is that there is no difference in bend to failure between the two samples.

(e) What can you use to model the population distribution?

The sample statistics for the non-boiled spaghetti could serve as the population under the Null Hypothesis, since this is, in essence, a control group.

(f) What can you use to model the sample distribution?

The sample statistics for the boiled spaghetti could serve as estimates of the population from which the treatment sample came from, since this is, in essence, a treatment group.

(g) What value serves as the center of the sampling distribution of the mean in this case?

You will see in the next chapter that we can use the difference in means, which we hypothesize to be zero under H_0, and estimate a sampling distribution of this difference.

9

2-Sample Statistics

At the very end of Chapter 8, I posed you a problem that was slightly different from the one-sample Z- and t-tests we had learned up to this point: To take two samples of spaghetti from a box, representing the population. Treat one sample of spaghetti by immersing it in boiling water for a few seconds, leaving the other sample untouched. Then, compare the mean displacement at which each strand of spaghetti fails, for each sample.

This scenario is one of the simplest, and best, ways to test hypotheses about treatment effects. Of course you have to substitute something important for the spaghetti, like the displacement of a (model) femur under stress induced by a robotic exoskeleton. But the idea that, even when you don't have information about a population, you can use two samples, both taken from the same population at random, to represent that population at the beginning of the study, is one of the fundamental building blocks of hypothesis testing, the Random Sample design discussed in Chapter 6. By treating only one sample and leaving the other untouched as a control, any difference between the samples' statistics can be considered an estimate of the treatment effect (see Figure 9.1).

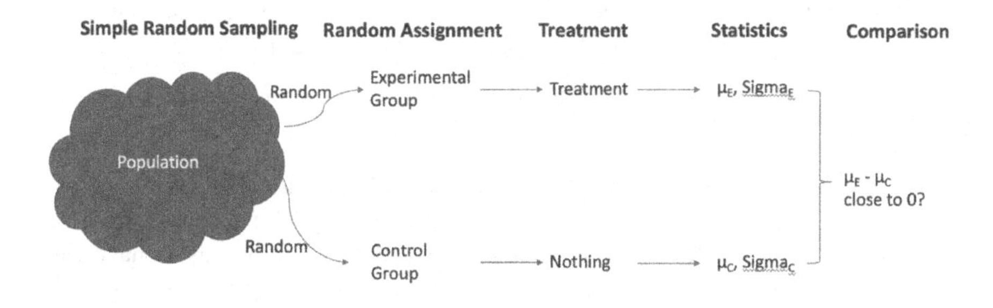

FIGURE 9.1: The random sample experimental design

The control group in a Random Sample design, because it does not undergo the treatment, still represents the population from which both samples were drawn at the beginning. The Experimental group may or may not represent a new population at the end of the treatment: The population of exquisitely designed exoskeletons, for example. If its mean is considerably different from the control group mean, this is evidence that the treatment made an effect: that the population mean of the treatment

DOI: 10.1201/9781003094227-9

group is likely different from the population mean for the control group, solely due to the effect of the treatment (plus a bit of random error). The Alternative Hypothesis for a study using this design looks something like this:

$$H_1 : \mu_1 \neq \mu_2, \text{two-tailed}$$

$$H_1 : \mu_1 > \mu_2, \text{right-tailed}$$

$$H_1 : \mu_1 < \mu_2, \text{left-tailed}$$

And the Null Hypothesis, like always, is the logical converse:

$$H_0 : \mu_1 = \mu_2, \text{two-tailed}$$

$$H_0 : \mu_1 \leq \mu_2, \text{right-tailed}$$

$$H_0 : \mu_1 \geq \mu_2, \text{left-tailed}$$

9.1 2-sample *t*-test. The Most Utilized Test in the world!

We will use $\overline{x}_1$ to estimates μ_1 and $\overline{x}_2$ to estimate μ_2, but our hypotheses are always about the populations from which our samples have been drawn. We usually assign the control group to μ_2, but because of the symmetry of the Z and t distributions, it doesn't really matter, so long as we keep our direction straight for one-tailed tests.

Just like the one-sample t-test, we don't know the population mean or standard deviation, so the t-distribution seems like the way to go. Now we can substitute our sample means in place of the population means in the one-sample t-test

$t_{(\alpha,df=n-1)} = \frac{\overline{x}-\mu}{\frac{s}{\sqrt{n}}}$ becomes $\frac{\overline{x}_1-\overline{x}_2}{\frac{s}{\sqrt{n}}}$, but there are a couple of problems staring back at us.

We have *two* samples, each with different standard deviations, s_1 and s_2, and *two* sample sizes, n_1 and n_2. We *could* restrict ourselves to only taking samples with the same number of observations, but this is often impractical. Moreover, as we have seen before, larger n gives our test more power—we want to use every last observation in our estimate of the true difference in means. Lastly, under H_0 we have assumed that both samples are drawn from the same population, and so they are estimating the variability in the same distribution. So, if the sample sizes are not too far off, we can take their weighted average as the best estimate of the population σ, calling it the **Pooled Standard Deviation**:

$$s_p^2 = \frac{(n_1 - 1)\, s_1^2 + (n_2 - 1)\, s_2^2}{n_1 + n_2 - 2}$$

$$s_p = \sqrt{\frac{(n_1 - 1)\, s_1^2 + (n_2 - 1)\, s_2^2}{n_1 + n_2 - 2}}$$

The standard error now becomes a function of both samples, and therefore we have to take into account the size of both samples when developing a good estimate. Again, the weighted average of the two sample sizes, if they are not too far off from each other, will be our best estimate:

$$SE_{\overline{x}} = s_p \sqrt{\frac{1}{n_1} + \frac{1}{n_2}}$$

The square of $SE_{\overline{x}}$ is often called the **Mean Squared Error Within Groups**, or Mean Square Within. It is the variance of both samples taken together, assuming the two groups are from the same population. Pooling the data into a single parent distribution, and computing the variance gives us the **Mean** (i.e. average), **Squared** (this is a variance estimate), **Error of estimation** (difference between from the mean and each data value), **Within** our pooled sample—Mean Square Within. You will see this statistic often in hypothesis tests using the t, χ^2, and F distributions (introduced later in this Chapter). Different statisticians use different terms to refer to this concept: Mean Square Within, Mean Square Error, and Mean Square Residual are the three most common.

$$s^2_{(\overline{x}_1 - \overline{x}_2)} = MS\,W = s_p^2 \left(\frac{1}{n_1} + \frac{1}{n_2} \right)$$

The pooled standard error becomes less precise as an estimate the more the variability of the two samples diverges, so we assume, in the t-test, that we have **homogeneity of variance**—that *the t-test only assesses the difference in means, not any differing variability* in the two samples.

> **The Pooled Standard Deviation** is a weighted mean of the standard deviations of two or more samples. We use s_p when we can assume that the samples being tested are drawn independently and randomly from the same population.

> **The Mean Squared Error Within Groups** is the variance of the sampling distribution of the difference between two means, taken from samples assumed to have been drawn from the same population. Like the variance of a population, the MSW is the second moment about the mean: the variance—of the sampling distribution. Its standard deviation is the standard error of the mean, sometimes called the Root Mean Square Error.

Now that we have a mean difference and a standard error, we can construct the test statistic. Just like the Z- and t-tests, the test statistic is just a difference in means divided by the standard error of the mean.

$$t_{(\alpha, df=n_1+n_2-2)} = \frac{\overline{x}_1 - \overline{x}_2}{s_p\sqrt{\left(\frac{1}{n_1} + \frac{1}{n_2}\right)}}$$

The degrees of freedom for the two-sample t-test is two less than the pooled sample size. We have $n_1 + n_2$ total observations in our pooled estimate, but we have computed two parameters, $\overline{x}_1$ and $\overline{x}_2$, and used those in our estimate. We thus lose two degrees of freedom when we fix $\overline{x}_1$ and $\overline{x}_2$, s_p becomes fixed at $n - 2$ iterations.

9.1.1 $E(x)$ of $\overline{x}_1 - \overline{x}_2$ Under the Null Hypothesis

Under H_0, the samples are drawn *from the same population*, so there is zero expected difference—the numerator of the t-statistic is expected, over the long haul, to be zero. Because of this, just like the 1-sample t-test, the distribution of t is centered on zero, and its unit values are comparable to standard deviation units due to the denominator of the test statistic. Because the t distribution is generally skinnier, but with fatter tails, the cumulative distribution function for t differs slightly from the Normal Distribution. The table of values for t provided in Chapter 8 will work for any t distribution, including the two-sample t-test. Confidence intervals are also calculated in the same manner (this one is two-tailed):

$$1 - \alpha\%CI = \overline{x}_1 - \overline{x}_2 \pm t_{(\frac{\alpha}{2}, df=n_1+n_2-2)} s_p\sqrt{\left(\frac{1}{n_1} + \frac{1}{n_2}\right)}$$

This is just a restatement of the test statistic, expressing the confidence interval of $\overline{x}_1 - \overline{x}_2$ in the original units as opposed to t-units. It represents the range, $p(\overline{x}_1 - \overline{x}_2) = 1 - \alpha$. In plain English, it is the region under the curve where we would expect the difference in sample means to fall $1 - \alpha$ percent of the time, assuming the Null Hypothesis is true.

The 2-sample *t*-test is a test of the difference in means between two samples, each estimating the common population from which they were drawn. The expected difference in sample means is zero under the Null Hypothesis. Its test statistic is computed by: $t_{(\frac{\alpha}{2}, df=n_1+n_2-2)} = \frac{\overline{x}_1 - \overline{x}_2}{s_p\sqrt{\left(\frac{1}{n_1} + \frac{1}{n_2}\right)}}$

And its $(1 - \alpha)$ % confidence interval is:

$$1 - \alpha\%CI = \overline{x}_1 - \overline{x}_2 \pm t_{(\frac{\alpha}{2}, df=n_1+n_2-2)} s_p\sqrt{\left(\frac{1}{n_1} + \frac{1}{n_2}\right)}$$

The assumptions of the t-test are:

1. The two samples are independent;
2. The parent population of the two samples follow normal distributions; and
3. The two samples have roughly the same variance (homogeneity of variance)

Let's do a problem to illustrate the logic of the 2-sample t-test.

1. You have designed a new release mechanism for step-in snowboard bindings and want to know if it is stronger under extreme stress, than the old mechanism. You have no information regarding the mean stress (in GPa) under which the old release mechanism will maintain integrity. So, you take two independent samples of bindings, and swap out the release mechanism on one sample, leaving the other sample original. Stress tests to failure indicate show the following sample statistics in Table 9.1:

Group	n	$\overline{x}$	s
Control	18	1.2 GPa	.32
Experimental	14	1.4 GPa	.40

TABLE 9.1: Summary statistics for snowboard binding experiment

(a) Should you run a 1-tailed or 2-tailed test for this question? What is H_0 and H_1?

(b) Given you are determining the strength of the new mechanism, and not yet ready to place the bindings with the new mechanism on the market, what is an appropriate Type I error rate? Justify your answer.

(c) Construct the 1- α % confidence interval for the difference in means for this analysis, given your Type I error rate in part a.

(d) What is the value of the test statistic?

(e) If the 2-sample t-test assumes normal parent populations, and homogeneity of variance among samples, can you reasonably justify this analysis?

For part a, we are trying to find out if our release mechanism is *stronger* all other things being equal, than the status quo mechanism. This calls for a 1-tailed, right-tailed test:

$$H_0 : \mu_1 \leq \mu_2$$

$$H_1 : \mu_1 > \mu_2$$

Where μ_2 is the control group and μ_1 is our experimental group.

For part b, a part failure could result in pretty serious injury if we design our bindings to accommodate a release mechanism that isn't as strong as we say it is. So ultimately, we have to be pretty darn sure that our release does in fact have the strength we report. But our test is probably the first among many of the upgraded binding, so it doesn't have to be super stringent. An α of about 0.01 or 0.05 is probably good enough to give us a strong test without breaking the bank. Let's select 0.01, because (I am making this up) this is our best selling binding, and if we are wrong, the company's reputation may go down the toilet.

In part c, this is just calculation. I will show it longhand here, but understand that once you get an understanding of the test and its assumptions, you will most likely use Matlab code, which I will provide later. Because this is a right tailed test, the confidence interval is to the left of the critical value of $t_{(0.01,df=33)}$.

$$1 - 0.01\%CI = 1.4 - 1.2 + t_{(0.01,df=33)} s_p \sqrt{\left(\frac{1}{14} + \frac{1}{18}\right)}$$

I look up the value of t from our table in Chapter 8, finding the critical value of $t_{(0.01,df=33)} = 2.457$.

$$99\%CI = 0.2 + 2.457 s_p \sqrt{\left(\frac{1}{14} + \frac{1}{18}\right)}$$

We now have to figure out our estimate for the standard error. The pooled standard deviation is straightforward (if you remember to square s_1 and s_2!).

$$s_p = \sqrt{\frac{(n_1 - 1)\, s_1^2 + (n_2 - 1)\, s_2^2}{n_1 + n_2 - 2}}$$

$$s_p = \sqrt{\frac{(14 - 1)\, 0.40^2 + (18 - 1)\, 0.32^2}{30}}$$

$$s_p = 0.3569$$

$$99\%CI = 0.2 + (2.457)\,(0.3569) \sqrt{\left(\frac{1}{14} + \frac{1}{18}\right)}$$

$$= 0.5125$$

This gives us the upper boundary of our 1-sided interval. Because the Null Hypothesis states $H_0: \mu_1 \leq \mu_2$, the 99% Confidence Interval for the difference in means is:

$$99\%CI = (-\infty, 0.5125)$$

Zero, the value of the difference in means, $\mu_{Experimental} - \mu_{Control}$ under the Null Hypothesis *is* found within this interval, so it is not so unlikely to have occurred that we would conclude that our redesigned release mechanism is significantly stronger than the old mechanism.

Part d, again, is a straightforward calculation. We can use the s_p we have already calculated.

$$t_{(0.01,df=33)} = \frac{1.4 - 1.2}{0.3569\sqrt{\left(\frac{1}{14} + \frac{1}{18}\right)}}$$

$$t_{(0.01,df=33)} = \frac{0.2}{.1272}$$

$$t_{(0.01,df=33)} = 1.5726$$

This value is to the left, towards the expected mean difference of zero, from the critical value, $t = 2.457$. This is consistent with the findings of the confidence interval method, that our mean difference isn't great enough to warrant claiming that our redesigned release mechanism is any stronger, on average, than the old binding.

For part d, we don't have a lot of information about the parent distribution of our bindings, both control and experimental. We assume they trend to normal or are very close, but this is only an assumption. If we had badly skewed parent populations, then our test would likely be under-powered because we have violated one of the assumptions of the *t*-test. The samples have different standard deviations as well. These are pretty close, but we don't yet have a way to figure out how close is considered "approximately equal" enough to claim that the populations have homogeneity of variance. We will take these assumptions up later and provide some heuristics for dealing with violations.

In the meantime, we just saved our company a ton of money, and perhaps a ruined reputation, by finding that our design is not yet up to snuff! Even though the mean value of our sample is higher than the control sample representing the status quo, it isn't a big enough effect to really be sure that the difference is systematically caused by our new release mechanism. Finding this out, we can advise the company to not use the new release yet, until it goes through a re-design phase to make it even stronger.

9.1.2 2- sample *t*-test: Effect Size

One of the issues we had in interpreting the 1-sample *Z*- and *t*-tests was, sometimes we find a difference that is significant statistically, meaning we reject H_0, but the actual difference between the sample mean and the population mean is so small it just isn't worth it to use the results of the test as an indicator that we have **practical significance**. In short, practical significance is that minimum real difference you would count as something action able, for example to retool your manufacturing plant, or to use a new release mechanism in your snowboard bindings.

This real difference is called an **Effect size**. There are a number of metrics assessing effect size, but the most common, and most easily understood is Cohen's d (d stands for "difference"). Cohen's d measures effect size as a difference between control and experimental means, scaled in units of the pooled standard deviation.

$$d = \frac{\overline{x}_1 - \overline{x}_2}{s_p}$$

Notice that it differs from the 2-sample *t*-test only in that it is *not* scaled by the size of the samples. For Problem 3, the effect size we found was:

$$d = \frac{0.2}{0.3569} = 0.56$$

This is about half a standard deviation difference, given our error of measurement s_p.

This may be big enough for us in practicality, but our test returned non-significant results. The only way to rectify this is to increase our power to detect a real difference of 0.56 standard deviation units. take a look at the denominator:

$$t = \frac{d}{\sqrt{\left(\frac{1}{n_1} + \frac{1}{n_2}\right)}}$$

You can see that, algebraically, for any given effect size, increasing our sample size, increases t. Just like we have learned with the one-sample *Z*- and *t*-tests, we can use this relationship to find a "sweet spot" that gives us adequate power to detect a real difference, without incurring the cost of taking an excessively large sample. This involves setting the minimum d we would count as practically significant, setting the power we want to detect this real difference ($1-\beta =\sim 0.80$ or above, usually), setting our Type I error rate, α, and then adjusting the sample size to meet these parameters.

The effect size we detected in Problem 1 is:

$$d = \frac{1.4 - 1.2}{0.3569}$$

$$d = 0.56$$

Effect size is a measure of the distance between the means of two samples. It is expressed in standard deviation units. The most common measure of effect size is Cohen's d.

$d = \frac{\bar{x}_1 - \bar{x}_2}{s_p}$

Used to determine the practical significance of an effect, the larger the effect size, the more we can expect an average difference of d in future design work.

Practical Significance is the assessment of whether the results found in a hypothesis test are of any value in a real-world application.

9.1.3 2-sample *t*-test: Dealing with Violations and Power

The 2-sample *t*-test is a powerful tool in the engineer's arsenal. It is widely applicable, and it allows us to make claims about experimental data without having to know the specific parameters of the population from which our samples are drawn. But it isn't without limitations. As stated earlier in this chapter, the assumptions of the *t*-test are:

(a) Independence of observations;
(b) Normalcy of the parent population(s); and
(c) Homogeneity of variance.

If, unbeknownst to us, one or more of these observations are violated, the *t*-test may not be robust, i.e., it may not return the proper decision to our hypothesis test. For example, if we violate the assumption of a normal population distribution, we tend to lose power. If our parent populations are skewed, for example, their samples will also tend to be skewed. As we learned in Chapter 2, in skewed distributions, our primary measure of center, the mean, is pulled away from the bulk of the data, being affected by the more extreme values in the skewed tail. It also increases the standard deviation of our samples, and thus the standard error will tend larger for such distributions. This decreases the power of our test.

If we violate the homogeneity assumption when our two samples have widely different widths the pooled standard deviation will be pulled towards that of the wider sample. This is generally not a problem if our sample sizes are large, but with small sample sizes, this asymmetry can increase our Type I error rate.

9.1.3.1 Assumption of Independence of Observations

Non-independence of observations also, can increase the Type I error rate if there is some systematic bias introduced to one sample over the other. For example, if I REALLY REALLY want my turbocharger turbine to perform better than my stock turbines, I may unwittingly lubricate its bearings better, observe its behavior more closely, or engage in any number of sub-conscious biased behaviors because I TRULY believe and really WANT the data to show an improvement. These wants and desires are good! They keep us in engineering through all the frustration and hard work, but we have to be very careful that our data are collected without bias, with each observation independent of all the others. Unfortunately, this assumption is the hardest to test for after the fact.

There are some corrections you can apply if you violate the Normalcy and Homogeneity assumptions of the *t*-test. We will not go into these corrections in this text, but they are pretty easy to look up and understand when you need them. However, it is important to be able to assess whether or not you might have violated them before you go Googling "what do I do now, when my data is not normal or homogeneous?" Here are a couple of good rules of thumb to go by:

9.1.3.2 Assumption of Normal Population Distribution(s)

The *t*-test is pretty robust to violations of Normalcy. This means that you typically won't increase your Type I error rate if your samples show a bit of skew, or if they are flat and not peaked in the middle. You will lose a bit of power, though.

To combat this, you can transform your data to fit the normal distribution, *normalizing it*, as they say, by transforming each datum into a z-score within its experimental condition, then computing a *t*-test between experimental conditions.

You may also use a *nonparametric* test, which do not have the same assumptions of the normal distribution. The Mann-Whitney U test is a test of ranked data that can be more powerful than t for skewed or flat distributions. In general, though, for distributions that are symmetric, or close to the normal distribution in shape, the *t*-test will give you the most power.

9.1.3.3 Assumption of Homogeneity of Variance

If your samples show widely different variability—if the variance of one sample is, say, 1.5 – 2 times that of the other–or if your ratio of sample sizes is 1.5 or greater, *and* the histograms of your samples show different widths, you run the risk of increasing your Type I error rate. To combat this, you can apply a simple correction that adjusts the standard error and degrees of freedom to account for these differences. This is called Welch's *t*-test for unequal variances. It is more powerful that Student's *t* when Homogeneity is violated, but is not as powerful when our sample means are fairly close to one another.

None of these corrections are worth the electrons with which they are computed if you don't first design your experiment correctly, account for Type I error, power, and effect size, and sample size up front, and *then* deal with any violations that might occur. Problems will still occur because not all natural phenomena follow a normal distribution, and not all sampling and testing can be done perfectly, so these corrections are good options to keep in mind in that eventuality. A simple check can be made using the following procedure:

(a) Graph your data and inspect the histograms!

(b) Look at your sample data and see if it is asymmetric, and/or if the variances are widely different;

(c) Check your power. If it is too low, you might need to get some more data;

(d) Based on 1 and 2, run either Student's *t*, or one of the appropriate corrections.

9.2 Paired Sample *t*-test: Testing Pre- and Post

Lots of times in engineering, we only have a few, really, really expensive machines with which to test (Rolls Royce aircraft engines, for example) or for ethical reasons, we can't randomly assign events to treatments (e.g., surgery versus no surgery for removing a cancerous tumor). In such cases, we can measure an event *before* we apply a treatment, (i.e., perform a Pre- or baseline-test), then apply the treatment and measure the event *after* (i.e., Post-test). In Chapter 6, we called this a Pre-post Randomized Sample, and we diagrammed it as shown in Figure 9.2:

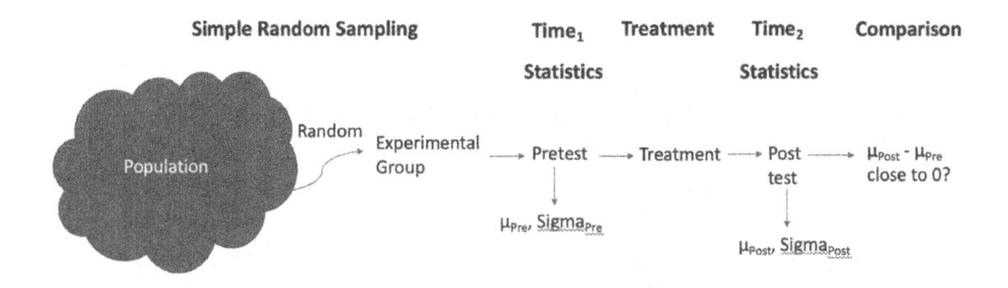

FIGURE 9.2: Pre-post experimental design

By subtracting the mean of the post-test from the mean of the pre-test for each event sampled, we get a sample of *gain* or *difference* scores. I perform this analysis every year when, going into the winter holidays, I am fit and trim, weighing 165 lbs (Pre-test). Then I apply a treatment of feasting and drinking and weigh myself afterwards (172 Post-test). This is a difference of seven pounds that can only be attributed to the treatment: Eating and drinking too much, and probably not exercising as much as when I am chasing after the assignments of my undergrads. If I were to sample people over the winter holidays, recording their pre-holiday weights and then post-holiday, I could find the mean weight gain—the average effect of the treatment:

$$\overline{d} = \sum_{i=1}^{n_d} \frac{d_i}{n_d}$$

Where d_i is each individual gain, and n_d is the number of paired events in the sample of Pre- and Post-measurements. So, if I sampled 100 people over the winter break n_d would be 100 gain scores, and $\overline{d}$ would be the mean of those gains. Some "gains" may be negative, so don't take the word too literally. That is why I use d to represent it. A difference can be either positive or negative.

The variability of our gain scores is simply their standard deviation:

$$s_d = \sqrt{\sum_{i=1}^{n_d} \frac{\left(d_i - \overline{d}\right)^2}{n_d - 1}}$$

The Null Hypothesis, if we assume the *treatment has no effect* must be that $\mu_{pre} = \mu_{post}$ which is the equivalent of saying $\mu_d = 0$. The Alternative Hypothesis is that the treatment had some effect:

$\mu_d \neq 0$ (two-tailed);
$\mu_d > 0$ (right tailed); or
$\mu_d < 0$ (left tailed)

Given these things, we can now construct a test statistic:

$$t_{(\alpha, df=n_d-1)} = \frac{\overline{d} - 0}{\frac{s_d}{\sqrt{n_d}}}$$

$$t_{(\alpha, df=n_d-1)} = \frac{\overline{d}}{\frac{s_d}{\sqrt{n_d}}}$$

Stop right there, Middleton (I know you are thinking it)! That looks suspiciously like the one-sample *t*-test! Well, you are right. It *is* the one-sample *t*-test! The population mean, if we assume there is no difference in pre- and post-test measurements, has to be zero, so we just perform a one-sample *t*-test on the sample of differences. The confidence interval for the **Paired-Sample *t*-test** is exactly the same as the one-sample *t*-test:

$$1 - \alpha\% CI = \overline{d} \pm t_{(\alpha, df=n_d-1)} \frac{s_d}{\sqrt{n_d}}$$

9.2.1 What Does the Paired-Sample *t*-test Buy You?

As alluded to when I first introduced the Pre-Post Randomized Sample design, there are many times when due to either cost, feasibility, or ethics, one cannot take multiple samples and subject them to different treatments. Taking a single sample and measuring its events twice reduces the overall cost of purchasing and preparing two or more samples, *and* it basically *guarantees* that the effect measured is due to the effect of the treatment, because each event serves as its own control. There *are* time-varying confounds, like test effects, fatigue and the like that you need to be aware of, but altogether, this design is extremely efficient and effective.

The Paired-sample *t*-test is a test of the mean of the differences between paired observations. The expected difference in Pre- and Post- measurements is zero under the Null Hypothesis. Its test statistic is computed by:

$$t_{(\alpha,df=n_d-1)} = \frac{\overline{d}}{\frac{s_d}{\sqrt{n_d}}}$$

And its $(1-\alpha)$ % confidence interval is:

$$1-\alpha\%CI = \overline{d} \pm t_{(\alpha,df=n_d-1)}\frac{s_d}{\sqrt{n_d}}$$

The assumptions of the Paired-sample *t*-test are:

(a) Data consists of independently-measured, paired observations;

(b) The parent population of the differences follows a normal distribution; and

(c) The distribution of differences has no obvious outliers

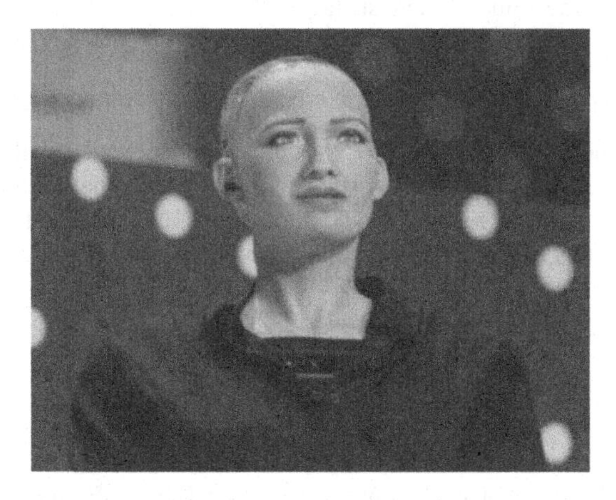

FIGURE 9.3: Robot capable of facial expression

2. Hegel, et al., 2008, studied the interaction of 20 male subjects with two different robot opponents, playing against them in a video game context: The control (Pre) condition was a purely functional robot, that typed on a keyboard and played the game. The experimental (Post) condition was a social robot that responded with facial expressions (Figure 9.3 shows a more contemporary social robot–still pretty scary looking to me!). All subjects interacted with both robots. They used a paired-sample t-test to compare the subjects' perceptions of the "friendliness" of each robot. The following data has been reconstructed from their analysis (See Table 9.2). Subjects' ratings range from 1 (not friendly at all) to 5 (very friendly):

Pre	Post	Pre	Post	Pre	Post	Pre	Post
5	5	1	4	2	2	3	5
1	4	4	5	2	4	1	3
4	5	2	5	4	5	4	5
2	5	3	5	2	5	2	5
2	5	1	4	2	5	2	4

TABLE 9.2: Pre/Post data reconstructed from Hegel et al's study of social robots.

(a) What are the Alternate and Null Hypotheses for this study? Should the authors use a one-tailed or two-tailed test?

(b) What is an appropriate Type I error rate for this study, given the results are used for research, and not for product creation?

(c) Construct the $1-\alpha$ % confidence interval for this analysis, given your Type I error rate in part c.

(d) What is the value of the test statistic?

(e) Should the researchers reject H_0, given this data?

(f) Were the assumptions of the paired-sample t-test met in this study?

For Part a, the Null and Alternative hypotheses are not well specified. The authors actually used a two-sided test, even though they felt that adding that hideous visage for an animatronic face would make the robot test friendlier with subjects. I would run this test as a one-tailed test for this reason. So here is what I would use:

$$H_0 : \mu_d \leq 0$$

$$H_1 : \mu_d > 0$$

In part b, there is not much onus for a Type I error. Further research would probably point out the error if it occurred. But we still don't want to make a mistake and waste our time in this and some future studies. I think $\alpha = 0.05$ is probably adequate, but you might go as low as $\alpha = 0.10$.

Part c is a straightforward calculation. I will use $\alpha = 0.10$, because we haven't used that yet. It will give you more experience looking up the appropriate critical value of *t:* $t_{(0.10,df=19)} = 1.328$.

$$1 - \alpha\%CI = 0.45 + 1.328\frac{1.1459}{\sqrt{20}}$$

$$1 - \alpha\%CI = 0.45 + 0.4457$$

$$1 - \alpha\%CI : \mu_d \leq 0.9957$$

Since $\mu_d = 0$ is not in the confidence interval, we have to conclude that it is less than 10% probable that we could have achieved a $\overline{d}$ of 0.45 or above just due to random chance. Instead, the test indicates that a value greater than zero (as predicted by the Alternative Hypothesis) is a better explanation for the measured differences in subjects' perceptions of the robot's friendliness.
Part d reiterates this finding:

$$t_{(\alpha,df=n_d-1)} = \frac{0.45}{\frac{1.1459}{\sqrt{20}}}$$

$$t_{(0.10,df=19)} = 1.7562$$

The critical value of $t_{(0.10,df=19)} = 1.328$. This was read directly off the *t*-table in Chapter 8. Since the computed value of *t* is greater than 1.328, we can conclude that the probability of getting a $\overline{d}$ of 0.45 or above is highly improbable (less than ($\alpha = 0.10$)). So we conclude that it is likely that the interaction with the social robot is a "friendlier" experience, on average, than the merely functional robot.

Part e is answered in our discussion of c and d.

For part f, let's look at the histogram see Figure 9.4 of the differences: The histogram is slightly skewed, but this might be due to the relatively restricted domain (*d* can only take on 11 discrete values, −5 to 5). Only 4 differences are negative and 16 are positive. This seems to be unlikely if there were no effect of the social robot. Lastly, I don't see any obvious outliers, so I would conclude that our test is probably valid.

You should understand the *t*-test, the one-sample, two-sample, and paired-sample cases and the logic of using mean differences to determine effects of an experimental treatment. This is basic knowledge that all engineers need to keep in their pocket, so to speak, for use interpreting and conducting tests of our designs. We will return to this logic again and again in this course, but more importantly, you will return to it again and again in practice.

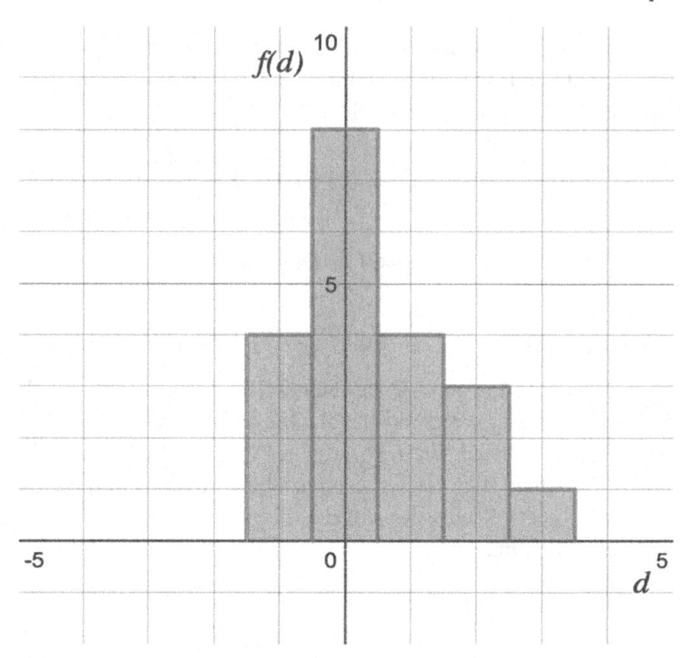

FIGURE 9.4: Histogram of pre- versus post-test differences in Hegel's social robot study

9.3 χ^2 Test of Independence: Testing the Independence of Proportions for Two or More Samples

You may recall back in Chapter 4, Introduction to Probability, that we can use conditional probabilities to test whether (or not) two variables are independent of each other. Well, I am going to let you in on a secret: Each of the hypothesis tests we have done, using the Z- and t- distributions, the Binomial, Hypergeometric, and Poisson distributions, is just a case of this test of independence. As a case in point, Problem 3 you were designing snowboard binding release mechanisms. The test you conducted was, under H_0, to find the probability that the mean stress to failure of the treated bindings would be less than or equal to the mean stress of the control bindings, *given* they both were drawn from the same population. The conditional probability we are testing is that $p(\mu_1 = \mu_2)$ will stay unchanged, given we treat our experimental sample. If this statement is true:

$$p(\mu_1 = \mu_2) = p(\mu_1 = \mu_2 | treatment)$$

Then binding release stress at failure is *independent* of the treatment—the application of the newly-designed release—the treatment doesn't make a difference. If this statement is true:

$$p\,(\mu_1 = \mu_2) \neq p\,(\mu_1 = \mu_2|treatment)$$

Then we conclude the treatment most likely caused the measured difference.

We use α as our cut-off point for $p\,(\mu_1 = \mu_2)$ under H_0. $p(\mu_1 = \mu_2|$ *treatment*) is symbolized by p, the actual, *experimental* probability that our treatment sample could have been drawn at random from the population of untreated bindings. Comparing p to α, (α being the *theoretical* probability that our treatment sample could have been drawn at random from the population), if $p < \alpha$, then we can conclude that our sample is unlikely to have been drawn from the population of untreated samples at random, i.e., release stress is NOT independent of treatment group–and we reject H_0. Using a test statistic and finding it further in the rejection region is a way we show that $p < \alpha$. We call this a **decision rule**.

If $p \geq \alpha$, then release stress and experimental group are either independent, or we have too little evidence to conclude that they are not independent, so we play it conservative and refuse to reject H_0. Using a test statistic and finding it closer towards the expected value of the sampling distribution than the critical value is a statement that $p \geq \alpha$ (See Table 9.5).

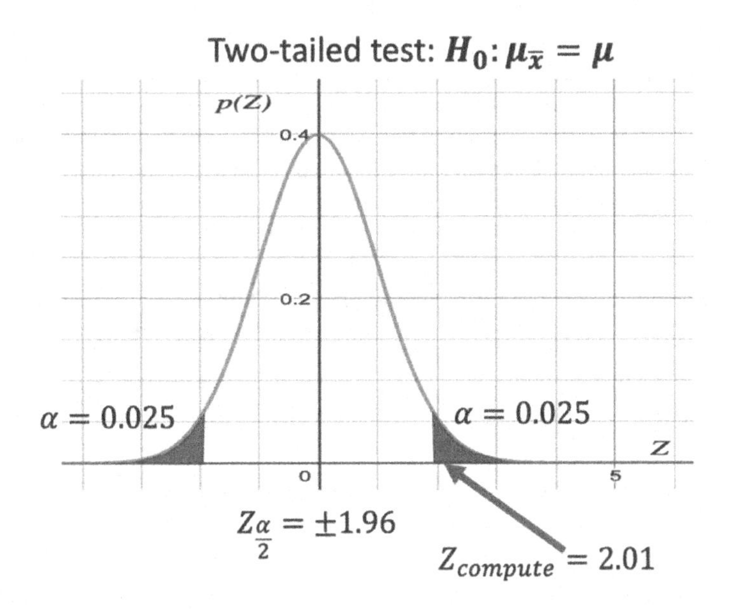

FIGURE 9.5: When the test statistic $\left(Z_{compute}\right)$ is further in the tail of the sampling distribution than the critical value $\left(Z_{\frac{\alpha}{2}}\right)$ for a 2-tailed test), $p < \alpha$

A Decision Rule is a statement that if

$$p(\mu_1 = \mu_2 | treatment) < \text{p}(\mu_1 = \mu_2)$$

there is good enough evidence to reject H_0 in favor of H_1. This conditional probability is typically written in terms of α and p:

If $p < \alpha$, reject H_0 in favor of H_1

Or in terms of a test statistic (this applies to any test statistic, not just t):

If $|t_{computed}| > |t_{criticalata\alpha}|$, reject H_0 in favor of H_1

These statements are algebraic identities to the conditional probability. They are functionally equivalent.

In Chapter 4, we used contingency tables to introduce conditional probabilities using nominal data. Now we are going to expand this discussion of nominal data to examine independence in a more formal way. Analysis of nominal data is no different than interval or ratio data in its logic: We examine our data and figure out its expected value under the Null Hypothesis, then we compare the expected value under the Null Hypothesis with the expected value of our actual data. If the probability is very small that we could have pulled our data, at random, out of the population, then we can reject H_0 in favor of H_1. The major difference with nominal data is that it is discrete, categorical data, and therefore we use a different sampling distribution to calculate probabilities: The χ^2 distribution.

3. Cheah & Ting (2005) studied the frequency of the use of function/cost ratio as a definition of value in 54 different construction firms Southeast Asia. This is called "Value Engineering." They classified companies into the two groups: Contractors vs. Non-contractors, and then recorded whether the use of function/cost ratio was used sometimes or frequently, versus never/rare. Table 9.1 shows their data.

	Type of Firm		
Frequency of VE Use	**Contractors**	**Non-Contractors**	**Total**
Never/Rare	19	10	29
Sometimes/Frequent	10	15	25
Total	29	25	54

TABLE 9.3: Frequency of use of value engineering in different types of companies in SE Asia

(a) Do Contractors use Value Engineering, on average, more often than Non-Contractors?

From Chapter 4, we know that, if Type of Firm and Frequency of Use are independent from each other, then:

$$p\,(\text{Frequency of Use}) = p\,(FrequencyofUse|TypeofFirm)$$

In other words, if the two variables are independent, the expected proportions of Frequency of Use will be the same for both Contractors and Non-Contractors. Right now they are pretty close. We can see that $\frac{29}{54} = 54\%$ of the firms never or rarely use VE. $\frac{25}{54} = 46\%$ of the firms use VE sometimes or frequently. If the two variables were independent, we would expect to see these ratios for each of the types of Firm. This involves simple proportions: Of the 29 firms who are Contractors, $\frac{29}{54}$ should use VE rarely or never. Likewise, of the 25 firms who are Non-Contractors, $\frac{29}{54}$ should use VE rarely or never.

$\frac{29}{54} \times 29 = 15.57$ Contractor firms who *should* use VE if the use is independent of firm

$\frac{29}{54} \times 25 = 13.43$ Non-Contractor firms who *should* use VE if the use is independent of firm

We call these the **Expected Frequencies** of the firms employing VE rarely or never. We can also find the expected frequencies of the firms employing VE sometimes/frequently:

$\frac{25}{54} \times 29 = 13.43$ Contractor firms who *don't* use VE if the use is independent of firm

$\frac{25}{54} \times 25 = 11.57$ Non-Contractor firms who *don't* use VE if the use is independent of firm

I am putting these numbers into the contingency table shown in Table 9.2 to organize them (see Table 9.5). The expected frequencies are in red:

	Type of Firm		
Frequency of VE Use	**Contractors**	**Non-Contractors**	**Total**
Never/Rare	19 (15.57)	10 (13.43)	29
Sometimes/Frequent	10 (13.43)	15 (11.57)	25
Total	29	25	54

TABLE 9.4: Observed versus expected fequency of use of value engineering in different types of companies in SE Asia

Now we can do our comparisons. It looks like fewer than expected Contractors use VE, while more Non-Contractors use VE than expected. Since these proportions are unequal, if this were a strict conditional probability problem, we would say that the two variables are *not independent* because:

$$P\,(Frequency\ of\ Use) \neq P\,(Frequency\ of\ Use|Type\ of\ Firm)$$

But these proportions are pretty close, to where I might be uncomfortable stating that they are systematic, and not just due to random chance. Enter one of the most famous papers in all of statistics, written by Karl Pearson. Pearson was one of W.S. Gosset's mentors and probably the most famous statistician in the world at the turn of the 20th Century. In 1900, Pearson used what he knew about conditional probabilities to show that the expected value of a categorical variable is just

$$E\,(f) = np$$

Where n is the column total of the independent variable, and p is the proportion that a row is of the total N. Designating rows i and columns j, the expected frequency of any cell in a contingency table, i, j becomes:

$$E\left(f_{ij}\right) = Np_{i.}\ p_{.j}$$

$$E\left(f_{ij}\right) = n_{.j}\ \frac{n_{i.}}{N}$$

or, in easier to remember notation.

$$f_{expected} = n_{col}\ \frac{n_{row}}{N}$$

We now have observed frequencies, $f_{observed}$, from our data. We also have expected frequencies, $f_{expected}$, that represent the values our data would take on if our two variables are independent. The variation of the data in each cell–the squared deviation between $f_{observed}$ and $f_{expected}$ is just like what we have always seen:

$$\left(f_{observed} - f_{expected}\right)^2$$

And so an estimate of the variance of the data in a cell is the proportion of this squared deviation in the population under a null hypothesis that assumes that the difference between $f_{observed}$ and $f_{expected}$ is zero. Therefore, n, the denominator of the variance estimate is $f_{expected}$.

$$\frac{\left(f_{observed} - f_{expected}\right)^2}{f_{expected}}$$

Using these cell variances, the variance of the whole set of data is the sum of the mean squared deviations across the table. This sum is distributed as χ^2 ("Chi-squared").

$$\chi^2 = \sum_{i=1}^{row} \sum_{j=1}^{col} \frac{\left(f_{observed} - f_{expected}\right)^2}{f_{expected}}$$

If this "χ^2" is very large, we see that there is a big difference between what we expect under the null hypothesis and what we got with our data. Just like the Normal or *t*-distributions, if χ^2 is large enough to be improbable to happen solely as a function of chance, we can reject H_0 and assume the two variables are not independent. A hypothesis test asks the question, how large does the value of our test statistic, χ^2, have to be to consider the proportions unlikely to have occurred solely as a function of chance?

9.3.1 Null and Alternative Hypotheses for Proportions

Like all hypothesis tests, when we are dealing with proportions, H_0 is a statement of the status quo: That the independent variable (our treatment) does not make a difference—the two variables are independent. The Alternative, H_1 is that the independent variable *does* make a difference—the value of the dependent variable is dependent on the value of the independent. Since these are proportions, they are expressed mathematically (using Phi as the expected value) as:

$$H_0 : \phi_1 \leq \phi_2$$

The proportions of my experimental treatment across the dependent variable, ϕ_1, is equal to (or less than) the proportions of my control group, ϕ_2.

The Alternative is the logical converse—what we really want to find out):

$$H_1 : \phi_1 > \phi_2$$

The χ^2 distribution is highly skewed for most applications, so, though it is possible to compute a CDF for small degrees of freedom, it is impractical given the relatively small, discrete number of categories to be tested. For that reason, we traditionally express the Null and Alternative hypotheses as one-tailed tests. Figure 9.6 shows how χ^2 has nearly vertical slope between $\chi^2 = 0$ and $\chi^2 = 3$.

Degrees of freedom for χ^2relate to the number of categories tested in each variable. For our example of Value Engineering in Southeast Asia, we have 2 categories in the independent variable, Type of Firm. We have 2 categories in the dependent variable, Frequency of Use. When we compute the $f_{expected}$ for one cell, this fixes the final 3 $f_{expected}$ in the table. So, we only have 1 degree of freedom in a 2 × 2 table. In general, once we fix $(row - 1)(column - 1)$ cells, the remaining cells are fixed, yielding (see Table 9.7):

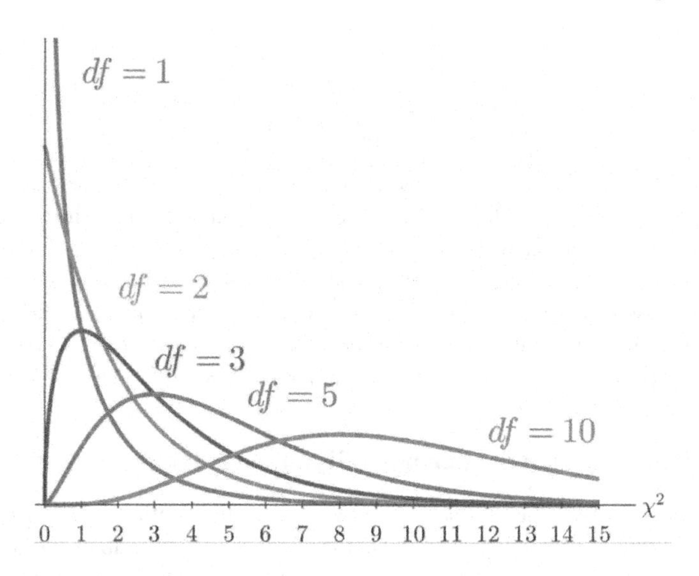

FIGURE 9.6: The χ^2 distribution changes shape dramatically as the number of categories grows

Degrees of Freedom	Type I Error Rate				
	$\alpha = 0.001$	$\alpha = 0.01$	$\alpha = 0.025$	$\alpha = 0.05$	$\alpha = 0.10$
1	10.8275662	6.6348966	5.02388619	3.84145882	2.70554345
2	13.8155106	9.21034037	7.37775891	5.99146455	4.60517019
3	16.2662362	11.3448667	9.3484036	7.8147279	6.25138863
4	18.466827	13.2767041	11.1432868	9.48772904	7.77944034
5	20.5150057	15.0862725	12.832502	11.0704977	9.2363569
6	22.4577445	16.8118938	14.4493753	12.5915872	10.6446407
7	24.3218863	18.4753069	16.0127643	14.0671404	12.0170366
8	26.1244816	20.090235	17.5345461	15.5073131	13.3615661
9	27.8771649	21.6659943	19.0227678	16.9189776	14.6836566
10	29.5882984	23.2092512	20.4831774	18.3070381	15.9871792

FIGURE 9.7: Values of the χ^2 for common degrees of freedom and Type I error rates

$$df = (row - 1)(col - 1)$$

Now we have all the information we need to go back to Problem 7 and see if use of Value Engineering appears to be independent of Firm Type or not.

i. Do Contractors use Value Engineering, on average, more often than Non-Contractors?
ii. What are the Null and Alternative Hypotheses for this test?

iii. What is an appropriate Type I error rate, given what you know about the potential consequences of being wrong?
iv. What is the value of the test statistic?

We will approach parts b, c, and d, to come up with an answer to our original question.

In part b, the authors wanted to know if the proportion of contracting firms using VE was different than non-contracting firms:

$$H_0 : \phi_{Contractor} \leq \phi_{Non-contractor}$$

$$H_1 : \phi_{Contractor} > \phi_{Non-contractor}$$

For Part c, I don't assess the consequences of a Type I error being too egregious. The authors may lose some confidence in the applicability of Value Engineering, but this will not really adversely affect anyone. I would put it at $\alpha = 0.05$, or even 0.10. I am feeling a bit brave today, so I will choose $\alpha = 0.10$.

For Part d, we have a straightforward calculation. The expected frequencies are shown in parenthesis in Table 9.5:

	Type of Firm		
Frequency of Use	**Contractors**	**Non-Contractors**	**Total**
Never/Rare	19 (15.57)	10 (13.43)	29
Sometimes/Frequent	10 (13.43)	15 (11.57)	25
Total	29	25	54

TABLE 9.5: Observed and expected frequencies for value engineering data

$$\chi^2 = \sum_{i=1}^{row} \sum_{j=1}^{col} \frac{\left(f_{observed} - f_{expected}\right)^2}{f_{expected}}$$

$$\chi^2 = \frac{(19 - 15.57)^2}{15.57} + \frac{(10 - 13.43)^2}{13.43} + \frac{(10 - 13.43)^2}{13.43} + \frac{(15 - 11.57)^2}{11.57}$$

$$\chi^2 = 3.514$$

Since this has 2 rows and 2 columns, our degrees of freedom are:

$$df = (2 - 1)(2 - 1)$$

$$df = 1$$

We can now look up the critical value of χ^2 at $df = 1$, and $\alpha = 0.10$ (Figure 9.8).

Degrees of Freedom	Type I Error Rate				
	$\alpha = 0.001$	$\alpha = 0.01$	$\alpha = 0.025$	$\alpha = 0.05$	$\alpha = 0.10$
1	10.8275662	6.6348966	5.02388619	3.84145882	2.70554345
2	13.8155106	9.21034037	7.37775891	5.99146455	4.60517019
3	16.2662362	11.3448667	9.3484036	7.8147279	6.25138863
4	18.466827	13.2767041	11.1432868	9.48772904	7.77944034
5	20.5150057	15.0862725	12.832502	11.0704977	9.2363569
6	22.4577445	16.8118938	14.4493753	12.5915872	10.6446407
7	24.3218863	18.4753069	16.0127643	14.0671404	12.0170366
8	26.1244816	20.090235	17.5345461	15.5073131	13.3615661
9	27.8771649	21.6659943	19.0227678	16.9189776	14.6836566
10	29.5882984	23.2092512	20.4831774	18.3070381	15.9871792

FIGURE 9.8: Critical value of χ^2 at 1 *df*, and α

Since our calculated value of χ^2, 3.514 is greater than the critical value, 2.705, we can conclude that $p < \alpha$, and we can reject H_0 in favor of H_1. FINALLY we have an answer to the problem: The evidence *does* suggest that Contracting firms tend to utilize Value Engineering practices more frequently than Non-Contracting firms.

9.3.2 Assumptions of the χ^2 Test of Independence

9.3.2.1 Independence

Since nominal data is so simple, there are few assumptions such as homogeneity of variance or shape of the sampling distribution of the proportion, that we have to contend with. Basically, we just have to be sure that our *data is collected independently*—this should be a no-brainer by this time. None of our trials should influence any further trials for any pair of observations. This reduces the chance of spurious systematic variation, and allows our model, which assumes just random variation within groups, to be valid.

9.3.2.2 Cell Frequencies Greater Than 5

The second assumption is that we have *adequate sample size* for each cell in the contingency table. As the number of categories grows for each variable, the degrees of freedom grow multiplicatively. With too few counts in a cell, our estimates of the $f_{expected}$ are unreliable. The general rule of thumb is that $f_{expected}$ should not be less than 5.

9.4 F-test of Equal Variances

Sometimes it is helpful to test whether or not one sample shows more consistency in its scores than another sample. When we want to make a decision regarding the quality of two different batteries for an exploratory vehicle for example. Batteries vary on their amp-hours—the time one can expect the battery to last under a constant load. The obviously better battery, all things being equal, is the one that provides the longest life, given the expected load. But not all batteries within a class, or manufactured by the same company, exhibit the same lifespan. There is variability among batteries.

1. You are trying to decide from which company to purchase batteries for your exploratory vehicle, which, if successful, may be included on the next spaceflight to Mars. The two companies provide you with the following data:

$$\bar{x}_{\text{company A}} = 1,598 \text{ hours at constant load}$$

$$\sigma_{\text{company A}} = \pm 80 \text{ hours}$$

$$n_{\text{company A}} = 31 \text{ batteries}$$

$$\bar{x}_{\text{company B}} = 1,602 \text{ hours at constant load}$$

$$\sigma_{\text{company B}} = \pm 120 \text{ hours}$$

$$n_{\text{company B}} = 25 \text{ batteries}$$

2. Which company do you choose to do business with?

   All things being equal, I would choose Company A, because its product has about the same lifespan as Company B, but they are far more consistent. If I am designing an exploratory vehicle to exacting standards, this consistency could mean the difference between collecting important data versus going back to a charging station and waiting until the charge is full enough to continue work. The most common test of such consistency is the F-test of equal variances. The F-test is very simple to conceptualize. It is just the ratio of the variances of two samples:

$$F_{(n_1-1,n_2-1)} = \frac{s_1^2}{s_2^2}$$

   If F is large, then the variation in sample 1 is greater than sample 2. If F is close to 1, there is little to no difference in the variation. For our example:

$$F_{(30,24)} = \frac{120^2}{80^2}$$

$$F_{(30,24)} = 2.25$$

Just like the Z, t, or χ^2 tests, we can compare this computed value of F to a critical value of the F distribution. F is actually a jointly-distributed χ^2 value, so it should be no surprise that its graph at lower degrees of freedom looks very much like χ^2. So, just like t and χ^2, we need to determine its value for a particular combination of degrees of freedom. Figure 9.9 shows values of F for commonly used degrees of freedom at $\alpha = 0.05$. Because there are two samples, each with a separate *df*, we have to make different tables for each value of α (making Matlab a much-coveted tool, when it can calculate F pretty much automatically).

	Values of the F distribution for $\alpha = 0.05$																
df.	**Degrees of Freedom in the Numerator**																
Denom.	**1**	**2**	**3**	**4**	**5**	**6**	**7**	**8**	**9**	**10**	**12**	**14**	**16**	**18**	**20**	**25**	**30**
1	161.5	199.5	215.7	224.6	230.2	234.0	236.8	238.9	240.5	241.9	243.9	245.4	246.5	247.3	248.0	249.3	250.1
2	18.1	19.0	19.2	19.3	19.3	19.3	19.4	19.4	19.4	19.4	19.4	19.4	19.4	19.4	19.5	19.5	19.5
3	10.1	9.6	9.3	9.1	9.0	8.9	8.89	8.85	8.81	8.79	8.74	8.71	8.69	8.67	8.66	8.63	8.62
4	7.71	6.94	6.59	6.39	6.26	6.16	6.09	6.04	6.00	5.96	5.91	5.87	5.84	5.82	5.80	5.77	5.75
5	6.61	5.79	5.41	5.19	5.05	4.95	4.88	4.82	4.77	4.74	4.68	4.64	4.60	4.58	4.56	4.52	4.50
6	5.99	5.14	4.76	4.53	4.39	4.28	4.21	4.15	4.10	4.06	4.00	3.96	3.92	3.90	3.87	3.83	3.81
7	5.59	4.74	4.35	4.12	3.97	3.87	3.79	3.73	3.68	3.64	3.57	3.53	3.49	3.47	3.44	3.40	3.38
8	5.32	4.46	4.07	3.84	3.69	3.58	3.50	3.44	3.39	3.35	3.28	3.24	3.20	3.17	3.15	3.11	3.08
9	5.12	4.26	3.86	3.63	3.48	3.37	3.29	3.23	3.18	3.14	3.07	3.03	2.99	2.96	2.94	2.89	2.86
10	4.96	4.10	3.71	3.48	3.33	3.22	3.14	3.07	3.02	2.98	2.91	2.86	2.83	2.80	2.77	2.73	2.70
11	4.84	3.98	3.59	3.36	3.20	3.09	3.01	2.95	2.90	2.85	2.79	2.74	2.70	2.67	2.65	2.60	2.57
12	4.75	3.89	3.49	3.26	3.11	3.00	2.91	2.85	2.80	2.75	2.69	2.64	2.60	2.57	2.54	2.50	2.47
13	4.67	3.81	3.41	3.18	3.03	2.92	2.83	2.77	2.71	2.67	2.60	2.55	2.51	2.48	2.46	2.41	2.38
14	4.60	3.74	3.34	3.11	2.96	2.85	2.76	2.70	2.65	2.60	2.53	2.48	2.44	2.41	2.39	2.34	2.31
15	4.54	3.68	3.29	3.06	2.90	2.79	2.71	2.64	2.59	2.54	2.48	2.42	2.38	2.35	2.33	2.28	2.25
16	4.49	3.63	3.24	3.01	2.85	2.74	2.66	2.59	2.54	2.49	2.42	2.37	2.33	2.30	2.28	2.23	2.19
17	4.45	3.59	3.20	2.96	2.81	2.70	2.61	2.55	2.49	2.45	2.38	2.33	2.29	2.26	2.23	2.18	2.15
18	4.41	3.55	3.16	2.93	2.77	2.66	2.58	2.51	2.46	2.41	2.34	2.29	2.25	2.22	2.19	2.14	2.11
19	4.38	3.52	3.13	2.90	2.74	2.63	2.54	2.48	2.42	2.38	2.31	2.26	2.21	2.18	2.16	2.11	2.07
20	4.35	3.49	3.10	2.87	2.71	2.60	2.51	2.45	2.39	2.35	2.28	2.22	2.18	2.15	2.12	2.07	2.04
21	4.32	3.47	3.07	2.84	2.68	2.57	2.49	2.42	2.37	2.32	2.25	2.20	2.16	2.12	2.10	2.05	2.01
22	4.30	3.44	3.05	2.82	2.66	2.55	2.46	2.40	2.34	2.30	2.23	2.17	2.13	2.10	2.07	2.02	1.98
23	4.28	3.42	3.03	2.80	2.64	2.53	2.44	2.37	2.32	2.27	2.20	2.15	2.11	2.08	2.05	2.00	1.96
24	4.26	3.40	3.01	2.78	2.62	2.51	2.42	2.36	2.30	2.25	2.18	2.13	2.09	2.05	2.03	1.97	1.94
25	4.24	3.39	2.99	2.76	2.60	2.49	2.40	2.34	2.28	2.24	2.16	2.11	2.07	2.04	2.01	1.96	1.92
26	4.23	3.37	2.98	2.74	2.59	2.47	2.39	2.32	2.27	2.22	2.15	2.09	2.05	2.02	1.99	1.94	1.90
27	4.21	3.35	2.96	2.73	2.57	2.46	2.37	2.31	2.25	2.20	2.13	2.08	2.04	2.00	1.97	1.92	1.88
28	4.20	3.34	2.95	2.71	2.56	2.45	2.36	2.29	2.24	2.19	2.12	2.06	2.02	1.99	1.96	1.91	1.87
29	4.18	3.33	2.93	2.70	2.55	2.43	2.35	2.28	2.22	2.18	2.10	2.05	2.01	1.97	1.94	1.89	1.85
30	4.17	3.32	2.92	2.69	2.53	2.42	2.33	2.27	2.21	2.16	2.09	2.04	1.99	1.96	1.93	1.88	1.84
40	4.08	3.23	2.84	2.61	2.45	2.34	2.25	2.18	2.12	2.08	2.00	1.95	1.90	1.87	1.84	1.78	1.74
60	4.00	3.15	2.76	2.53	2.37	2.25	2.17	2.10	2.04	1.99	1.92	1.86	1.82	1.78	1.75	1.69	1.65
100	3.94	3.09	2.70	2.46	2.31	2.19	2.10	2.03	1.97	1.93	1.85	1.79	1.75	1.71	1.68	1.62	1.57
∞	3.84	3.00	2.60	2.37	2.21	2.10	2.01	1.94	1.88	1.83	1.75	1.69	1.64	1.60	1.57	1.51	1.46

FIGURE 9.9: Values of the F-distribution for $\alpha = 0.05$. Degrees of freedom for the numerator are listed across the top of the table. Degrees of freedom for the denominator are listed down the side

Looking up the critical value of in F, we have to coordinate both degrees of freedom, keeping track of which sample variance represents our numerator, and which variance represents the denominator. For our example, we look down the column representing 30 *df* in Company B, and then look across the row representing 24 *df* in Company A. The critical value of F is 1.94 (Figure 9.10).

df. Denom.	Values of the F distribution for $\alpha = 0.05$ — Degrees of Freedom in the Numerator																
	1	**2**	**3**	**4**	**5**	**6**	**7**	**8**	**9**	**10**	**12**	**14**	**16**	**18**	**20**	**25**	**30**
1	161.5	199.5	215.7	224.6	230.2	234.0	236.8	238.9	240.5	241.9	243.9	245.4	246.5	247.3	248.0	249.3	250.1
2	18.1	19.0	19.2	19.3	19.3	19.3	19.4	19.4	19.4	19.4	19.4	19.4	19.4	19.4	19.5	19.5	19.5
3	10.1	9.6	9.3	9.1	9.0	8.9	8.89	8.85	8.81	8.79	8.74	8.71	8.69	8.67	8.66	8.63	8.62
4	7.71	6.94	6.59	6.39	6.26	6.16	6.09	6.04	6.00	5.96	5.91	5.87	5.84	5.82	5.80	5.77	5.75
5	6.61	5.79	5.41	5.19	5.05	4.95	4.88	4.82	4.77	4.74	4.68	4.64	4.60	4.58	4.56	4.52	4.50
6	5.99	5.14	4.76	4.53	4.39	4.28	4.21	4.15	4.10	4.06	4.00	3.96	3.92	3.90	3.87	3.83	3.81
7	5.59	4.74	4.35	4.12	3.97	3.87	3.79	3.73	3.68	3.64	3.57	3.53	3.49	3.47	3.44	3.40	3.38
8	5.32	4.46	4.07	3.84	3.69	3.58	3.50	3.44	3.39	3.35	3.28	3.24	3.20	3.17	3.15	3.11	3.08
9	5.12	4.26	3.86	3.63	3.48	3.37	3.29	3.23	3.18	3.14	3.07	3.03	2.99	2.96	2.94	2.89	2.86
10	4.96	4.10	3.71	3.48	3.33	3.22	3.14	3.07	3.02	2.98	2.91	2.86	2.83	2.80	2.77	2.73	2.70
11	4.84	3.98	3.59	3.36	3.20	3.09	3.01	2.95	2.90	2.85	2.79	2.74	2.70	2.67	2.65	2.60	2.57
12	4.75	3.89	3.49	3.26	3.11	3.00	2.91	2.85	2.80	2.75	2.69	2.64	2.60	2.57	2.54	2.50	2.47
13	4.67	3.81	3.41	3.18	3.03	2.92	2.83	2.77	2.71	2.67	2.60	2.55	2.51	2.48	2.46	2.41	2.38
14	4.60	3.74	3.34	3.11	2.96	2.85	2.76	2.70	2.65	2.60	2.53	2.48	2.44	2.41	2.39	2.34	2.31
15	4.54	3.68	3.29	3.06	2.90	2.79	2.71	2.64	2.59	2.54	2.48	2.42	2.38	2.35	2.33	2.28	2.25
16	4.49	3.63	3.24	3.01	2.85	2.74	2.66	2.59	2.54	2.49	2.42	2.37	2.33	2.30	2.28	2.23	2.19
17	4.45	3.59	3.20	2.96	2.81	2.70	2.61	2.55	2.49	2.45	2.38	2.33	2.29	2.26	2.23	2.18	2.15
18	4.41	3.55	3.16	2.93	2.77	2.66	2.58	2.51	2.46	2.41	2.34	2.29	2.25	2.22	2.19	2.14	2.11
19	4.38	3.52	3.13	2.90	2.74	2.63	2.54	2.48	2.42	2.38	2.31	2.26	2.21	2.18	2.16	2.11	2.07
20	4.35	3.49	3.10	2.87	2.71	2.60	2.51	2.45	2.39	2.35	2.28	2.22	2.18	2.15	2.12	2.07	2.04
21	4.32	3.47	3.07	2.84	2.68	2.57	2.49	2.42	2.37	2.32	2.25	2.20	2.16	2.12	2.10	2.05	2.01
22	4.30	3.44	3.05	2.82	2.66	2.55	2.46	2.40	2.34	2.30	2.23	2.17	2.13	2.10	2.07	2.02	1.98
23	4.28	3.42	3.03	2.80	2.64	2.53	2.44	2.37	2.32	2.27	2.20	2.15	2.11	2.08	2.05	2.00	1.96
24	4.26	3.40	3.01	2.78	2.62	2.51	2.42	2.36	2.30	2.25	2.18	2.13	2.09	2.05	2.03	1.97	1.94
25	4.24	3.39	2.99	2.76	2.60	2.49	2.40	2.34	2.28	2.24	2.16	2.11	2.07	2.04	2.01	1.96	1.92
26	4.23	3.37	2.98	2.74	2.59	2.47	2.39	2.32	2.27	2.22	2.15	2.09	2.05	2.02	1.99	1.94	1.90
27	4.21	3.35	2.96	2.73	2.57	2.46	2.37	2.31	2.25	2.20	2.13	2.08	2.04	2.00	1.97	1.92	1.88
28	4.20	3.34	2.95	2.71	2.56	2.45	2.36	2.29	2.24	2.19	2.12	2.06	2.02	1.99	1.96	1.91	1.87
29	4.18	3.33	2.93	2.70	2.55	2.43	2.35	2.28	2.22	2.18	2.10	2.05	2.01	1.97	1.94	1.89	1.85
30	4.17	3.32	2.92	2.69	2.53	2.42	2.33	2.27	2.21	2.16	2.09	2.04	1.99	1.96	1.93	1.88	1.84
40	4.08	3.23	2.84	2.61	2.45	2.34	2.25	2.18	2.12	2.08	2.00	1.95	1.90	1.87	1.84	1.78	1.74
60	4.00	3.15	2.76	2.53	2.37	2.25	2.17	2.10	2.04	1.99	1.92	1.86	1.82	1.78	1.75	1.69	1.65
100	3.94	3.09	2.70	2.46	2.31	2.19	2.10	2.03	1.97	1.93	1.85	1.79	1.75	1.71	1.68	1.62	1.57
∞	3.84	3.00	2.60	2.37	2.21	2.10	2.01	1.94	1.88	1.83	1.75	1.69	1.64	1.60	1.57	1.51	1.46

FIGURE 9.10: Finding the critical value of F at $df = (30{,}24)$

Our calculated value of $F_{(30,24)}$, 2.25, is further in the right tail of the distribution than F_{critical}, 1.94. So, we can conclude that $p(2.25) < \alpha$.

9.4.1 Assumptions of the F-test

The assumptions of the F-test are, like its parent the χ^2 test, pretty unrestrictive:

Independence Both parent distributions are independent. Again, this is just an assumption of any hypothesis test. We have to be pretty sure we don't have systematic bias in our sampling and measurement procedures.

Normalcy Both parent distributions are assumed to be normal or close to normal. Violations of this assumption wreak havoc with F. The Variance of a skewed distribution is highly affected by the extreme values, so a distribution that is asymmetric will tend to underestimate one, the other, or both values in the test statistic.

9.5 Summary of 2-sample Statistics

The most basic model of an experiment, where we do not know the parameters of the populations from which our experimental groups have been sampled, involves taking two random samples, applying a treatment to one of the samples, and then comparing the behavior of the treated sample with that of the control. This **Random Sample Design** is a general model that applies to nominal (and ordinal) as well as interval and ratio data.

The testing of two-sample hypotheses starts with the assumption that both samples are, for all intents and purposes, equivalent except for random error. Armed with this assumption, when we compare the behavior of one sample to the other, we can either take the difference between a sample statistic and its expected value $(\overline{x}_1 - \overline{x}_2)$, $\overline{d}$, and $\left(f_{observed} - f_{expected}\right)$, or we can take their ratio: $\frac{s_1^2}{s_2^2}$. If these comparisons are unlikely, under the Null Hypothesis, we consider that evidence that our experimental treatment *caused* the observed difference.

The t, χ^2, and F distributions are the most common and most useful tests in the engineer's arsenal. Even when our data gets complex, with multiple experimental conditions across several dependent variables, the basic model of the **Random Sample Design**, and its associated hypothesis tests is the core conceptual model for all empirical exploration and verification.

9.6 References

Boneau, C. A. (1960). The effects of violations of assumptions underlying the t test. *Psychological bulletin*, *57*(1), 49-64.

Cheah, C. Y., & Ting, S. K. (2005). Appraisal of value engineering in construction in Southeast Asia. *International Journal of Project Management*, *23*(2), 151-158.

Hegel, F., Krach, S., Kircher, T., Wrede, B., & Sagerer, G. (2008, August). Understanding social robots: A user study on anthropomorphism. In *RO-MAN 2008-The 17th IEEE International Symposium on Robot and Human Interactive Communication* (pp. 574-579). IEEE.

Sawilowsky, S. S., & Blair, R. C. (1992). A more realistic look at the robustness and Type II error properties of the t test to departures from population normality. Psychological Bulletin, 111(2), 352-360.

Snaterse, M. (2014). chi2cont. Matlab code retrieved from https://uk.mathworks.com/matlabcentral/mlc-downloads/downloads/submissions/45203/versions/1/previews/chi2cont.m/index.html

Uttley, J. (2019). Power Analysis, Sample Size, and Assessment of Statistical Assumptions—Improving the Evidential Value of Lighting Research. *Leukos*, *15*(2-3), 143-162.

9.7 Study Problems for Chapter 9

1. Discuss the logic of conditional probability as it pertains to hypothesis testing. How does this apply to the notion of independence of two samples?

   We know from basic probability that two events are independent if $p(B|A) = p(B)$. That is, the probability of B occurring has nothing to do with A occurring. With hypothesis testing using samples, the null hypothesis is a statement of no effect—a statement of independence of the population(s) from which the sample(s) were drawn. So, a hypothesis test is estimating whether or not the probability of a sample being drawn from a population, *given* a treatment, is the same as a control sample, which was drawn from the population with no treatment.

2. Choose a topic that you would be interested in studying (it could be a capstone project, term paper, or just something you are interested in). Develop a hypothesis about the effect of an independent variable. Now describe a study with a Random Sample Design that could be used to test your hypothesis.

i. How will you determine your experimental groups?
ii. How will you insure your groups are functionally equivalent before treatment?
iii. What scale of measure will you use for the dependent variable?
iv. What are the Null and Alternative Hypotheses?
v. For your dependent variable, what is the appropriate test of your hypothesis?
vi. For a given Type I error, what is the critical value of the test statistic for a given sample size?

Answers will vary depending on your interests.

A. Experimental groups need to be thought out carefully so that they are as equivalent as possible so that the only difference between treatment conditions is the treatment itself (plus insignificant systematic error, plus random error).

B. How will you insure your groups are functionally equivalent before treatment?

Random assignment will guarantee that the only difference in groups is random. One can also use blocking (Chapter 6) to equilibrate samples across important pre-existing variables.

C. What scale of measure will you use for the dependent variable?

Answers will vary depending on the context.

D. What are the Null and Alternative Hypotheses?

The Null Hypothesis should be a statement of no effect. The Alternative Hypothesis is what you really expect to find. A two-tailed hypothesis posits no directional effect. A one-tailed hypothesis posits a directional effect.

E. For your dependent variable, what is the appropriate test of your hypothesis?

Right now we have a few options! The two-sample t-test should be considered if the dependent variable is continuous (interval or ratio data); The Chi-square test of independence should be considered if the dependent variable is frequency; The F-test of equal variances should be considered if the dependent variable is continuous and if you are interested in comparing the spread of two distributions.

F. For a given Type I error, what is the critical value of the test statistic for a given sample size?

Answers will vary. Make sure you keep your degrees of freedom straight!

3. What is Effect Size? Why is it important to consider Effect Size prior to conducting an experiment?

Effect size is the distance between sample means, scaled by their pooled standard deviation. It is important, because there will always be some non-zero difference in sample means. The effect size helps us determine when a difference is large enough to be considered worth following up on after a hypothesis test, it tells us the extent to which our treatment made an effect.

4. **The test statistic for a hypothesis test and the probability value computed for that test statistic are inversely related. State the logic of Decision Rules using both test statistics compared to critical values, and p versus α.**
When a test statistic is large, beyond the critical value of a distribution, the probability that any sample, drawn randomly from the sampling distribution could have the test statistic value or greater, is small. Likewise, when its magnitude is large, but its sign is negative, the probability of drawing a sample randomly from the population, with a value further in the tail of the distribution is small. If this probability is smaller than the Type I error rate agreed upon a priori, then we consider the difference measured to be significant—very unlikely due to chance, and therefore more likely due to the effect posited by the Alternative Hypothesis.
5. **Why is Practical Significance different from Statistical Significance?**
If you choose a sample size that is extremely large, the power of the test will be very high. That means that even very small treatment effects can be considered significant, from a probabilistic perspective. These very small differences may be so small as to not be worth devoting extra time, money, personnel, to pursue in the process of improving a design. So we want both adequate power, but also a decent effect size to insure that the differences we do detect have practical significance.
6. **What are the advantages and disadvantages of using a Paired Sample t-test versus a 2-sample t-test? Under what conditions is one more appropriate than the other?**
The paired-sample t-test has a great advantage of being able to use a single sample with a relatively small sample size to determine the effect of a treatment by measuring it twice: Pre-test versus Post-test. This is cost-effective, and can account for naturally occurring change. The 2-sample t-test has an advantage in its ability to compare two functionally different samples to determine the effect of a treatment applied to only 1. Because the two samples are functionally the same, the difference measured post-treatment can be attributed to the treatment. This type of test can be applied in situations where objects cannot be measured twice (due to fatigue or test effects). Its disadvantage is that the overall number of measured phenomena is large relative to the paired-sample t-test.
7. **A researcher hypothesizes that electrical pulses, appearing at random intervals around a food source will result in a decrease in food intake in rats. She wants to design a means of reducing the depredation caused by vermin on grain stores. Ten rats (kept at 80% body weight) are tested for the number of times they approach a target food source in a**

10 minute period of time first with and then without the electric pulses turned on. The testing conditions are counter balanced.

Rat	No Pulse	Pulse
1	12	8
2	7	7
3	3	4
4	11	14
5	8	6
6	5	7
7	14	12
8	7	5
9	9	5
10	10	8

vii. **What is the appropriate test to run on this data?**
Because the same rats are tested twice, this is a pre-post design. The Paired-sample *t*-test is an appropriate analysis technique.

viii. **Should you run a 1-tailed or 2-tailed test for this question? What is H_0 and H_1?**
Because I know a bit about the impact of punishment on behavior, I think there will be an initial reduction in the number of times rats approach food when they have been buzzed by electricity. This is a one-tailed, right-tailed test.

$$H_0 : \mu_{\sim pulse} - \mu_{pulse} \leq 0$$

$$H_1 : \mu_{\sim pulse} - \mu_{pulse} > 0$$

ix. **What is an appropriate Type I error rate? Justify your answer.**
The results of this experiment, if we find a significant effect of the pulses, would be to install equipment in our grain storage facilities. This could cost a lot of money, so if we are making a Type I error, it could cause a lot of financial loss with no improvement in curtailing the depredation. For this reason, I might use a moderate α of 0.05 or 0.01. But depending on the consequences you envision, this may change to reflect the actual context in which you find yourself.

x. **Construct the 1− α % confidence interval for the difference in means for this analysis, given your Type I error rate in part a.**
Because this is a 1-tailed, right-tailed hypothesis, the appropriate confidence interval looks like this:

$$1 - \alpha\%CI = \bar{d} - t_{(\alpha,df=n_d-1)}\frac{s_d}{\sqrt{n_d}}$$

I can use Matlab to help me compute this confidence interval. First, I enter the two columns of data as two column vectors. I can then use the Matlab function:
[h,p,ci,stats] = ttest(nopulse1,pulse1,'Alpha',0.01, 'tail','right');
Matlab gives me a 99% confidence interval of $(-1.446, \infty)$
What is the value of the test statistic?
Matlab returns a value of $t = 1.3156$

xi. **Interpret the result of your hypothesis test**
Because zero, the hypothesized value of $\overline{d}$ under the null hypothesis, is in this interval, we fail to reject the null hypothesis. It looks like the pulses didn't do a great job of discouraging those darn rats! This is consistent with a test statistic that is below the critical value at $\alpha = 0.01$

xii. **What is the power of this test as conducted? Should the researcher have sampled more rats?**

Because this is a paired-sample test, you need to use the function for a 1-sample t-test, where the 1-sample is our sample of pre-post differences. The following Matlab code provides an estimate of the power of the test using the population mean $\mu_d = 0$, the sample mean $\overline{x}_d = 1$, and the sample standard deviation, $s = 2.4037$. $n = 10$ pairs of differences pre-post.

«power = sampsizepwr('t',[0 2.4037], [1],[],10,'Alpha',0.01,'Tail','right');

The computed power is 0.1154. I could fiddle with the code, substituting new values for n to see what sample size would give a reasonably powerful test, or I could alter the code a bit thusly, making n a variable, and the power at 0.8, a strong test. You may use any value for power that you feel is the appropriate probability you want of making a correct decision:

» n = sampsizepwr('t',[0 2.4037], [1],[0.8],[],'Alpha',0.01,'Tail','right');

This gives an estimate of 61 pairs of measurements that is needed to detect a mean difference of 1 time approaching the food. Looks like rats are hard to deter!

8. You take a random sample of 100 automobiles from an assembly line and outfit half of them with new fire-injectors you have designed. The other half is outfitted with standard spark-plugs. You test the difference in performance for both samples, recording if there was an increase, no change, or a decrease in performance. Your data is summarized below:

	Type of Plug		
	Experimental	**Regular**	**Total**
Increase	23	17	40
No Change	14	17	31
Decrease	13	16	29
Total	50	50	100

(a) **What is the appropriate test to run on this data?**

We have two categorical variables, the independent variable, Type of Plug (Experimental versus Regular), and Performance (Increase, No Change, Decrease). This is appropriate for a χ^2 test of independence.

(b) **Should you run a 1-tailed or 2-tailed test for this question? What is H_0 and H_1?**

I have to make some assumptions here, but as engineers, we don't create products with the notion that they will perform more poorly than the status quo products. We use the physics and fluid dynamics to predict the performance of our products. Because of this we have a very good idea what direction the effect of a treatment will be. So, a one-tailed test here is appropriate. If I had no idea how the experimental plugs would perform, say in the very beginning of their design, I might run a 2-tailed test.

$$H_0 : \Phi_{experimental} \leq \Phi_{regular}$$

$$H_1 : \Phi_{experimental} > \Phi_{regular}$$

(c) **What is an appropriate Type I error rate? Justify your answer.**

For this, because it is unlikely to cause loss of life and limb, and because the unit costs of the plugs is fairly low, I would go with $\alpha = 0.05$, but of course the actual alpha one uses must be determined against the costs of making a Type I error.

(d) **What is the value of the test statistic?**

Using the following Matlab script, I input the data as a matrix with 2 columns and 3 rows (I don't compute the marginal totals). Then the following code will compute the χ^2 test statistic for me.. **You are welcome to cut-and-paste this script into Matlab, because it has particularly poor routines for the χ^2 test of independence.**

```
function [h,p,X2,df,X2crit] = chi2cont(x,varargin)
if isempty(varargin)
        alpha = 0.05;
else
        alpha = varargin1;
end
e = sum(x,2)*sum(x)/sum(x(:));
X2 = (x-e).^2./e;
X2 = sum(X2(:));
df=prod(size(x)-[1 1]);
p = 1-chi2cdf(X2,df);
X2crit=icdf('chisquare',[1-alpha],df);
h = double(p<=alpha);
```

Now typing in the chi2cont function: «[h,p,X2] = chi2cont(x, alpha);
A χ^2 value (2 df) of 1.5007 is returned, $p = 0.4722$

(e) Interpret the result of your hypothesis test

Since the χ^2 computed is less than the critical value of 6.635, and p is greater than $\alpha = 0.05$, I fail to reject the null hypothesis. I need to do more work on the design of my experimental plugs.

(f) Did the researcher account for all the assumptions of this test?

There is clearly adequate sample size. The number of plugs in each of the cells of the table indicate that the expected frequencies, f_e cannot be less than 5. Because the sample is random, I assume that the measurements are independent. So it looks like the researcher did her job well in design of this experiment.

9 An HVAC engineer studied the efficiency of two different heat pumps: One for which she had installed anhydrous ammonia as a refrigerant (Experiment). The other used R134a hydrofluorocarbon refrigerant (Control). She wanted to know if the experimental refrigerant improved the efficiency of the heat pump. She took two random samples of heat pumps, and installed the experimental refrigerant in 11 of them. Here is her data.

Control	Experimental
.546	0.556
.547	0.874
.774	0.554
.465	0.635
.459	0.672
.665	0.754
.456	0.558
.539	0.574
.528	0.664
	0.586
	0.578

(a) What is the appropriate test to run on this data?

Because this is a Random Sample Design, with a continuous dependent variable (Ratio data), we can use the 2-sample t-test.

(b) Should you run a 1-tailed or 2-tailed test for this question? What is H_0 and H_1?

Because the experimental question is about *improving* the efficiency. A 1-tailed test, right-tailed test is most appropriate.

$$H_o : \mu_{experimental} \leq \mu_{control}$$

$$H_1 : \mu_{experimental} > \mu_{control}$$

(c) **What is an appropriate Type I error rate? Justify your answer.**

This is testing a new refrigerant. Overall, I don't think the stakes are high enough to have a tiny Type I error rate. $\alpha = 0.05$ is a good, conservative figure. But if the cost of making a Type I error were very high (e.g., the dangers of anhydrous ammonia, or the cost of installing new condensers to handle the different temperatures and flow rates), one might be even more conservative and choose $\alpha = 0.01$.

(d) **What is the value of the test statistic?**

First, I enter my two columns of data as column vectors, experimental and control. The pooled standard error accounts for minor differences in sample sizes, so the vectors don't have to be the same length. Then I use the Matlab ttest2 function.

«[h,p,ci,stats]=ttest2(experimental, control, 'alpha',0.05,'tail','right');

The stats structure returns a test statistic with a value of $t = 1.8111$

(e) **Compute the $1 - \alpha\%$ confidence interval**

ci returns the interval $(0.0036, \infty)$.

(f) **Interpret the result of your hypothesis test**

Because the calculated value of t, 1.811 is larger than the critical value of $t_{18} = 1.734$, and p is less than $\alpha = 0.05$, we reject the null hypothesis and assume that the anhydrous ammonia increased the efficiency of the heat exchanger. The confidence interval does not contain 0, the difference in means hypothesized under H_o, consistent with the test statistic.

(g) **What is the power of this test? Is it ok, underpowered or overpowered?**

«power = sampsizepwr('t2',[mean(control) 0.1027], [mean(exp)], [], 9, 'Ratio', 1.22222, 'Alpha', 0.05, 'Tail', 'right');

Returns a power of 0.5388. This is pretty low, I would want it closer to 0.75 or 0.80, but in our case, we did find a significant difference despite the low power.

(h) **Did the researcher account for all the assumptions of this test?**

Histograms of the two samples reveal distributions that aren't very normal. The experimental sample, in particular looks skewed. So, we may be violating the assumption of normalcy of parent populations.

The variance of the experimental sample, $s^2_{\text{exp}} = 0.0102$, and the variance of the control sample, $s^2_{\text{exp}} = 0.0110$. These are very close, so it appears that homogeneity of variance is accounted for.

10. **The same researcher as in Problem 8 wanted to see if the variation in her experimental data was somehow greater (less consistent) or less (more consistent) than the control group.**

(a) **What is the appropriate test to run on this data?**

Because we are looking at the precision of the samples, we should use the F-test of equal variances.

(b) **What is an appropriate Type I error rate? Justify your answer.**

Again, I think that $\alpha = 0.05$ is appropriate for the same reasons as above.

(c) **What is the value of the test statistic?**

$$F = \frac{s^2_{control}}{s^2_{\text{exp}}} = \frac{0.0110}{0.0102} = 1.078$$

(d) **Interpret the result of your hypothesis test**

The degrees of freedom used to fix the F-distribution here are $9 - 1 = 8$ for the control group, and $11 - 1 = 10$ for the experimental group. I can then look up the critical value, $F_{(8,10)} = 3.07$.

(e) **Did the researcher account for all the assumptions of this test?**

The two samples appear to be independent, given the experimental design.

The assumption of normalcy may be violated because of the skew of (particularly) the experimental sample.

10

Simple Linear Regression

Thus far, we have studied some simple experimental designs that test whether or not two expected values of a variable, that are supposed to be the same, are different enough that we can't really attribute that difference to random error. This, in essence, is determining the relationship of a categorical independent variable (Experimental Condition) with a dependent variable which can be either continuous (our set of measurements for Z and t), or categorical (frequencies of each categorical outcome for the binomial, hypergeometric, Poisson, and χ^2 distributions).

Many times, we don't have the luxury of experimentally manipulating phenomena. Additionally, many of the relationships we want to know about are between two continuous variables. Take the following set of data for example, taken from a report on the relationship between the specific gravity of wood species and how it varies as a function of the distance the wood sample is taken from the pith of the tree: The pith, generally, being less dense, and the heartwood, generally, being more dense (see Figure 10.1):

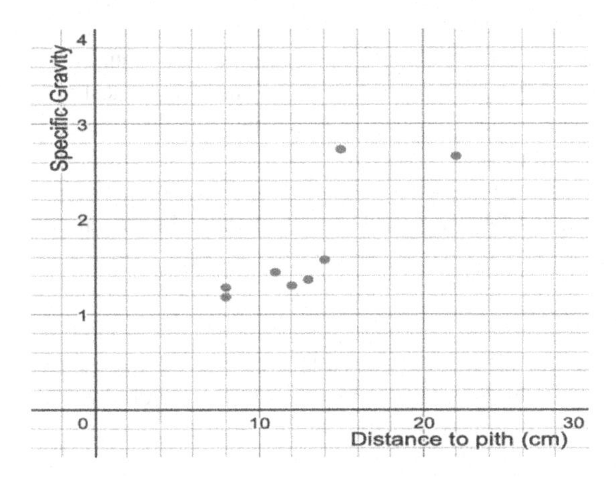

FIGURE 10.1: Relationship between specific gravity of *Cecropia Obtusifolia* by distance to pith

DOI: 10.1201/9781003094227-10

The researchers in this study wanted to be able to compute biomass of a plantation of trees, and needed to create a model that would describe, given the diameter of a tree, the way in which the wood density varies across its cross section.

1. Given the data in Figure 10.1, what is a reasonable estimate of the value of Specific Gravity, given a Distance to pith of 12 cm? Given a distance of 20 cm?

   This question seems pretty straightforward, but there is more hidden underneath the surface. There are a lot of assumptions you have to make to be able to answer this question. You might play it safe and say that the data points are the best guess, because they are real values of the relationship. So, SG at 12 cm is about 1.3. For 20 cm, or other values that are not represented in the data set, you have to make an assumption about the shape of the relationship, and the ways in which the function you ascribe to that shape relates to the variability of the data.

   Most students, straight out of linear algebra, would automatically try to fit a line to the cloud of points. But is this relationship linear? Why would you think so? Could it be quadratic, but just not show a lot of curve in the range of the independent variable within which the data were collected? Why not exponential (there are good reasons), or logistic (many ecological examples are logistic as the system being measured reaches some natural limit of energy)? This question is important if you really want to know the shape of the underlying pattern. If you only want to be able to predict points within the contextual domain, it may not be as important—linear approximation of a curve is often good enough.

   The data collected in Figure 10.1 also shows variability. The points are more densely packed between 8 and 14 cm than they are between 14 and 22 cm. Does this difference in density affect your estimate? Do you trust the data more at the lower end of the domain, or the upper end? This question is also important to tackle, because once you decide upon the shape of your model, you have to fit it to the cloud of data, trying to maximize its fit by minimizing the error of prediction. Suppose I *do* assume the relationship is linear, or very close to linear. That means the model I am going to use for this relationship has the form:

$$\hat{y}_i = \beta_0 + \beta_1 x$$

   The little "hat" over the letter denoting the dependent variable, y in this case, is used to indicate that the values in the model are *predicted* values, not necessarily actual values. Statisticians actually call the predicted value "y-hat."

   One could reasonably fit any number of linear functions to this data:

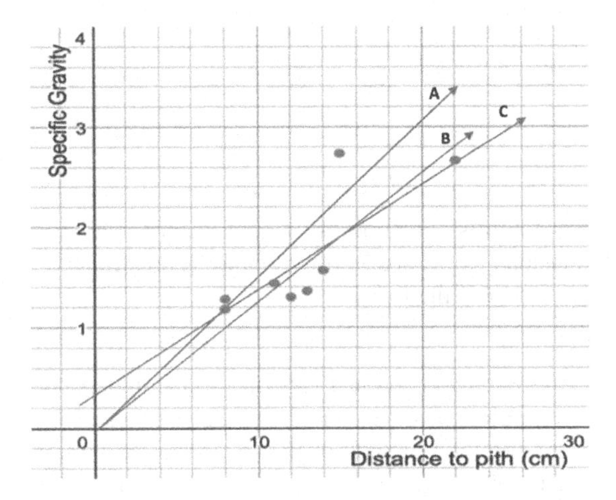

FIGURE 10.2: Potential models that "fit" the given data

Which one has the best fit? By what criteria do we judge, "best?" Two of the models in Figure 10.2 go through the origin. Since we know that both specific gravity and distance have absolute zeros (they exist on a ratio scale), the best fit line should, theoretically, pass through the origin. But, given the actual data, the y-intercept of the line may or may not correspond exactly to the origin, and line C appears to balance out the data points a bit better than either line A or line B.

To determine "best" we have to define the error of measurement in some way. It could be the absolute distances of each measured point, from the plotted function. This is entirely reasonable, and is used in practice in a number of fields. It could also be a Euclidean distance, using the Pythagorean theorem, or some kind of *Maximum Likelihood* function, which deals with the probabilities of the points appearing in a bivariate sample and parameters that describe the line of best fit. Both of these methods are used in practice.

Just like a "univariate" sample, we can assess the total Euclidean distance from a central value as the sum of the squared distances from the mean, and the average distance as this sum divided by the degrees of freedom (the 2nd moment of x around $\overline{x}$). In **Linear Regression**, we first fit a line to bivariate data. We can then define the error of prediction as the squared difference between the predicted values, $\hat{y}$ –those that lie on the prediction line–and the actual measured values of our dependent variable, y_i:

$$\varepsilon_i = (y_i - \hat{y}_i)^2$$

This is a Euclidean distance between these points in the model. The total distance is just the sum of the individual distances:

$$SS_{\varepsilon} = \sum_{i=1}^{n} (y_i - \hat{y}_i)^2$$

Linear Regression is a method of fitting a linear function to a set of data. In two dimensions, the function truly is line, geometrically. It is expressed as a linear equation of the form:

$y_i = \beta_0 + \beta_1 x_i + \varepsilon_i$

where y is an observed dependent variable, x is an independent variable, ε_i is the error of prediction, and β_0 and β_1 are the parameters that fix the values of y. The most common method of linear regression is the **Least Squares** method, where the squared error of prediction is minimized, yielding the line of best fit to the data:

$\hat{y}_i = \beta_0 + \beta_1 x_i$

Where $\hat{y}_i$ is the predicted value of y_i.

The assumptions of linear regression are:

(a) Independence of the observations

(b) The true relationship between independent and dependent variables is linear in its parameters;

(c) Errors of prediction are normally distributed around zero; and

(d) Errors of prediction are equally distributed around the line of best fit.

Let's see how this plays out for those three arbitrary lines fit to the data in Problem 1.

The equations of the three prediction lines in Figure 10.2 are:

$$A: y_i = 0 + 0.1538x_i$$

$$B: y_i = 0 + 0.1273x_i$$

$$C: y_i = 0.36 + .1228x_i$$

Here are the actual measurements taken from the study along with the predicted values from the three prediction lines, and their individual error of prediction (Table 10.1):

Distance to Pith	Specific Gravity	A $\hat{y}$	A $(y_i - \hat{y})^2$	B $\hat{y}$	B $(y_i - \hat{y})^2$	C $\hat{y}$	C $(y_i - \hat{y})^2$
22	2.66	3.38	0.5184	2.80	0.0196	2.62	0.0016
15	2.73	2.31	0.1764	1.91	0.6724	1.90	0.6889
14	1.57	2.15	0.3364	1.78	0.0441	1.80	0.0529
11	1.44	1.69	0.0625	1.40	0.0016	1.49	0.0025
13	1.36	2.00	0.4096	1.65	0.0841	1.70	0.1156
12	1.3	1.85	0.3025	1.53	0.0529	1.59	0.0841
8	1.28	1.23	0.0025	1.02	0.0676	1.18	0.01
8	1.18	1.23	0.0025	1.02	0.0676	1.18	0.01
$SS_\varepsilon = \sum(y_i - \hat{y})^2$			**1.8108**		**1.0099**		**0.9656**

TABLE 10.1: SS_ε of prediction calculated for three different models

Line C line is the best fit of the three (see Table 10.1), because its SS_ε is lower than either line A or line B. We can illustrate how this "Sum of Squares" is used with a geometric example. The errors of prediction are shown as vertical distances from a measured point to the line of best fit.

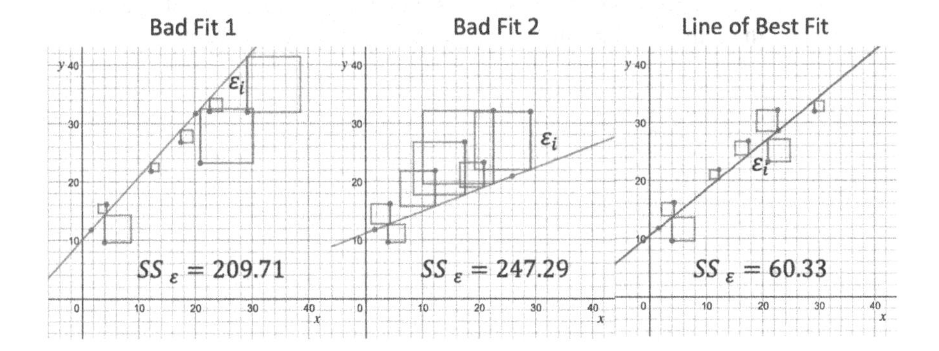

FIGURE 10.3: The sum of the squares of the ε_i for three fit lines are literally sums of squares, geometrically speaking

Figure 10.3 above shows that, if you compute the area of squares with side length ε_i, the sum of these squared residual values can be minimized. All of these lines have been drawn on the same data, but you can see that the error for each predicted value of y is quite different for the three models. The line of best fit, shown in the far right panel, is the model with the minimum SS_ε of any line that can be plotted. But how can we get the best line of fit the first try? There are, after all, infinitely many lines one could draw on the graph of our data, only one of which will fit best. It would take forever to compute all of them. This is where regression comes in.

10.1 Finding the Line of Best Fit

The key thing to remember when fitting a line to data is that the line of best fit will pass through the point in the plane that has the highest probability of being selected at random. Let's think about this carefully. If you were to pick the value of the independent variable, x, that is most probable, what value would you pick? If you are like me, you would go back to our discussion of probability density functions and say to yourself, "Self, my best guess of the expected value of a random variable is its mean, $\overline{x}$." Likewise, if you are going to pick the value of the dependent variable, y, you might say to yourself, "Self, I am going with $\overline{y}$ for the same reasons as I explained to you earlier," and hope that your self agrees with your iron-clad logic. This allows us to write our prediction equation for at least one point:

$$\overline{y} = \beta_0 + \beta_1 \overline{x}$$

The value of $(\overline{x}, \overline{y})$ for our Specific Gravity data is (12.875, 1.7025). Figure 10.4 illustrates that, if all potential lines of best fit pass through $(\overline{x}, \overline{y})$, all we need to do is find the right value for the intercept, β_0 to solve for the slope, β_1.

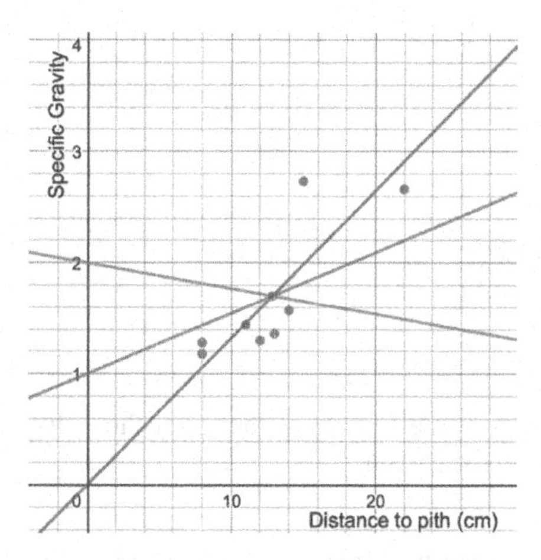

FIGURE 10.4: There are infinitely many lines that pass through $(\overline{x}, \overline{y})$

One method for determining β_0 is to just choose an arbitrary value, solve for β_1, calculate the SS_ε, and do this many times, iterating until the value of SS_ε converges to a minimum with a level of tolerance you are willing to live with. This will approximate the true equation for the line of best fit good enough for nearly any application. We know this because the function that pairs SS_ε with values of β_0 is a quadratic (Sum of the *Squared* Errors of Prediction). There *has* to be a unique minimum value. But the mathematician in all of us is dissatisfied with a practical, sensible and easily conceptualized method. Instead, we need to find a conceptually dense, computationally efficient, formula that gives the exact answer...

Here goes...

Let's start with our assumptions. First, we define the total error in the system as the SS_{ε} . Everything else depends on this assumption. If you change your definition of the total error, the regression equation will change to minimize the function you choose.

$$SS_{\varepsilon} = \sum_{i=1}^{n} (y_i - \hat{y})^2$$

From the regression equation,

$$\hat{y}_i = \beta_0 + \beta_1 x_i$$

We can rewrite SS_{ε} as:

$$SS_{\varepsilon} = \sum_{i=1}^{n} (y_i - \beta_0 + \beta_1 x_i)^2$$

From first semester calculus, we know we can minimize SS_{ε} for values of β_0 and β_1 by taking the derivative of SS_{ε} and setting it equal to zero. But we have two variables, x_i, and y_i, so we need to take partial derivatives of both and set both equal to zero. This will then give us a system of two equations and two unknowns, yielding one, unique solution. Let's start by minimizing SS_{ε} with respect to β_0 :

$$\frac{\partial SS_{\varepsilon}}{\partial \beta_0} = \sum_{i=1}^{n} -2\,(y_i - \beta_0 - \beta_1 x_i) = 0$$

$$\frac{\partial SS_{\varepsilon}}{\partial \beta_0} = \sum_{i=1}^{n} 2\left(n\beta_0 + \beta_1 \sum_{i=1}^{n} x_i - \sum_{i=1}^{n} y_i\right) = 0$$

Solving for β_0, we get an equation we have seen before:

$$\beta_0 = \overline{y} - \beta_1 \overline{x}$$

or more familiarly...

$$\overline{y} = \beta_0 + \beta_1 \overline{x}$$

Now, let's minimize SS_{ε} with respect to β_1 :

$$\frac{\partial SS_{\varepsilon}}{\partial \beta_1} = \sum_{i=1}^{n} -2x_i\left(y_i - \beta_0 - \beta_1 x_i\right) = 0$$

$$\frac{\partial SS_{\varepsilon}}{\partial \beta_1} = \sum_{i=1}^{n} -2\left(x_i y_i - \beta_0 x_i - \beta_1 x_i^2\right) = 0$$

Substituting $\beta_0 = \overline{y} - \beta_1\overline{x}$, we get the following:

$$\frac{\partial SS_{\varepsilon}}{\partial \beta_1} = \sum_{i=1}^{n} \left(x_i y_i - x_i\overline{y}_i + \beta_1 x_i\overline{x}_i - \beta_1 x_i^2\right) = 0$$

Now we can solve for β_1 :

$$\frac{\partial SS_{\varepsilon}}{\partial \beta_1} = \frac{\sum_{i=1}^{n}\left(x_i y_i - x_i\overline{y}_i\right)}{\sum_{i=1}^{n}\left(x_i^2 - x_i\overline{x}_i\right)} = 0$$

Simplifying, we find:

$$\beta_1 = \frac{\sum_{i=1}^{n}\left(x_i - \overline{x}_i\right)\left(y_i - \overline{y}_i\right)}{\sum_{i=1}^{n}\left(x_i - \overline{x}_i\right)^2}$$

With both β_0 and β_1 expressed solely in terms of x and y, we can now calculate the exact line of best fit for our data. The progression of calculations is shown in Table 10.2:

Distance to Pith (x)	**Specific Gravity (y)**	$(x_i - \overline{x}_i)$	$(y_i - \overline{y}_i)$	$(x_i - \overline{x}_i)(y_i - \overline{y}_i)$	$(x_i - \overline{x}_i)^2$	$\beta_1 = \frac{\sum_{i=1}^{n}(x_i-\overline{x}_i)(y_i-\overline{y}_i)}{\sum_{i=1}^{n}(x_i-\overline{x}_i)^2}$
22	2.66	9.125	0.9575	8.7371	83.2656	$\beta_1 = 0.11139$
15	2.73	2.125	1.0275	2.1834	4.5156	
14	1.57	1.125	-0.1325	-0.1491	1.2656	
11	1.44	-1.875	-0.2625	0.4922	3.5156	
13	1.36	0.125	-0.3425	-0.0428	0.0156	
12	1.3	-0.875	-0.4025	0.3522	0.7656	
8	1.28	-4.875	-0.4225	2.0597	23.7656	
8	1.18	-4.875	-0.4225	2.0596	23.7656	
$\overline{x}_i$ = 12.875	$\overline{y}_i$ = 1.7025	**Sum**		**15.6925**	**140.875**	

TABLE 10.2: Full calculations of β_1 for specific gravity data

With β_1 fixed, we now can solve for β_0 :

$$\beta_0 = 1.7025 - 0.11139 \cdot 12.8750$$

$$\beta_0 = 0.2683$$

... and... drumroll... our final, exact (within rounding error) line of best fit is... (see Figure 10.5)

$$\hat{y}_i = 0.2683 + 0.11139x_i$$

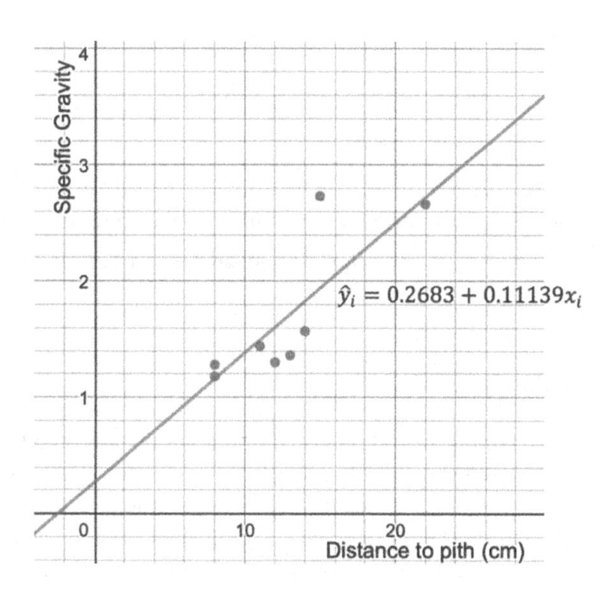

FIGURE 10.5: Least squares regression line for specific gravity data

To show that our model of the Specific Gravity Data is indeed the best, I have plotted the SS_ε for several different lines that pass through $(\overline{x}, \overline{y})$ but that have different slopes and intercepts. You can see that the shape of the function $\beta_0 = f(SS_\varepsilon)$ looks parabolic, and that the value of SS_ε at $\beta_0 = 0.2683$ appears to be at the vertex. The proof above shows that this is the case, but somehow, looking at the picture is more convincing to me. (see Figure 10.6)

It is important to remember that the values predicted from our linear model, $\hat{y}_i$, is a set of points, not the linear model itself. The real power of regression lies in the fact that the model can be used predictively–that ***we can take a finite sample of bivariate data, and make a prediction about values that are not in our sample***. Similar to the Normal Distribution for univariate data, we can use this function as a model that serves as a proxy for the average values we would respect in the population of bivariate points.

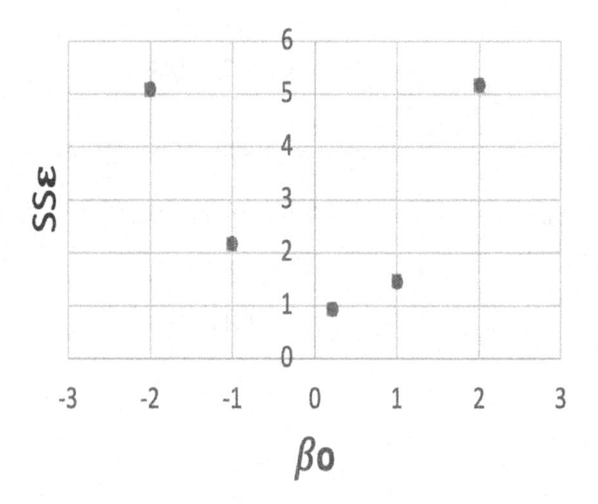

FIGURE 10.6: The SS_ε is minimized at the solution to the regression equation where $\beta_1 = \frac{\sum_{i=1}^{n}(x_i-\overline{x}_i)(y_i-\overline{y}_i)}{\sum_{i=1}^{n}(x_i-\overline{x}_i)^2}$ and $\beta_0 = \overline{y} - \beta_1\overline{x}$. This example is where $\beta_1 = 0.11139$ and $\beta_0 = 0.2683$

10.1.1 Goodness of Fit

Not all models resulting from linear regression are equally good models. Take a look at the two scatterplots in Figure 10.7. This is data modeling the degree to which acid rain degrades the surface of historical tombstones, typically made of limestone Meierding(1993). Mechanical engineers are interested in designing means of preserving these artifacts without having to spray them with chemicals to neutralize the acid.

2. Of the two models in Figure 10.7, which one do you trust more, Model A or Model B?

   Just like univariate statistics, all things being equal, we tend to trust the sample that has the least variation (i.e., the one that is more *precise*). In Figure 10.7, we can see that the values of the independent variable are exactly the same for both Models. The values of the *dependent* variable, however, are more tightly packed around the line of best fit in model A, than those in Model B, so we would say Model A has better *fit* than Model B. **Goodness of fit**, therefore, is not just minimizing the error of prediction within a data set. It involves comparison of two or more models against their overall error in prediction. We can do this by comparing the SS_ε for the two models:

   $$SS_{\varepsilon(ModelA)} = 2.5315 \qquad SS_{\varepsilon(ModelB)} = 7.1352$$

   Model A shows a lot less total variation in its predicted values, compared to

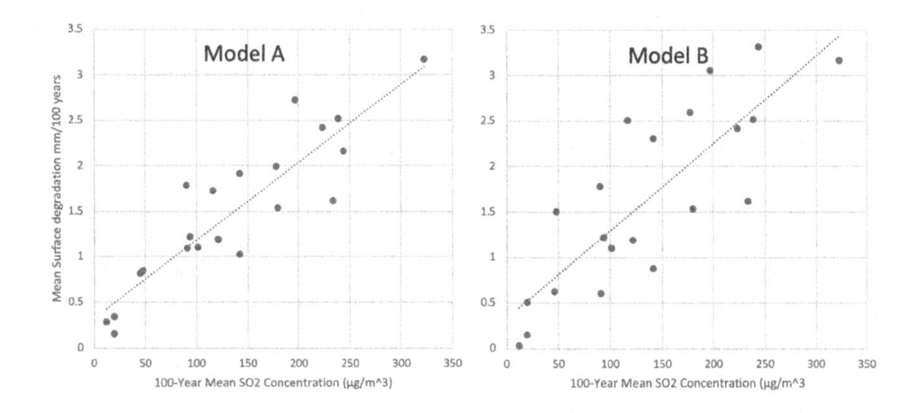

FIGURE 10.7: Two data sets, measuring the same phenomena, but with different error of measurement can produce a line of best fit with the same estimated parameters

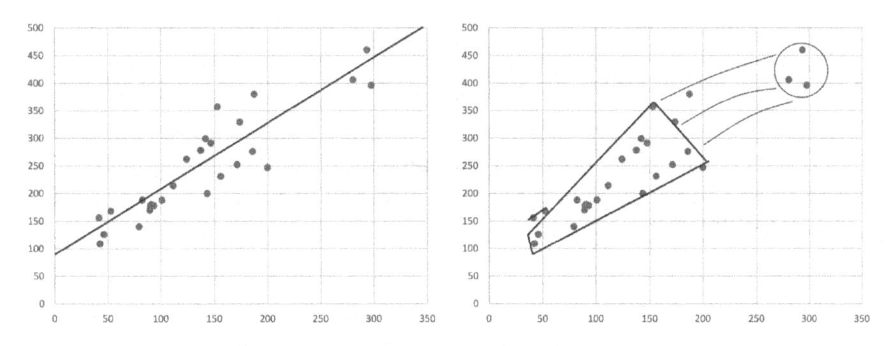

FIGURE 10.8: Two different models of the same data. One has better fit than the other adapted

the actual values. It is like an NFL quarterback is squeezing the football-shaped cloud of data a lot more, compared to Model B. This makes perfect sense. A perfect model would show all predicted values, $\hat{y}_i$, equivalent to the actual values, y_i —*all values of* $\hat{y}_i$, *would lie on the line of best fit*, with no variation around it. In that case $SS_\varepsilon = 0$.

This works great for two models that just happen to have the same n and the same prediction equation. But this is unlikely if we take two samples of bivariate data and compare their goodness of fit. The two samples would most likely result in different variability, just due to random chance, making the intercept and slope estimates somewhat different. To get around this problem, we use a measure of goodness of fit that accounts for the *proportion of variability each model is able to explain*, relative to the total variability within each sample of points.

10.1.1.1 R^2: The Coefficient of Determination

The total amount of variation in the dependent variable, y_i, is the numerator of y's variance, $\sum_{i=1}^{n}(y_i - \bar{y})^2$. This quantity which we will call SS_{total}, is just the sum of the individual differences of our measured values of y compared to the value we would predict without taking the independent variable, x into account ($\bar{y}$).

$$SS_{total} = \sum_{i=1}^{n}(y_i - \bar{y})^2$$

We will introduce another measure of the variability in our data that accounts for how good each data point in the linear *model* predicts $\bar{y}$.

$$SS_{regression} = \sum_{i=1}^{n}(\hat{y}_i - \bar{y})^2$$

Notice that $SS_{regression}$ is **exactly the same as** SS_{total} except that $SS_{regression}$ assesses the variability of our **model** around $\bar{y}$. If $SS_{regression} = SS_{total}$, then $\hat{y}_i = y_i$ for all i. If this is the case,

$$\frac{SS_{regression}}{SS_{total}} = 1$$

A perfect fit of data to model. Unfortunately, if your data is like mine, the model I use to predict values of y_i is *not* perfect. There is variation around the line of best fit. This variation, as we have shown earlier is SS_{ε}.

$$SS_{total} = SS_{regression} + SS_{\varepsilon}$$

This equation means that the total amount of variation in y_i is equal to the amount of variation accounted for by the regression model, plus the error of prediction—that variation we can't account for with our linear model. The proportion of the total variation accounted for by our model, $\boldsymbol{R^2} = \frac{SS_{regression}}{SS_{total}}$. This is our measure of goodness of fit. If our model accounts for 100% of the total variation, it is a perfect fit, and R^2 will be 1. If it is horrible, not accounting for any of the variation, R^2 will be 0.

$$Goodness\ of\ Fit = R^2 = \frac{\sum_{i=1}^{n} (\hat{y}_i - \bar{y})^2}{\sum_{i=1}^{n} (y_i - \bar{y})^2} = \frac{SS_{regression}}{SS_{total}}$$

Because $SS_{total} = SS_{regression} + SS_{\varepsilon}$, R^2 can also be written as 1 – the proportion of variation that we attribute to random error, SS_{ε}:

$$R^2 = 1 - \frac{\sum_{i=1}^{n} (\hat{y}_i - y_i)^2}{\sum_{i=1}^{n} (y_i - \bar{y})^2} = 1 - \frac{SS_{\varepsilon}}{SS_{total}}$$

The Coefficient of Determination, R^2 is the proportion of the total variation in the independent variable that is accounted for by a linear model. R^2 is often used as a measure of goodness of fit of a model to the data from which it was generated. It is defined by the following equations:

$R^2 = \frac{\sum_{i=1}^{n}(\hat{y}_i - \bar{y})^2}{\sum_{i=1}^{n}(y_i - \bar{y})^2} = \frac{SS_{regression}}{SS_{total}}$

or

$R^2 = 1 - \frac{\sum_{i=1}^{n}(\hat{y}_i - y_i)^2}{\sum_{i=1}^{n}(y_i - \bar{y})^2} = 1 - \frac{SS_{\varepsilon}}{SS_{total}}$

Where variance components of the model are additive:

$SS_{total} = SS_{regression} + SS_{\varepsilon}$

$\sum_{i=1}^{n} (y_i - \bar{y})^2 = \sum_{i=1}^{n} (\hat{y}_i - \bar{y})^2 + \sum_{i=1}^{n} (y_i - \hat{y}_i)^2$

As a proportion, R^2 ranges from 0, where the fit of the model is completely random, to 1, where the model perfectly fits the data.

3. What is the goodness of fit of our linear model for the Specific Gravity Data?

If you remember back a couple of sections in this chapter, the line of best fit for the relationship between distance to pith of wood samples to their measured specific gravity was

$$\hat{y}_i = 0.2683 + 0.11139x_i$$

I have placed the original data, the predicted values of specific gravity from the model, and the individual errors of prediction, ε (See Table 10.3).

Distance to Pith (cm)	Specific Gravity	$\hat{y}$	ε
22	2.66	2.74	−0.08
15	2.73	1.93	0.80
14	1.57	1.82	−0.25
11	1.44	1.47	−0.03
13	1.36	1.70	−0.34
12	1.3	1.59	−0.29
8	1.28	1.13	0.15
8	1.18	1.13	0.05

TABLE 10.3: Predicted values and errors of prediction for specific gravity data

With this data, we can calculate the coefficient of determination, R^2. For this model, $R^2 = 0.6664$. About 2/3 of the variation in Specific Gravity of our wood samples is determined by the distance to pith at which the samples were taken. That is not too shabby! Now, we still don't know anything about the other 33% of the variation, but I could begin using the relationship we modeled to predict future values of the density of the wood, just given a length measurement! That is a powerful modeling tool.

10.1.2 When is a Linear Model NOT Appropriate?

So far we have fit a model to our data, assuming the underlying relationship between our independent and dependent variables was indeed linear. But so many of the relationships we use in engineering are *not* linear. They might be polynomial, as in projectile motion, power functions as in radiation intensity, or exponential (nearly everything, but especially Thermodynamics). I briefly mentioned that, to trust a linear model, you have to make some assumptions about the data. The assumptions of linear regression are:

1. Independence of the observations
2. The true relationship between independent and dependent variables is linear *in its parameters*;
3. Errors of prediction are normally distributed around zero; and
4. Errors of prediction are equally distributed around the line of best fit.

The first assumption is the same as any data gathering enterprise, we have to reduce the bias in our sampling and measurement to be pretty sure that the measurements we do analyze actually reflect the values of the variables they are measuring.

The 2nd assumption allows us to take non-linear data and perform regression on it if it can be expressed as $f(x) = Ax + By + Cz + \ldots + \varepsilon$. If you remember back in your linear algebra class, we could model polynomials (definitely non-linear) functions using a *linear system.* In such a system, the coefficients of the model are the parameters. $f(x) = Ax^2 + Bx + C$ is linear in its parameters because neither A, nor B or C is multiplied or divided by another parameter, nor is used as an exponent. If the data you are trying to fit has an underlying structure that is non-linear, sometimes you can transform the data, to linearize it if you will, so that you can perform linear regression on the transformed data.

The 4th and 5th assumptions have to do with whether or not the error of prediction is truly random. And this is tied up with whether or not our model is appropriate. If we model a quadratic with a straight line, our error of prediction will have a distinct, non-random pattern. Let's explore this notion more carefully.

The basic idea of transforming your data relies on the identity:

$$f^{-1}(f(x)) = x$$

Multiplying a function by its inverse returns the original function. A simple quadratic example will illustrate this:

$$f(x) = x^2$$

$$f^{-1}(x) = x^{\frac{1}{2}}$$

Figure 10.9, below, shows a scatterplot of a set of data that has an underlying quadratic structure. There is some random variation, but looking at the graph, I would say it either could be modeled with a linear equation, or with a 2nd degree polynomial. It could also be exponential. The thing to keep in mind is that we don't go into a data modeling study blind. We generally have some idea, by reviewing the appropriate scientific literature, what the basic structure of the data will be. If this is the case, we should use that theory to guide our analyses.

If I suspect the data does indeed have an underlying 2nd degree polynomial model, I can transform it to be linear by taking its square root (Table 10.4):

Plotting $\sqrt{y}$ versus x gives us a linearized dataset. Using the procedures for Least Squares Regression, I have fitted a line to the data (see Figure 10.10).

Now, this is all well and good, but we still don't know the quadratic equation. To get this in quadratic form, we have to *undo* the transformation for the values of $\hat{y}$, and then use what we know about quadratics to compute an appropriate curve as the best fit.

Since we linearized the data, we now have to "undo" the linearization, applying the inverse function to our line of best fit. We used $\sqrt{y_i}$ to transform the data into a linear model, so we now have to use $\widehat{\sqrt{y_i}}^{\,2}$ to transform the linearized *model* back

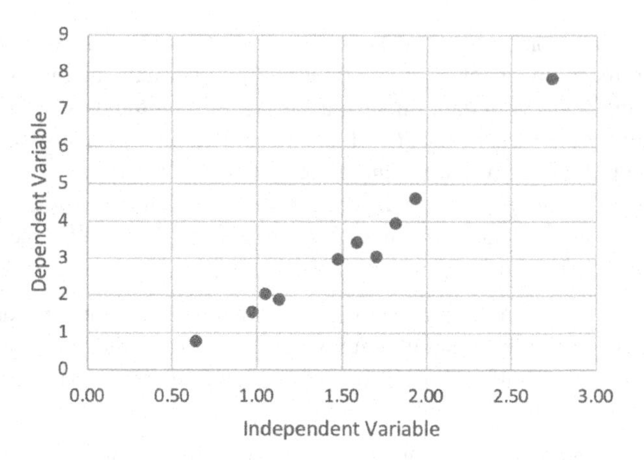

FIGURE 10.9: Bivariate scatterplot displaying an underlying quadratic structure

x_i	y_i	$\sqrt{y_i}$
2.74	8.04	2.84
1.93	4.52	2.13
1.82	4.04	2.01
1.47	2.56	1.60
1.70	3.58	1.89
1.59	3.47	1.86
1.13	1.80	1.34
1.05	1.55	1.24
0.97	1.58	1.26
0.64	1.31	1.15

TABLE 10.4: Data transformed by taking the square root of the dependent variable

into the original units of the *data*. Rather than computing each value of $\hat{y}$ from the data, we can use our line of best fit, and transform that function to get the curve that best fits the original data (see Figure 10.11).

$$\hat{y}_i = (0.8646x + 0.4304)^2$$

$$\hat{y}_i = 0.7475x^2 + 0.7442x + 0.1852$$

Assessing the goodness of fit of our model to the original data, we use the same index, R^2. Because we are assessing the goodness of our curve of best fit compared to the original data, we need to use those values, x_i , y_i , and $\hat{y}_i$ to compute R^2, *not* our transformed data. R^2 of our quadratic model is about 0.95. Ninety-five percent of all the variation in our dependent variable is accounted for by the independent variable. Only about 5% is considered random error. This is an exceptionally good fitting model.

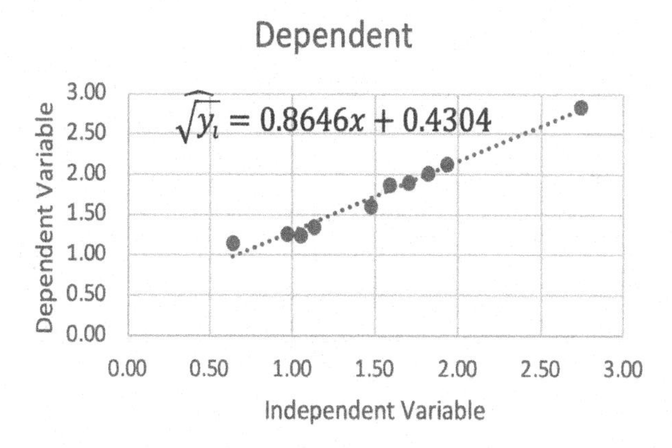

FIGURE 10.10: Line fit to linearized data: I transform the data by the inverse function

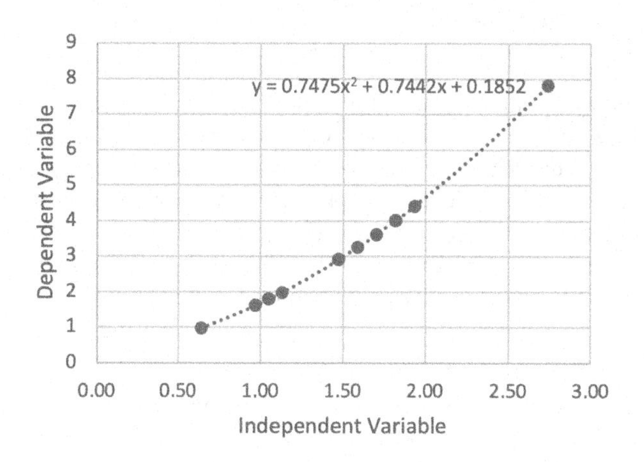

FIGURE 10.11: Curve of best fit transformed back into original units. Undo the linearization by applying the inverse of $\hat{y}$

The key takeaway here is, since Least Squares Regression requires the relationship being modeled to be linear in its parameters, a nonlinear set of data can be made linear by applying the inverse function of the one you think really is underlying the data. Regression on the linearized (transformed) data will give a line of best fit, then you can apply the inverse function to the line of best fit to transform the model back into the units of the original data. Table 10.5 provides the transformations for the most commonly used functions in engineering.

Function	Transformation(s)	Regression equation	Predicted value ($\hat{y}$)
Linear	None	$\hat{y}_i = \beta_0 + \beta_1 x_i$	$\hat{y}_i = \beta_0 + \beta_1 x_i$
Exponential	$DV = ln\,(y)$	$ln\,(y) = \beta_0 + \beta_1 x_i$	$\hat{y} = 10^{\beta_0 + \beta_1 x_i}$
Quadratic	$DV = \sqrt{y}$	$\sqrt{y} = \beta_0 + \beta_1 x_i$	$\hat{y}_i = (\beta_0 + \beta_1 x_i)^2$
Inv. Prop.	$DV = \frac{1}{y}$	$\frac{1}{y} = \beta_0 + \beta_1 x_i$	$\hat{y}_i = (\beta_0 + \beta_1 x_i)^{-1}$
Log	$IV = ln\,(x)$	$y = \beta_0 + \beta_1 ln\,(x_i)$	$\hat{y}_i = \beta_0 + \beta_1 ln\,(x_i)$
Power	$DV = ln\,(y)$ $IV = ln\,(x)$	$ln\,(y) = \beta_0 + \beta_1 ln\,(x_i)$	$\hat{y}_i = 10^{\beta_0 + \beta_1 ln(x_i)}$

TABLE 10.5: Linearization transformations for commonly encountered relationships among bivariate data. ∗Note: Here x and y are independent and dependent variables of the *original* function. IV means independent variable and DV means dependent variable of the *transformed* function

Linear regression has the added assumptions (Assumptions 3 and 4) that the residual errors ε_i are random—that there is no systematic bias in our prediction equation, and that the distribution of the ε_i is Normal. The randomness assumption is pretty easy to understand. We have had this tacit assumption every time we use the Normal distribution as a model—it is a property of the CLT. But the Normal distribution of the errors is a little less obvious. This will become clearer in the next chapter, when we look at Regression as a projection of a vector from a 3d basis onto a 2d basis.

But for now, assume that the population distribution of y_i for any fixed value of x_i, should be normal. This makes sense, in that, if y and x are both continuous random variables, and if we do a good job of random sampling, the long term expected value of each will tend to their mean. If we fix x_i at a single value, the resulting distribution of y_i should be normal, with the mean of that distribution falling on the prediction line at $\hat{y}_i$. This is most obvious at $(\overline{x}, \overline{y})$. If these cross-sectional distributions are NOT normal, then the skew will pull the mean away from $\hat{y}_i$, making some or all of the predicted values to be in error.

10.2 Residual Analysis

There is a nice way to test for the violation of these two assumptions: **Residual Analysis**. Residual analysis involves plotting the values of our residual error, ε_i against the values of the independent variable, x_i. See Figure 10.12 for a plot of the residuals of our Specific Gravity data.

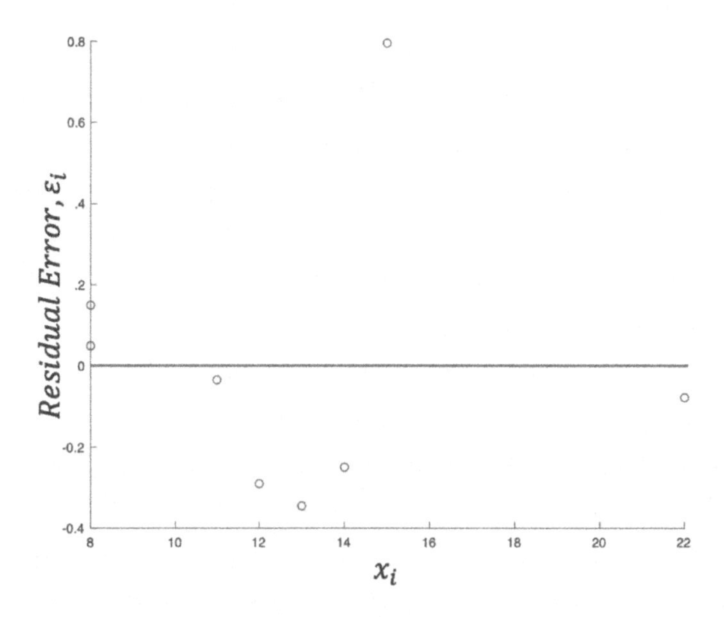

FIGURE 10.12: Residual plot for specific gravity model

What we are looking for in this plot is some obvious pattern in the data. Ideally, the residual error should be evenly and randomly distributed about $\varepsilon_i = 0$. This would be an indication that the errors of prediction are truly random, and that there is little systematic bias in the data. For our data, we see no obvious pattern, but the data are not evenly distributed about $\varepsilon_i = 0$. This could be an indication that our data is skewed or otherwise non-normal.

If we suspect that our assumption of non-normal distribution of errors is violated, we can check it using a Quantile-Quantile plot. This is a plot of the standardized residuals (the Actual Z for each ε_i) versus their standard normal equivalent values (the values predicted by the Normal Distribution). Figure 10.13 shows the Q-Q plot for our Specific Gravity data:

If you look closely at Figure 10.13, you can see a small, + at about $(-1.5, -0.8)$., right on the lower boundary of the window. This indicates that our lowest residual value is way out of line (literally) if the ε_i were truly normally distributed. The rest of the data show little deviation from the Q-Q line. We might choose, based on this finding, to eliminate the lowest value as an outlier, reducing the overall skew of the data. Or we might examine whether or not the data might really reflect some other function than a line. Reading the literature of wood density, it looks like a linear model is appropriate, so I would most likely choose to drop one of the measurements at $x_i = 8$, assuming it is an error of measurement.

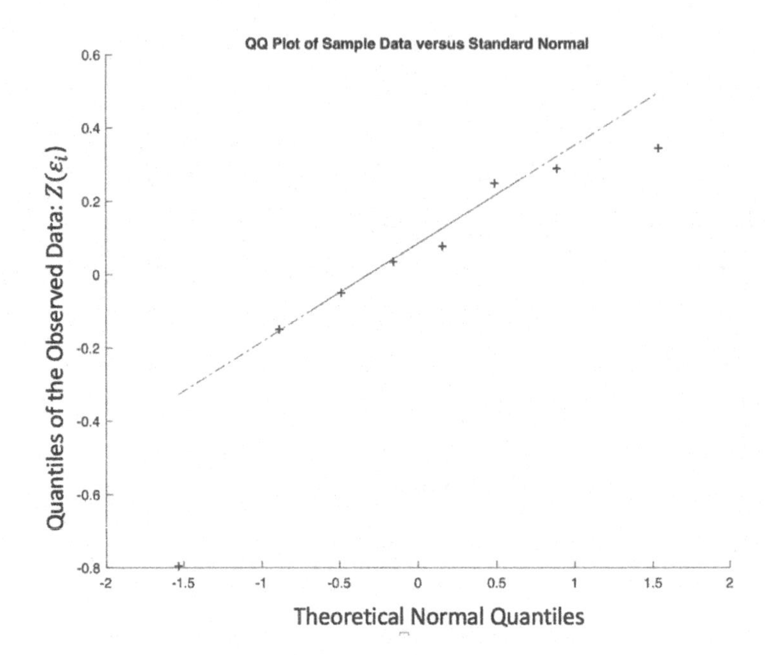

FIGURE 10.13: Quantile-Quantile plot of residual error

Taken together, the residual plot tells us that our error is pretty random, but isn't evenly distributed about the prediction line. The Q-Q plot tells us that we have one or two points that are skewing the distribution, and by cleaning up the data, we have a pretty good model.

Residual Analysis is a method of examining the distribution of the errors of prediction of a model to determine if they are random, and normally distributed.

1. Draw a scatterplot of the prediction errors, ε_i versus the values of the independent variable, x_i. There should be no obvious pattern in the position of the errors on the plot. The errors should appear to be randomly distributed about the central value of zero. If there *is* some obvious pattern, the shape of that pattern is a good clue to the potential of an underlying non-linear relationship among the variables, and that transformation may be a good option;

2. Draw a Quantile-Quantile plot. This graph plots the standardized value of the error, $Z(\varepsilon_i)$ versus its predicted normal curve equivalent Z-score. If the ε_i are normally distributed, the scatterplot will appear as a straight line.

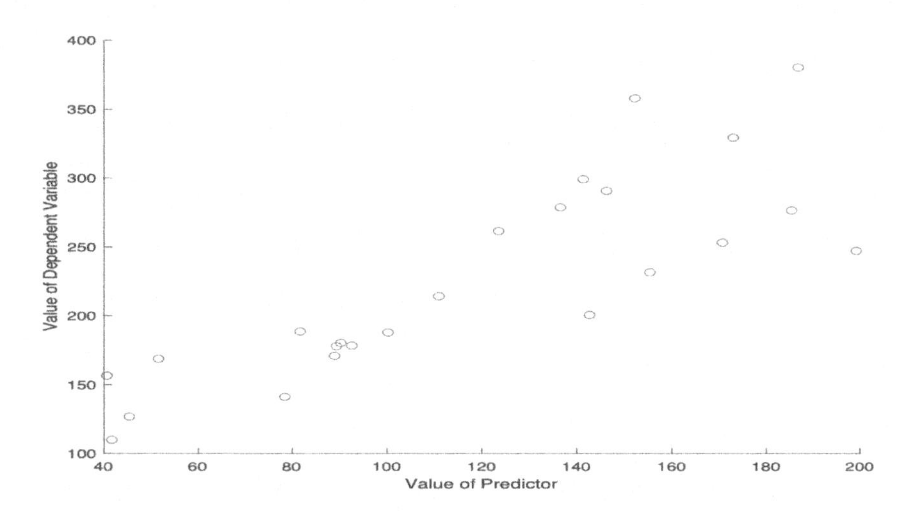

FIGURE 10.14: "Heteroscedasticity." The variance of dependent variable, y, grows as a function of the independent variable, x

10.2.1 Heteroscedasticity

One last mention regarding the distribution of errors. Remember, we have this assumption that the errors are randomly and evenly distributed about zero. Our residual plot showed that, for our Specific Gravity data, we probably have some issues with the even distribution of our prediction error. This unevenness is called, "heteroscedasticity," and it will get you extra credit on the final if you remember it and can spell it correctly. What this means is that the variance of y should not vary with x. The variances of all the little cross-sectional distributions of y for each value of x should not be very different from each other. The classic violation of this assumption is what is called "fan-spread" (See Table 10.14).

This kind of non-homogeneous variance tends to underestimate the standard error of our estimate of the slope of $\hat{y}_i$. So, when we test the hypothesis that our fitted model is better than a chance model, we tend to inflate the Type I error rate. Least Squares regression will fit a line to any cloud of data, but it does not detect the different variances along the line of best fit. So checking for heteroscedasticity using residual analysis is always a good practice. Figure 10.15 shows the residual plot for this fan-spread data. Notice how the fan-shape is apparent with the residuals, just like it was with the raw data. This is one of the key patterns to look for when doing a residual analysis.

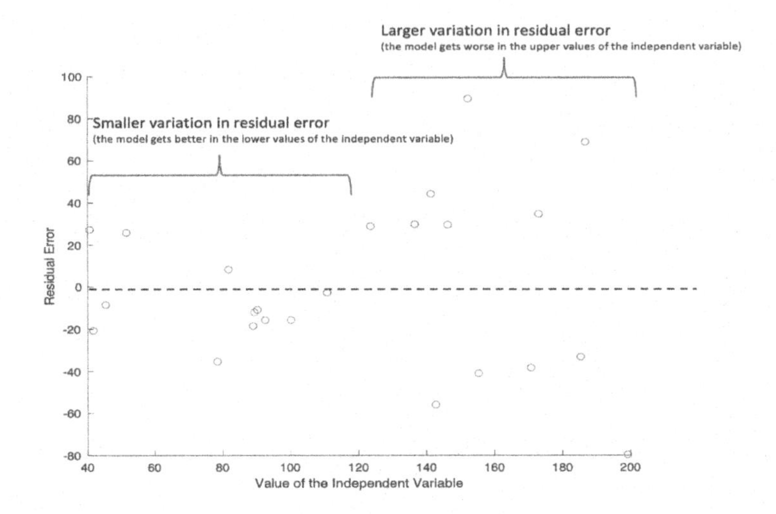

FIGURE 10.15: Residual plot for "Fan-Spread" data

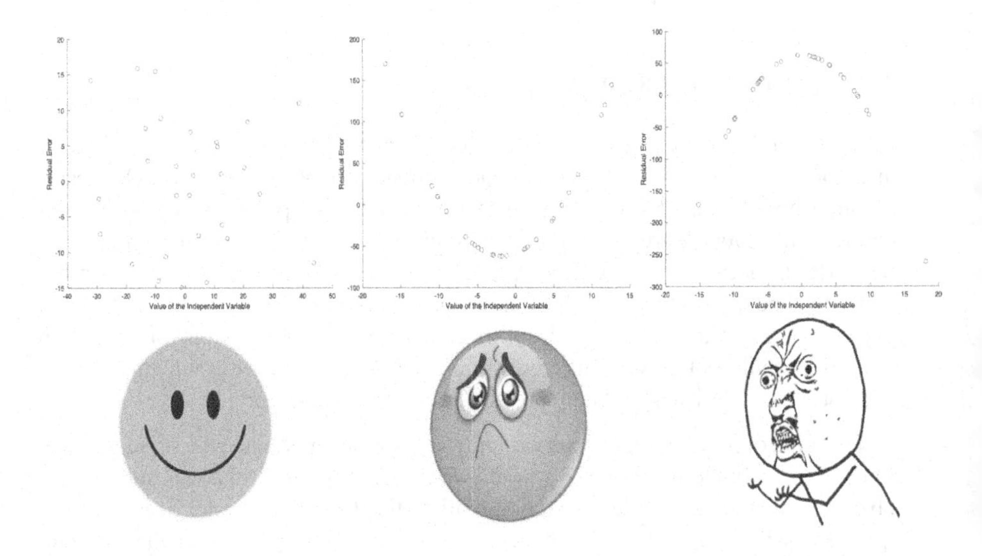

FIGURE 10.16: Only one of these residual plots makes me happy...

Heteroscedasticity is when the variability of a dependent variable is unequal across the range of values of the independent variable. One of the classic signs of heteroscedasticity is the "fan shape" of the scatterplot.

10.3 Hypothesis Testing in Regression: *t*-test of the Slope

After we have fitted a line to our data, after we have checked the assumptions and made any transformations and adjustments, we still have to determine whether or not our linear model is better than a chance model: That the probability we could have gotten our prediction equation is highly unlikely just due to random chance. After all we suspect there *is* some kind of systematic relationship between our independent and dependent variable, so we would like to have evidence that the model we have generated is significantly different from chance. This hypothesis essentially is asking the question, "Is the slope of my line significantly different from $\beta_1 = 0$?" A slope of zero is the average slope of all lines that can be drawn through the point $(\overline{x}, \overline{y})$ and so it is the expected value of β_1 over the long haul.

The null hypothesis and alternative hypotheses for a regression analysis are:

$$H_0 : \beta_1 = \mu_{\beta_1} = 0 \ \textbf{Two-tailed}$$

$$H_0 : \beta_1 \geq \mu_{\beta_1} \ \textbf{Left-tailed}$$

$$H_0 : \beta_1 \leq \mu_{\beta_1} \ \textbf{Right-tailed}$$

$$H_1 : \beta_1 \neq \mu_{\beta_1} \ \textbf{Two-tailed}$$

$$H_1 : \beta_1 < \mu_{\beta_1} \ \textbf{Left-tailed}$$

$$H_1 : \beta_1 > \mu_{\beta_1} \ \textbf{Right-tailed}$$

We now know β_1 and its expected value under the null hypothesis, $\mu_{\beta_1} = 0$. If we just had a measure of the standard error of β_1, we could construct a test statistic:

$$t = \frac{\beta_1 - \mu_{\beta_1}}{SE_{\beta_1}} = \frac{\beta_1}{SE_{\beta_1}}$$

Just like the one-sample *t*-distribution we studied back in Chapter 8, we can think of the SE_{β_1} as the error variation in the sampling distribution of β_1 —the width of the sampling distribution. From our work fitting a line, we know that the error variation assessed vertically about the line is $SS_\varepsilon = \sum_{i=1}^{n} (y_i - \hat{y}_i)^2$. We also know that the error variation in the horizontal direction is the variation of the independent variable, x, so $SS_x = \sum_{i=1}^{n} (x_i - \overline{x}_i)^2$. Taking these quantities, we can construct the error of the slope:

$$s_{\beta_1} = \sqrt{\frac{\sum_{i=1}^{n} (y_i - \hat{y}_i)^2}{\sum_{i=1}^{n} (x_i - \overline{x}_i)^2}}$$

So, this is the average error of β_1 in our sample. But our sample had n pairs of observations. The standard error of the sampling distribution of β_1 must, therefore be this error quantity divided by the degrees of freedom in the sampling distribution. Because we are using two parameters to estimate s_{β_1}, the degrees of freedom for our standard error must be $(n - 2)$.

$$SE_{\beta_1} = \sqrt{\frac{1}{(n-2)} \cdot \frac{\sum_{i=1}^{n}(y_i - \hat{y}_i)^2}{\sum_{i=1}^{n}(x_i - \overline{x}_i)^2}}$$

Now the ***t*-test of the Slope** can be written as:

$$t_{(n-2,\alpha)} = \frac{\beta_1}{SE_{\beta_1}} = \frac{\beta_1}{\sqrt{\frac{1}{(n-2)} \cdot \frac{\sum_{i=1}^{n}(y_i-\hat{y}_i)^2}{\sum_{i=1}^{n}(x_i-\overline{x}_i)^2}}}$$

Applying this test to our Specific Gravity data, assuming $\alpha = 0.05$, we get

$$t_{(6,0.05)} = \frac{\beta_1}{SE_{\beta_1}} = \frac{0.11139}{\sqrt{\frac{1}{(6)} \cdot \frac{0.9303}{140.875}}}$$

$$t_{(6,0.05)} = 3.358$$

The critical value of t at 6 degrees of freedom and $\alpha = 0.05$ is 1.943 (one tailed). Since our calculated value of $t_{(6,0.05)} = 3.358$ is further in the tail than the critical value, we can conclude that it is highly unlikely, $p < 0.05$, that the relationship between Specific Gravity and Distance from Pith in our species of wood is due solely to random chance. Since it is so unlikely to have happened as a function of chance, we reject H_0 and accept the alternative hypothesis, H_1 as the best explanation for the pattern we found in our data.

10.4 General Procedure for Performing Regression Analyses

Procedure for Performing Linear Regression

1. Read up about the relationship you are modeling. Is it likely to be linear? Quadratic? Do your homework!
2. Determine if you are making a 1-tailed or 2-tailed hypothesis;
3. Express the Alternative Hypothesis and then the Null Hypothesis;

4. Select the Type I error rate;
5. Collect your Data;
6. Make a scatterplot of the data;
7. Examine the scatter for obvious signs of non-linearity;
8. Transform the data if necessary
9. Compute the line of best fit
10. Compute the residuals
11. Perform a residual analysis, transform the data or eliminate outliers if necessary
12. Perform the t-test of the slope to determine if the relationship modeled is significantly different from that expected due to random chance
13. Figure out what to do with the results of your analysis and have a beer!

One of the best historical examples of using curve-fitting (regression hadn't been invented yet in the annals of mathematics) is when Galileo modeled how far brass balls (no jokes here involving monkeys) would travel in the horizontal direction, when they were allowed to roll down an inclined plane, and released from the end of the inclined plane from fixed heights. He really didn't drop balls off the Leaning Tower of Pisa. That is apocryphal. His experiments were much more subtle. He suspected there was some relationship between the height from which the ball was dropped, and the horizontal distance it would travel. If all balls were traveling at the same speed when they left the ramp, then the horizontal distance should increase as a function of the height at which the balls were released from the ramp. Here is some of his real, actual data (Table 10.6)! (see Table 10.6)

distance	**height**
573	1000
534	800
495	600
451	450
395	300
337	200
253	100

TABLE 10.6: Some of Galileo's real, original, experimental data!

Because his work was entirely exploratory, I would probably use a Type I error rate of 0.10, to give me some flexibility with imprecise measurements and random error given the balls might not be perfectly round or all be exactly the same mass and diameter. I might also use a Two-tailed test for this, but since I have observed rocks when I throw them, I know that rocks tend to go farther the higher they are thrown. This would push me to do a 1-tailed (a right-tailed, test means that I believe β_1 is greater than zero).

So now I have the Null and Alternative Hypotheses settled:

$$H_0 : \beta_1 \leq \mu_{\beta_1} \text{ Right-tailed}$$

$$H_1 : \beta_1 > \mu_{\beta_1} \text{ Right-tailed}$$

Think a bit, if you were Galileo and hadn't studied projectile motion in your Calculus and Physics classes (those subjects still not having been invented), how would you try to figure out this relationship given this data? He didn't even have the luxury of graphs, that coming after Descartes, who was Galileo's contemporary. It is really difficult to imagine the work of creativity and genius that came up with the idea that the acceleration of gravity is a constant! I am going to go through this data analysis in way too much detail. I want you to see how a data analyst thinks as s/he walks through some data, where s/he doesn't know what the answer is going to be (I don't. I have an idea of the physics, but the actual model and underlying formula is not one I have committed to memory). So here goes!

In the twenty-first century, I would first sketch the scatterplot of distance versus height (see Figure 10.17). This will help me determine if the relationship might be linear (or not). It will also give me an indication of what the slope might be as well as the intercept.

What I see here is something that looks distinctly non-linear! We have to figure this out! Following plotting the data, I might have, if I hadn't reviewed the scatter plot, computed the line of best fit. But now, I have to figure out a proper transformation, given what I know about the shape of the curve, and the physics of projectile motion. It might be a polynomial function of height— 3rd or 4th degree polynomial, but we only see part of its domain. It might also be a bounded exponential or a logistic curve, approaching some upper asymptote. But given I know that vertical position versus time of a ball in freefall is quadratic, I might assume that the distance traveled in the horizontal direction is a square-root (power) function. The functions, in this case, would be inverses, indicating a constant acceleration of gravity that governed both.

There is only one problem with my thinking at this point: The y-intercept for my graph is not even close to zero. Just looking at the curve, it appears that, if the trend continues, there may not even **be** an intercept, I might want to take this into account. If I square this non-zero intercept, I will get an even worse intercept in my model!. There are a number of ways in which statisticians deal with this, but what I decided to do was to convert the whole problem into a familiar form.

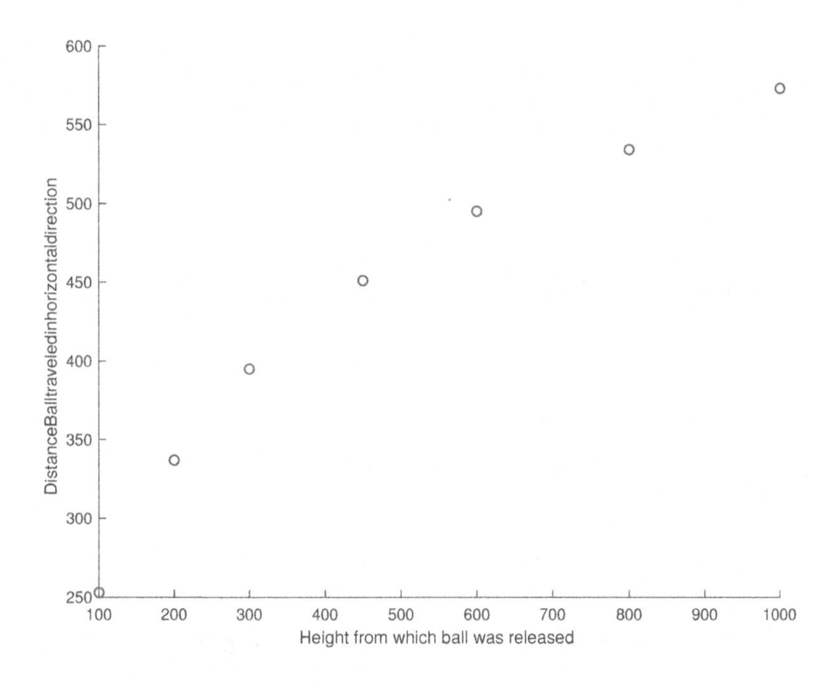

FIGURE 10.17: Scatterplot of Galileo's data

Back in high school, you dealt with function transformation of radical functions by making x a function of y, turning the function into a quadratic. This makes the whole transformation scheme easier. With a quadratic model, we can transform the data into a square-root function, model it, and then transform back to the quadratic to assess the fit. Because the function is one-to-one we can *re-express* this function, $height = f(distance\ traveled)$ as its inverse: $distance\ traveled = f(height)$ to get a prediction equation. Note, there are some problems with this method, namely that some relationships we model have a clear independent variable, and dependent variable, and reversing their roles in the function may make the function not one-to-one, and therefore, not invertible.

Plotting height as a function of distance give us a familiar graph (see Figure 10.18):

Now we can transform height to linearize the function (Table 10.7):

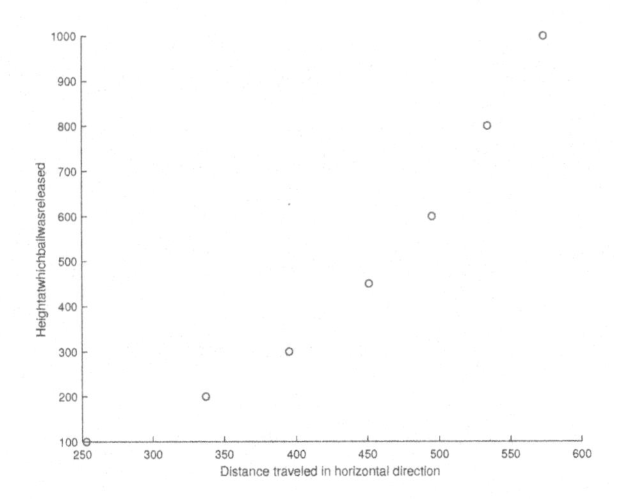

FIGURE 10.18: Scatterplot of height versus distance data

distance	**height**	**Sqrt(Height)**
573	1000	31.62
534	800	28.28
495	600	24.49
451	450	21.21
395	300	17.32
337	200	14.14
253	100	10

TABLE 10.7: Galileo's data, "Linearized"

and regressing the square root of height on distance returns the following data for our line of best fit:

$$\beta_0 = -8.4109$$

$$\beta_1 = 0.0678$$

The scatter plot of the fitted data against the transformed data now looks very, **very** good. The transformed data and the model fitted to the transformed data show very close values (see Figure 10.19):

The Q-Q plot of the transformed data now looks better (see Figure 10.20). The bulk of the data in the center of the model are even closer to the line, and looking at the quantile scale for the residuals, the overall model is tighter.

Lastly, $R^2 = 0.9840$ is a great fit!

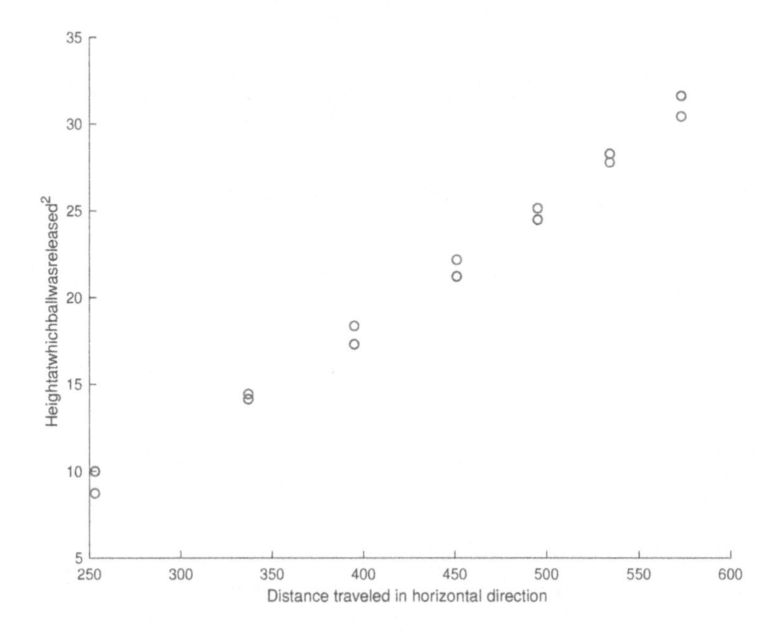

FIGURE 10.19: Scatterplot of transformed height versus distance data, with predicted values

We still have the problem that there is a definite pattern to the residuals. This, given what we know right now will have to be an error in our model. This is because we haven't really dealt with the coefficient of the second term in the quadratic, B. In future courses, you will learn how to handle issues of intercept error versus slope error for more complex models. But for now we can take our prediction equation:

$$\sqrt{height} = -8.4109 + 0.0678\left(\hat{d}\right)$$

and re-express it as a function of distance.

$$\frac{\sqrt{height} + 8.4109}{0.0678} = \hat{d}$$

As a final check, I have plotted our predicted model versus the original data (see Figure 10.21). I am pretty satisfied with this model. It fits the physics and it fits the data well. But, now that I am finished modeling, I need to see if this model is better than just random chance. Matlab returns

$$t_{(df=5,\alpha=0.05)} = 17.54$$

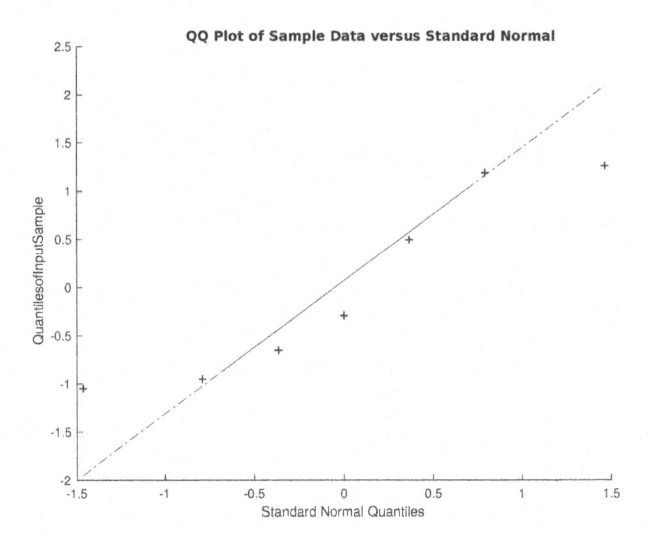

FIGURE 10.20: Q-Q plot of the residuals of the transformed model

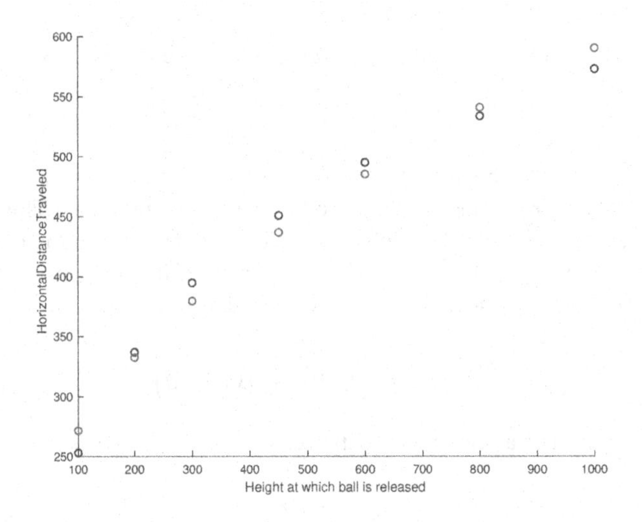

FIGURE 10.21: Scatterplot of model and original data values in original units

and

$$p(t) = 1.038 \times 10^{-5}$$

This value means that, if Galileo repeated this study 100,000 times, collecting 6 data points each time, we would expect our model results about 1 time in that 100,000 just due to random chance. Them's good odds! We can now conclude that our model fits well, and is likely due to some systematic relationship between the height at which a ball is rolled from a ramp, and the horizontal distance it will travel before hitting the ground. Whew! Time for a beer (to honor Gossett's contribution to the analysis).

10.5 Summary of Simple Linear Regression

Wow! That was a long and detailed chapter. I really want you to study this approach to curve fitting. Not only is it the most commonly used method for fitting models to data, it is also the mathematical basis for nearly all the advanced statistics you are likely to encounter. Starting in Chapter 11, we will be extending bivariate regression to predicting the relationship between values of a dependent variable and multiple independent variables that can be measured on any scale of measure, nominal, ordinal, interval or ratio.

The important concepts to remember include the fact that our "best" model is the one that minimizes the squared error of prediction. This basic assumption about the nature of error drives all of the subsequent mathematics. If we don't buy this assumption, then we have to develop a wholly different approach to fitting curves to data. *With* this assumption, we can model the total variation in our dependent variable as a simple sum of the variation of the line of best fit around $E(y)$, which we call $SS_{regression}$ plus the variation that our model cannot account for the residual error of prediction, SS_{ε}:

$$SS_{total} = SS_{regression} + SS_{\varepsilon}$$

This basic equation will be rewritten in subsequent chapters in what will be called "The General Linear Model" that underlies all subsequent statistics you are likely to encounter.

With this understanding of the different components of variation and the fact that they add together, we can define the goodness of fit of a model. The coefficient of determination, R^2, is the proportion of the total variation in our dependent variable that our model accounts for. Good fitting models have higher R^2 than poor-fitting models. But R^2 is not the only determinant of "good." We also want to make sure that the model we do settle on both accurately reflects the underlying relationship, and that it is significantly different from what we might come up with due to random chance. For this reason, it is critical to examine the residual error (ε) to insure that we have random error of prediction (or as close to random error as we can get).

Residual analysis should involve both a residual plot, and a Quantile-Quantile plot so that the nature of any departure from normally-distributed, random errors can be assessed and perhaps rectified. This might involve transforming our data by an inverse function to linearize it, and then fitting the linear model, assessing its goodness, and then transforming the predicted values back into the original units.

Always, always, always get to know your data before you start fitting lines to it! The extended example using Galileo's dataset should illustrate the issues associated with trying to figure out the right function to fit to the data, and the problems associated with just barging ahead with a linear model when the underlying pattern of the data is non-linear. Two equally reasonable transformations can result in different prediction equations and different goodness of fit. A thorough understanding of the data and the science behind it is a necessary component of any curve-fitting enterprise.

Lastly, the t-test of the slope is a statistical method for determining if the relationship you have modeled is due to just random chance or some probable *true* relationship. The t-test doesn't tell you your model is correct, only that, *given that the model does reflect the data*, the relationship is unlikely to appear that way just due to random chance.

10.6 References

Chamberlain, A. (2016). The Linear Algebra View of Least-Squares Regression.

Dickey, D.A., and Arnold, J.T. (1995), Teaching Statistics With Data of Historic Significance: Galileo's Gravity and Motion Experiments, *Journal of Statistics Education, 3, No. 1.* Retrieved online: https://amstat.tandfonline.com/doi/full/10.1080/10691898.1995.11910483.

Hastie, T., & Tibshirani, R. & Friedman, J. (2017). *The elements of statistical learning (12th printing).* New York: *Springer.*

Helmut Spaeth (1991). Mathematical Algorithms for Linear Regression. New York: Academic Press.

Mackowiak, P. A., Wasserman, S. S., and Levine, M. M. (1992), A Critical Appraisal of 98.6 Degrees F, the Upper Limit of the Normal Body Temperature, and Other Legacies of Carl Reinhold August Wunderlich, *Journal of the American Medical Association, 268*, 1578-1580.

T.C.Meierding(1993).Marble Tombstone Weathering and Air Pollution in

North America, *Annals of the Association of American Geographers, 83*(4), 568-588.

Wiemann, Michael C.; Williamson, G. Bruce. 2014. Wood specific gravity variation with height and its implications for biomass estimation. Research Paper FPL-RP-677. Madison, WI: U.S. Department of Agriculture, Forest Service, Forest Products Laboratory.

10.7 Study Problems for Chapter 10

1. Examine each of the following assumptions of Simple Linear Regression.

   What are the consequences of violation of these assumptions? Which ones are *really* critical to make sure your analysis doesn't violate?

   (a) Independence of the observations

   This is an essential assumption of all probabilistic data modeling. If the observations are not independent, there will be some unaccounted for bias in the model resulting from the data, leading to potentially erroneous decisions made from the data.

   (b) The true relationship between independent and dependent variables is linear in its parameters;

   This is critical! Because the underlying mathematics is linear algebra, if the real relationship between the data is nonlinear, any model created from it will not well reflect the true relationship. That is why we transform nonlinear data if we want to use Least Squares Regression to create a model.

   (c) Errors of prediction are normally distributed around zero; and

   If the errors of prediction are not normal, then there is some non-random (i.e., systematic) error in the model. If they are not equally distributed around zero, then there is some problem with model, having some systematic bias in its prediction.

   (d) Errors of prediction are equally distributed around the line of best fit.

   If the errors of prediction are not equally distributed around the line of best fit, the line isn't a best-fit line, or the underlying data is nonlinear.

2. The relationship between the Pitch-to-Diameter ratio and Static Thrust of aircraft propellers is being studied. Your professor has downloaded data from a NASA report. The data is presented below:

P/D (X)	Static Thrust (Y)
0.50	8.10
0.50	8.20
0.50	8.30
0.50	8.40
0.50	8.50
0.70	6.80
0.70	7.10
0.70	7.40
0.70	7.60
0.90	6.20
0.90	6.30
0.90	6.60
1.10	5.00
1.10	5.10

Describe how would you go about determining the relationship that exists (if any) between P/D and Static Thrust for this data. Write down the steps in the process, you don't have to write out a regression equation if you don't need to in order to construct your answer.

First I would do my homework, studying this relationship. The relationship should show static thrust reducing as P/D increases. I don't know right now (yes I do) what the shape of the relationship should be. This means I would be making a 1-tailed, left tailed hypothesis.

$$H_0 : \beta_1 \geq \mu_{\beta_1}$$

$$H_1 : \beta_1 < \mu_{\beta_1}$$

For this, given I am designing aircraft propellers, I would want to be fairly conservative. If this is in the later stages of design, I would probably select $\alpha = 0.01$. After collecting this data, I would want to make a scatterplot:

The data look fairly linear overall, despite the fact that there are only 4 values of the independent variable. I don't think I need to transform the data, but I will use a residual analysis after the regression model is computed to double-check.

Next, I will compute the line of best fit.

```
»stats=regstats(x,y,'linear');
```

Matlab returns the parameters of my regression model:

$$\beta_0 = 2.0391$$

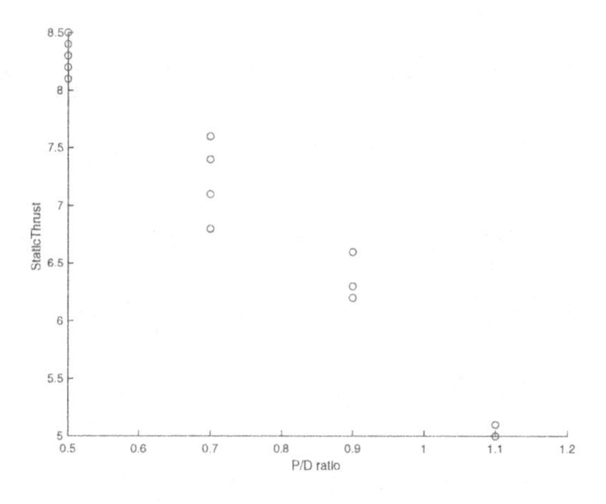

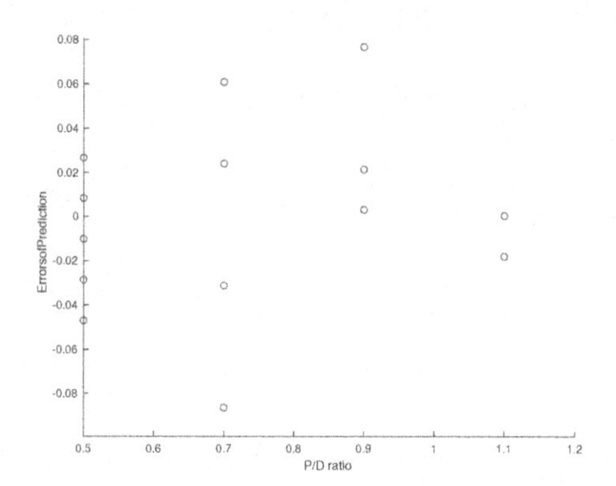

$$\beta_1 = -0.1842$$

I will use the vector of residuals to perform a residual analysis:

»scatter(x,stats.r);

This residual plot shows no distinct—good!

» qqplot(stats.r);

The actual quantiles of the data are pretty close to the expected quantiles from the Standard Normal Distribution. None of the values appears to lie far away from y=x. I don't see any outliers. Looks like a linear model is appropriate.

Now I will perform the *t*-test of the slope to determine if the relationship modeled is significantly different from that expected due to random chance.

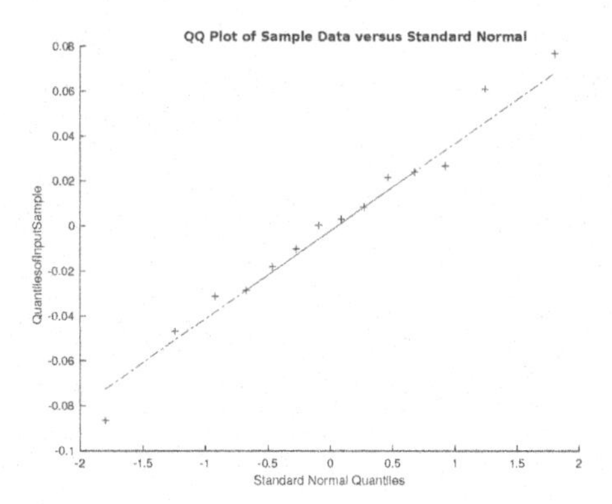

$$t = -17.6471$$

The critical value of t at 12 degrees of freedom, $\alpha = 0.01$, is $t_{12} = -2.6810$

Since the calculated value of t is greater in magnitude (further in the tail) than the critical value, I conclude that the relationship I have modeled is unlikely to be due to random chance, it is likely a true relationship.

1. Why does the line of best fit have to pass through $(\overline{x}, \overline{y})$?

   Because these values are the E(x) for the independent and dependent variable, respectively, the point that they make up is the E(x) of the bivariate distribution (x,y).

2. The table below shows the body temperature and heart rate of 130 individuals (Mackowiak, et al., 1992). The military is interested in looking at new ways of cooling infantrymen and women, particularly in situations where they experience physical exertion. Knowing the relationship between body temperature and heart rate, therefore, is an important factor to consider in designing new cooling apparatus for suits and vehicles.

Body Temperature	Heart rate	Body Temperature	Heart rate	Body Temperature	Heart rate	Body Temperature	Heart rate	Body Temperature	Heart rate
96.3	70	97.4	68	97.8	73	98	78	98.4	70
96.7	71	97.4	72	97.8	65	98.1	73	98.4	82
96.9	74	97.4	78	97.8	74	98.1	67	98.4	84
97	80	97.5	70	97.9	76	98.2	66	98.5	68
97.1	73	97.5	75	97.9	72	98.2	64	98.5	71

continued on next page

97.1	75	97.6	74	98	78	98.2	71	98.6	77
97.1	82	97.6	69	98	71	98.2	72	98.6	78
97.2	64	97.6	73	98	74	98.3	86	98.6	83
97.3	69	97.7	77	98	67	98.3	72	98.6	66
97.4	70	97.8	58	98	64	98.4	68	98.6	70
98.6	82	99.1	71	97.4	57	98	76	98.2	69
98.7	73	99.2	83	97.6	61	98	87	98.2	57
98.7	78	99.3	63	97.7	84	98	78	98.3	79
98.8	78	99.4	70	97.7	61	98	73	98.3	78
98.8	81	99.5	75	97.8	77	98	89	98.3	80
98.8	78	96.4	69	97.8	62	98.1	81	98.4	79
98.9	80	96.7	62	97.8	71	98.2	73	98.4	81
99	75	96.8	75	97.9	68	98.2	64	98.4	73
99	79	97.2	66	97.9	69	98.2	65	98.4	74
99	81	97.2	68	97.9	79	98.2	73	98.4	84
98.5	83	98.7	82	99	81	98.8	73	99.4	77
98.6	82	98.8	64	99.1	80	98.8	84	99.9	79
98.6	85	98.8	70	99.1	74	98.9	76	100	78
98.6	86	98.8	83	99.2	77	99	79	100.8	77
98.6	77	98.8	89	99.2	66	98.7	64	98.7	79
98.7	72	98.8	69	99.3	68	98.7	65	98.7	59

(a) i. What are the Null and Alternate Hypotheses you would use?
I assume this is a positive relationship. Therefore it is a 1-tailed, right-tailed test.
$H_0 : \beta_1 \leq \mu_{\beta_1}$
$H_1 : \beta_1 > \mu_{\beta_1}$

ii. What is the Type I error rate you would use for this context?
Here, I think a Type I error is not too problematic. It is likely to not heat up infantry more than their usual kit, so I will hedge on a more liberal Type I error rate of $\alpha = 0.10$

iii. Is a linear model appropriate?
The scatterplot of the data (not pictured here) looks like there is a lot of random variation, but I don't see any obviously nonlinear structure.

iv. Given the data, what is the relationship between heart rate and predicted temperature?
Matlab returns the linear parameters of the model as:

$$\beta_0 = 96.3068$$

$$\beta_1 = 0.0263$$

$$R^2 = 0.0643$$

v. Are the errors random and Normally distributed?
The Q-Q plots of both variables, and the residuals look very tight around y=x. The residual plot, however, shows a pretty linear pattern in the residuals. So, it looks like there may be some systematic error in the model. Unfortunately, with so much random error (plot the (x,y) scatterplot of the original data), there doesn't seem to be much guidance for what this relationship should be (if there really is one).

vi. Is this relationship significant—is the slope of the line of best fit significantly greater than zero?
R^2 is exceptionally small. Only 6% of the variation in heart rate is explained by body temperature.

$$t = 2.97$$

The critical value of t at 128 degrees of freedom, $\alpha = 0.01$, is $t_{128} = 1.282$.
Since the calculated value of t is greater in magnitude (further in the tail) than the critical value, I conclude that the relationship I have modeled is unlikely to be due to random chance it is likely a true relationship.
So I have a dilemma. The t-test of the slope says there is a relationship, but R^2 says that that relationship is poorly modeled. I would conclude that, because of this, any relationship I have found is small enough in magnitude to be impractical for use in designing new cooling apparatus.

3. Why does R^2 follow from this relationship: $SS_{total} = SS_{regression} + SS_{\varepsilon}$?

(a) i. What meaning does R^2 have in an analysis?
$R^2 = \frac{SS_{regression}}{SS_{total}}$. This means that it is the proportion of the total variation in the dependent variable that is explained by values of the independent variable. It tells us the goodness of fit of our model.

ii. What values of R^2 would you feel comfortable in the kinds of work that you do as an engineer?
The higher the better, generally. For early work, I might be happy with low values of R^2. But for later stages in design, I might want a much more precise model and want R^2 in the 0.75 to 0.90 range.

iii. Why are there no hard and fast rules for assessing goodness of fit?
Different applications require different precision of our models.

4. **Why should a Q-Q plot fall on a straight line?**

   If we assume that the residuals are normally distributed, we can plot their actual quantile against the normal distribution. If they are exactly the same, then the line made up of points, $(Q_{data_i}, Q_{Normal_i})$ will all be points where the ordinate and abscissa are exactly the same, falling on the line $y = x$.

5. **What is the problem with heteroscedasticity?**

   (a) i. **How can we detect it?**
   Heteroscedasticity is where the variability in the dependent variable is not approximately equal for all values of the independent variable. We can detect this using scatterplots (fan spread), and residual analysis (non-random distribution of residuals).

   ii. **How can we account for it in our analyses?**
   We may have to partition the dependent variable up into regions and estimate several linear models.

6. Each of your kidneys has a ton of tiny blood vessels, called nephrons, that filter your blood to keep it clean of toxins. This filtration system can be damaged by high concentrations of blood sugar—too much glucose can cause the kidneys to filter too much blood. The stress put on the nephrons can damage them causing diabetic kidney disease. You are trying to create a new dialysis machine for patients who suffer from diabetic kidney disease. Some of the data you collect include the concentration of glycated hemoglobin (HBA1C)–a form of hemoglobin that is chemically linked to glucose, and random blood glucose (RBG) measurements of the amount of glucose at random times during the day. Patients who have consistently elevated RBG, and HBA1C, are at a greater risk of kidney failure. You need to assess the relationship between RBG and HBA1C, so that you can predict HBA1C, given a simple blood test.

   You have collected data from 350 patients. This data is contained in the file hba1c_vs_rbg_regression.xlsx.

   (a) i. What are the Null and Alternate Hypotheses you would use, given what you know about the context?
   We assume that there is a positive relationship (slope-wise) to the data. This is a one-tailed, right-tailed hypothesis:
   $H_0 : \beta_1 \leq \mu_{\beta_1}$
   $H_1 : \beta_1 > \mu_{\beta_1}$

   ii. What is the Type I error rate you would use for this context?
   Because this has potentially serious consequences for people's health and wellbeing, I would use $\alpha = 0.01$

   iii. Is a linear model appropriate?
   The scatterplot of the data appears to be nice and football shaped with no apparent curvilinearity in the overall pattern. I will go ahead with a linear model.

iv. Given the data, what is the relationship between RBG and HBA1C?
» stats=regstats(rbg,hbalc,'linear');
Matlab returns parameters of:

$$\beta_0 = -0.0841$$

$$\beta_1 = 1.4053$$

and goodness of fit:

$$R^2 = 0.4470$$

This tells me that, for every unit change in HBA1C results in a change in estimated red blood glucose of about 1.4. The goodness of fit is moderate, about 45% of the total variation in red blood glucose being accounted for solely by HBA1C concentration.

v. Are the errors random and Normally distributed?
» scatter(hbalc,stats.r);

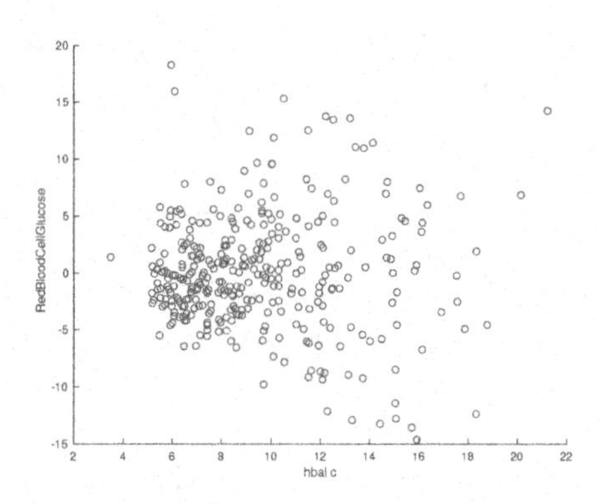

» qqplot(stats.r);

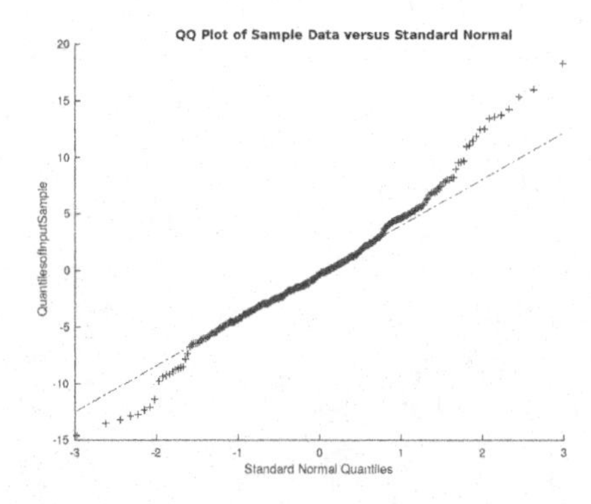

The residual plot looks great. There doesn't appear to be any definitive pattern in the data. Both the Q-Q plot and the residual plot show a bit of heteroscedatiscity with greater density towards the lower values of HBA1C, and less density at upper values. But the deviation from what is expected in the Q-Q plot (the y=x line), is fairly small, and only located at the tails of the distribution. I would conclude that a linear model is appropriate, and that the residual errors are fairly random and pretty much Normally distributed.

vi. Is this relationship significant—is the slope of the line of best fit significantly greater than zero?

$t = 16.7461$, $p = 1.4938 \times 10^{-46}$

This value of t is very far into the right tail of the distribution. So far, in fact, that its probability is nearly infinitesimally small. We can therefore reject the null hypothesis that the slope of our relationship is zero, and accept the alternative, that the relationship is a significant, positive relationship.

7. How does the t-test test the hypothesis that the slope of a fitted line is (or is not) greater than zero?

The t-test of the slope is essentially a one-sample t-test. The slope represents a difference in the variation in the y-direction, as a function of the variation in the x-direction. If the slope is zero, none of the variation in y is uniquely predicted by x. Therefore, the further the slope is from zero, scaled by how much it is expected to vary $\left(SE_{\beta_1}\right)$ can be modeled by a t-distribution. If the slope is much greater than zero relative to its variability, then t will be large in magnitude, further in the tails of the t-distribution, and less probable to have occurred solely as a function of chance.

8. The dataset in Groundwater_evap_regression.xlsx contains records of the minimum and maximum ground temperatures, and associated min and max air temperatures for 25 sites. The dependent variable measured was daily quantity of evaporated water. You are trying to design a means of recapturing the evaporated water from such sites to help residents obtain clean drinking water.

(a) i. What are the Null and Alternate Hypotheses you would use, given what you know about the context?

Here the dependent variable is evaporated groundwater. The temperature conditions are the independent variables. You might choose to combine some of the independent variables into composites (e.g., subtracting or taking their ratio). But the essential question relates to whether or not there is a relationship between independent variables and the dependent variable, and what the nature of that relationship might be.

Generally, I think that temperature, whenever it is measured would be positively related to evaporation of water, just given chemistry and thermodynamics. So I would posit a one-tailed, right-tailed hypothesis:

$H_0 : \beta_1 \leq \mu_{\beta_1}$
$H_1 : \beta_1 > \mu_{\beta_1}$
But this may not be your understanding. Try to conceptualize the problem and make your own hypotheses.

ii. What is the Type I error rate you would use for this context?
The consequences of this model are not immediately dangerous or costly, so I will use $\alpha = 0.05$

iii. Is a linear model appropriate?
I will provide my analysis for maximum ground temperature only. The procedure will be the same for the other variables, though the conclusions may be different!

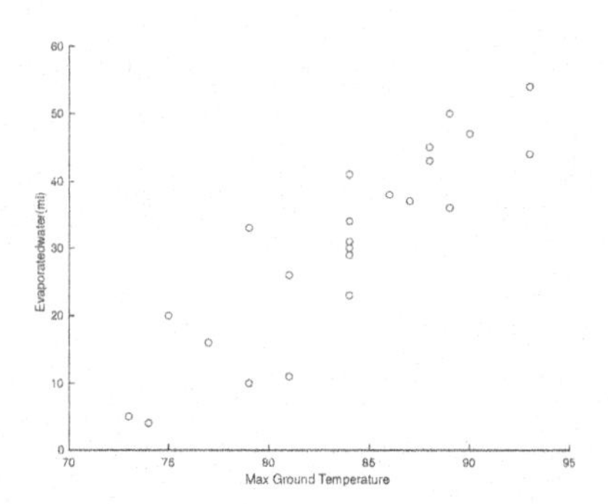

The scatterplot of the data looks nice and football shaped. No apparent outliers. I think a linear model is appropriate.

iv. Given the data, what is the relationship between the independent variables and the amount of evaporated water?
» stats=regstats(evap,temp,'linear');
Matlab returns parameters of:

$$\beta_0 = -160.4133$$

$$\beta_1 = 2.2842$$

and goodness of fit:

$$R^2 = 0.7980$$

First of all, the y-intercept is not possible in our world. That clues me in to my dusty memories that, when water freezes, it doesn't evaporate. Thus the global relationship cannot be linear. But for the conceptual domain (73, 93), it is close enough that a linear model appears to be reasonable. With a goodness of fit of about .80, I am pretty happy with the model and its predictive value within the conceptual domain.

v. Are the errors random and Normally distributed?

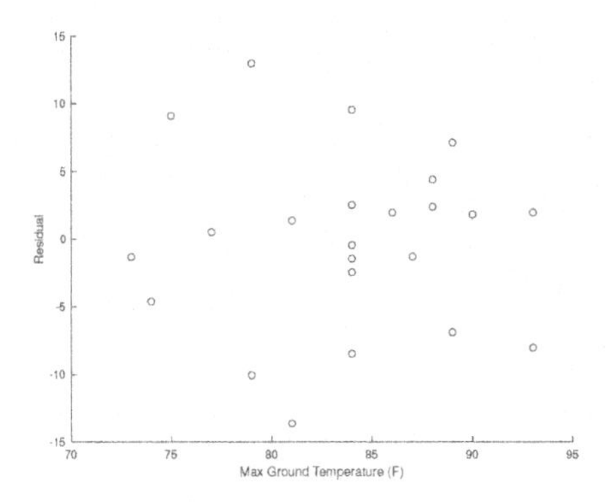

The residual plot looks beautifully random, evenly spread about zero.

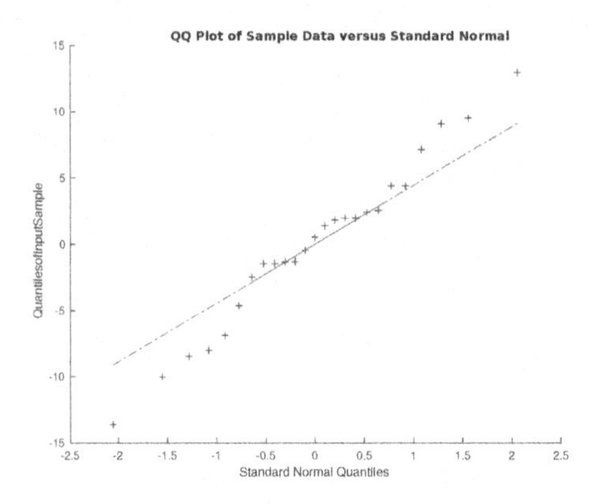

The Q-Q plot shows a pretty linear structure, even if it deviates from $y=x$ a bit at the tails. I don't see a big reason to assume that the distribution of errors varies widely from normal.

vi. Is the relationship significant—is the slope of the line of best fit significantly greater than zero for each of the variables?
$t = 9.5319, p = 1.8739 \times 10^{-9}$
This value of t is very far into the right tail of the distribution. Its probability is about 2 in a billion. We can therefore reject the null hypothesis that the slope of our relationship is zero, and accept the alternative, that the relationship is a significant, positive relationship.

vii. Given what you know about linear regression, which of the independent variables seem to make the biggest impact on the daily amount of water evaporation?

I will leave this up to you to argue. Look at three indices:

β_1 : This is the size and direction of the effect of the independent variable on the dependent variable;

R^2 : This is an indication of the relative "goodness" of the model in terms of the amount of variation in the dependent variable the independent variable explains;

t_{β_1} : This is an indication that the linear relationship modeled is unlikely to be random, and therefore more likely to be some systematic (i.e., true) relationship.

In the next Chapter, we will learn how to account for all of these independent variables simultaneously multidimensionally.

Part III

Applications of the General Linear Model

11

The General Linear Model: Regression with Multiple Predictors

As we saw in the previous chapter, fitting a model to a set of data is an important technique for being able to describe and predict relationships among variables. Curve-fitting is easy when points lie on a perfect function that can be described by a closed-form equation. But the typical approach most students have been taught doesn't account for the inherent variability in data. Instead, we have to deal with the fact that multiple potential functions could be fit to the data, and so we have to figure out which of those potential models fits the best. That is where a statistical approach, layered onto the linear algebra becomes necessary. I introduced this in Chapter 10, to set the stage. Now, we will take this basic understanding and show how it can form the basis for a general model that will undergird nearly all statistical relationships you will encounter over your career. It is best to start with an applied example in 2 dimensions and then generalize this to any number of independent variables you wish to model.

Satellites in low earth orbit (LEO) operate between 250 and 1500 km above the ground. Because Earth's atmosphere extends hundreds of miles into space, LEOs eventually experience enough friction that they fall back to earth and burn up (http://image.gsfc.nasa.gov/poetry/). NASA Goddard Space Flight Center has provided data for the number of sunspots each year from 1969 through 2004, along with the number of satellites that re-enter the earth atmosphere. Here is what they say about the phenomenon (in the file Satellite_failures_regression.xlsx data file):

Satellites experience drag as they move through the outer reaches of Earth's atmosphere, a large region of hot gas known as the thermosphere. Like a marshmallow held over a campfire, the thermosphere puffs up when heated by solar ultraviolet and x-radiation. The more the thermosphere swells, the more drag satellites experience. For satellites at lower inclinations and at low latitudes near the equator, this increase in energy mostly comes from the bright area surrounding sunspots and solar flares. The number of photons at the higher energies can increase by up to 100 times or more within a few minutes due to a single flare, and can then last up to a day before returning to pre-flare levels.

https://www.nasa.gov/topics/solarsystem/features/solar-effects.html

DOI: 10.1201/9781003094227-11

Here are the first few years-worth of data to use as an example to get us started (see Table 11.1). We will expand this data once we get the linear models figured out.

Number of Sunspots	**Number of Satellites Re-entering Earth's Orbit**
105	26
107	25
66	19

TABLE 11.1: Three years of paired data

If we consider the number of sunspots the independent variable, x, and number of satellites entering Earth's orbit the dependent variable, y, then we can create a simple linear system:

$$\beta_0 + \beta_1 x_i = y_i$$

$$\beta_0 + \beta_1 105 = 26$$

$$\beta_0 + \beta_1 107 = 25$$

$$\beta_0 + \beta_1 66 = 19$$

The problem with this system is that there are no lines that fit the data perfectly in the two dimensions, (x,y). If we assume there ***is*** a one-to-one, linear function that best models the data, then we can model it by the following:

$$\beta_0 + \beta_1 x_i = \hat{y}_i$$

Where $\hat{y}_i$ is an estimation of the assumed true value of y_i. The little "hat" above the y indicates that this is a predicted value of y. Statisticians call this value "y-hat". By this model, any value of x maps, one-to-one, to a single value of $\hat{y}$, even though the three actual data points do not lie on the line itself. Because $\hat{y}_i$ is estimated, the true value of y_i differs from $\hat{y}$ by the error of prediction, the residual ε_i. Let's represent the two variables, x and y, as column vectors:

$$X = \begin{bmatrix} 105 \\ 107 \\ 66 \end{bmatrix} \quad \text{and } \hat{Y} = Y - \varepsilon = \begin{bmatrix} 26 - \varepsilon_1 \\ 25 - \varepsilon_2 \\ 19 - \varepsilon_3 \end{bmatrix}$$

Our prediction equation now becomes

$$X\beta = \hat{Y}$$

where β is a vector of the parameters β_0 and β_1.

Unfortunately, a 3x1 matrix cannot be multiplied by a 2x1 matrix without some help.

$$\begin{bmatrix} 1 & x_1 \\ 1 & x_2 \\ 1 & x_3 \end{bmatrix} \begin{bmatrix} \beta_0 \\ \beta_1 \end{bmatrix} = \begin{bmatrix} y_1 \\ y_2 \\ y_3 \end{bmatrix} - \begin{bmatrix} \varepsilon_1 \\ \varepsilon_2 \\ \varepsilon_3 \end{bmatrix}$$

$$X \times \beta = Y - \varepsilon$$

To insure we have a constant term in the linear equation, X is an augmented matrix, where the first column is made up of 1s. Performing the multiplication left to right, we can see that $(1 \times \beta_0) + (x_1 \times \beta_1) = \hat{y}_1$, $(1 \times \beta_0) + (x_2 \times \beta_1) = \hat{y}_2$, and $(1 \times \beta_0) + (x_3 \times \beta_1) = \hat{y}_3$, preserving our linear system.[1] In sum, the column of 1s gives us the intercept term in our prediction equation, β_0.

Our error of prediction, ε, for each value of Y is just the actual value, y_i minus its predicted value $\hat{y}_i$ —exactly the same as we learned in Chapter 10.

$$\begin{bmatrix} \varepsilon_1 \\ \varepsilon_2 \\ \varepsilon_3 \end{bmatrix} = \begin{bmatrix} y_1 \\ y_2 \\ y_3 \end{bmatrix} - \begin{bmatrix} \hat{y}_1 \\ \hat{y}_2 \\ \hat{y}_3 \end{bmatrix}$$

$$\varepsilon = Y - \hat{Y}$$

The vector $\hat{Y}$ is the predicted values of the dependent variable Y. As such, they contain some error, represented by the vector ε. Like all measurements, the predicted value plus the error of measurement equals the true value of the variable, y_i.

$$\begin{bmatrix} 1 & 22 \\ 1 & 15 \\ 1 & 14 \end{bmatrix} \begin{bmatrix} \beta_0 \\ \beta_1 \end{bmatrix} = \begin{bmatrix} 2.66 - \varepsilon_1 \\ 2.73 - \varepsilon_2 \\ 1.57 - \varepsilon_3 \end{bmatrix}$$

$$X \times \beta = \hat{Y}$$

11.1 Linear Algebra Approach to Regression

There are a number of ways to conceptualize fitting a line to a cloud of data. I will provide two methods, the combination of which, I believe, gives us a pretty deep understanding of linear regression, and what will be called, **The General Linear Model**, a model that applies to most advanced approaches to hypothesis testing as well as curve fitting.

[1]The column of 1s is also a way, in matrix algebra, to be able to count the length of a column vector. It becomes useful in statistics, because we always have to deal with the degrees of freedom of a vector space (i.e., we consider n is important variable).

The first, and most assumption-free approach is that provided by linear algebra. The logic is this: Now that we know the structure of our linear system, including the variability in measurements (ε), we can begin to solve for the unique vector of parameters, β, that will project Y onto X creating a one-to-one function $\hat{Y} = f(X) = X\beta$.

Let's look at this geometrically. X exists in a n-dimensional space, where n is the number of observations—the column length of X. That means that X_1 and X_2 can be represented as a vector in this space. These two vectors, (1,1,1), and (22, 15, 14) will lie in a plane[2] that denotes the set of all possible linear combinations of Xs—column vectors called the *column space of* X, denoted *C(X)*.

Y, because its length is also *n*, exists in the same *n*-dimensional space as X. However, looking at our values, we can see that *Y* is not in the plane defined by X_1 and X_2–it is not a perfect linear combination of the two. We hope that our dependent values would be close to this plane, but we now have to figure out what linear combination in *C(X)* is the closest approximation to Y. This is the set of points that **Minimize the Error of Prediction.**

In a space defined by orthogonal dimensions (i.e., dimensions that are not linear combinations of each other) we can imagine a *second* plane that intersects the column space of X at right angles. This plane is defined by Y and $\hat{Y}$—the orthogonally-projected vector of Y onto *C(X)*. (see Figure 11.1)

Minimizing Error of Prediction. *The key to fitting a best fit line to data is to minimize our error of prediction, ε, by choosing a matrix of parameters, β, so that the residual vector $\varepsilon = Y - \hat{Y}$ is orthogonal (i.e., the shortest distance) to C(X).*

Because it is orthogonal, the dot-product between X and ε must be zero:

$$X \cdot \varepsilon = 0$$

We also know (or can work out if we think a bit), that since $Y - \hat{Y} = \varepsilon$, and since $\hat{Y} = X\beta$,

$$X^T (Y - \varepsilon) = X^T X\beta$$

$$X^T (Y) - 0 = X^T X\beta$$

then,

[2]We are restricting our example to 3-dimensions. Later, we will see that X can have unlimited dimensions in both row and column, in which case, each column forms a vector that together define an *n*-1 dimensional hyperplane.

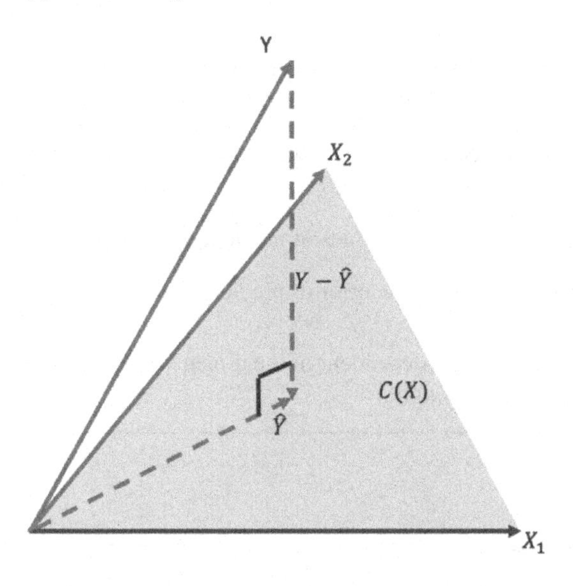

FIGURE 11.1: The projection of the dependent variable, Y onto the plane spanned by the independent variables X_1 and X_2. $\hat{Y}$ is the vector of predicted values of Y. $Y - \hat{Y} = \varepsilon$, the residual (error) vector, is orthogonal to $C(X)$

$$\beta = \left(X^T X\right)^{-1} X^T Y$$

Substituting this into our original linear model, $X\beta = \hat{Y}$, we get the prediction equation that minimizes our error of prediction:

$$\hat{Y} = X\left(X^T X\right)^{-1} X^T Y$$

Multiple Linear Regression: The four equations that govern linear regression are:

$$\hat{Y} = X\beta$$

$$\beta = \left(X^T X\right)^{-1} X^T Y$$

$$\hat{Y} = X\left(X^T X\right)^{-1} X^T Y$$

and

$$\varepsilon = Y - \hat{Y}$$

Where

- $\hat{Y}$ is a vector of predicted values of the dependent variable;

- β is a vector of parameters, β_0 and β_1, corresponding to the intercept and slope of a linear equation;
- X is an augmented matrix of predictor (i.e., independent variable) values; and
- Y is a vector of measured values of the dependent variable; and
- ε is a vector of residua, the error of prediction in $\hat{Y}$

These equations are independent of the dimensionality of X, so long as both X and Y have the same length.

11.2 Calculus Approach to Regression

The more traditional approach to linear regression is what is called **Least Squares Regression**. This approach also involves minimizing the error of prediction, but it uses calculus, and so has some assumptions that the linear algebra approach does not. Namely, there has to be a model function that is differentiable, by which we can minimize the residual error. Error of prediction is conceived of just like any variation in statistics: As the difference between some predicted value $\hat{y}_i$, and a measured value y_i:

$$\varepsilon_i = y_i - \hat{y}_i$$

just like we found when computing the variance of a sample, the sum of the errors of prediction will be zero. So, we look to the Euclidean distance of each residual value, and sum the square of the differences and try to minimize that value.

$$\varepsilon_i^2 = (y_i - \hat{y}_i)^2$$

$$\sum_{i-1}^{n} \varepsilon_i^2 = \sum_{i=1}^{n} (y_i - \hat{y}_i)^2$$

Notice that this quantity, the sum of the squared residual values, is the same as the *numerator of the variance of a sample*. Indeed it *is* the numerator of the variance of residual values around the regression line. Minimizing this quantity insures that the line of best fit is the linear function producing the minimum error of prediction.

$$\sum_{i-1}^{n} \varepsilon_i^2 = \sum_{i=1}^{n} (y_i - \hat{y}_i)^2$$

$$= \sum_{i=1}^{n} \left(y_i - (\beta_0 + \beta_1 x_i)^2\right)$$

This equation is quadratic, but it would be helpful to have it in a more easily differentiable form, so we can find the minimum:

$$= (Y - X\beta)^T (Y - X\beta)$$

The minimum of this quadratic exists at the point the first derivative with respect to β is equal to zero:

$$\frac{\partial \varepsilon}{\partial \beta} = -2X^T (Y - X\beta)$$

$$-2X^T (Y - X\beta) = 0$$

Now solving for β, we get the equation that minimizes the squared residual error:

$$\beta = \left(X^T X\right)^{-1} X^T Y$$

It is no coincidence that this equation, and all the others derived from it are exactly the same as we found using the simpler linear algebra method.

11.3 Fitting a Line

Now that we have a good model for understanding why, out of all the lines I might try to fit to my data, I can select the one with the least error of prediction and be sure that it is the **only** line that minimizes the error. With this established I can begin modeling data. So, let's try the real data that we introduced the linear algebra with: The relationship between number of sunspots, and the number of satellites reentering Earth's Orbit. This is the first 8 rows of X and Y (see Table 11.2). The entire dataset has 36 rows having 36 years of data. You may download the Excel file and follow along.

Number of Sunspots	Satellites Re-entering Earth's Orbit
105	26
107	25
66	19
67	12
37	14
32	21
14	15
12	16

TABLE 11.2: The first 8 rows of the sunspot data

Just like with univariate data, the first thing one should do when analyzing data is examine it visually. Looking at the scatterplot in Figure 11.2, I see something that *could*, kind of vaguely, represent a linear relationship. Most of the points seem to be clustered around the center of the graph. Few large values of Number of Satellites appear where Number of Sunspots is small, and few small values of Number of Satellites appear where Number of Sunspots is large. What I am looking for is a kind of football shape: An ellipse with a definite long axis and a definite short axis:

If the graph had appeared like someone had taken a shotgun to it, where the points appeared randomly distributed in all directions and distances, I might conclude that, if there is a relationship, it probably isn't linear. Also, if I saw a definite curvilinear pattern in the data, I might pause and speculate about whether a linear model is appropriate. We will discuss this situation later on, but for now, I don't think it would hurt to fit a line and see how good it is as a model.

I need to designate my X matrix-my matrix of predictor values. In this case X is Sunspots, and Y, my outcome variable, is Satellites. So I know:

$$\beta = \left(X^T X\right)^{-1} X^T Y$$

$$\begin{bmatrix} 1 & 105 \\ 1 & 107 \\ \vdots & \vdots \\ 1 & 14 \\ 1 & 12 \end{bmatrix} \qquad \begin{bmatrix} 25 \\ 26 \\ \vdots \\ 15 \\ 16 \end{bmatrix}$$

Where the actual vectors have length 36.

$$\left(X^T X\right) = \begin{bmatrix} 36 & 2671 \\ 2761 & 280527 \end{bmatrix}$$

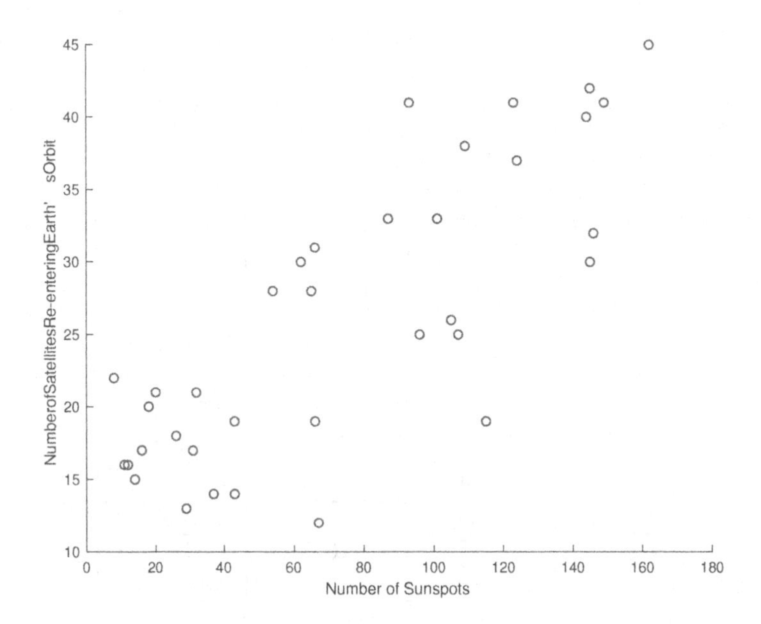

FIGURE 11.2: Scatterplot of sunspot data

Notice how $X^T X$ (1,1) is n, the number of paired observations. This is a result of adding the column of 1s.

$$\left(X^T X\right)^{-1} = \begin{bmatrix} 1.3017 & -0.0914 \\ -0.0914 & 0.0071 \end{bmatrix}$$

This inverting allows us to "divide" the matrices (multiplying by the inverse of $X^T X$).

$$X^T Y = \begin{bmatrix} 939 \\ 83134 \end{bmatrix}$$

Now we can find our β -vector, our estimated parameters of the line of best fit:

$$\beta = \left(X^T X\right)^{-1} X^T Y = \begin{bmatrix} 13.9520 \\ 0.1635 \end{bmatrix}$$

With our β vector defined, we now have the equation for our line of best fit:

$$\hat{Y} = X\beta$$

$$\hat{y}_i = 13.9520 + 0.1635x_i$$

Substituting in all the values of X into the equation, we get the following predicted values in Table 11.3:

Number of Sunspots	Satellites Re-entering Earth's Orbit	$\hat{Y}$	ε
105	26	31.1203	−5.1203
107	25	31.4473	−6.4473
66	19	24.7435	−5.7435
67	12	24.9070	−12.9070
37	14	20.0018	−6.0018
32	21	19.1842	1.8158
14	15	16.2411	−1.2411
12	16	15.9141	0.0859

Table 11.3. The first 8 rows of sunspot data with predicted and residual values

Plotting $\hat{Y}$ versus X, and plotting the prediction equation in the graph, we can see two important features:

1. The vector $\hat{Y}$ is a vector of values, not the continuous function!
2. The function $\hat{Y} = f(X)$ allows us to predict values of f that aren't in our original data set! This is the real power of regression, in that we can take a finite sample of bivariate data, and make predictions about the overall structure of the data, and the relationship(s) among the variables.

Once we have the line of best fit fitted (See Figure 11.3), all the relationships we studied in Chapter 10 are exactly the same. The total variation is just the sum of the variation accounted for by our model and the variation due to random error of prediction.

$$SS_{total} = SS_{regression} + SS_{\varepsilon}$$

$$\sum_{i=1}^{n}(y_i - \bar{y})^2 = \sum_{i=1}^{n}(\hat{y}_i - \bar{y})^2 + \sum_{i=1}^{n}(y_i - \hat{y}_i)^2$$

and

$$R^2 = \frac{\sum_{i=1}^{n}(\hat{y}_i - \bar{y})^2}{\sum_{i=1}^{n}(y_i - \bar{y})^2} = \frac{SS_{regression}}{SS_{total}}$$

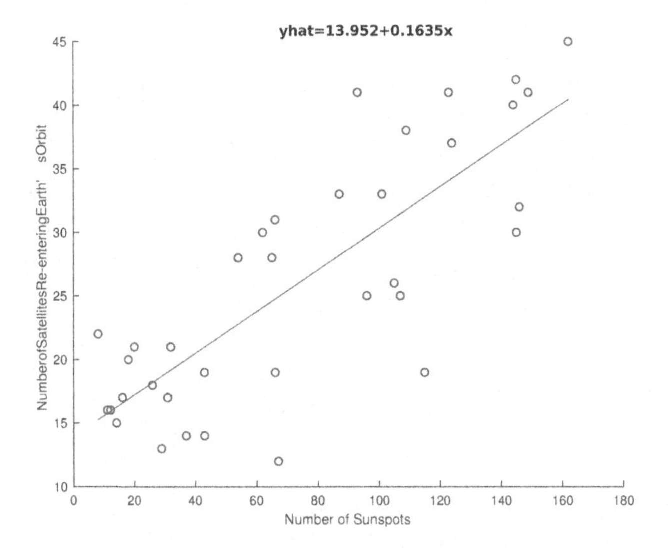

FIGURE 11.3: Line of best fit overlaid on sunspot data

So, if $\hat{Y} = X\beta$, then

$$SS_{\varepsilon} = (Y - X\beta)^{-1}(Y - X\beta)$$

The mean of a column vector using matrix algebra can be computed by

$$\overline{Y}^T = 1^T Y\left(1^T 1\right)^{-1}$$

Where $\overline{Y}^T$ is a row vector of the means of the columns of Y, and 1 is a column vector of ones with the same length as Y. With this,

$$SS_{Total} = \left(Y - \overline{Y}\right)^T\left(Y - \overline{Y}\right)$$

and

$$SS_{regression} = \left(\hat{Y} - \overline{Y}\right)^T\left(\hat{Y} - \overline{Y}\right)$$

$$SS_{total} = SS_{regression} + SS_{\varepsilon}$$

can then be written

$$\left(Y-\overline{Y}\right)^T\left(Y-\overline{Y}\right)=\left(\hat{Y}-\overline{Y}\right)^T\left(\hat{Y}-\overline{Y}\right)+\left(Y-\hat{Y}\right)^T\left(Y-\hat{Y}\right)$$

Walking through this identity, you can see pretty clearly that it is equivalent to the identity written in summation notation:

$$\sum_{i=1}^{n}(y_i-\bar{y})^2=\sum_{i=1}^{n}(\hat{y}_i-\bar{y})^2+\sum_{i=1}^{n}(y_i-\hat{y}_i)^2$$

So,

$$R^2=\left[\left(Y-\overline{Y}\right)^T\left(Y-\overline{Y}\right)\right]^{-1}\left[\left(\hat{Y}-\overline{Y}\right)^T\left(\hat{Y}-\overline{Y}\right)\right]$$

$$=\frac{SS_{regression}}{SS_{total}}$$

11.4 Expanding to Multiple Predictor Variables: Multiple Linear Regression

What I have just done is spend 10 pages, hoping to show that the *linear algebra approach is equivalent to the simple linear regression we learned in Chapter 10.* Everything is the same! We are just using the equivalent operations using column vectors to represent our variables. But this approach is much more powerful. After all, it is pretty simple to compute the line of best fit with only 2 variables. Why go through the trouble of learning the matrix algebra?

The power of this approach is that it can be expanded to incorporate any number of independent, predictor variables as we want! What's that? Yes, Sparky, we can now examine a set of predictor variables, and determine their individual and collective impact on a dependent, outcome variable. Let's start with an example:

Hydrocarbon vapors escaping during the process of filling an automobile, and through poorly sealed gas caps are a significant source of hydrocarbon pollution in the US. For this reason, most states have instituted gas cap testing. The following data (see Table 11.3) was collected in the 1980s (yes, back when we used stone tires and our feet to power the car). The dependent variable is the amount of hydrocarbons, in grams, emitted during refueling for 32 automobiles. The independent variables are the temperature of the fuel and the pressure of the fuel. The first three rows of this data are provided in Table 11.3 below to illustrate the data structure (see Table 11.3):

Gas Temp (C)	**Gas Pressure (KPa)**	**Escaped Hydrocarbons (g)**
11.67	23.58	29.00
2.22	22.48	24.00
10.56	21.93	26.00
⋮	⋮	⋮

TABLE 11.3: Data embodying a potential relationship between escaped hydrocarbons and the temperature and pressure at which Gasoline is pumped during automobile fueling

Given these different conditions at refueling, can we reasonably predict the amount of hydrocarbons escaping? Which of the four conditions makes the most impact on hydrocarbon emissions? These are the questions that **Multiple Linear Regression** will help us answer.

Procedure for Performing Linear Regression

1. Read up about the relationship you are modeling. Is it likely to be linear? Quadratic? Do your homework!
2. Determine if you are making a 1-tailed or 2-tailed hypothesis;
3. Express the Alternative Hypothesis and then the Null Hypothesis;
4. Select the Type I error rate;
5. Collect your Data;
6. Make a scatterplot of the data;
7. Examine the scatter for obvious signs of non-linearity;
8. Transform the data if necessary;
9. Compute the model of best fit;
10. Compute the residuals;
11. Perform a residual analysis, transform the data or eliminate outliers if necessary;
12. Perform the F-test of the "slope" to determine if the relationship modeled is significantly different from that expected due to random chance;
13. Figure out what to do with the results of your analysis and have a beer!

The procedures for Multiple Regression, as it is often called in shorthand, are the same as Simple Linear Regression. The only exception is # 12, where we have to change the t-test to an F-test to handle multiple dimensions. The only substantive difference between Multiple Regression and Simple Linear Regression is trying to understand all the extra dimensions in the model. We also have a few more assumptions we have to buy into if our model is to be trusted.

Going through this procedure, we need to think about the problem a bit: First, by the Ideal Gas Law, temperature and pressure are related proportionally. So, as temperature increases, so does pressure. Both of these should cause several things to occur that may impact the rate at which hydrocarbon vapors escape the gas tank: 1) as T and P increase, more hydrocarbon molecules will begin zipping around at faster speeds (this is just the definition of Heating); 2) however, as pressure increases in a closed system, fewer molecules escape the surface tension of the liquid gasoline; but in an open system, we expect hotter gasoline to release more hydrocarbon vapor. A gas tank is not completely closed when refueling. But the opening is very small. I really don't know what might happen at different T and P levels, so my hypothesis might have to be two-tailed.

Like all Regression analyses, our null hypothesis is that the "slope" of the "line" of best fit is equal to zero. Here, with 3 dimensions, we actually plot a plane of best fit best fit linear model with the form:

$$\hat{Y} = \beta X$$

Which yields the closed-form equation:

$$\hat{y}_i = \beta_0 + \beta_1 X_{1i} + \beta_2 X_{2i} + \ldots + \beta_k X_{ki}$$

Where X_k are column vectors in X, our augmented matrix of predictors, and β_k is its partial slope. We also call β_k the **regression coefficient** of X_k. Interpreted, β_k is the rate of change between a predictor variable, X_k, and the outcome variable Y that is independent of the other predictor variables. In differential equations terms, it can be thought of as the slope of $\hat{Y}$ in the two-dimensional cross section defined by X_k, $\hat{Y}$.

Regression Coefficient. The β -weights in a linear model are the partial slopes of an independent variable with respect to the dependent variable. They can be interpreted as the rate of change of X_k with respect to Y while holding all other X constant. The larger the β -weight, the more influence a variable has on the value of the outcome, Y.

The Null Hypotheses (plural) for multiple predictor variables are that each of these partial slopes is equal to zero. The Alternative is that at least one of these slopes is not equal to zero:

$$H_{0k} : \beta_k = 0$$

$$H_{1k} : \beta_k \neq 0$$

for all independent variables, *k*.

I hope you noticed that the prediction equation, expressed in matrix form is no different for multiple dimensions than it was for our two-dimensional example of the relationship between sunspots and failed satellites. *That is the most important idea in this Chapter: Once we have this general model, we can test multiple hypotheses simultaneously.*

Ok. So now we have our Hypotheses established, we need to set a Type I error rate. For this problem, I think we can fudge a bit on the high side. If we erroneously find that one of the variables contributes to escaping hydrocarbons, it doesn't hurt much. If we were going to enact legislation that made it mandatory for consumers to make extremely expensive repairs on their vehicles, we might be more restrictive. Let's use a default $\alpha = 0.05$ for our analyses.

The data is already collected, so we need to make a scatterplot. *Whoa!* Too many dimensions! I can only really visualize 2 dimensions, or 3 if the data is pretty simple. Since I can't rotate the 3d scatterplot in this book, I will do the next best thing—two 2-d scatterplots, one for each of the independent variables.

Examination of the scatterplots (Figure 11.4), I see what looks to be nice, football shapes with a clear long axis and short axis—iconic graphs for linear regression. This makes me feel pretty comfortable to run the full model as a linear combination of the variables, untransformed.

Computing the matrix algebra in Matlab, I get the following β vector:

$$\begin{bmatrix} 8.9802 \\ 0.4994 \\ 0.5225 \\ \vdots \end{bmatrix}$$

Translating that into a more familiar form, I get the prediction equation:

$$\hat{y}_i = 8.9802 + 0.4994\,(Gas\ Temp) + 0.5225\,(Gas\ Pressure)$$

Interpreting this, I expect that every 1 degree increase in temperature of my gasoline will result in about 0.4994 grams of hydrocarbons lost when I refuel my tank. For every 1 KPa of gas pressure that increases, I expect about 0.5225 grams of hydrocarbons to be lost. The intercept, 8.8902 means that we expect about 9 grams of hydrocarbons to be lost regardless of temperature and pressure values. Now I need to

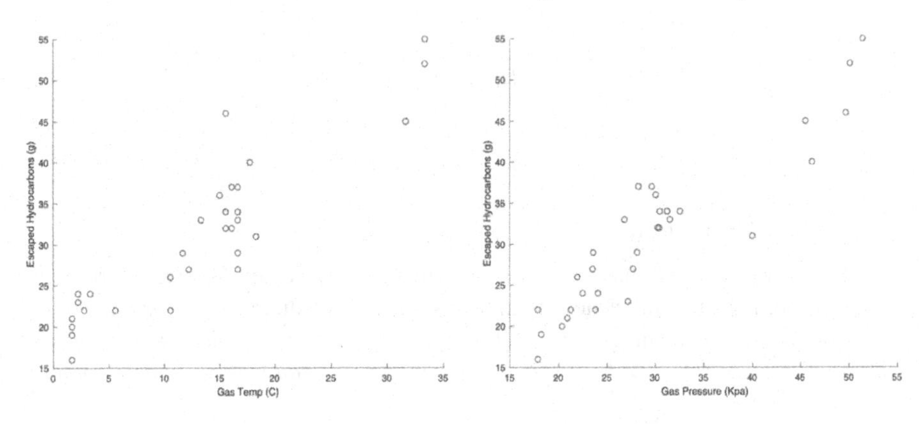

FIGURE 11.4: Scatterplots of escaped hydrocarbons (Y) versus each independent variable in X

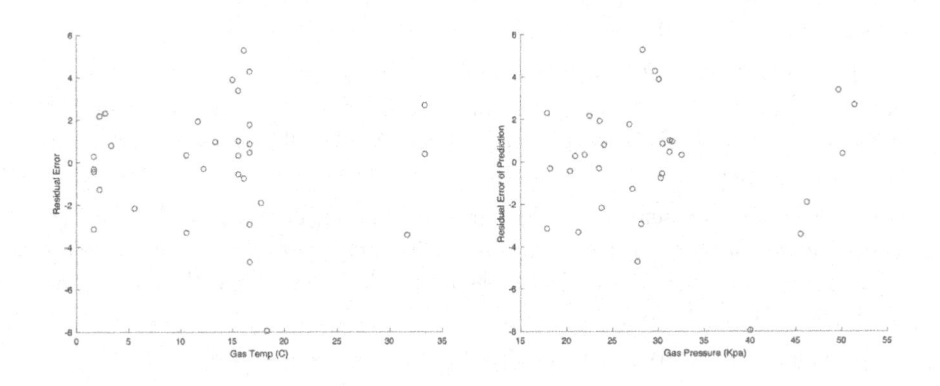

FIGURE 11.5: Residual plots of the error of prediction versus each independent variable in X

compute the residuals and perform a residual analysis. Again, because there are two independent variables, I will need to do two residual plots (See Figure 11.5).

These look excellent! The error of prediction shows no apparent pattern for each of the two predictor variables. It is fairly evenly distributed about zero—just what we want. I would conclude that there appears to be no problem with our linear model. Now let's look at the Q-Q plot presented in Figure 11.6 to see if our assumption of normalcy is violated:

This also looks great! The data lie very close to the line, and there are few that deviate much at all. I can conclude that the errors are both random and distributed close to normal. I won't have to transform the data. Now it is time to check on our

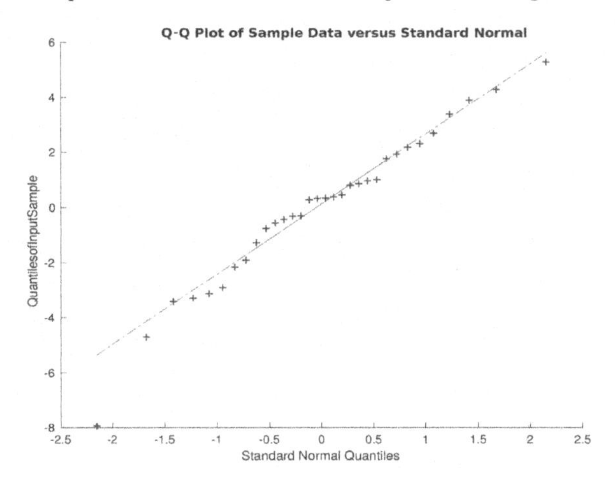

FIGURE 11.6: Q-Q plot of the residual error of prediction

goodness of fit. R^2 is interpreted just the same as before. It is the proportion of the total variation in our outcome variable, y, accounted for by our model. It uses the same formulas, even as we add more and more predictors:

$$R^2 = \frac{SS_{regression}}{SS_{total}}$$

$R^2 = 0.9124$, meaning that our model accounts for over 90% of the variation in hydrocarbon loss when refueling. In other words, by knowing only the temperature and the pressure of the gasoline we are pumping into our tanks, we have a model that is 90% accurate in predicting the amount of hydrocarbons we are inadvertently releasing into the atmosphere.

We established in Chapter 10 that we could have a high R^2 but still have a relationship that is not differentiated from any old random chance model. We should now determine if the partial slopes of the regression line are significantly different from zero. We have two β_k values. We could run two t-tests of the slope, but this is both cumbersome, and reduces our Type I error rate without applying some correction for conducting two non-independent analyses. So, we turn to the analog that we use for any number of independent variables, not just one. And it is our old friend, the F-test. Let's see how it works:

Remember, with the F-test, we are testing whether the variance of one sample is greater than another sample. If the ratio of F is much greater than 1, then it indicates that the variance of sample 1 is greater than that of sample 2.

$$F_{(df_1, df_2)} = \frac{\sigma_1^2}{\sigma_2^2}$$

Here, we are looking at the variance estimated by our regression line as our σ_1^2, while the residual error variance is our σ_2^2. If $\sigma^2_{regression} > \sigma^2_{\varepsilon}$, then our model explains more variation than it doesn't explain. Or we can say the error in our model is very small compared to the "truth." If you remember the formula for the variance of in a population, it is just this:

$$\sigma^2 = \frac{\sum_{i=1}^{n} (y_i - \mu)^2}{N}$$

The average of the squared deviations of y from its mean. We estimate σ^2 with the sample variance, s^2. It is the same thing, except we take into account we are estimating one parameter after first estimating the sample mean, so we divide by the degrees of freedom, the sample size minus 1[3]:

$$s^2 = \frac{\sum_{i=1}^{n} (y_i - \bar{y})^2}{n-1} = \frac{SS_{total}}{df_{total}}$$

The numerator of this formula should be looking familiar. It is the SS_{total} of our outcome variable, y. So, the SS divided by the degrees of freedom is an estimate of a population variance. If we want to know the estimate of the variance of our regression model over the long-haul, it would then be

$$MS_{regression} = \frac{\sum_{i=1}^{n} (\hat{y}_i - \bar{y})^2}{k-1} = \frac{SS_{regression}}{df_{regression}}$$

$MS_{regression}$ here means "Mean-Square" of the regression model. If you look at the formula, it is the Mean of the Squared deviations from the expected value of y. "Mean-Square" therefore, just means an estimate of a population variance. I don't know why statisticians use all these different names for the same thing, but they do.

By the same logic, the MS_{ε} is the $\frac{SS_{\varepsilon}}{df_{\varepsilon}}$.

$$MS_{\varepsilon} = \frac{\sum_{i=1}^{n} (y_i - \hat{y}_i)^2}{n-k} = \frac{SS_{\varepsilon}}{df_{\varepsilon}}$$

Where k is the number of independent variables being modeled, and n is the total number of paired observations. Notice that Sums of Squares add in the general model, and so do the degrees of freedom.

$$SS_{regression} + SS_{\varepsilon} = SS_{total}$$

[3]In Linear Algebra, the degrees of freedom of a system is the number of independent ways in which the system can be specified, without putting additional constraints on it. In easier terms, degrees of freedom is the minimum number of independent coordinates that can specify the system completely. Once you add constraints, like computed parameters, the degrees of freedom reduces as a function of the number of constraints imposed.

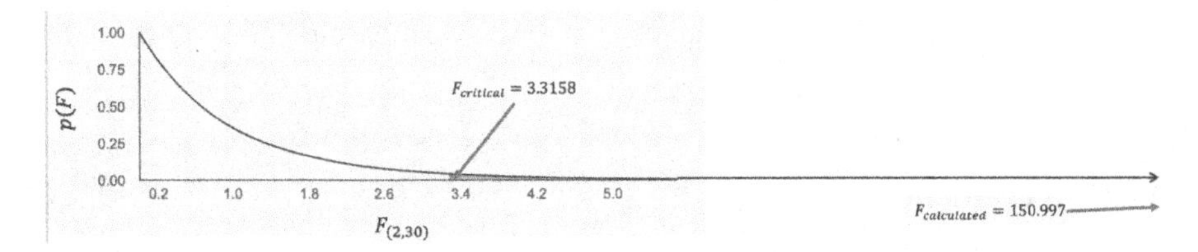

FIGURE 11.7: F-distribution at 2 *df* for the numerator and 30 *df* for the denominator

$$df_{regression} + df_{\varepsilon} = df_{Total}$$

With this information, we can now compute our F ratio of the slope to see if either of the β_k in our model, our partial slopes, are significantly different from zero.

$$F_{(k-1,n-k)} = \frac{MS_{regression}}{MS_{\varepsilon}}$$

$$F_{(2,30)} = 150.997$$

$$p\,(F \geq 150.997) = 4.65 \times 10^{-16}$$

Figure 11.7 shows the F-distribution for 2 degrees of freedom for $MS_{regression}$, and 30 degrees of freedom for MS_{ε}. The critical value, *where the cumulative probability that the model could have β -weights significantly different from zero* (at $\alpha = 0.05$), is $F_{(2,30)} = 3.3158$.

This F-value is so huge (I can't even fit it on the graph), and its p-value so incredibly small (the area under the curve to the right of $F_{calculated}$ is almost infinitesimally small), that we will reject H_0 and conclude that our β -weights are significantly different than zero. Because of this, the temperature and pressure of the gasoline we pour into our tanks do appear to contribute significantly to increased hydrocarbons escaping and polluting the atmosphere.

In English, we would say that the *chance* of obtaining an $F_{calculated}$ so incredibly huge given we have a sample of 32 automobiles, *is so far less* than $\alpha = 0.05$, that we have to conclude that the relationship we modeled is not a function of chance, but instead, represents some systematic (i.e., *real*) relationship among the variables. The prediction equation,

$$\hat{y}_i = 8.9802 + 0.4994\,(Gas\ Temp) + 0.5225\,(Gas\ Pressure)$$

is our best estimate of the shape and location of this relationship in the vector space made up of our data values.

11.4.1 Prediction

With our model established, I can now use it to predict *future* hydrocarbon pollution given just two values: The temperature of the gasoline, and the pressure at which it is pumped. Wow! When you think about it, that is really powerful. I now have a model that can be used to design new hydrocarbon containment and recovery systems that may significantly reduce air pollution. Of course, this extended example is just a model itself—a model of how to think and go about discerning such a relationship.

11.4.2 Extrapolation

It is easy to see how to predict such values: just substitute values for gasoline temperature and pressure into the equation and bing! Out comes our best guess of what the mass of the escaping hydrocarbons is. Because our model is so good, our error of prediction will generally be quite small, *if we pay attention and only make predictions within the contextual domains of the independent variables*! This is critically important to keep in mind. When we try to extrapolate *beyond* the values used to develop the model, our error of prediction grows exponentially. If you remember the notion of confidence interval, the $(1-\alpha)\%$ confidence interval of the mean is that region of the reference distribution where we would expect $(1-\alpha)\%$ of all the means of samples in a sampling distribution to fall if we just choose them randomly. Anything outside this range is considered very unlikely. The narrower the sampling distribution, the more precise our estimates are.

The same is true for the sampling distribution of our predicted values about $\hat{y}$. At $(\overline{x}, \overline{y})$, the most highly probable value for both x and y, the confidence interval is narrowest. As we move along the line of best fit, we can see that our samples for any vertical cross section of the "football" have generally fewer data points. This widens its sampling distribution—the area where we expect $(1-\alpha)\%$ of $\hat{y}_i$ to fall if we repeated the experiment. As an illustration, I have plotted the 95% confidence interval of $\hat{y}_i$ for Hydrocarbons vs. Gas Pressure.

Notice in Figure 11.8, how, in the middle of the model (at $(\overline{x}, \overline{y})$), the Confidence Interval is narrowest, and gets successively wider as the independent variable gets larger or smaller. Beyond our domain of the independent variable, the Confidence Interval gets wider at an exponential rate. The moral of this story, my children, is... be very cautious if you choose to extrapolate. You might wind up in this (reasonably common) situation (see Figure 11.9).

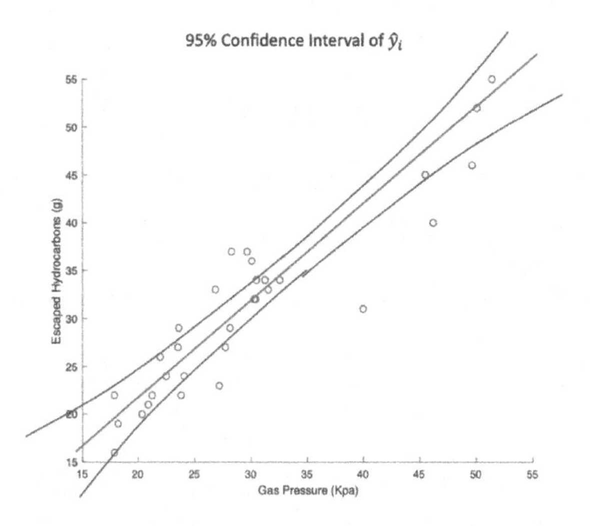

FIGURE 11.8: Confidence interval of $\hat{y}_i$, showing decreasing precision as we extrapolate beyond the contextual domain of the model

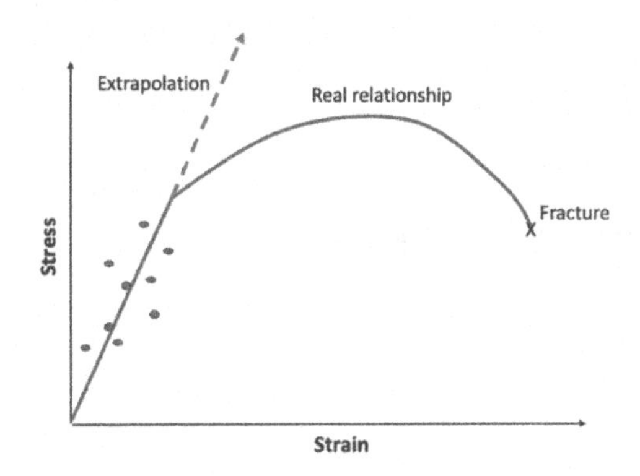

FIGURE 11.9: Extrapolation beyond the contextual domain may lead to erroneous conclusions!

11.4.3 Assumptions of Multiple Regression

Since we are just expanding our linear model to multiple dimensions of the independent variable, it should be no surprise that the assumptions we have to make for our model to be valid are the same as those for Simple Linear Regression. To recap:

The assumptions of linear regression are:

1. Independence of the observations
2. The true relationship between independent and dependent variables is linear in its parameters;
3. Errors of prediction are normally distributed around zero; and
4. Errors of prediction are equally distributed around the line of best fit.

   To these assumptions, we add a 5th, dealing with the fact that our relationship is now not just *between* two variables, but potentially *among* the multiple predictor variables in the model.

5. The independent predictors should not display multicollinearity.

This is just a way to say that, since our model assumes orthogonal bases, there should be no correlation between any of our independent variables. Each should be related to the dependent variable, but none should significantly predict the others. Think about the fact that we are projecting a vector onto the column space of X (beginning of Chapter 11). When the Y and X axes of our linear system are at 90 degrees from each other (i.e., the variables are orthogonal), the projection gives us the line of best fit. If the axes are at acute angles with each other, the projection becomes elongated, no longer a perfect linear combination of the dependent variable with any co-related independent variables. So, what does it mean that two independent variables are "correlated?"

11.4.4 Covariance and Correlation

Conceptually, when two variables are correlated, that means the slope of their line of best fit doesn't equal zero. So, correlation is some function of the slope of the bivariate regression line, β_k. Take a look at the numerator of β_1:

$$\beta_1 = \frac{\sum_{i=1}^{n}(x_i - \overline{x}_i)(y_i - \overline{y}_i)}{\sum_{i=1}^{n}(x_i - \overline{x}_i)^2}$$

This quantity looks suspiciously like a sum of squares (compare with the SS_x) in the denominator. The only difference is that it really is a sum of "rectangles," with length $(x_i - \overline{x}_i)$ and width $(y_i - \overline{y}_i)$. If we divided this sum by our degrees of freedom, we get a kind of variance *between* x and y. We call this the **Covariance of x with y**, symbolized Cov_{xy} or more simply s_{xy}. It is the average size of the rectangle made by deviations between any point in the bivariate distribution and the expected value of the bivariate distribution, $(\overline{x}, \overline{y})$:

$$s_{xy} = \frac{\sum_{i=1}^{n}(x_i - \overline{x}_i)(y_i - \overline{y}_i)}{n - 1}$$

It is useful for me to think of this quantity as the amount of shared variation between x and y taken as a single (bivariate) distribution. The larger the magnitude of s_{xy}, the more the two variables are related to each other. The matrix algebra to compute the covariance between any two variables is just an extension of that to find the variance of a single variable:

$$s_{xy} = \frac{1}{n-1}\left(X - \overline{X}\right)^T\left(Y - \overline{Y}\right)$$

$$s_{xy} = \left(X_1^T X_1 - 1\right)^{-1}\left(X - \overline{X}\right)^T\left(Y - \overline{Y}\right)$$

The denominator of β_1

is just the variance of x,

$$s_x^2 = \frac{\sum_{i=1}^{n}(x_i - \overline{x}_i)^2}{n-1}$$

$$s_x^2 = \left(X_1^T X_1 - 1\right)^{-1}\left(X - \overline{X}\right)^T\left(X - \overline{X}\right)$$

So the slope of the bivariate regression line is the rate of change of the covariance, with respect to the variance of X:

$$\beta_1 = \frac{s_{xy}}{s_x^2}$$

Again, we can think of the covariance as a measure of the relationship between the two variables, X and Y. Because of this, there are a couple of nice relationships that fall out.

The Coefficient of Determination, R^2 can also be expressed as a standardization of β_1 as the rate of change of the regression slope with respect to the variation in Y. This is yet another way to say that R^2 is the amount of variation of Y that is explained by the regression line.

$$R^2 = \frac{\beta_1}{s_y}$$

Karl Pearson (yes, he keeps turning up in all of this), showed that the square root of the Coefficient of Determination, symbolized r, retains the directionality of the relationship between X and Y, and can be a nice, dimensionless measure. We now call this measure **The Pearson-Product Moment Correlation Coefficient.**

$$r = \frac{s_{xy}}{s_x s_y} = \sqrt{R^2}$$

and therefore, β_1 can be expressed as a function of *r*.

$$\beta_1 = r\frac{s_y}{s_x}$$

"What does this all mean, professor?" I am glad you asked, Sparky.

1. *The covariance between any two variables is an indicator of the relationship between the two;*
2. *Like R^2, the magnitude of r indicates the strength of relationship between two variables. But r ranges from −1 to 1, preserving the direction of the relationship;*
3. *The slope of the regression line, β_1 is the parameter that fixes this relationship in the vector space of X, Y so that the values of the dependent variable can be 1) predicted from the values of the independent variable, and 2) the amount of variation about β_1 can be used to describe the strength of the relationship.*

11.4.5 Collinearity: Covariance Among Independent Variables

We can assess the degree of collinearity between any two variables in a regression analysis. It is not just restricted to the relationship between independent variable and dependent variable. For our matrix of k independent variables,

$$S_{xx} = \frac{1}{n-1}\left(X - \overline{X}\right)^T\left(X - \overline{X}\right)$$

$$S_{xx} = \left(X_1^T X_1 - 1\right)^{-1}\left(X - \overline{X}\right)^T\left(X - \overline{X}\right)$$

yields a *matrix* wherein the covariance between each X_k are represented in the cells. Notice how the diagonal of the covariance matrix, $s^2_{X_kX_k}$, is the variance of each independent variable, while the off-diagonal cells are the pairwise covariances. Not surprisingly, statisticians call this the **Variance-Covariance Matrix of X**, or just the Covariance Matrix of X (see Table 11.4).

	X_1	X_2	X_3	...	X_k
X_1	$s^2_{X_1X_1}$	$s^2_{X_1X_2}$	$s^2_{X_1X_3}$	...	$s^2_{X_1X_k}$
X_2	$s^2_{X_2X_1}$	$s^2_{X_2X_2}$	$s^2_{X_2X_3}$	...	$s^2_{X_2X_k}$
X_3	$s^2_{X_3X_1}$	$s^2_{X_3X_2}$	$s^2_{X_3X_3}$	...	$s^2_{X_3X_k}$
⋮	⋮	⋮	⋮	...	⋮
X_k	$s^2_{X_kX_1}$	$s^2_{X_kX_2}$	$s^2_{X_kX_3}$	...	$s^2_{X_kX_k}$

TABLE 11.4: Covariance matrix of X on itself

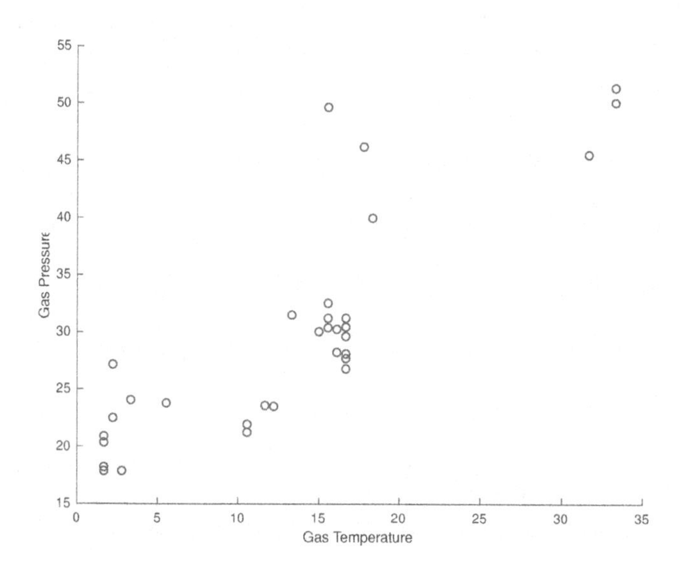

FIGURE 11.10: Scatterplot of bivariate relationship between gas pressure as a function of temperature

For our two Independent variables in our extended example, Gas Temperature and Gas Pressure, the covariance matrix looks like this (Table 11.5):

	X_1	X_2 **Gas Temp**	X_3 **Gas Pressure**
X_1	0	0	0
X_2 **Gas Temp**	0	76.38	70.23
X_3 **Gas Pressure**	0	70.23	92.07

TABLE 11.5: Covariance matrix for the independent variables in the gas pressure data

Remember, that X_1 is just a column of 1s, that we use to compute the intercept of the regression line. Because there is no variation in X_1, it can have no variance, nor any linear relationship with any of the other independent variables. The other variables, Gas Temperature and Gas Pressure, unfortunately, *do* appear to have a relationship. If X_2 and X_3 were completely independent—orthogonal vectors, their covariance would be zero. We can confirm this by looking at their scatterplot (Figure 11.10):

This looks like a pretty close relationship! To further confirm, we can look at the correlation coefficient. $r = 0.8375$, and its big sister, $R^2 = 0.7014$. Interpreted, we see that just about 70% of the variation in the two independent variables is held in common! This is a prime example of Collinearity.

Back to Collinearity (Assumption 5). Because s_{xy}, r, and R^2 are so closely related, if we want to test to see if our independent variables are collinear, we can look at the covariance, s_{xy}. If the magnitude of the covariance is large, the two variables have some dependence relation, they are not orthogonal to each other. Or, we can look at the correlation coefficient. If the magnitude of r is large, then the two variables have some dependence relation to each other. And because $R^2 = r^2$, if the proportion of variance accounted for in the bivariate linear model is large, then the two variables are not-independent. They have some degree of collinearity.

Covariance is the average amount that any two variables vary together. Precisely, it is the mean of the product of the deviations of two variables from their respective means. The sample variance, s^2 is the covariance of a variable with itself.

In a covariance matrix, the diagonal represents the variance of each of the column vectors in the matrix of predictors, while the off-diagonals represent their covariances.

The covariance is a relationship that draws together β_k, r, and R^2 into a single concept: The direction and strength of relationship between any two variables. For a bivariate relationship, s_{xy} is:

$$s_{xy} = \frac{\sum_{i=1}^{n}(x_i - \bar{x}_i)(y_i - \bar{y}_i)}{n-1}$$

or in Linear Algebra terms, the **variance-covariance matrix** for any number of variables is:

$$S_{xx} = \left(X_1^T X_1 - 1\right)^{-1} \left(X - \overline{X}\right)^T \left(X - \overline{X}\right)$$

Pearson Product-Moment Correlation, or just correlation coefficient, is a measure of the strength of the linear association between two variables. r is a measure of the direction of the relationship, and how far away all the data points are the line of best fit in a regression equation.

R^2, the coefficient of determination, is the square of the correlation coefficient. It represents the goodness of fit of a linear model.

What do we do if our independent variables ARE collinear? The answer, unfortunately, is "not much." Sometimes, when we are modeling curvilinear relationships, a transformation will fix the problem. Sometimes, we can apply a "Box-Cox" power transformation to the data to make each of the independent variables normal in shape.

This *may* help. But in my experience, you really need to examine the predictor variables you are using, and look at their theoretical relationship. In our case, we would expect gas temperature and gas pressure to be proportionally related due to the ideal gas law. If we *didn't* get this relationship I would be surprised! Unfortunately, that means that the β—weights in our model are not accurately accounting for the rate of change of each variable. The overall model has a very high R^2, it is just our parameter estimates that are a little off.

When you encounter such non-collinearity problems, there are two rules of thumb I would recommend:

1. For the collinear relationships, when you report your analysis, provide a bivariate scatterplot of the culprits, along with the variance/covariance matrix or R^2. This will alert the reader of your report to the fact that the estimates of the collinear variables β_k will be a bit off;
2. Either report the collinearity and admit that there are some problems with your model and that it violates the assumption of non-collinearity; or
3. Examine the bivariate relationship between each collinear variable, and choose the "best" one, the independent variable with the best fit, significant t-test of the slope, and best theoretical justification as the one you will keep in the model. Now run the model again with the reduced number of independent variables.

11.5 The General Linear Model

It turns out that Multiple Linear Regression is THE GENERAL MODEL for all subsequent statistical topics in this course. Everything we do from now on, until we get to computational methods like Bootstrapping, is just regression. To review, the general regression equation is:

$$y_i = \beta_0 + \beta_1 x_1 i + ... + \beta_k x_{ki} + \varepsilon_i$$

Where the outcome variable y_i is modeled by a linear combination of predictor variables x_{ki}, $k = 1, \ldots, k$ plus the error of prediction, ε_i. The model is linear in its parameters, so this is perfectly fine to model:

$$yi = \beta_0 + \beta_1 x_{1i} + \beta_2 x_{2i}^2 + \beta_3 e^{x_{3i}} + \ldots \beta_k x_{ki} + \varepsilon_i$$

but this is not:

$$y_i = \beta_0 + \beta_1 x_{1i} + \beta_2 x_{2i}^2 + x_{3i} e^{\beta_3} + \ldots \beta_k x_{ki} + \varepsilon_i$$

We assume the following:

1. All our measurements, x_{ki}, are independent;
2. The true relationship between y_i and x_{ki} variables is linear in its parameters;
3. Errors of prediction are normally distributed around zero;
4. Errors of prediction do not display heteroscedasticity; and
5. The independent predictors should not display (much) multicollinearity.

Notice that we have not put any restrictions on the level of measurement for either independent or dependent variables. We could choose a dependent variable that is continuous (i.e., interval or ratio scale), and independent variables that are categorical (i.e., nominal). This form of regression is what is known as the Analysis of Variance.

We could likewise choose a categorical dependent variable with continuous independent variables. This is what is called logistic regression. Or, we could choose a categorical dependent variable and categorical independent variable. This is called multinomial logistic regression. The point is, from here on out, it is *all* regression: Trying to predict an outcome variable from a set of independent predictors. We can even mix it up and regress an outcome (of any type) on a mixture of categorical and continuous independent variables. No matter what the type of analysis, it is defined by the same identities that define multiple regression of a continuous variable on a set of continuous dependent variables:

$$\hat{Y} = X\beta$$

$$\beta = \left(X^T X\right)^{-1} X^T Y$$

$$\hat{Y} = X\left(X^T X\right)^{-1} X^T Y$$

and

$$\varepsilon = Y - \hat{Y}$$

For this reason, it is called **The General Linear Model**. Different statisticians symbolize it in different ways, but it is all regression as I have described it in this chapter.

The General Linear Model or GLM, refers to linear regression fit by least squares and weighted least squares estimation. The GLM is restricted to one dependent variable, but may have many independent predictors of different scales of measurement.

Because it is not restricted to level of measurement, nor dimensionality, the GLM can be seen as the underlying mathematics of inferential statistics. In conventional terms, each value of the dependent variable y is a linear combination of parameters, β_0, an intercept, an initial value of y when all other variables are equal to zero–and the individual contributions of a set of predictors $\beta_k x$ plus some error of estimation, ε:

$$y_i = \beta_0 + \beta_1 x_{1i} + ... + \beta_k x_{ki} + \varepsilon_i$$

In matrix form, the GLM is greatly simplified:

$$Y = X\beta + \varepsilon$$

Where Y is a vector of outcome measurements, X is a matrix of predictor variables, β is a vector of parameters corresponding to the individual contributions of the variables in X to the value of Y, and ε is a vector of individual errors of estimation. This makes the estimated values of Y:

$$\hat{Y} = X\beta$$

Where

$$\varepsilon = \hat{Y} - Y$$

11.6 Extended Example

The best way to learn how to analyze models using multiple independent predictors and interpret results is to use an applied example. The dataset we will use is from Yeh (1998), who studied the impact different components of concrete had on its compressive strength. I have provided the full data set in in the file Concrete_data.xslx. The first few rows of the data matrix are provided below in Table 11.6.

Cement $\frac{kg}{m^3}$	**Water** $\frac{kg}{m^3}$	**Super plasticizer** $\frac{kg}{m^3}$	**Age days**	**Compressive strength Mpa**
540.0	162.0	2.5	28	79.99
540.0	162.0	2.5	28	61.89
332.5	228.0	0.0	270	40.27
⋮	⋮	⋮	⋮	⋮

TABLE 11.6: Structure of the concrete hardening data from Yeh (1998)

Yeh challenged a long-held rule in the concrete industry that water-to-cement ratio (W/C) is the most important factor in determining the strength of high-performance concrete. An increase in the W/C decreases compressive strength, whereas a decrease in the W/C ratio increases strength. The implication, therefore, is that the strengths of various but comparable concrete are identical as long as their W/C ratios remain the same, regardless of the remainder of the composition. He acknowledged in his study, which used neural networks to predict strength, compared to Multiple Regression, that some of the relationships among the data might be non-linear.

1. What kind of hypothesis should we run to examine Yeh's question of interest?

   Given only the paragraph above, I have little to go on. Also, I know only a little about concrete formulations and the potential effects of the different components. So, with the exception of W/C ratio, which the literature tells me should be a one-tailed (right-tailed) hypothesis, the other hypotheses should probably be two-tailed. This distinguishes an *omnibus test*, a test whether at least one of many independent variables contributes significant predictive variation to the model, versus *individual tests* of each independent variable separately. We haven't done this yet as a procedure, but it is simple enough: First do the omnibus F-test to see if any of the variables have a significant relationship with the dependent variable. Then do separate t-tests *post hoc*, starting with the largest Beta-weights—the variables that contribute most to the equation, working your way down the list until you find non-significant results. To specify this, I separated the Null and Alternative Hypotheses for W/C from the omnibus test.

$$H_{0k} : \beta_k = 0$$

   For W/C:

$$H_{1k} : \beta_k \neq 0$$

$$H_{0\,\text{w/c}} : \beta_{w/c} \leq 0$$

   For Super Plasticizer and Age:

$$H_{1\,\text{w/c}} : \beta_{w/c} > 0$$

2. What should the Type I error rate be?

   Yeh didn't report any Type I error rate. He was only interested in the prediction equation. BOO! This can lead to *overfitting*—the fact that, as you add more variables, R^2 will grow, even if the variables do not contribute significant predictive variation, they may give erroneous results. For an exploratory study like this, an α of 0.05 or even 0.10 is appropriate. There is no severe consequence of making a Type I error. I will choose $\alpha = .05$ as a kind of default.

3. What do the scatterplots of the data tell you about the potential relationships between the individual independent variables and the dependent variable?

   First, you need to enter your data into a statistics package of choice. Here, I used Matlab. Then, compute the W/C ratio, and eliminate the two columns for Cement and Water from the analysis, since they are clearly collinear with their ratio. Finally, insert a column of ones to make our augmented matrix for analysis.

   This gives us an augmented X-matrix as shown in Table 11.7:

Ones	**W/C Ratio** $\frac{kg}{m^3}$	**Super plasticizer** $\frac{kg}{m^3}$	**Age days**
1	0.300	2.50	28
1	0.300	2.50	28
1	0.422	0	270
⋮	⋮	⋮	⋮

TABLE 11.7: Augmented matrix of predictors

   and our vector of outcomes, Y (Table 11.8):

Compressive strength *MPa*
79.990
61.890
40.270
⋮

TABLE 11.8: Vector of values of the dependent variable, compressive strength

   Now we can make a few scatterplots (See Figures 11.11, 11.12, and 11.13):

   Examining the shape of these graphs (Tables 11.11 to 11.13), I can see that there are a some variables that look like they have a tighter relationship than the others. W/C (X_2) looks pretty good. It looks, like Yeh surmised, that the greater the W/C content, the lower the compressive strength of the concrete. This, not surprisingly has no values of zero, and the small column at W/C = 0.23 $\frac{kg}{m^3}$ may indicate a point where, at lower concentrations of water, the concreted doesn't set up properly, leading to weaker compressive strength. It does look, however like it might be curvilinear. This also is not surprising, as we could take the ratio to an extreme and really dilute the concentration, wherein we would expect very little, even negligible compressive strength.

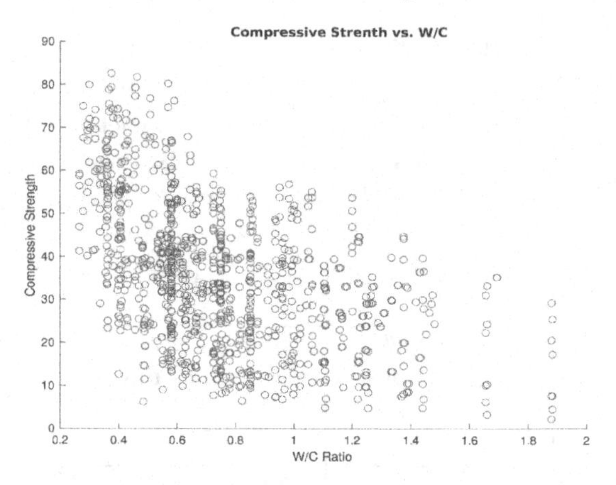

FIGURE 11.11: Scatterplot of compressive strength versus water content

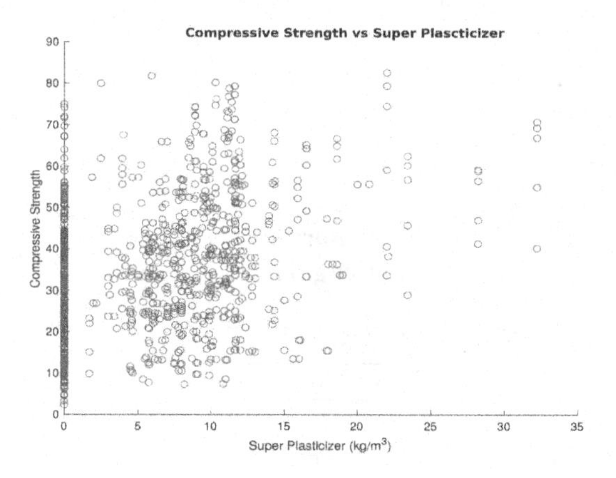

FIGURE 11.12: Scatterplot of compressive strength versus "Super" plasticizer

The super plasticizer (X_3) looks like the larger the concentration, the greater the compressive strength of the concrete, though there is lots of variation, and we have a problem with the greater variability at the lower values compared to higher values—heteroscedasticity. The cure time, the Age of the concrete at the time of testing (X_4), looks like it has a fairly strong, asymptotic relationship—that is definitely non-linear!

4. How should we deal with the curvilinear relationship that appears between W/C, and Age and Compressive Strength?

   For each of these, I need to transform my values to linearize them with respect to Compressive Strength. For W/C, this appears to be an inverse proportion. So, if I transform by the function:

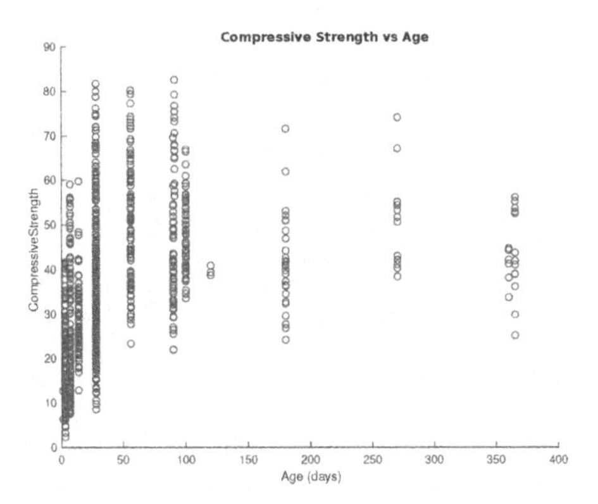

FIGURE 11.13: Scatterplot of compressive strength versus age

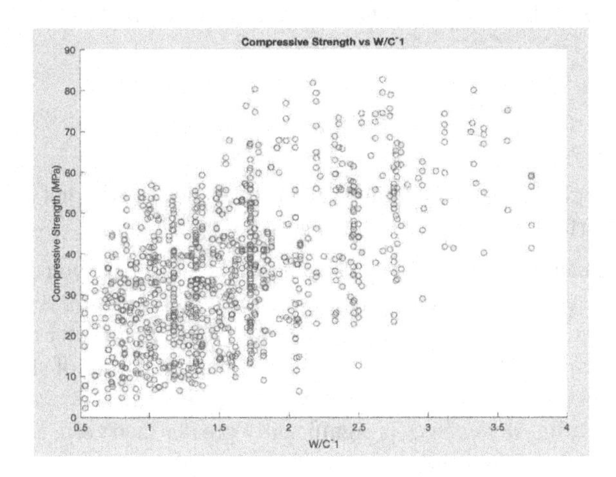

FIGURE 11.14: Scatterplot of transformed values of W/C with Compressive Strength

$$y = x^{-1}$$

I should get a more football shaped scatterplot (Figure 11.14):

Now THAT is more like it!

For Age, the form of this curing time curve should be a bounded exponential:

$$y = C\,(1 - e^{-x})$$

Ignoring the constants and taking the inverse with respect to x, we get a transformation to linearize the relationship (see Figure 11.15):

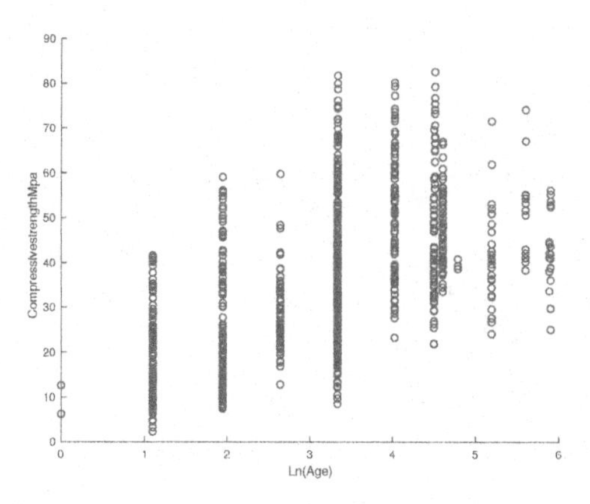

FIGURE 11.15: Scatterplot of Ln(Age) with compressive strength

$$y^{-1} = ln\,(x)$$

Whew! That looks like something I can use! A nice (somewhat squashed) football. Not perfect, but as close as I can get!

Finally, we can compute the line of best fit. Rather than have you do all the matrix algebra, I have included my Matlab results using the function regress. It returns the β -matrix, the $(1 - \alpha)\,\%$ confidence interval for β, a vector of the residuals, the $(1 - \alpha)\,\%$ confidence interval for the residuals, and a set of statistics: R^2, the computed F-test of the regression model, the *p-value* the probability that at least one of the partial slopes in β is significantly greater than zero, and the *MSE* the estimate of the residual variance $\left(SS\epsilon^2\right)$.

I don't want you to be concerned about the confidence intervals of β and r at this time. Suffice it to say that they show the expected range of each parameter in the model (see Figure 11.8). We have enough to deal with as it is. Here is the β -matrix, and the important statistics we can use to make sense of the model as it currently stands.

Parameter	**Coefficient**
β_0, Intercept	−15.2356
β_1, W/C^{-1}	13.3441
β_2, "Super" Plasticizer	0.6227
β_3, ln(Age)	8.2548

TABLE 11.9: Regression coefficients for each of the transformed predictor variables

$$R^2 = 0.6962$$

$$F_{(2,1027)} = 783.810$$

$$p = 7.6424 \times 10^{-265}$$

$$MSE = 85.028$$

5. How do you interpret this analysis?

   (a) How good is the fit?

   (b) Is the regression model significantly different from random chance?

   (c) Do you understand the meanings of the partial slopes and the regression equation?

The next thing I notice is that, just as Yeh surmised, W/C ratio appears to have, by far, the largest effect of all the predictors on compressive strength of concrete. Its β -weight is nearly 2 times the magnitude of the next best predictor, ln(Age). For its part, ln(Age) does appear to have an effect, but what 8.3 ln(days) means, I really don't know. For both W/C and ln(Age) I will have to transform them back into the original units to really interpret when I calculate future concrete recipes. "Super" Plasticizer may not be so super. Its β -weight is so low relative to the other two variables, I may choose to ignore its effect in my recipe, taking Yeh's advice to really concentrate (see what I did there?) on W/C and age.

Looking at the stats, we have a good predictive model. $R^2 = 0.6962$ means that with only 3 predictors, I can account for about 70% of the variability in the compressive strength of concrete. In other words, I can use my regression solution to get an optimal mixture for the right strength needed. This is born out when looking at the F-test. F is so large, and p is so small relative to my α, I feel pretty confident that the relationship modeled is systematic and not just random.

6. Does this analysis violate any assumptions of Multiple Linear Regression? If so, what should be done about them?

Now to determine if I am violating any assumptions, and to see if any of my variables really aren't significant predictors. A Q-Q plot of the residuals shows a beautiful linear pattern with the exceptions of the very tail of the residual distribution. I am very happy that our errors appear to be normally distributed.

Now I will plot the scatterplots of the residual error against each of the independent variables in the model. Let's start with W/C Ratio^{-1}.

This is exactly what I am looking for no distinct pattern, and fairly evenly distributed about zero.

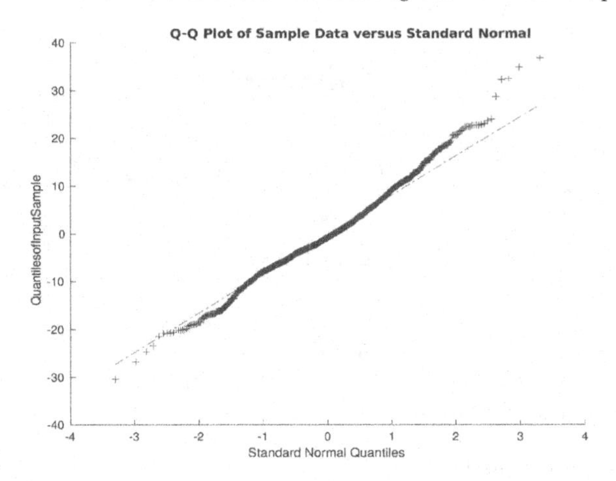

FIGURE 11.16: Q-Q plot of residuals in concrete analysis

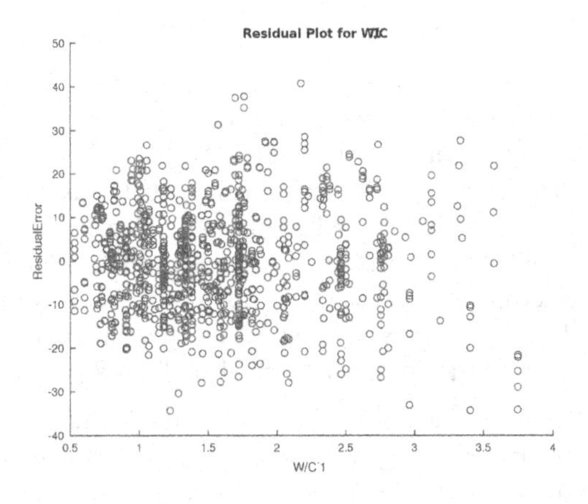

FIGURE 11.17: Residual plot for transformed values of W/C

This is not so great. I can see that all those zeroes appear to be problematic, and values above about 25 or so appear to have lower overall error than the bulk of the data. I may choose to eliminate those values from the analysis as outliers, because they will tend to overestimate β_2. But, since β_2 is so tiny, relative to β_1 for W/C, and β_2 for ln(Age), I am beginning to think that the effect of this variable is negligible.

ln(Age) shows a reasonable residual plot. There may be some curve-down in the pattern, indicating that my transformation didn't completely linearize the relationship. But, it is much better than it would have been prior to the transformation, so I will stay with it.

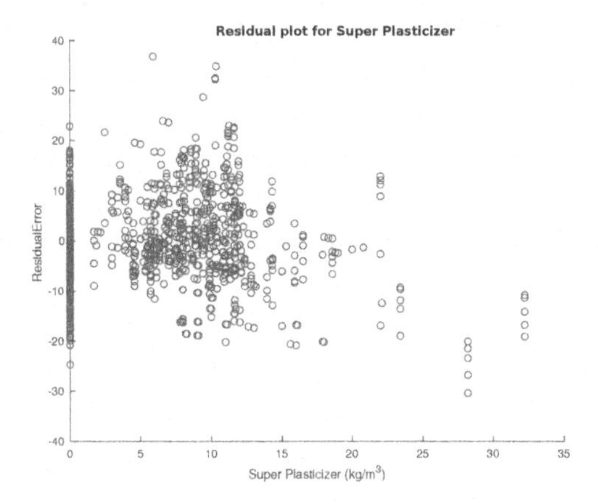

FIGURE 11.18: Residual plot of values for super plasticizer

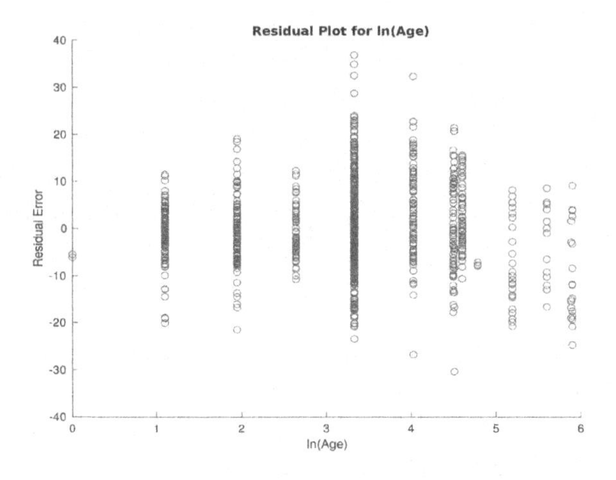

FIGURE 11.19: Residual plot of ln(age)

Looking at these 3 bivariate plots, aside from a few outliers and a not-quite random plot for ln(Age), there is no obvious pattern that I could ascribe to some nonlinear function that may be acting on one or more of the variables. To that end, I am very relieved! But it does reiterate the dangers of extrapolation—the model will get more "off" as we get to the tails of the independent variables and beyond their contextual domains.

We are almost done! But first, let's take a look at the potential for multicollinearity among our predictors. Here is the covariance matrix of X in Table 11.10:

	W/C Ratio^{-1}	"Super" Plasticizer	ln(Age)
W/C Ratio	0.420	1.263	−0.040
"Super" Plasticizer	1.263	35.683	−0.306
ln(Age)	−0.040	−0.306	1.420

TABLE 11.10: Covariance matrix of predictor variables

This shows that most of the variables have small covariance among them (!). Compared to the variance of the residuals ($MSE = 85.028$), none of the covariances in the table appear to be problematic.

The highest of the correlations is between W/C^{-1} and "Super" Plasticizer

$$r_{(W/CRatio\hat{}-1,"Super"Plasticizer)} = 0.3261$$

$$R^2 = 0.3261^2 = 0.11$$

They share only about 11% of their variance. This is acceptable for most analyses.

7. How do we test to see, especially the variables with the smaller β -weights, if these partial slopes are different than what we would expect from random chance?

We know that this omnibus test is significantly different than what we would expect by random chance. But we still don't know if our small β_k are ALL significant. So, we have to use the t-test of the slope to help us.

$$t_{(n-2,\alpha)} = \frac{\beta_k}{SE_{\beta_k}} = \frac{\beta_k}{\sqrt{\frac{1}{(n-2)} \cdot \frac{\sum_{i=1}^{n}(y_i-\hat{y}_i)^2}{\sum_{i=1}^{n}(x_{ki}-\bar{x}_{ki})^2}}}$$

$$t_{W/C^{-1}(1028,0.05)} = \frac{\beta_{W/C^{-1}}}{\sqrt{\frac{1}{(n-2)} \cdot \frac{\sum_{i=1}^{n}(y_i-\hat{y}_i)^2}{\sum_{i=1}^{n}\left(x_{W/Ci^{-1}}-\bar{x}_{W/Ci^{-1}}\right)^2}}}$$

$$= \frac{\beta_{W/C^{-1}}}{\sqrt{\frac{1}{(n-2)} \cdot \frac{MSE}{S^2_{W/C^{-1}}}}}$$

$$= \frac{13.343}{\sqrt{\frac{1}{(1030-2)} \cdot \frac{85.028}{0.4200}}}$$

$$= \frac{13.343}{\sqrt{0.1969}}$$

$$= 30.070$$

This is way, way, *way* further in the tail of the t-distribution than the critical value, $t_{(1028,\,0.05)} = 1.9623$. We do the same thing for the rest of our independent variables.

	t	α	$t_{critical}$
β_1 **W/C^{-1}**	30.070	0.05, one-tailed	1.6463
β_2" **Super" plasticizer**	4.709	0.25, two-tailed	1.9623
β_3 ln (Age)	21.241	0.25, two-tailed	1.9623

All of these tests of the individual, partial slopes, are much larger than their critical value. So, each of the variables, given our large sample size (n=1,030), significantly predict compression strength greater than what we would expect by chance ($\alpha = 0.05$). As a consequence, we can retain each of them in our final model.

The final thing we have to do is to specify our prediction equation that is linear in its parameters:

$$\hat{y}_i = -15.232 + 13.343\left(W/C_i^{-1}\right) + 0.623\,(\text{"}Super\text{"}Plasticizer_i) + 8.255\,(ln\,(Age)_i)$$

In summary, it looks like, for the most part, W/C *is* by far the most important factor in determining the compressive strength of concrete. Next, of the studied variables, we have to take the cure time, Age into account. Finally, there is a small, but significant, effect of "Super" Plasticizer.

This is a typical analysis in the real world! You won't know what the real relationship among the variables are, though you are smart and wouldn't be doing the analysis if you didn't have some idea. You will also have some idea what the direction of the relationships might be, but you might not know their shape. The power of Multiple Linear Regression is that, with only a few pieces of concrete, knowing the composition of their mixture, we can predict with pretty good accuracy, values of *future* mixtures we haven't tried yet.

11.7 Summary

The General Linear Model is a powerful, all-purpose Leatherman (TM) Multi-tool that every engineer should keep in their pocket, so to speak, to use whenever we need to efficiently predict future values of an outcome variable, given a small set of predictors. It is also useful to determine, of those predictors, which ones contribute the most predictive value, versus those whose contribution may be less predictive, or even confounding.

$$y_i = \beta_0 + \beta_1 x_{1i} + ... + \beta_k x_{ki} + \varepsilon_i$$

The model is simple, and the mathematics you need to know to understand it is fairly straightforward we are just plotting lines in space!

Like all mathematical models, when we actually apply them to the real world, we have to make some basic assumptions about the mathematical properties of the phenomena we are dealing with, and with the nature of the data we extract from them. The GLM relies on the basic assumption that our predictor variables and outcome are distributed multivariate normal. This means that each orthogonal cross section in the k+2 dimensional space (k independent variables, one dependent variable, and the frequencies as a function of all the points in the space) is normal. When that is not the case, for example, when some variables are related in a non-linear manner, or when predictor variables are not independent—not orthogonal to each other, we have to jump through some hoops so that the GLM will do work for us, transforming the data so that it *does* conform to this multivariate normal shape.

Navigating any analysis that is based on the GLM requires that we examine our regression equation for signs that one or more assumptions have been violated. Analysis of the raw data, using bivariate plots, analyzing the residual error to make sure it is (close to) normal and (close to) randomly distributed about zero are chief among these. Both qualitative examination and quantitative confirmation (e.g., computing the covariance of X), are key skills to develop.

The most important thing to keep in mind about the General Linear Model is this basic fact: The GLM is mathematically identical to multiple regression analysis. However, the "general" part means that we use this mathematical model for examining the relationship between and among multiple independent variables regardless of whether or not they are continuous (interval or ratio scales), or discrete (ordinal or nominal scales). The single outcome, the dependent variable can also be continuous or discrete, incorporating any of the scales data comes in. Because of this flexibility, the GLM is the most widely used mathematical tool for inferential statistics. Those simple tools we have already examined, the Pearson χ^2 test of independence, and the t-test, are just basic applications of this powerful mathematical concept. And it all boils down to something so simple and universal it almost boggles the mind:

$$\hat{Y} = X\beta$$

$$\beta = \left(X^T X\right)^{-1} X^T Y$$

$$\hat{Y} = X\left(X^T X\right)^{-1} X^T Y$$

$$\varepsilon = Y - \hat{Y}$$

11.8 References

Brooks, T. F., Pope, D. S., & Marcolini, A. M. (1989). *Airfoil self-noise and prediction*. Technical report, NASA RP-1218.

Chamberlain, A. (2016). The Linear Algebra View of Least-Squares Regression.

Grossman, Y. L., Ustin, S. L., Jacquemoud, S., Sanderson, E. W., Schmuck, G., & Verdebout, J. (1996). Critique of stepwise multiple linear regression for the extraction of leaf biochemistry information from leaf reflectance data. Remote Sensing of Environment, 56(3), 182-193.

Hastie, T., & Tibshirani, R. & Friedman, J. (2017). *The elements of statistical learning (12th printing)*. New York: *Springer*.

Li, X., Ono, T., Wang, Y., & Esashi, M. (2003). Ultrathin single-crystalline-silicon cantilever resonators: fabrication technology and significant specimen size effect on Young's modulus. *Applied Physics Letters*, *83*(15), 3081-3083.

Sakurai, T., Yuasa, S., Ando, H., Kitagawa, K., & Shimada, T. (2016). Performance and regression rate characteristics of 5-kN swirling-oxidizer-flow-type hybrid rocket engine. *Journal of Propulsion and Power*, *33*(4), 891-901.

Spaeth, H. (1991). *Mathematical Algorithms for Linear Regression*. New York: Academic Press.

Wiemann, Michael C.; Williamson, G. Bruce. 2014. Wood specific gravity variation with height and its implications for biomass estimation. Research

Paper FPL-RP-677. Madison, WI: U.S. Department of Agriculture, Forest Service, Forest Products Laboratory.

Yeh, I-C. (1998). Modeling of strength of high performance concrete using artificial neural networks, *Cement and Concrete Research,* 28(12), 1797-1808.

11.9 Chapter 11 Study Problems

1. Grossman, and colleagues (1996) provided analyses that allow us to compare the cellulose and lignin content of leaves as functions of the carbon, nitrogen, and water concentrations in the leaves. These factors are related to environmental water scarcity. Suppose you were part of a team trying to create technology for improving the water absorption of plants. Use the data I recreated from this study to examine the relationship between Carbon and Nitrogen concentrations and Water content on the development of both dependent variables: Leaf Cellulose and Lignin content. Use this data file: Leaf_Cellulose_and_Lignin_Multiple_Regression.xlsx.

   (a) Would you make a 1-tailed or 2-tailed hypothesis?

   I know little about Cellulose production in leaves, especially about the impact of these two variables, so I would use a 2-tailed hypothesis.

   (b) Express the Alternative Hypothesis and the Null Hypothesis.

$$H_0 : All\, \beta = 0$$

$$H_1 : At\ least\ 1\, \beta \neq 0$$

   (c) Justify your Type I error rate;

   Since I am just learning about this relationship, and I don't have any products or applications on the line, I can be less conservative. I might use $\alpha = 0.05$.

   For the remainder of the questions, I have written a well-documented Matlab code to guide you:

```
% After entering the data into Matlab, create column vectors for each
% dependent variable:
cellulose=data(:,5);
lignin=data(:,6);
% Now create the augmented matrix of predictor variables:
X=data(:,1:4);
```

```
% Examine scatterplots of each predictor with respect to the dependent variable:
scatter(X(:,2),cellulose);
figure
scatter(X(:,3),cellulose);
figure
scatter(X(:,4),cellulose);
% If the bivariate scatterplots do not look non-linear, go ahead and run
% your regression:
[b,bint,r,rint,stats]=regress(cellulose,X);
% Make residual plots for each independent variable:
figure
scatter(X(:,2),r);
figure
scatter(X(:,3),r);
figure
scatter(X(:,4),r);
% If these look nice and randomly scattered about zero, draw the QQ plot of
% the residuals:
figure
qqplot(r);
% Interpret the Beta-weights. These will be in the b vector
% Now interpret R^2 and the F-test of the "slope." These are in the stats structure.
% Finally, express the regression equation. This will give you the
% predicted values of the dependent variable.
y_hat=X* b;
% repeat for the process for all dependent variables
```

(d) What do the scatterplots of the data reveal about the nature of the relationship(s) (if any) that exist between independent and dependent variables.

None of the scatterplots show any distinct curvilinearity. All appear to be appropriately modeled with a linear function

(e) Do any variables need to be transformed?

Not at this stage of the analysis.

(f) Compute the parameters of the best fitting model and other important statistics.

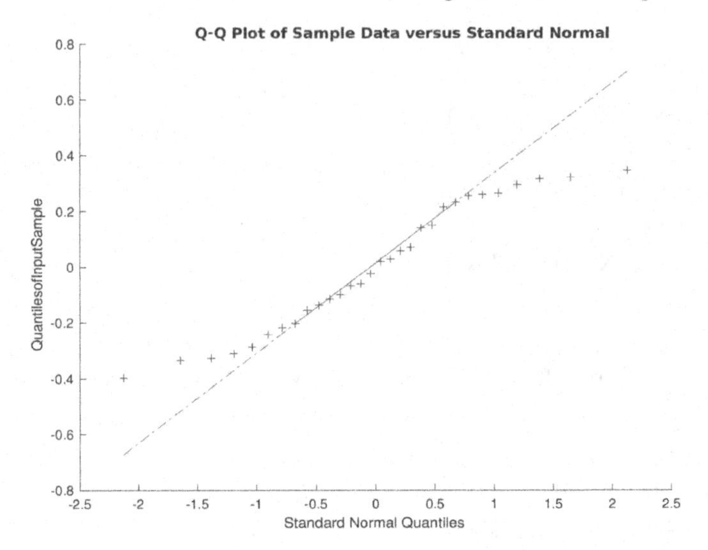

	β
Intercept	1.379
Carbon	0.003
Nitrogen	−0.013
Water	−0.005

R^2	F	p	MSE
0.328	4.232	0.015	0.059

These are the b-vector and Stats structures in Matlab

(g) **Which, of the independent variables, have the most impact on cellulose and lignin content?**

From the β -weights, we can see that nitrogen concentration, with a partial slope of −0.013, contributes most to the cellulose content of the leaf. The lower the nitrogen concentration, the higher the cellulose content. Carbon and Water content contribute much less than nitrogen.

(h) **Perform a residual analysis. Are there any violations of the assumptions of Multiple Linear Regression? What, if anything, should you do?**

None of the residual plots show a distinct pattern. All of the error values are nicely randomly scattered about 0. The Q-Q plot shows some deviation at the tails, indicating some non-normality:

This is quite a bit of deviation. I might go look at the independent variables to see which ones appear to be distributed non-normally. I could then convert them to z-scores (i.e., standardization) and re-analyze using the z-scores. This, however, makes the resulting model more difficult to understand and use as a prediction equation.

(i) **Perform the F-test of the slope to determine if the relationship modeled is significantly different from that expected due to random chance.**

F is 4.232, with a p-value less than 0.05, so I must reject the null hypothesis and conclude that my model has revealed a real relationship among the variables.

(j) **State the final model you would use to compute new values of cellulose and lignin from the significant predictors.**

Given the β -weights, I can express the prediction equation as:

$$\widehat{Cellulose} = 1.379 + 0.003\,(Carbon\,) - 0.013\,(Nitrogen\,) - 0.005\,(Water)$$

I would then repeat this procedure to analyze the contribution of these variables to the Lignin content in leaves.

2. **Li et al., (2003) studied the resonant frequency of ultrathin (in *nm*!) cantilevered silicon beams as a function of Temperature and Length. They were characterizing the material for use in sensors that could be used in a variety of contexts, including ultrasensitive electron microscopes, space exploration, and threat detection. Recreated data is listed below:**

Length (μm)	**Temp (K)**	**Resonant Frequency (kHz)**
17	343.120612	114
18	342.64533	102
20.5	342.31532	80
22	342.19918	70.5
24	341.843292	60
26	341.700146	54
28	341.656033	50

If you were the engineers, given what you know about beams and bending moment:

(a) **Would you make a 1-tailed or 2-tailed hypothesis?**

For this analysis, I would make a 1-tailed hypothesis. We would expect, from my thermodynamics notes, that temperature should be positively related to resonant frequency, and that length would be negatively related.

(b) **Express the Alternative Hypothesis and the Null Hypothesis.**

As hypotheses get more complicated, it is more efficient to express them in

English than in strict mathematical form. But remember that the null and alternative hypotheses must be mutually exclusive and collectively exhaustive!

$$H_0 : \beta_{Length} \geq 0$$

$$H_0 : \beta_{Temp} \leq 0$$

$$H_1 : \beta_{Length} < 0$$

$$H_1 : \beta_{Temp} > 0$$

(c) Justify your Type I error rate;

If I were at the beginning of modeling these relationships, I might use $\alpha = 0.05$. But, I am going to assume that I am well into designing a threat assessment system, so we really need to be sure about our model. So, I will choose $\alpha = 0.01$

(d) **What do the scatterplots of the data reveal about the nature of the relationship(s) (if any) that exist between independent and dependent variables.**

The scatterplot of temperature vs. resonant frequency looks beautifully linear. No problems anticipated there! However, the scatterplot of length versus resonant frequency looks like an inverse proportion. I think I will have to transform this data:

(e) Do any variables need to be transformed?

I used the following transformation in Matlab:

»lengthtransformed=data(:,1)^ −1. This resulted in a much straighter curve (though there still is some small curvature at the ends of the domain). This looks pretty good to create my model.

(f) Compute the parameters of the best fitting model and other important statistics.

R^2	F	p	MSE
0.991	229.19	0.0000748	7.747

	β
Intercept	−5922.41
Length^{-1}	1759.17
Temperature	17.29

(g) **Which, of the independent variables, have the most impact on resonant frequency?**

Looking at the β -vector, I can see that Length^{-1} has much more of an effect than temperature. I have to be careful, because if I want the units to be in length units so I can determine the appropriate size of the cantilevered beams for manufacturing, I have to transform them back to μm.

(h) **Perform a residual analysis. Are there any violations of the assumptions of Multiple Linear Regression? What, if anything, should you do?**

Both residual plots show no pattern in the scatter of the errors. The Q-Q plot looks acceptable. There is only one point that differs much from the expected pattern, y = x: This point is at the upper end of the residual values, and might represent a point where our model begins to break down. We don't know for sure. But, overall, it appears to be a good linear model.

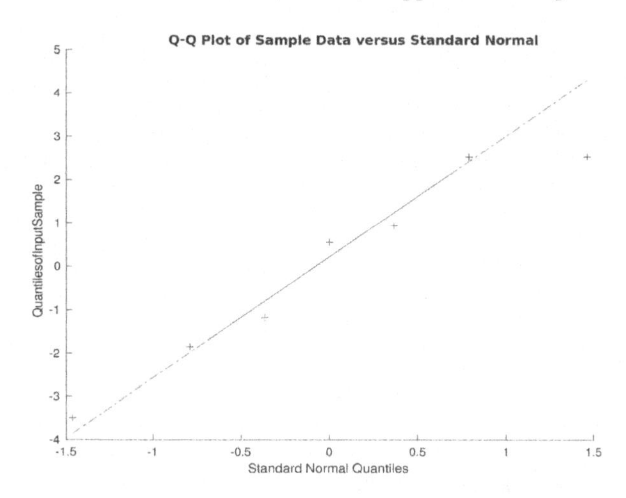

(i) **Perform the *F*-test of the slope to determine if the relationship modeled is significantly different from that expected due to random chance. Compare these results to R^2 . What issues will you have to grapple with?**

Both F and R^2 are very high. Over 99% of the variation in resonant frequency is accounted for by our two predictor variables, and it is highly unlikely ($p = 0.0000748$ is way less than alpha!) that the relationship is due just to random chance.

(j) **State the final model you would use to compute new values of Frequency from the two predictors.**

Taking my two parameters, making sure I have my length in the original units, I get the following:

$$ResonantFrequency = -5922.41 + 1759.17\left(Length^{-1}\right) + 17.29\,(Temp)$$

3. **What are the assumptions of Multiple Linear Regression, and why are each important to be able to build a case for a trustworthy prediction model?**

   The assumptions of linear regression are:

   (a) Independence of the observations. This is important so that we have a truly random distribution.

   (b) The true relationship between independent and dependent variables is linear in its parameters. Because the model plots a first degree equation, there cannot be logarithmic, exponential, or other non-linear relationships underlying the data.

   (c) Errors of prediction are normally distributed around zero. This assumptions accounts for the fact that we are using the normal distribution to estimate the probabilities of the parameters over the long-haul;

   (d) Errors of prediction are equally distributed around the line of best fit; This accounts for heteroscedascity—when the values of the dependent variables do not have uniform distribution across all values of the independent variable(s). If the relationship is heteroscedastic, the Mean Square Error will be poorly estimated because there will be a pattern in the errors.

   (e) Absence of multicollinearity (Independent variables are, in fact independent and not linearly dependent upon one another). Because we are projecting a vector onto a space of predictor variables, the relationships among all independent variables must be orthogonal to minimize the distance accurately the errors must not be distorted by angles between axes that are significantly greater than or less than 90 degrees. Beta-weights will include interactions among variables, and therefore will inaccurately estimate the unique contribution of an independent variable, on the dependent variable.

4. **Why is the formula for the coefficient of determination exactly the same for Multiple Linear Regression as it was for Simple Linear Regression?**

   Because each predictor variable is independent, they each contribute unique explained variation to the overall prediction equation. R^2 is equivalent to the true variation divided by the total variation. The true variation in a model is always the total minus the error variation, therefore $R^2 = \frac{SS_{Total} - SS_{Regression}}{SS_{Total}}$.

5. **Explain the relationship between the covariance and the correlation coefficient between two variables.**

   (a) **What does the covariation matrix tell us?**

   The covariance between two variables tells us the degree to which they hold variation in common, when other variables are not taken into account. In multiple linear regression, the covariance matrix can be used to examine whether or not the independent variables have a high or low degree of collinearity, and the direction of their relationship.

(b) What does the correlation coefficient tell us that the covariance doesn't?

The correlation is the covariance divided by the product of the sample standard deviations of the two variables being compared. As such, it is a standardized measure of the relationship between two variables that retains the direction of the relationship. Because of this standardization, the correlation coefficient can be used to judge both the direction and strength of the relationship between two variables.

6. Why is multicollinearity such a problem when trying to piece together a model among several potentially confounding independent variables?

(a) What are your options when you find two or more variables are collinear?

Collinearity can cause the Beta-weights in our prediction model to be wildly off. If we have collinearity in our matrix of predictors, we can transform our variables to be orthogonal, risking a lack of interpretability of the findings, we can eliminate problem variables from the analysis, or we can keep them in, and acknowledge that the estimated partial slope parameters may be misleading.

7. Explain how transformation works in the multiple regression context.

Because all independent variables are orthogonal, we can transform them separately with respect to the outcome variable. We have to remember that the prediction equation, being linear in its parameters, uses the transformed variables as predicters, and that we have to transform back into the original units to record their true values.

8. Sakurai et al., (2016) published performance data of Swirling-Oxidizer-Flow-Type Hybrid Rocket Engines. They characterized the Thrust (Force, measured in kN) of the engines as a function of, among many variables, pressure in the combustion chamber, the flow rate of the liquid oxygen, and the burn time. Their data is presented below:

Run	F (kN)	Pressure (MPa)	O_2 Mass-Flow Rate (kg/s)	Burn Time (s)
1	0.36	0.51	0.154	1.6
2	1.658	1.62	0.501	4.6
3	1.4	1.43	0.459	8.8
4	1.3	1.26	0.417	19.6
5	4.152	2.76	1.228	1.5
6	4.438	2.58	1.357	4.5
7	4.323	2.68	1.208	4.4
8	3.91	2.13	1.14	9.3

If you were the engineers, given what you know about rocket engines:

(a) Would you make a 1-tailed or 2-tailed hypothesis for this regression analysis?

I don't know a whole lot about this type of hybrid engine, so I would have to do a 2-tailed test for these ranges of independent variables. Ordinarily, somebody performing this type of study will be an expert and know the intended impact of the independent variables, and should use a one-tailed test. But, I am not them, so I will use a 2-tailed test for this example.

(b) Express the Alternative Hypothesis and the Null Hypothesis.

$$H_0 : All\beta = 0$$

$$H_1 : At\ least\ 1\beta \neq 0$$

(c) Justify your Type I error rate if these rockets were to be used to send sensitive equipment into space;

Here, the consequences of making a false positive might lead to catastrophic and expensive errors. I am going to use $\alpha = 0.001$. This is a very large data set ($n = 1{,}503$), so we should have enough power to detect a significant difference at this level of precision.

(d) What do the scatterplots of the data reveal about the nature of the relationship(s) (if any) that exist between the independent and the dependent variables.

The scatterplots for Pressure and Oxygen mass flow rate are beautifully linear. For Burn Time, however, it looks like little to no relationship with Thrust:

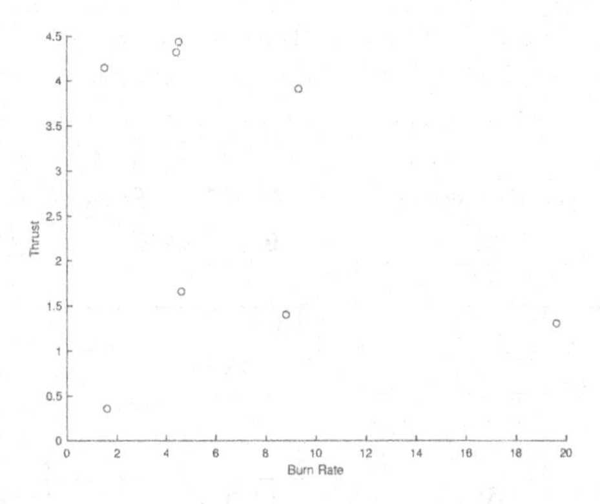

I will go ahead and keep Burn Rate in as a predictor for this example, but really, I might eliminate it because it will not add anything to the predictive model.

(e) **Do any variables need to be transformed?**

I see no curvilinearity in the scatterplots, so I will run the regression analysis as is and return to this when I look at the residuals.

(f) **Compute the line of best fit and other important statistics.**

R^2	F	p	*MSE*
0.996	313.221	3.36×10^{-5}	0.0206

	β
Intercept	−0.2653
Pressure	0.1835
Oxygen Mass Flow Rate	3.237
Burn Rate	−0.0002

(g) **Which, of the independent variables, have the most impact on thrust?**

In this analysis, it looks like everything depends on Oxygen Mass Flow Rate. Its β -weight is about **18 times** greater than the next closest predictor, Pressure. For its part, Burn Rate appears to be so close to zero as to have negligible effect (just what we would expect from its bivariate scatterplot with Thrust).

(h) **Perform a residual analysis. Are there any violations of the assumptions of Multiple Linear Regression? What, if anything, should you do?**

The scatterplots all look beautifully random. No problems with using a linear model. The Q-Q plot is also really nice:

So it looks like the residuals are distributed as normal.

(i) **Perform the F-test of the slope to determine if the relationship modeled is significantly different from that expected due to random chance. Compare these results to R^2 . What issues will you have to grapple with?**

$R^2 = 0.996$. The model predicts way over 99% of the variation in Thrust. F is HUGE! The resulting probability that we could have gotten this relationship just due to random chance is near zero.

(j) **State the final model you would use to compute new values of thrust from the three predictors.**

$$Thrust = -0.2653 + 0.1835\,(Pressure) + 3.237\,(Oxygen) - 0.0002\,(BurnRate)$$

9. Back in 1989, Brooks and colleagues studied the sound pressure produced by airfoils in wind tunnel tests. They recorded the frequency of the vibrations of the airfoils, the angle of attack, the chord length of the airfoil, the free stream velocity

of the wind tunnel, and the airfoil's suction-side displacement thickness. Sound pressure was measured in decibels. That data is available in the following file: Airfoil_Sound_Multiple_Regression.xlsx.

(a) **Following good regression procedure, what is your best estimate of the relationship between the independent variables and sound pressure of the studied airfoils?**

Everything looked okay to use a linear model with no variables requiring transformation. The resulting parameters and statistics are as follows:

R^2	F	p	MSE
0.516	318.824	1.148 × 10 − 232	0.002

	β
Intercept	1.3283
Frequency	−1.2822
Angle	−0.0042
Chord Length	−0.3569
Velocity	0.0010
Suction	−1.4730

(b) **How sure are you of your model?**

With $R^2 = 0.516$, this is not a great fit. There appears to be a lot of variation in each of the independent variables. BUT, it is likely that our model does reflect a true relationship among the variables. F is very high, and its associated probability is extremely low that the model we developed is due to random chance (so close to zero as to be negligible). This is much smaller than our Type I error rate, $\alpha = 0.001$.

(c) **If you were the engineers, how would you report your findings to your CEO (who is not an engineer), if the company has received complaints about the noise caused by your airfoils?**

I am very confident in the overall model. As far as individual variables are concerned, Frequency, Chord Length and Suction are definitely variables I would want to look at. Angle of Attack and Velocity of the air stream are much smaller in their effect.

(d) **How would you report your findings to other engineers who need to use your results to design new airfoils?**

Pay attention to Frequency, Chord Length, and Suction when designing new airfoils. We can't eliminate angle of attack or velocity, but their effect is so much smaller for the contextual domain here, that, for airfoils in this range of application, I would ignore them in the design.

10. What are the key characteristics you would look for in an analysis to use in evaluating the results of a study that utilized Multiple Linear Regression?

    (a) What characteristics of the data and data gathering are most important to know?

    I always look at whether the independent and dependent variables have been adequately defined. Their range and scale, in particular are important to ensure that we get a good picture of the relationship among the variables. In addition, there is no excuse for not doing your homework. You should be making explicit hypotheses about why independent variables are included and how they potentially impact the dependent variable. I don't want to see variables just thrown into the pot to see what happens. Because of collinearity, this can result in erroneous models and Type I and Type II errors.

    (b) What characteristics of the analysis and results are most important to know?

ALL OF THEM! In particular, paying attention to the initial "eyeballing" and subsequent residual analysis is what helps make the case that the model you develop is one that can be used to make decisions based on your analyses.

12

The GLM with Categorical Independent Variables: The Analysis of Variance

Throughout Chapter 11, I kept hinting that the General Linear Model really is *general*, serving as the central mathematical concept that governs how we think and apply inferential statistics. The history of this revelation stems back to our old friend, William Seeley Gossett, he of the t-test fame (and Guinness Brewer extraordinaire). His work, integrated with that of Karl Pearson, inspired Sir Ronald Fisher to create a general approach to manipulation of treatment conditions on experimental units.

Fisher applied most of his work in the area of biology, especially agriculture, where he basically codified most of the experimental designs we studied in Chapter 6. Fisher's mathematical approach to data was... non-standard... to many. Most people had a difficult time following his logic. His writings are considered the cornerstone of experimental statistics, but, like today, people had to translate them into more 'user friendly, applicable articles before they became widely used. As an interesting coincidence, his wife's name was Ruth Guinness (no relation).

In a true stroke of genius, Fisher in his now classic *Statistical Methods for Research Workers* (1925), showed that one can test for *differences in means*, by treating the differences, squared, as *variances between groups*. Because sums of squares (i.e., variances) are additive by the rules of Multiple Linear Regression, the two-sample t-test, Fisher argued, could be expanded to include any number of experimental groups, it was not limited to just two. This revelation was termed the **Analysis of Variance**, and is one of the hallmarks of statistical thinking of all time—on par with the LLN, the Central Limit Theorem, Student's t, and linear regression.

12.1 The 2-sample t-test as Regression

Let's do what Fisher did back in the roaring '20s (I suspect under the influence of port, which he particularly loved), and show that the 2-sample t-test is just a disguised regression problem.

To review, you may recall that the 2-sample t-test is a test of the difference in means between two samples, each estimating the center of the common population

DOI: 10.1201/9781003094227-12

from which they were drawn. The expected difference in sample means is zero under the Null Hypothesis, and its test statistic is computed by:

$$t_{(\alpha, df=n_1+n_0-2)} = \frac{\overline{x}_1 - \overline{x}_0}{s_p\sqrt{\left(\frac{1}{n_1} + \frac{1}{n_0}\right)}}$$

$\overline{x}_1$ and $\overline{x}_2$ are parameter estimates of the mean of population from which both of their samples were drawn. The two estimates will differ a bit just due to random chance. But over the long haul, the means of all samples drawn from this common population will average out to the same value, μ.

The pooled standard deviation assumes that these two samples are taken from the same population, and uses their combined variances to estimate the common width of the population. The pooled standard error just scales this variation to the sampling distribution with degrees of freedom equal to the average of the two sample sizes.

To represent these two data vectors as a matrix of predictors, we can treat the experimental condition as an independent variable, *X*, with two values: 0 corresponding to the control group, and 1 corresponding to the experimental group. So a two-sample *t*-test would yield a predictor matrix with this structure:

$$\begin{bmatrix} 1 & 0 \\ 1 & 0 \\ 1 & 0 \\ 1 & 1 \\ 1 & 1 \\ 1 & 1 \end{bmatrix}$$

Just like we learned about multiple regression, X_1 is a column of ones so that the intercept of the prediction equation can be computed, and X_2 is the values of our independent variable. The ONLY difference to what we have done before is that these values are **dummy codes**, meaning that they code our different experimental groups into mutually exclusive values. *Y*, our vector of outcome values is just a continuous random variable measured on an interval or ratio scale. The vector below is pretty lame, but the values are simple so that the process of regression is made more clear. Please forgive me for my lack of creativity here in picking an exciting topic...

$$\begin{bmatrix} 1 \\ 2 \\ 3 \\ 4 \\ 5 \\ 6 \end{bmatrix}$$

Now let's see how the regression plays out. Here are our defining equations:

$$\hat{Y} = X\beta$$

$$\beta = \left(X^T X\right)^{-1} X^T Y$$

$$\hat{Y} = X\left(X^T X\right)^{-1} X^T Y$$

$$\varepsilon = Y - \hat{Y}$$

When we estimate β, we get values of

$$\beta = \begin{bmatrix} 2 \\ 3 \end{bmatrix}$$

and therefore the prediction equation is:

$$\hat{y}_i = 2 + 3x_i$$

Substituting each value of X_2 into this equation we get the following predicted values:

$$\hat{Y} = \begin{bmatrix} 2 \\ 2 \\ 2 \\ 5 \\ 5 \\ 5 \end{bmatrix}$$

These are the means of our two groups—their expected values. In other words, our line of best fit passes through the points (0, 2) and (1, 5). The slope of the line, therefore is:

$$\frac{(\bar{y}_1 - \bar{y}_0)}{(\bar{x}_1 - \bar{x}_0)} = 3$$

Since our cleverly chosen dummy variables make the denominator 1 (See Figure 12.1).

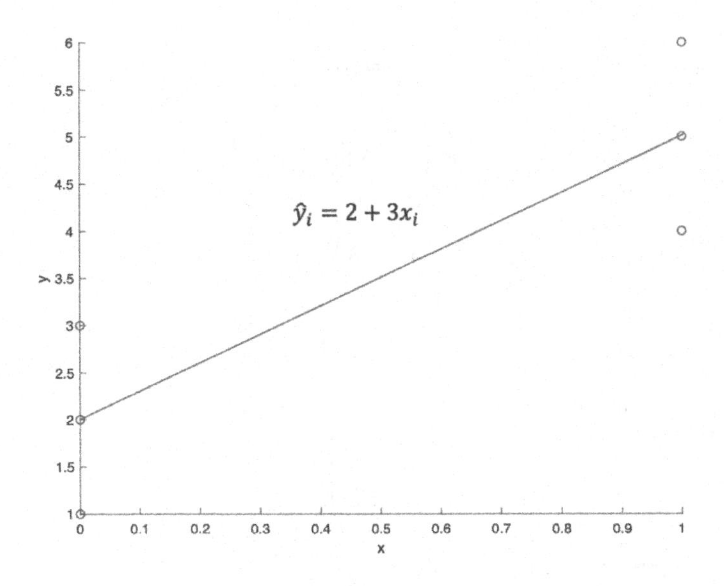

FIGURE 12.1: Line of best fit goes through $(\overline{x}_0, \overline{y}_0)$, and $(\overline{x}_1, \overline{y}_1)$, with a slope of $\overline{y}_1 - \overline{y}_0$

Now let's look at the t-test of the slope of a regression line:

$$t_{(n-2,\alpha)} = \frac{\beta_1}{SE_{\beta_1}} = \frac{\beta_1}{\sqrt{\frac{1}{(n-2)} \cdot \frac{\sum_{i=1}^{n}(y_i-\hat{y}_i)^2}{\sum_{i=1}^{n}(x_i-\overline{x}_i)^2}}}$$

You can see that the numerator now is just the difference in sample means, and the denominator reduces to the Standard Error of the Mean for the 2-sample t-test!

$$t_{(n-2,\alpha)} = \frac{\overline{y}_1 - \overline{y}_0}{\sqrt{\frac{1}{(n-2)} \cdot \frac{\sum_{i=1}^{n}(y_i-\hat{y}_i)^2}{\sum_{i=1}^{n}(x_i-\overline{x}_i)^2}}}$$

The standard error of the slope can be calculated easily if we have our vector of residuals: $(y_i - \hat{y}_i)$.

$$\begin{aligned} 1-2 &= -1 \\ 2-2 &= 0 \\ 3-2 &= 1 \\ 4-5 &= -1 \\ 5-5 &= 0 \\ 6-5 &= 1 \end{aligned}$$

$$SE_{\beta_1}\sqrt{\frac{1}{(6-2)}\cdot\frac{-1^2+0^2+1^2+-1^2+0^2+1^2}{\left(\frac{1}{2}\right)^2+\left(\frac{1}{2}\right)^2+\left(\frac{1}{2}\right)^2+\left(-\frac{1}{2}\right)^2+\left(-\frac{1}{2}\right)^2+\left(-\frac{1}{2}\right)^2}}$$

$$SE_{\beta_1}=\sqrt{\frac{1}{4}\cdot\frac{4}{\frac{6}{4}}}$$

$$SE_{\beta_1}=\sqrt{\frac{2}{3}}=0.8615$$

so

$$t_{(4,\alpha)}=\frac{3}{0.8615}=3.6742$$

The test statistic for the 2-sample *t*-test for this same data is:

$$t_{(\alpha,df=n_1+n_2-2)}=\frac{\overline{x}_1-\overline{x}_0}{s_p\sqrt{\left(\frac{1}{n_1}+\frac{1}{n_0}\right)}}$$

$$t_{(\alpha,df=4)}=\frac{5-2}{s_p\sqrt{\left(\frac{1}{n_1}+\frac{1}{n_0}\right)}}$$

$$s_p^2=\frac{(n_1-1)\,s_1^2+(n_2-1)\,s_0^2}{n_1+n_0-2}$$

$$s_p^2=\frac{2\,(1)+2\,(1)}{4}=1$$

$$t_{(\alpha,df=4)}=\frac{5-2}{1\sqrt{\left(\frac{1}{3}+\frac{1}{3}\right)}}=\frac{3}{1\,(.8165)}=3.6742$$

Check. And Mate... the *t*-test *is* just a simple case of linear regression where the independent variable, groups, is a nominal variable and the dependent variable is measured on an interval or ratio scale.

12.2 Expanding the *t*-test to *n* Groups

Fisher's contribution to modern data analysis was expanding this connection to account for differences in means for *multiple groups*. We don't need to limit ourselves to just a treatment group and a control group. We can now, for example look at different dosages (a medical term that has stuck in a lot of statistical circles) of the treatment, or even qualitatively different treatments. In airfoil design, for example, the National Advisory Committee for Aeronautics (NACA), which later became NASA, created the basic system we use to characterize the design of airfoil sections. Their original system denotes different key geometric features of an airfoil using 4 numbers corresponding to: 1) the maximum camber, $\mathbf{C_{max}}$, as a percent of the chord; 2) the *position* on the chord, of the maximum camber, $\mathbf{X_{Cmax}}$, in tenths of the chord; and maximum thickness of the airfoil section, **t**, as a percent of the chord, expressed in tenths and hundredths (the 3rd and 4th numbers in the system) (Ladson, & Brooks, 1975). Figure 12.2 shows a picture of three different airfoils that differ on the first number, chord-length, but are equivalent on the last three:

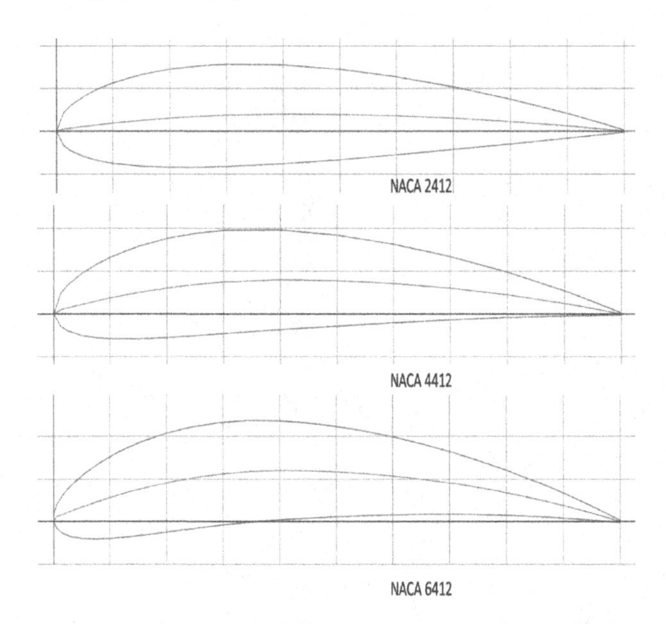

FIGURE 12.2: Three different airfoils, differing only on chord length

These airfoils will, obviously, display very different aerodynamic performance. In particular lift/drag ratio at the same angles of attack (angle with respect to airflow). How could we compare the performance of samples of these three airfoils on lift/drag ratio? This is the kind of question that a regression approach to hypothesis testing affords.

Suppose we use the NACA 2412 airfoil, the most symmetric, as our referent. How do the others that differ on chord length compare to the 2412 in lift/drag ratio at a low angle of attack (+2%) and low airspeed (150 ft/sec)? Let's look at some data in Table 12.1:

NACA 2412	NACA 4412	NACA 6412
50	55	57
54	53	54
48	49	52
48	50	51

TABLE 12.1: $\frac{l}{d}$ Ratio for three different airfoils at low angle of attack, and low airspeed

It looks like the asymmetric airfoils show a higher *l/d* ratio for these conditions, but is it *really* higher, or is this data just due to random chance? We could run 3 different *t*-tests contrasting each pair of sample means, but there are some complications with this method, namely, that doing 3-different *t*-tests in succession greatly increases our Type I error rate. To compensate for this, we would have to divide our chosen α by the number of *t*-tests performed–in this case, 3. If our $\alpha = 0.05$, it would end up being $\alpha = \frac{0.05}{3} = 0.0167$ for each test. For a small number of contrasts, this can be fine, but as the number of classes of airfoils grows, or if other independent variables are added to the analysis, the combinations get so large that it is both impractical, from a time perspective, and unwise, as α can quickly get so small that we would have to sample many, many, very expensive airfoils to have any power at all.[1]

A more universally utilized and practical method is to use regression to test the differences of all the means in one go—use the omnibus test, just like in Chapter 11. We would first make a vector of our outcome variable, *l/d* ratio and call it *Y*, then make an augmented matrix of our predictor variable, and call it *X* for some reason. If we regress Y on X, then we get the following relationship, shown in Figure 12.3:

All the relationships we found for projecting *Y* onto the column space of *X* still hold:

$$SS_{Total} = \left(Y - \overline{Y}\right)^T \left(Y - \overline{Y}\right) = 92.25$$

$$SS_{regression} = \left(\hat{Y} - \overline{Y}\right)^T \left(\hat{Y} - \overline{Y}\right) = 24.50$$

$$SS_{\varepsilon} = (Y - X\beta)^T (Y - X\beta) = 67.75$$

$$R^2 = \frac{SS_{regression}}{SS_{total}} = 0.2656$$

[1]This procedure is known as the Dunn-Bonferroni correction (Bitnun, 2009). Simple, you just divide the Type I error rate by the number of samples you are comparing, and use the critical value of t for the reduced for each *t*-test you run.

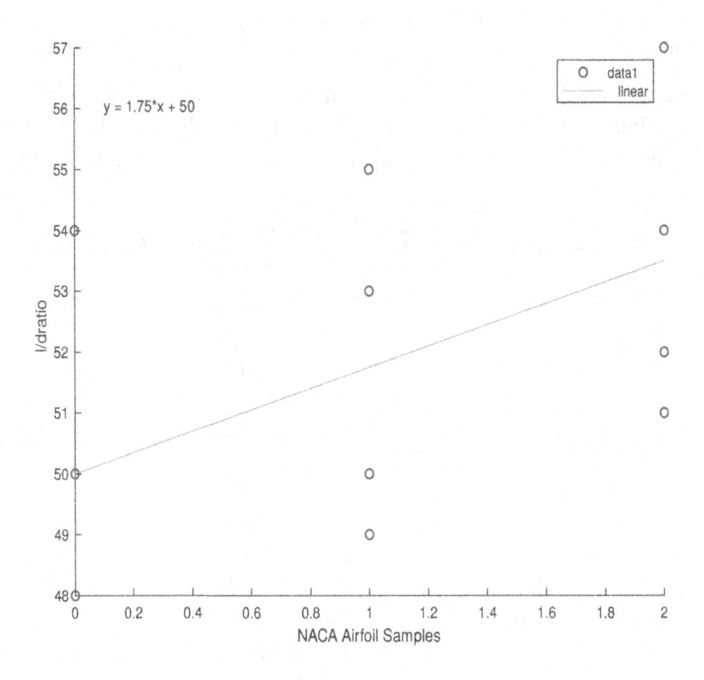

FIGURE 12.3: Regression line of best fit for 3 categorical variables. $\beta_{10} = 50, \beta_1 = 1.75$

1. What do you make of the relationship between these three airfoils that differ only on their chord length? In other words, does chord-length make a difference in the *l/d* ratio of these airfoils?

As I look at the relationship, given we only have 4 measurements in each sample. It looks like there might be a slight relationship. I see NACA 6412, which has the longest Chord length, appears to have the highest *l/d* ratio, while NACA 2412 has the least. But the *l/d* ratio is pretty close for all samples, and with such a small n, I am loathe to make a conclusion without testing whether or not this relationship could have happened, in all probability, just due to random chance. In the 2-sample case, we would have just used the t-test of the slope. But since we have three samples, which could have been put into the X-matrix in any order, due to the fact that they are nominal in scale, we need to try something a bit different—an omnibus test that accounts for this fact—the F-test.

The F-ratio, named after Sir Ronald Fisher, as we established in Chapter 9, is just a test of whether one distribution is wider than another, i.e., whether the variance in the distribution represented in the numerator is wider, or narrower than the distribution in the denominator. In regression, we use it to estimate if the variation we are accounting for in our regression model is significantly greater than the random error variation that our model does not account for:

$$F_{(k-1,n-k)} = \frac{MS_{regression}}{MS_{\varepsilon}}$$

Where the degrees of freedom for the numerator is the number of regression parameters being estimated in the model (minus 1), and the degrees of freedom in the denominator is the sample size minus the number of parameters being estimated.

$$MS_{regression} = \frac{\sum_{i=1}^{n}(\hat{y}_i - \bar{y})^2}{k-1} = \frac{SS_{regression}}{df_{regression}} = \frac{24.5}{2} = 12.25$$

$$MS_{\varepsilon} = \frac{\sum_{i=1}^{n}(y_i - \hat{y}_i)^2}{n-k} = \frac{SS_{\varepsilon}}{df_{\varepsilon}} = \frac{67.75}{9} = 7.53$$

If F is much greater than 1, we begin to feel that our model is pretty good: That the amount of variation the best fit equation explains is much greater than we would expect due to random chance.

$$F_{(2,9)} = \frac{MS_{regression}}{MS_{\varepsilon}} = 1.63, p = .249$$

The degrees of freedom in the numerator is the number of categories, airfoil types, minus 1. The degrees of freedom for the denominator is 12 total measurements, minus 3: the three parameters corresponding to each of the airfoil types, equalling 9. Interpreted, this F for our NACA airfoils means that the variation we explain as a function of the different groups is greater than the random variation within the groups, but given we only have 4 measurements per group, this difference is not too unlikely to be just a function of random chance ($p = .249$).

"This makes plenty of sense," I hear you thinking, "but I don't really get a feel for how we can regress a continuous variable on a nominal variable given categories are not ordered."

Any two samples taken *from the same population*, will exhibit some difference in means as illustrated in Figure 12.4. The distance between any sample mean and the grand mean of the two samples, $\bar{\bar{y}} - \bar{y}_{X_1}$, can then be added to the distance between the other sample and the grand mean $\bar{y}_{X_2} - \bar{\bar{y}}$ to yield the total distance between the samples: $\bar{y}_{X_2} - \bar{y}_{X_1}$. But, the sum of the squared deviations between each sample mean and the grand mean gives us the numerator of the variance, the 2nd moment about the grand mean, with respect to the sample means.

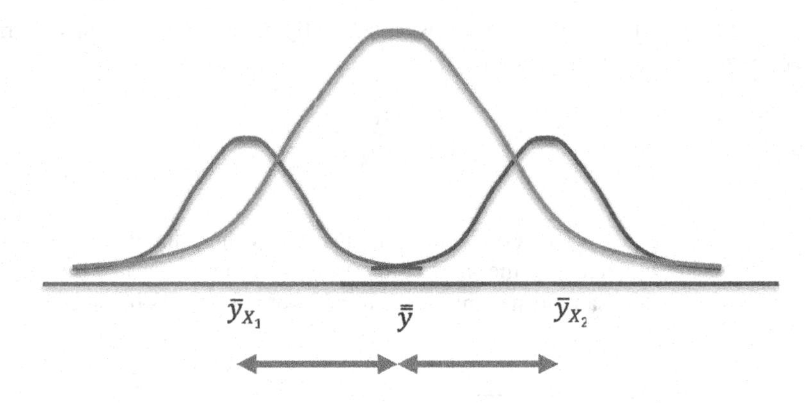

FIGURE 12.4: Two samples, X_1 and X_2 taken from a single population

$$SS_{\text{Between Groups}} = \sum_{k=1}^{K} \left(\bar{y}_{X_k} - \bar{\bar{y}}\right)^2$$

This is remarkably similar to the $SS_{regression}$

$$SS_{regression} = \sum_{i=1}^{n} (\hat{y}_i - \bar{y})^2$$

Where $\bar{y}$ is the mean of the dependent variable, and $\hat{y}_i$ is the expected value of any *i*th value of *X*. Really think about this when *X* is now a set of samples, each denoted with a dummy code (0, 1, 2, etc). For a continuous dependent variable, *X*, $\hat{y}_i$ is the mean of the distribution at X=*i*. *SOOoooo,*

$$\hat{y}_i = \bar{y}_{X_k}$$

The predicted value of y_i, $\hat{y}_i$ is the sample mean for each airfoil category. That means the total distance between our groups is a measure of the variation between groups: $SS_{\text{Between Groups}}$, and this distance is estimated by $SS_{regression}$. For the rest of your life, you need to think of these two concepts as synonymous.

"Fine." I hear you thinking. This works for two samples, but you already showed me that with the 2-sample *t*-test. Get on with it man!"

Ok. Getting on with it... you can see in Figure 12.5 that we have just added another experimental condition, X_3. Its distance is determined in exactly the same way.

$$\left(\bar{y}_{X_3} - \bar{\bar{y}}\right)^2$$

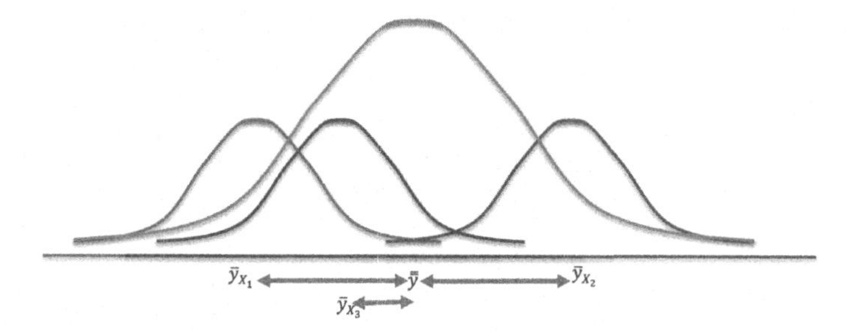

FIGURE 12.5: Adding more experimental conditions doesn't change anything!

Its contribution to the total distances of the means is just the same as the other two groups:

$$SS_{\text{Between Groups}} = \sum_{k=1}^{K} \left(\bar{y}_{X_k} - \bar{y}\right)^2$$

The formula doesn't change at all. As more groups are added, the total variation they contribute to the regression equation is just the sum of their individual squared distances. The total variation in the dependent variable Y is exactly the same:

$$SS_{Total} = \sum_{i=1}^{n} (y_i - \bar{y})^2$$

and the error, by the general linear model is $SS_{Total} - SS_{\text{Between Groups}} = SS_{\varepsilon}$

$$SS_{\varepsilon} = \sum_{i=1}^{n} \left(y_i - \bar{y}_{X_{ki}}\right)^2$$

The point is... regressing a continuous dependent variable on a categorical independent variable is still just regression! The matrices are the same, the algebra is the same, the formulas are the same.

Now that we have our regression taken care of, we can now determine the strength of the relationship between Y and X, and determine if this relationship is significantly different from that which we would expect under the null hypothesis: Zero differences in means.

Using the data we got from our NACA airfoils, we can perform the **(ANOVA):**

$$SS_{Total} = \left(Y - \bar{Y}\right)^T \left(Y - \bar{Y}\right) = 92.25$$

Analysis of Variance

Source	Sum Sq.	d.f.	Mean Sq.	F	Prob>F
X1	24.5	2	12.25	1.63	0.2493
Error	67.75	9	7.5278		
Total	92.25	11			

Hierarchical (Type II) sums of squares.

FIGURE 12.6: ANOVA table for independent variable with 3 levels (NACA 2412, NACA 4412, and NACA 6412), with 4 measurements per level for a total sample size of 12

$$SS_{regression} = \left(\hat{Y} - \overline{Y}\right)^T \left(\hat{Y} - \overline{Y}\right) = 24.50$$

$$SS_{\varepsilon} = (Y - X\beta)^{-1} (Y - X\beta) = 67.75$$

$$R^2 = \frac{SS_{regression}}{SS_{total}} = 0.2656$$

$$MS_{regression} = \frac{\sum_{i=1}^{n} (\hat{y}_i - \bar{y})^2}{k - 1} = \frac{SS_{regression}}{df_{regression}} = \frac{24.5}{2} = 12.25$$

$$MS_{\varepsilon} = \frac{\sum_{i=1}^{n} (y_i - \hat{y}_i)^2}{n - k} = \frac{SS_{\varepsilon}}{df_{\varepsilon}} = \frac{67.75}{9} = 7.53$$

$$F_{(2,9)} = \frac{MS_{regression}}{MS_{\varepsilon}} = 1.63, p = 0.2493$$

This analysis is usually displayed in an ANOVA table. This is a convenient display that, being common to all ANOVA analyses, makes it easy to see all these relationships in one go. Here is the ANOVA table presented in Figure 12.14 for our NACA airfoil data.

The sources of variation are listed by name in the first column. X1 is the name of the independent variable. The equation that shows the relationship among sources of variation, is displayed in the second column. Notice that the $SS_{Total} = SS_{X1} + SS_{Error}$. This is just a restatement of the now-familiar $SS_{Total} = SS_{Regression} +$

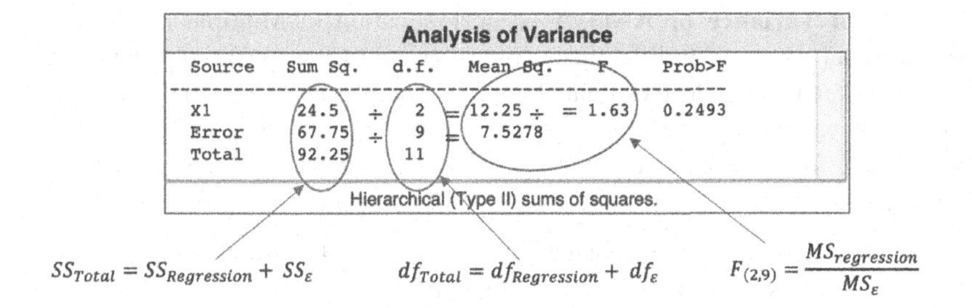

FIGURE 12.7: The ANOVA table shows the regression model in ways that make all the calculations explicit

SS_{ε}, or as we have noted it for this chapter $SS_{Total} = SS_{\text{Between Groups}} + SS_{\text{Within Groups}}$. If you have done the calculations correctly this equation should all add up.

The degrees of freedom for each variance component estimated is listed in the 3rd column of the table. We had 3 groups, so the degrees of freedom for estimating the average distance between the groups is $3 - 1 = 2$. We had 12 total observations, so to estimate the variance of all the observations (the numerator of which is SS_{Total}) is $12 - 1 = 11$. The degrees of freedom for the residual error can be found knowing that *df* adds up to the total: 2+? = 11 so $df_{\varepsilon} = 9$.

To compute the F-ratio testing whether the slope of the regression line is significantly different from zero, we divide the *SS* for each variance component by its degrees of freedom.

$$F_{(2,9)} = \frac{MS_{regression}}{MS_{\varepsilon}} = 1.63, p = 0.2493$$

All of these relationships are shown explicitly in Figure 12.7.

The final column in an ANOVA table is the p-value, the probability that the between groups variation—the distance between the means of each sample, given the overall variability in the data—is a function of random chance. Looking at this value, it appears our samples of NACA airfoils are not significantly different from each other, i.e., the model did not account for enough variation between groups relative to the error in the model. In fact, we would expect this result just due to random chance about $\frac{1}{4}$ of the time. This is not low enough for me to chance designing multimillion-dollar aircraft based on the results of my experiment.

Analysis of Variance or ANOVA is an application of Multiple Linear Regression that models the differences among group means as variance between groups. When this variance estimate between groups is compared to the residual (i.e., error) variance, the probability that the samples being compared could have been drawn from the same, common population, completely at random, can be computed.

As a case of the General Linear Model, ANOVA is the orthogonal projection of a continuous vector, Y, onto the column space of a matrix of predictors, X. Because the projection is orthogonal to Xs column space, it is the linear model with the least error (i.e., has the error vector with least magnitude).

$$SS_{\text{Between Groups}} = \sum_{k=1}^{K} \left(\bar{y}_{X_k} - \bar{\bar{y}}\right)^2$$

$$SS_{\varepsilon} = \sum_{i=1}^{n} \left(y_i - \bar{y}_{X_{ki}}\right)^2$$

$$SS_{Total} = \sum_{i=1}^{n} \left(y_i - \bar{\bar{y}}\right)^2$$

The omnibus test of mean differences in ANOVA is the F-test. The F-ratio compares the $MS_{\text{Between Groups}}$ the variance estimator of the regression model, with MS_{ε} , the variance estimator of the residual error.

$$MS_{\text{Between Groups}} = \frac{\sum_{k=1}^{K}\left(\bar{y}_{X_k} - \bar{\bar{y}}\right)^2}{k-1} = \frac{SS_{\text{Between Groups}}}{df_{\text{Between Groups}}}$$

$$MS_{\varepsilon} = \frac{\sum_{i=1}^{n}\left(y_i - \bar{y}_{X_{ki}}\right)^2}{n-k} = \frac{SS_{\varepsilon}}{df_{\varepsilon}}$$

$$F_{(df_{\text{Between Groups}}, df_{\varepsilon})} = \frac{MS_{\text{Between Groups}}}{MS_{\varepsilon}}$$

R^2, the coefficient of determination is computed in the same manner:

$$R^2 = \frac{SS_{\text{Between Groups}}}{SS_{total}}$$

It is important to know all of these terms, because different statisticians use different terms and notation to denote the same concepts. Even in this chapter, when I have switched from talking about Multiple Linear Regression as a general model to ANOVA as a specific example, I have used different subscripts for these terms. Here is a handy cheat sheet (Figure 12.8):

12.2.1 Residual Analysis

Just like any other regression analysis based on the General Linear Model, the assumptions include normally distributed residuals, randomly spread about zero.

It is a little more difficult to discern because of their discrete nature, but a residual analysis for the NACA airfoils in Figure 12.9 shows no obvious pattern. The values

What I Prefer	Alternative, Equivalent Terms		
Independent Variable	Factor	Predictor Variable	
Dependent Variable	Response Variable	Criterion Variable	Outcome Variable
$SS_{Regression}$	$SS_{Between\ Groups}$	SS_{Factor}	
$SS_{Residual}$	$SS_{Within\ Groups}$	SS_{ε}	SS_{error}
$MS_{Regression}$	$MS_{Between\ Groups}$	MS_{Factor}	
$MS_{Residual}$	$MS_{Within\ Groups}$	MS_{ε}	MS_{error}
R^2	η^2		

FIGURE 12.8: Common terms statisticians use for the same concepts

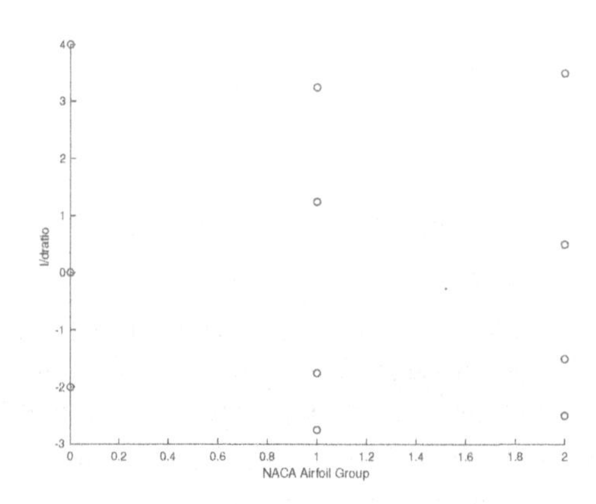

FIGURE 12.9: Residual analysis for NACA airfoils

of $y_{residual}$ appear to be fairly randomly distributed about zero. A Q-Q plot in Figure 12.10, however, looks a bit off in the lower tail. With so few data points, again, it is hard to tell, but we might want to look at the mathematical relationship of chord length to *l/d* ratio to see if there might be an appropriate transformation for the data. From my understanding of wing dynamics, chord length has a linear relationship to *l/d* ratio for fairly low speeds like those done in our testing. Given this, I don't think a transformation is necessary, but I might want to collect more data in a more controlled environment to reduce error of measurement. That might give me a narrow enough standard error to detect a difference that I still suspect exists.

12.2.2 Multiple Comparisons: What to Do if You Find Significant Results in an ANOVA

Unfortunately, our comparison of NACA airfoils returned non-significant results for an effect of chord length on *l/d* ratio for the three models tested. But what if we had found a significant difference? How would we test to see which of the models outperformed one or more of the others? Just like in Multiple Linear Regression, we

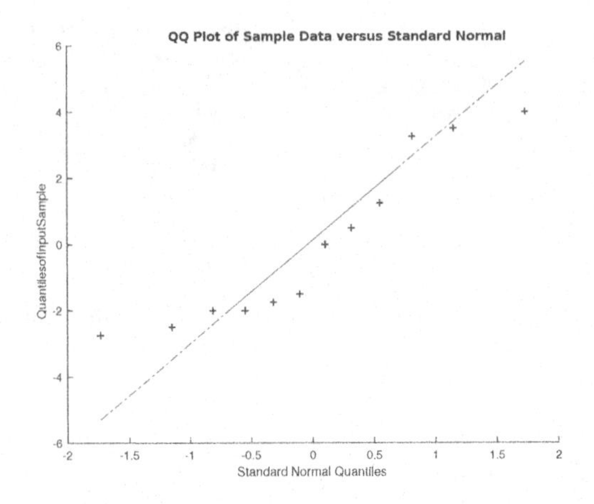

FIGURE 12.10: Quantile-Quantile plot of NACA airfoils residuals

can run the ANOVA *omnibus* test, and then, if we find a significant difference there, we know that at least one pairwise comparison of means is significantly different. So, we can run a *post hoc* comparison of the group means to pinpoint the difference.

There are several methods for post hoc comparisons[2]. We will focus on two of the most robust: The Dunn-Bonferroni correction and the Scheffé test. These are conservative tests, being very careful to not inflate alpha, at the sacrifice of the power of the tests. The differences between them are subtle, not easily understood without a practical example. So, let's grab some data!

Dakeev (2014) presented a nice experimental analysis of an experimental small-scale wind turbine attachment that directs wind from the "dead-space" near the hub, towards the airfoil tips. This attachment was hypothesized to increase the wind-speed at the tips of the airfoils. He outfitted a number of turbines with three different attachment conditions: 1) A cone-shaped attachment on the experimental turbine, 2) the experimental turbine without the attachment, and 3) a control condition with a bare hub and no airfoils attached. The data for this analysis is reconstructed from his results. It is available in the following file: Wind_Turbine_Oneway_ANOVA.xlsx. There are 37 observations.

[2]I have chosen the Dunn-Bonferroni correction and Scheffé as the two post hoc methods to include in this volume because I tend to be conservative with post hoc tests. My reasoning is thus: If we really don't have a good hypothesis about which conditions will perform significantly differently from the others, then we don't want to just go fishing to find any result that appears significant, reporting them all. Instead, we should be as sure as we can be that we are not inflating our Type I error rate, even if it means sacrificing power. Dunn-Bonferroni and Scheffé are both moderately (D-B) to very (Scheffé) conservative. If I use these methods I can be pretty sure that the results of my post hoc analysis are as likely to be true as my Type I error rate will allow.

Setting up the data, Y has to be a column *vector* of dependent variable values. X is a column *vector* of dummy codes for each of the experimental groups. Any statistical package you are using converts it to a typical X-matrix when it does the regression for us. If you are following along, you will need to enter this data. You may want to put real variable names in their places if it helps you keep things straight). The dummy codes are: Control condition = 0; Turbine with Attachment = 1; Turbine with No Attachment = 2.

Just like any regression analysis, we need to state our Null and Alternative Hypotheses.

Dakeev stated a 1-tailed (upper) Alternative Hypothesis. But because we have several groups we are comparing, the hypotheses are a bit more involved. If you are anal retentive, you should specify each of them separately:

$$H_0 : \mu_{X_1} \leq \mu_{X_0}$$

$$H_0 : \mu_{X_1} \leq \mu_{X_2}$$

$$H_0 : \mu_{X_2} = \mu_{X_3}$$

$$H_1 : \mu_{X_1} > \mu_{X_0}$$

$$H_1 : \mu_{X_1} > \mu_{X_2}$$

$$H_1 : \mu_{X_2} \neq \mu_{X_3}$$

Or, you can be lazy like me and state in English:

H_0: All group means are equal

H_1: At least one mean is greater than one of the others

This analysis is not too dangerous in terms of making a Type I error. We are not yet ready to build the new turbines. We are testing the effect of a new prototype. A Type I error rate of 0.05 is just fine. With the data entered and these preliminaries specified, we can perform our Analysis of Variance (See Figure 12.11).

If I am Dakeev, I am getting pretty excited. The F-test is signficant. F is clearly higher than the critical value of 3.1404 (one tailed), and its associated probability is about 6.045 x 10^{-10} (in the 10 billionths). $R^2 = \frac{SS_{\text{Between Groups}}}{SS_{total}} = \frac{18.8096}{38.7931} = 47\%$. This is not super, but given I only have three categorical groups, about half of all the variation in wind speed is accounted for by the difference in experimental groups.

Analysis of Variance

```
Source    Sum Sq.   d.f.   Mean Sq.      F        Prob>F
------------------------------------------------------------
X1        18.8096     2    9.40478    30.12   6.04467e-10
Error     19.9835    64    0.31224
Total     38.7931    66
```

Hierarchical (Type II) sums of squares.

FIGURE 12.11: ANOVA table for Dakeev's wind turbine data

That isn't too bad for prototyping. When I examine the summary statistics for my samples, I get even more excited. It really looks like my experimental attachment is working... But I still don't know if my experimental turbine with the prototype attachment significantly outperforms the other groups. All I know is that ONE of the experimental conditions is DIFFERENT from at least one other. That is where the post hoc tests will come in.

12.2.2.1 Dunn-Bonferroni Correction

This procedure, as I hinted early on in the chapter is to divide the Type I error rate by the number of samples you are comparing, and use the critical value of t for the reduced α for each t-test you run. The reason you need to divide α by the number of comparisons, is that these post hoc tests are not independent of each other. In fact, they are sequential: If you run 3 post hoc t-tests at $\alpha = 0.05$, you inflate your alpha to the point where your probability that at least one of those post hoc tests is a Type I error gets up to about 15%. If $\alpha = 0.05$ really is the highest chance you will take for making a false-positive decision, then $\alpha = 0.15$ is unacceptable.

The big takeaway is that, to control for this inflation of the Type I error rate, you have to account for it by reducing the overall α for each post hoc t-test so that the total Type I error rates for each pairwise test add up to your chosen α. Luckily, any decent statistical package has a nice function that will compute all these combinations for you. YOU have to know what they mean, and specify your alpha correctly!

Here are our results of the Dunn-Bonferroni procedure for Dakevs analysis are shown in Table 12.2, with means and standard errors for the different factors presented in Table 12.3:

	Mean	**SE**
Control	3.50746105	0.12494854
Attachment	4.27274082	0.11913376
No Attachment	3.00902703	0.11175737

TABLE 12.3: Means of groups tested

Group$_1$	Group$_2$	95% CI Lower Bound	Mean Difference	95% CI Upper Bound	$p(t > t_{critical})$
Control	Attachment	−1.1897239	−0.7652798	−0.3408356	0.0001118
Control	No Attachment	0.08629525	0.49843402	0.91057279	0.01244841
Attachment	No Attachment	0.8621174	1.26371379	1.66531018	2.82E–10

TABLE 12.2: Multiple comparisons using Bonferroni's method

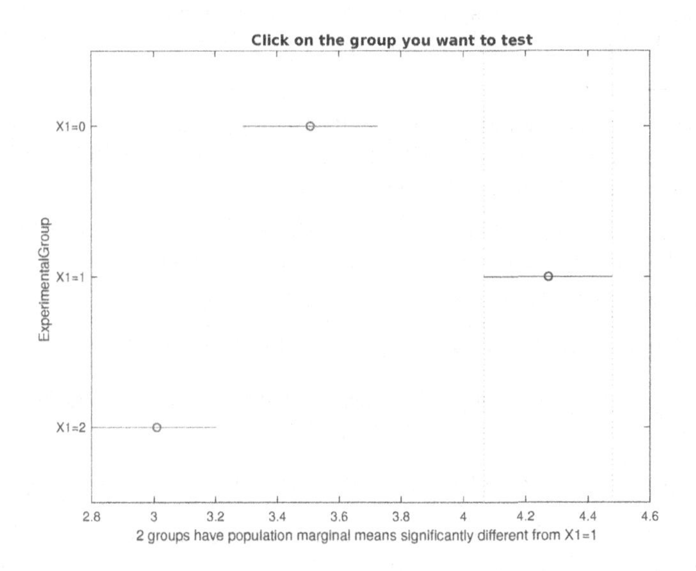

FIGURE 12.12: Multiple comparisons using Dunn-Bonferroni method. width of bars represent the standard error of the estimate

Matlab, the package I am using to write this book, also returns a lovely tool, an interactive graph that shows the differences in means graphically (see Figure 12.12). You can click on each mean and check to see each of the other groups that have a mean significantly different from it.

Looking at the p column, we can see that the innovative attachment does seem to result in wind speed at the tips of the airfoils to be greater than either the control or the no-attachment condition. Each of the contrasts show that the attachment returns a difference great enough to be considered highly improbable. Additionally, each of the confidence intervals do not contain 0, the hypothesized value of the difference in means.

All of this is great, but the thing to keep in mind is that you need to divide your α by the number of comparisons you wish to examine post hoc. Then compare the p-value in the results matrix with that reduced p-value. If you get too many comparisons, say if you had 15 groups to compare, you would have to run $\binom{15}{2} = 105$ t-tests! Type I error rate skyrockets to nearly 100% under such conditions. A good rule of thumb is to choose the specific contrasts that make theoretical sense ahead of time, then only partition your α by that number. So, if you had 15 groups, but were only interested in 10 contrasts, divide your Type I error rate by 10 instead of 105. This will increase your overall power to detect a real difference without having to collect a huge data sample just to account for a large number of comparisons that don't make theoretical sense.

The **Dunn-Bonferroni Correction for Pairwise Comparisons** is a procedure for testing, following an Analysis of Variance omnibus test, the differences among group means. The logic is simple:

1. Divide the experiment-wise Type I error rate, α, by the number of comparisons to be made post hoc;
2. Perform an Analysis of Variance as an omnibus test for all groups;
3. If the Analysis of Variance shows significant results for your groups, compute pairwise t-tests between group means;
4. Compare the computed t-statistic with the critical value at $\frac{\alpha}{\#comparisons}$;
5. If the t-statistic is further in the tail of the distribution than the critical value, reject H_0 for that comparison. If t is towards the middle of the distribution compared to the critical value, fail to reject H_0.
6. A good rule of thumb is to choose only those comparisons you need to make, theoretically, before performing the Dunn-Bonferroni procedure. Reducing the overall number of comparisons boosts the overall power of each.

12.2.2.2 Scheffé Test

Henry Scheffé's (he was American, so he spelled his first name, Henry, not Henri, as did his French ancestors) contribution to the Analysis of Variance was to create a test that increases the critical value of F for the post hoc tests (as opposed to reducing α), by multiplying it by the between groups degrees of freedom:

$$CriticalValue = (k-1)\,F_{(k-1,N-k,\alpha)}$$

With this critical value, one then computes a test statistic that is essentially an F-statistic for two samples:

$$F_{Scheff\acute{e}} = \frac{\left(\bar{y}_{X_1} - \bar{y}_{X_2}\right)^2}{MS_{\text{Within Groups}}\left(\frac{1}{n_{X_1}} + \frac{1}{n_{X_2}}\right)}$$

Like all statistical tests, we compare $F_{Scheff\acute{e}}$ with the critical value. If $F_{Scheff\acute{e}}$ is larger than the critical value, then we say the comparison is significant.

Scheffé is similar in computation to the Dunn-Bonferroni procedure. If you look at $F_{Scheff\acute{e}}$, you can see it is just the square of a 2-sample t-test. So it is not the test statistic that Scheffé manipulated, but the critical value to which it is compared. The major difference between the two, computationally, is the means by which the critical value is determined. Of these two post hoc procedures, the Dunn-Bonferroni method is more generous, giving you more power, but is limited to only pairwise post hoc comparisons. Scheffé's method, which is one of the most conservative of all post hoc tests, accounts for pairwise comparisons, interactions among two or more groups and other more sophisticated comparisons, but does reduce your overall power to reject a Null Hypothesis. If you choose to utilize Scheffé, then you should work ahead of time to develop the power you need by increasing your sample size accordingly.

Here are the results of our post hoc comparisons using the Scheffé method (Table 12.4):

Group$_1$	**Group$_2$**	**95% CI Lower Bound**	**Mean Difference**	**95% CI Upper Bound**	**$p(F_{Scheff\acute{e}} > F_{critical})$**
Control	Attachment	−1.1979476	−0.7652798	−0.332612	0.00019008
Control	No Attachment	0.07831	0.49843402	0.91855804	0.01591518
Attachment	No Attachment	0.85433641	1.26371379	1.67309117	6.68E–10

TABLE 12.4: Multiple Comparisons using the Scheffé test

Again, we can look at the p column and see that Dakeev's innovative, wind-focusing attachment, makes a difference. Each of the contrasts show that the attachment returns a difference great enough to be considered highly improbable. Additionally, each of the confidence intervals do not contain 0, the hypothesized value of the difference in means. The means matrix and the graph of the contrasts are exactly the same as before. This confirms our analysis using Dunn-Bonferroni.

One thing you may have noticed is that the confidence intervals for the Scheffé test are *wider* than those computed with Dunn-Bonferroni, and the probabilities, *larger*. Using the same data with the same number of pairwise contrasts, this will always be the case.

The **Scheffé Test for Pairwise Comparisons** is a procedure for testing, following an Analysis of Variance omnibus test, the differences among group means. Unlike the Dunn-Bonferroni Correction, the Scheffé Test computes a critical value that is increased by the degrees of freedom between groups, and compares its test statistic, $F_{Scheff\acute{e}}$, to this value.

$$Critical\ Value = (k-1)\,F_{(k-1,N-k,\alpha)}$$

$$F_{Scheff\acute{e}} = \frac{\left(\bar{y}_{X_1} - \bar{y}_{X_2}\right)^2}{MS_{\text{Within Groups}}\left(\frac{1}{n_{X_1}} + \frac{1}{n_{X_2}}\right)}$$

The procedure for using the Scheffé Test to make post hoc comparisons among group means is as follows:

1. Perform an Analysis of Variance as an omnibus test for all groups;
2. If the Analysis of Variance shows significant results for your groups, compute the critical value and $F_{Scheff\acute{e}}$ between group means;
3. If $F_{Scheff\acute{e}}$ is larger than the critical value, then we can reject H_0. If it is less than the critical value, we fail to reject H_0.

Scheffé is very conservative, and is best used in exploratory situations where one does not have the ability to choose, before-hand, the comparisons of group means, one would like to examine.

12.2.3 Assumptions of the ANOVA

Like all models, the ANOVA is just an approximation of the real relationship among variables. As a model we have to make some basic assumptions about its mathematical properties for our model to hold. Luckily, those assumptions are minimal, making the ANOVA a robust, widely applicable technique that all engineers should have in their toolkit.

Assumptions of Analysis of Variance are the same as that of any Multiple Linear Regression analysis:

1. Normality of the parent populations from which samples are drawn;
2. Independence of observations; and

3. Homogeneity of variance. The sample(s) must display equal variances across levels of the independent variable. This the extension of Homogeneity of variance for the 2-sample *t*-test to the test of multiple group means. It is the analog to Homoscedasticity for Multiple Regression with a continuous independent variable.

Like the *t*-test, Analysis of Variance is robust to violations of these assumptions, generally losing power if samples have widely differing variance, or if the population distributions are widely non-normal. Lack of independence, however, is something even ANOVA can't recover from.

12.3 Extended Example

The American Society of Testing and Materials (ASTM) published a study of the stress corrosion cracking undergone by sample aluminum alloys using a slow strain-rate technique (Ugiansky, & Payer, 1979). The data for this study is included in the file Corrosion_Anova.xlsx. This particular study examined the days to failure exhibited by 35 aluminum bars of 2124 Al, each tested in one of 7 laboratories (n=5 measurements per lab). We are investigating whether one or more of the labs shows poor performance in their testing, by exhibiting significantly faster time to failure than the other labs. After entering the data into our chosen statistical package, we can first look at the scatterplot shown in Figure 12.13:

This looks like there is a great deal of consistency in labs 1, 2, 5, 6, and 7. Lab 3 looks very different, both lower in average time to failure of the sample, and very little variation among measurements. Lab 4 may be a bit lower in average time to failure, but they have much greater variability than the other labs. I will keep this in mind as we check assumptions later on.

Now, what is a good overall α to use? I think I might go with the default (α = 0.05) if this were just to look at consistency across testing labs. But if we were considering shutting down a lab that showed bad performance, I might use $\alpha = 0.01$. For sake of this problem, let's use $\alpha = 0.01$. It may affect our post hoc tests, but that is a good learning opportunity here. I would use a 2-tailed test, because we don't really have an experimental treatment. I don't know, *a priori*, which labs will be better or worse than the others. With these specified, I can write the Null and Alternative Hypotheses for our analysis:

$$H_0\text{:All means are equal}$$

$$H_1\text{:At least one mean is different from the others}$$

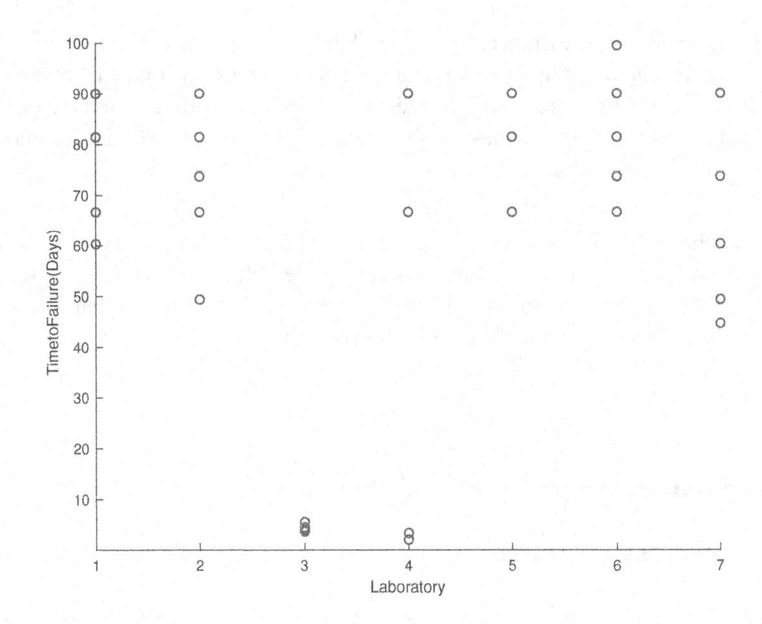

FIGURE 12.13: Scatterplot of time to failure versus laboratory

Analysis of Variance

Source	Sum Sq.	d.f.	Mean Sq.	F	Prob>F
X1	26416.3	6	4402.71	11.1	2.49843e-06
Error	11109.5	28	396.77		
Total	37525.8	34			

Hierarchical (Type II) sums of squares.

FIGURE 12.14: ANOVA table for time to failure analysis

See how nice that is, to specify it in English? I don't want to write out all $\binom{7}{2} = 21$ statements!

With this out of the way, we can run our data. The ANOVA table presented in Figure 12.14 shows us that we have a significant result. As we suspected from looking at the scatterplot, at least one of the laboratories has a different mean time-to-failure for its Al sample. We can see this because F is very high ($F_{critical(6,28)} = 2.79$), corresponding to a p-value that is very small the probability we could have gotten our results solely as a function of random chance is about 2.5 in 1 million.

R^2 of our regression model is 0.703. Over 70% of the variation in time to failure of our Al sample is accounted for by this analysis. This is quite high, so I

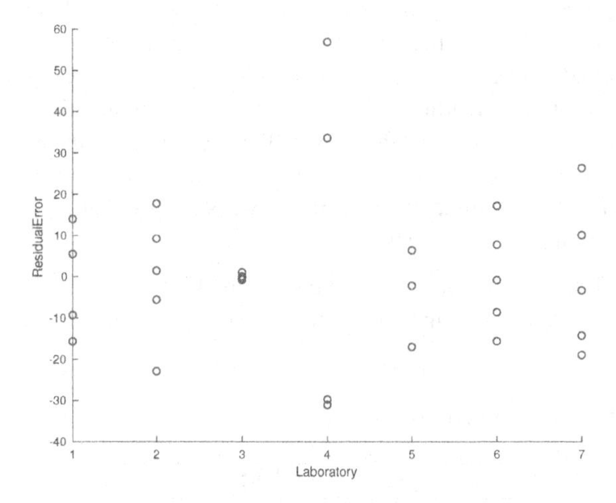

FIGURE 12.15: Residual plot for time to failure analysis

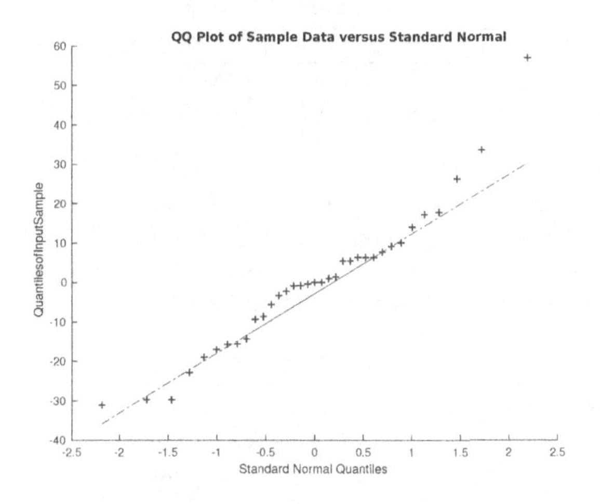

FIGURE 12.16: Q-Q plot for time to failure analysis

am feeling pretty confident in our results already. But what about our assumptions? I need to do a residual analysis (See Figure 12.15):

The residual plot shows that each of the samples appear to be fairly evenly distributed about zero, but the wide variation in widths makes me worried that we may have violated our assumption of equal variances. Lab 3 and Lab 4 are the only ones that appear greatly out of line.

The Quantile-Quantile plot of the residuals (Figure 12.16) shows a couple of modest deviations. First, in the middle we have a bulge right around where I would expect the residual for Lab 4 to follow. It is wider than the other distributions, so it would have a measured quantile greater than the predicted quantile causing a bulge higher than the fit line. But worse than this is the upper end of the plot. I see several points that are quite a ways from the line. That might warrant transformation of the data, or I might have to eliminate one or more points if I find I am not able to discern any differences in the post hoc analyses.

Now, I will do the post hoc analyses. I really didn't have any specific hypotheses about which labs might be better or worse, so I would need to perform $\binom{7}{2} = 21$ contrasts. I really can't reduce the number of contrasts to justify the Dunn-Bonferroni correction, so I choose Scheffé:

Lab	Mean	SE
1	75.9890975	8.90805726
2	72.2513148	8.90805726
3	4.43236443	8.90805726
4	33.0714898	8.90805726
5	83.6377187	8.90805726
6	82.2676881	8.90805726
7	63.6321692	8.90805726

The Results matrix shows us 5 of the 21 contrasts are significant, retaining our overall alpha of 0.01. The big culprit, of course, is Lab 3, with its extremely low value for time to failure of its Al bars, it is significantly different from Labs 1, 2, 5, 6, and 7 ($p < 0.01$). For its part, Lab 4 is not significantly different in its mean value, but its variability is still a concern.

Group_1	Group_2	95% CI Lower Bound	Mean Difference	95% CI Upper Bound	$p(F_{Scheffé} > F_{critical})$
1	2	−44.516525	3.73778269	51.9920906	0.99998328
1	3	23.3024251	71.556733	119.811041	0.00084329
1	4	−5.3367002	42.9176077	91.1719156	0.10999729
1	5	−55.902929	−7.6486213	40.6056867	0.99891778
1	6	−54.532899	−6.2785906	41.9757173	0.99965059
1	7	−35.89738	12.3569282	60.6112361	0.98516011
2	3	19.5646424	67.8189503	116.073258	0.00168885
2	4	−9.0744829	39.179825	87.4341329	0.18080391
2	5	−59.640712	−11.386404	36.867904	0.99032772

2	6	-58.270681	-10.016373	38.2379346	0.99514177
2	7	-39.635162	8.61914552	56.8734534	0.99787965
3	4	-76.893433	-28.639125	19.6151826	0.53497842
3	5	-127.45966	-79.205354	-30.951046	0.00020056
3	6	-126.08963	-77.835324	-29.581016	0.0002596
3	7	-107.45411	-59.199805	-10.945497	0.00804119
4	5	-98.820537	-50.566229	-2.311921	0.03468781
4	6	-97.450506	-49.196198	-0.9418904	0.04314804
4	7	-78.814987	-30.560679	17.6936285	0.4566477
5	6	-46.884277	1.37003065	49.6243386	0.99999996
5	7	-28.248758	20.0055495	68.2598574	0.85918114
6	7	-29.618789	18.6355188	66.8898268	0.89501144

So, overall, it appears I have two problem Labs: Lab 3 and Lab 4. Lab 4 appears to be primarily inconsistent with its wide variation though its mean value is about the same as the other labs. Lab 3 is more of a problem. It is very consistent, but its measurements are way off where they should be if we want our Labs to produce equivalent results. Is this a systematic problem with Lab 3? The results suggest it is. If I were the ASTM, I would examine the procedures and equipment in both labs, providing help to Lab 4, as their issues don't appear to be too bad, but Lab 3? I might have to look deeply at how measurements are done, and if there is something wrong that is not fairly easy to rectify, make some hard decisions about its viability, or the capacity of its personnel.

12.4 Summary

The Analysis of Variance is a flexible approach to testing hypotheses about differences in means. When we take two or more samples randomly from the same population, we expect there to be a slight differences among the sample means. Most should be relatively close to each other in magnitude, somewhere close to the population mean. Any variation in sample means is assumed to be due to random chance. But if, after we select our samples from the population, their means differ by a wide margin, we might ask the question, "I wonder what happened?" Did some systematic thing happen to one or more of the samples to cause them to be so different from the others?

This is the essential question that the Random Sample Design seeks to answer. If there are only two samples selected from a population, we can use the 2-sample t-test to help us determine the chance that the difference between the means is due to random chance. But if we have more than two groups to compare—more than two samples of measurements, the t-test just doesn't cut it, after all, you can't take the difference between 3 or more sample means without first establishing their order. Analysis of Variance to the Rescue!

Analysis of Variance takes each of the differences between each group mean and the overall sample mean, and develops a measure of between groups variation, the $MS_{\text{Between Groups}}$. If that measure is compared to the variation found within groups, $MS_{\text{Within Groups}}$, the larger their ratio, the more likely it is that the distance between the groups is large enough to be considered unlikely, just due to random chance. The F-test is a probability distribution that determines the probability that this ratio, $\frac{MS_{\text{Between Groups}}}{MS_{\text{Within Groups}}}$ is due to random chance

Because the Analysis of Variance is an omnibus test, detecting whether any difference in means is significant or not, we must follow up with a post hoc comparison to determine which means are indeed different from the others. Post hoc comparisons account for the inflation of Type I error rate due to successive tests of a single sample of data.

But the most important concept to remember about the Analysis of Variance is *that it is just Regression!* ANOVA regresses a continuous outcome variable on a matrix of dependent variables, which may have any number of comparable groups. By the General Linear Model, the goodness of fit of ANOVA is measured with R^2 the coefficient of determination.

With this basic knowledge of the Analysis of Variance, engineers can add more independent variables, and develop more sophisticated experimental designs than just the Random Sample Design described thus far. Next, we turn to ANOVA models applied when we have multiple independent variables, *each* with several potential groups.

12.5 References

Abbott, I. H., Von Doenhoff, A. E., & Stivers, L. S., Jr. (1945). *Summary of Airfoil Data. NACARep.* 824, (SupersedesNACAWRL-560).

Aschan, C., Hirvonen, M., Rajamäki, E., & Mannelin, T. (2005). Slip resistance of oil resistant and non-oil resistant footwear outsoles in winter conditions. *Safety science*, *43*(7), 373-389.

Bitnun, A. (2009). Bonferroni Correction. *J Clin Viro. l*, *44*, 262-7.

Bouacha, K., Yallese, M. A., Mabrouki, T., & Rigal, J. F. (2010). Statistical analysis of surface roughness and cutting forces using response surface methodology in hard turning of AISI 52100 bearing steel with CBN tool. *International Journal of Refractory Metals and Hard Materials*, *28*(3), 349-361.

Box, G. E. P. & Cox, D. R. (1964) An analysis of transformations, *Journal of the Royal Statistical Society, Series B*, *26*, 211-252.

Covassin, T. Swanik, C.B., Sachs, M.L. (2003).Sex Differences and the Incidence of Concussions Among Collegiate Athletes, Journal of Athletic Training,(38)3, 238-244.

Dakeev, U. (2014). Analysis of wind power generation with application of Wind tunnel attachment. Proceedings of the 121st Annual Conference and Exposition of the IEEE, Indianapolis, IA.

Fisher, R.A. (1925) Statistical Methods for Research Workers. Oliver & Boyd, Edinburgh.

Ladson, C. L., & Brooks Jr, C. W. (1975). Development of a computer program to obtain ordinates for NACA 4-digit, 4-digit modified, 5-digit, and 16 series airfoils.

Lee, S., & Lee, D. K. (2018). What is the proper way to apply the multiple comparison test?. *Korean journal of anesthesiology*, *71*(5), 353-360.

Ndaro, M. S., Jin, X. Y., Chen, T., & Yu, C. W. (2007). Splitting of islands-in-the-sea fibers (PA6/COPET) during hydroentangling of nonwovens. *Journal of engineered fibers and fabrics*, *2*(4), 1-9.

Owen, A. R. G. (1962) An Appreciation of the Life and Work of Sir Ronald Aylmer Fisher *Statistician, 12*, 313-319.

Ugiansky, G. M., & Payer, J. H. (Eds.). (1979). *Stress Corrosion Cracking: The Slow Strain-Rate Technique* (Vol. 665). ASTM International.

12.6 Study Problems for Chapter 12

1. Aschan et al., (2005) reported the coefficient of dynamic friction for two different sole types for work boots: Those with oil resistant soles versus those without oil resistant soles. Their data is listed below:

Non-Oil Resistant Soles		**Oil Resistant Soles**	
Oil Res Code	Steel w/Glycerine Dynamic Coeff. Friction	Oil Res Code	Steel w/Glycerine Dynamic Coeff. Friction
0	0.186	1	0.131
0	0.163	1	0.072
0	0.106	1	0.112
0	0.147	1	0.109
0	0.161	1	0.141
0	0.096	1	0.134
0	0.078	1	0.115
0	0.155	1	0.133
0	0.161	1	0.115
0	0.142		
0	0.159		
0	0.156		
0	0.116		
0	0.2		

(a) Using what you know about the Analysis of Variance, is there any difference in friction that differentiates the two different types of soles from each other?

i. What Hypotheses are you testing?

We are testing to see if the oil-resistant soles has different friction than soles without the oil-resistant formula. This is a 2-tailed hypothesis, because I don't know which direction the analysis should predict (I really do, but I am going to do this as a 2-tailed test. You might want to run it as a 1-tailed test to see the difference).

ii. What alpha should you choose for this analysis?

This doesn't seem to have any big problems should I make a Type I error, so I will choose $\alpha = 0.05$.

Analysis of Variance

Source	Sum Sq.	d.f.	Mean Sq.	F	Prob>F
X1	0.00391	1	0.00391	4.4	0.0482
Error	0.01865	21	0.00089		
Total	0.02256	22			

Hierarchical (Type II) sums of squares.

The ANOVA table returned from Matlab shows that the two groups are significantly different from each other in their grip on the steel with glycerine test surface.

	$\overline{x}$	s
Not Oil Resist	0.1447	0.0342
Oil Resist	0.1180	0.0207

We can see from the means, that the soles without the oil resistant formula have a significantly greater coefficient of dynamic friction than the oil resistant soles. The effect size is quite large: about 1 standard deviation.

(b) Are there any violations of the assumptions of ANOVA in this analysis?

i. If so, how do you propose to deal with those violations?

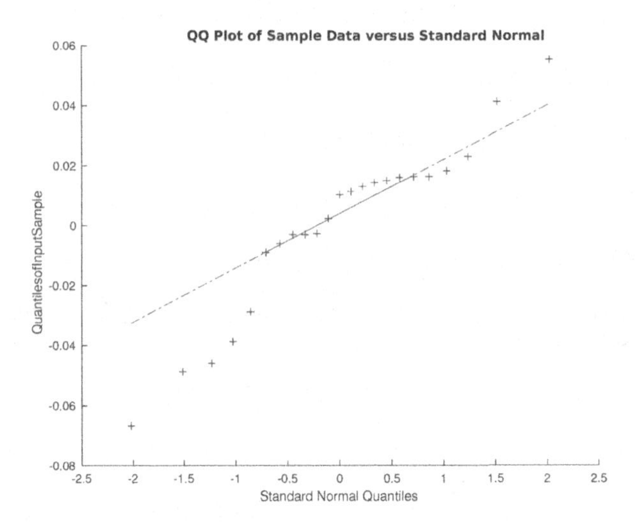

This doesn't look very normal! The input samples' quantiles appear to be lower in the lower region of the Q-Q plot, than predicted using the Standard Normal Distribution. What to do about it is another thing. I might want to transform the data in some way, but I have no good understanding of what the appropriate transformation function should be. Plus, ANOVA is robust to violations of the assumption of normality, so I am confident that I haven't committed a Type I error. Further, the standard deviations of the samples appear to be about the same, so homogeneity of variance looks to be okay. I think, in this case, I will go with my analysis as-is.

(c) **What is the goodness of fit of your analysis?**

i. **Does this analysis explain a lot of the variation in dynamic coefficient of friction?**

$R^2 = \frac{SS_{\text{between groups}}}{SS_\varepsilon} = \frac{0.1447}{0.1180} = 0.1733.$ This is not a lot of the total variation in dynamic friction. Only 17%. But, it is more than I could predict had I not done the analysis. Perhaps, if I were wanting to design oil-resistant soles that didn't lose their grip on slick, oily surfaces, I might look at the tread design, or other design factors.

2. **Explain what is meant by a single independent variable with two *levels* for the Analysis of Variance.**

Levels refer to the number of groups within a categorical variable. In the boot sole analysis, we had one independent variable: Type of Sole. That variable had two levels: 1) Without oil resistance, and 2) With oil resistance.

3. **Describe how we can use estimates of variance to test differences in group means.**

Variance is just a measure of distance. So, if we have two different group means (or as many as we want to analyze), we can just measure their difference with the grand mean, square that difference, and divide by the degrees of freedom, making a variance a measure of the distance between groups. Comparing that measure against a measure of the variation we expect under the Null Hypothesis, MS_ε the random error of prediction using The F-ratio, we can tell if the distance measured is significantly different from random chance.

4. **In the ANOVA table below, fill in the missing information.**

(a) **As you work through the table, try to understand how the calculations make sense from the General Linear Model.**

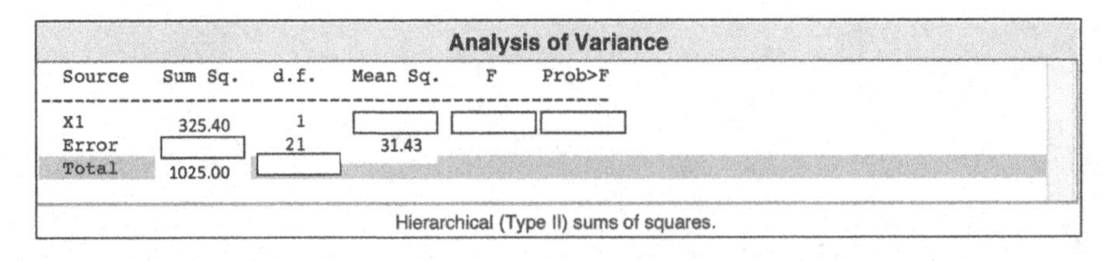

Analysis of Variance

Source	Sum Sq.	d.f.	Mean Sq.	F	Prob>F
X1	325.40	1			
Error		21	31.43		
Total	1025.00				

Hierarchical (Type II) sums of squares.

$SS_\varepsilon = 1025 - 325.40 = 699.60$

$df_{total} = 1 + 21 = 22$

$MS_{\text{between groups}} = \frac{325.40}{1} = 325.40$

$F : \frac{325.40}{31.43} = 10.353$

$p(F_{(1,21)} \geq 10.353)$

5. The risk of concussion in college and professional sports is a growing concern. To this end, you are studying the incidence of concussion in college athletes who are not football players. Covassin, et al., (2003) collected the following data over three years. Both men and women athletes were studied across 5 sports, and the number of concussions was recorded.

Sport	**# Concussions**	**Sport**	**# Concussions**
Soccer	51	Lacrosse	12
Soccer	47	Lacrosse	7
Soccer	60	Lacrosse	7
Soccer	34	Lacrosse	19
Soccer	27	Lacrosse	15
Soccer	40	Lacrosse	17
Basketball	16	Soft/Baseball	9
Basketball	30	Soft/Baseball	10
Basketball	26	Soft/Baseball	28
Basketball	8	Soft/Baseball	22
Basketball	21	Soft/Baseball	6
Basketball	20	Soft/Baseball	25
Gymnastics	1	Gymnastics	0
Gymnastics	0	Gymnastics	0
Gymnastics	0	Gymnastics	0

(a) What Hypothesis makes the most sense for this study?

For this one, I don't have a good way to order all of the sports in terms of their violence to the head. So I will opt for 2-tailed hypotheses.

(b) What alpha should you choose for this analysis, given the results will be used to design protective equipment and policies to protect athletes?

I would choose a fairly conservative Type I error rate: $\alpha = 0.01/6 = 0.0017$.

(c) Compute the ANOVA table.

i. What does the degrees of freedom between groups mean?

ii. Why is the degrees of freedom of the residual error only 25?

Analysis of Variance

Source	Sum Sq.	d.f.	Mean Sq.	F	Prob>F
X1	5896.53	4	1474.13	23.26	4.07477e-08
Error	1584.67	25	63.39		
Total	7481.2	29			

Hierarchical (Type II) sums of squares.

The degrees of freedom between groups is 4. This represents the 5 different sports analyzed, minus 1. The df_{ε} is equal to 25, because there are 30 total observations, and 5 sports being analyzed: $(30 - 1) - (5 - 1) = 25$

(d) Are there any violations of the assumptions of ANOVA in this analysis?

i. If so, how do you propose to deal with those violations?

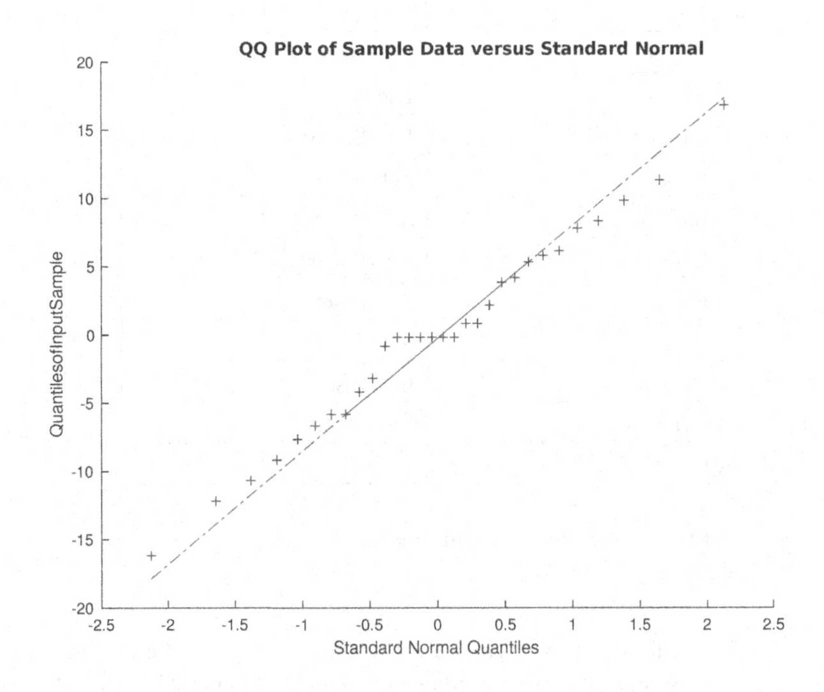

The Q-Q plot of the residual errors looks just about perfect. I am happy that I haven't egregiously violated the assumption of normality. The standard deviations of the different samples look like this:

Soccer	Basketball	Gymnastics	Lacrosse	Soft/Baseball
11.9568669	7.704	0.408	5.076	9.416

Gymnastics looks way out of bounds here. Looking at the data, in 6 years, there was only one recorded concussion. If people were honest in their reporting, this is a very small number, so we might consider the data for gymnastics to violate homogeneity of variance when compared to the other sports. We could eliminate it, knowing it is an anomaly. Redoing the analysis without Gymnastics yields the following ANOVA table:

Analysis of Variance

Source	Sum Sq.	d.f.	Mean Sq.	F	Prob>F
X1	3348.12	3	1116.04	14.09	3.63143e-05
Error	1583.83	20	79.19		
Total	4931.96	23			

Hierarchical (Type II) sums of squares.

Eliminating gymnastics makes our overall model have worse fit than when we had included it. Its mean is so much lower than the others (in the tail of the distribution), it served as a kind of "anchor" that pulled the overall model in its direction. We know that gymnastics is a special case, so by eliminating it, we can look at the more subtle distinctions among the other sports without violating homogeneity of variance. The Q-Q plot of the residuals looks very similar to the one reported earlier. I am still happy that my distributions are fairly normally distributed.

(e) What is the goodness of fit of your analysis?

i. Does this analysis explain a lot of the variation in the incidence of concussion?
$R^2 = \frac{3{,}348.12}{4{,}931.96} = 0.679$. Over 2/3 of the variation in concussion frequency is accounted for by the sport in which one plays.

(f) Compute post hoc comparisons for the different sports. Use both Bonferroni and Scheffé.

i. Are there any differences in the comparisons? Does one method have a more conservative confidence interval than the other?

ii. Given these results, which of the sports would you choose to focus on in developing protective equipment? Which appear to be fairly safe with respect to the potential of concussion?

iii. Given what you know about college athletics, what could be potential sources of error for reporting concussions for these sports?
Here are the results using Dunn-Bonferroni (Table 12.4):

Group$_1$	Group$_2$	99% CI Lower Bound	Mean Difference	99% CI Upper Bound	p
1	2	7.961	23.000	38.039	0.001
1	3	15.294	30.333	45.372	5.369×10^{-5}
1	4	11.461	26.500	41.539	0.000
2	3	−7.706	7.333	22.372	1
2	4	−11.539	3.500	18.539	1
3	4	−18.872	−3.833	11.206	1

We can see that Group 1 (Soccer) has a significant difference with the three other sports: Group 2 (Basketball), Group 3 (Lacrosse), and Group 4 (Soft/Baseball). These other three sports do not differ significantly from each other in the number of concussions reported per year. How do I know this? First, the confidence intervals for Soccer, compared with each of the other sports do not span zero, the expected value of the difference in means under the null hypothesis. Second, I can look at the probability for each comparison and see that the values for Soccer, compared with each of the other sports is less than $\alpha = 0.01/6 = 0.0017$.

Here are the results using Scheffé:

Group$_1$	Group$_2$	99%CI Lower Bound	Mean Difference	99%CI Upper Bound	p
1	2	7.336	23.000	38.664	0.003
1	3	14.669	30.333	45.998	1×10^{-4}
1	4	10.836	26.500	42.164	0.001
2	3	−8.331	7.333	22.998	0.575
2	4	−12.164	3.500	19.164	0.925
3	4	−19.498	−3.833	11.831	0.905

You can see that the decision you would make if you used Scheffé is the same in this instance. But the confidence intervals are wider, and the p-values larger, making it more conservative than Dunn-Bonferroni.

6. **Islands-in-the-sea is a type of bi-component fiber where strands of one polymer are dispersed in the matrix of another polymer. The strands are known as islands and the matrix is the sea. Manufacturers use water jet pressure to aid in the hydroentanglement process.** Ndaro et al., (2007) studied this process for 6 different levels of water pressure (60, 80, 100, 120, 150, and 200 bars). They recorded the tensile strength of Islands-in-the-sea fibers (in N/5 cm) for each of these levels. They analyzed the data using analysis of variance. Reconstructed data from their study is presented below:

Water Pressure (bar)	Tensile Strength (N/5 cm)	Water Pressure (bar)	Tensile Strength (N/5 cm)
60	225.6	120	234.05
60	189.25	120	293.08
60	245.86	120	299.33
60	284.25	120	319.85
60	281.34	120	300.79
80	294.22	150	265.53
80	250.71	150	262.88
80	272.36	150	367.48
80	287.13	150	280.29
80	262.89	150	274.13
100	318.21	200	278.55
100	249.14	200	360.15
100	238.34	200	323.82
100	298.36	200	373.39
100	312.46	200	273.9

(a) Critique their use of Analysis of Variance as opposed to treating Water Pressure as a continuous variable.

i. What do they gain by using ANOVA?

If the company were only interested in these fixed levels of water pressure (e.g., their machines only put out those pressures), then they could determine which might be optimal.

ii. What do they lose?

Because the levels of water pressure are not ordered, in ANOVA the distances between them cannot be taken into account like it would be if they were treated as a continuous random variable as we did with Multiple Regression. So by using ANOVA, they lose a bit of sensitivity if their intent is to gain insight into the global relationship between water pressure and tensile strength. It is not inappropriate to use ANOVA like this, it is robust with respect to Type I error rate. It just loses a bit of power when the independent variable is collapsed from a ratio/interval scale to a nominal scale. In this case, the ANOVA returns a non-significant result ($F_{(5,24)} = 2.25, p = 0.08$). Multiple regression returns a significant overall linear model ($F_{(5,24)} = 11.09, p = 0.002$).

(b) What Hypothesis makes the most sense for this study?

I would run this as a Regression model to take advantage of the continuous nature of the independent variable. If the researchers are only interested in the fixed levels of the independent variable, then a hypothesis that uses ANOVA would be:

H_0 : All levels of water pressure show the same tensile strength

H_1 : At least one of the levels of water pressure has a different tensile strength than one of the others.

(c) **What alpha should you choose for this analysis, given the results will be used to design protective equipment and policies to protect athletes?**

I would run this at $\alpha = 0.01$ because a Type I error may end up with equipment that doesn't improve the performance of the safety equipment.

(d) **Should you use a 1-tailed or 2-tailed ANOVA given what you know about these kinds of fibers?**

I would use a two-tailed test here. But a person who knew more about the dynamics of composite layout might use a 1-tailed test.

(e) **Compute the ANOVA table.**

Analysis of Variance

Source	Sum Sq.	d.f.	Mean Sq.	F	Prob>F
X1	15597.6	5	3119.51	2.25	0.0822
Error	33295.9	24	1387.33		
Total	48893.4	29			

Hierarchical (Type II) sums of squares.

(f) **Are there any violations of the assumptions of ANOVA in this analysis?**

i. **If so, how do you propose to deal with those violations?**
The Q-Q plot looks very nice and linear. The standard deviations of the samples are not widely divergent, so I think we are okay.

(g) **What is the goodness of fit of your analysis?**

i. **Does this analysis explain a lot of the variation in the tensile strength of the fibers?**
$R^2 = 0.319$. About 32% of the variation in tensile strength is explained by the levels of water pressure. This is not very much, and with the small sample sizes, we may be underpowered in this test. Incidentally, using Multiple Regression did not improve the goodness of fit. It merely was more sensitive in determining the overall F-test of the slope.

(h) **Compute post hoc comparisons for the different sports. Use both Bonferroni and Scheffé.**

i. Are there any differences in the comparisons? Does this matter in this analysis?

ii. Given these results, and the results of your omnibus test, what conclusion do you have about the impact of water pressure on tensile strength of islands-in-the-sea fibers?

Normally, because the omnibus test returned non-significant results, we wouldn't follow up with post hoc contrasts. But for purposes of teaching, it is useful to confirm the findings of the omnibus test with contrasts.

Here are the results using Dunn-Bonferroni:

Group_1	Group_2	99% CI Lower Bound	Mean Difference	99% CI Upper Bound	p
1	2	−104.96	−28.202	48.56	1
1	3	−114.80	−38.042	38.72	1
1	4	−120.92	−44.16	32.60	1
1	5	−121.56	−44.802	31.96	1
1	6	−153.46	−76.702	0.06	0.05
2	3	−86.60	−9.84	66.92	1
2	4	−92.72	−15.958	60.80	1
2	5	−93.36	−16.6	60.16	1
2	6	−125.26	−48.5	28.26	0.76
3	4	−82.88	−6.118	70.64	1
3	5	−83.52	−6.76	70.00	1
3	6	−115.42	−38.66	38.10	1
4	5	−77.40	−0.642	76.12	1
4	6	−109.30	−32.542	44.22	1
5	6	−108.66	−31.9	44.86	1

We can see that all 99% confidence intervals span zero, and that all probabilities are above $\frac{\alpha}{2} = 0.025$ for our two-tailed test. Just doing the omnibus test first eliminates the need to follow up if non-significant results are found. One contrast, that between 60 bar and 200 bar pressures has the narrowest confidence interval. If I were to continue this investigation, I would definitely increase the domain of the pressure variable to include values over 200 bar, keeping the lower values in this analysis (e.g., 60, 80, 100 bar).

Here are the results using Scheffé:

$Group_1$	$Group_2$	99% CI Lower Bound	Mean Difference	99% CI Upper Bound	p
1.00	2.00	−113.47	−28.20	57.07	0.92
1.00	3.00	−123.31	−38.04	47.23	0.76
1.00	4.00	−129.43	−44.16	41.11	0.63
1.00	5.00	−130.07	−44.80	40.47	0.61
1.00	6.00	−161.97	−76.70	8.57	0.10
2.00	3.00	−95.11	−9.84	75.43	1.00
2.00	4.00	−101.23	−15.96	69.31	0.99
2.00	5.00	−101.87	−16.60	68.67	0.99
2.00	6.00	−133.77	−48.50	36.77	0.53
3.00	4.00	−91.39	−6.12	79.15	1.00
3.00	5.00	−92.03	−6.76	78.51	1.00
3.00	6.00	−123.93	−38.66	46.61	0.75
4.00	5.00	−85.91	−0.64	84.63	1.00
4.00	6.00	−117.81	−32.54	52.73	0.86
5.00	6.00	−117.17	−31.90	53.37	0.87

Like before, the results of this analysis are consistent with the omnibus test. The confidence intervals are wider than Dunn-Bonferroni, indicating that it is more conservative, typically.

7. The following analysis was taken from a study of car collisions and the pressure experienced by crash-test dummies to the back of the head for small, mid-sized, and large sedans.

ANOVA

Source of Variation	*SS*	*df*	*MS*	*F*	*P-value*
Between Groups	86049.55556	2	43024.78	25.17541	0.001207
Within Groups	10254	6	1709		
Total	96303.55556	8			

(a) What does 2 degrees of freedom Between Groups mean?

There are three different sedan sizes. So there are 3 − 1 = 2 degrees of freedom.

(b) What does 8 total degrees of freedom mean?

9 total cars were evaluated. 9 − 1 = 8 total degrees of freedom.

(c) **Interpret the MS Within Groups in this table.**

The MS_{within} is an estimate of the residual (error) variance of the dependent variable, pressure. It is found by dividing the SS_{within} by df_{within}.

(d) **How much of the total variation in pressure to the back of the head, do the different experimental conditions explain?**

$$R^2 = \frac{SS_{between}}{SS_{total}} = \frac{86049.56}{96303.56} = 0.894$$

This means that about 90% of the variation in pressure is accounted for by the variation between the three experimental groups. This is a good fitting analysis.

(e) **What does the F-ratio mean?**

$F_{(2,6)} = 25.18$ is very high. It means that the variation explained by the distance between groups is about 25 times the error variation. This is an indication that the between groups variation is wide enough to be considered unlikely due to random chance. We can confirm this by looking at the tiny p-value, $p = 0.001$ So one would expect, if I performed this experiment 1000 times, only 1 time would I expect to get mean differences this large or larger.

(f) **Should you compute post hoc comparisons for this experiment? How do you know yes or no?**

Yes. Because the omnibus test returned a significant value of F, that means at least one of my experimental groups is different from one of the others.

8. Describe what the $SS_{\text{Between Groups}}$ means.

(a) **Explain the computational formula**

$SS_{\text{Between Groups}}$ is the total distance groups are from the grand mean (the mean of all the values of the dependent variable). It can be thought of as the "effect" of the groups on the expression of the dependent variable. The computational formula: $\sum_{k=1}^{K}\left(\overline{y}_{X_k} - \overline{\overline{y}}\right)^2$ shows that it is the numerator of a variance. If $\overline{\overline{y}}$ is the expected value of the whole vector of values of the dependent variable, then each deviation from that grand mean can be considered the variation of a single group.

9. Describe what the $SS_{\text{Within Groups}}$ means.

(a) **Explain the computational formula**

$SS_{\text{Within Groups}}$ is the residual error of the ANOVA model. It is the amount of variation in the dependent variable that we are not able to account for with the dependent variable. Its computational formula $SS_{\varepsilon} = \sum_{i=1}^{n}\left(y_i - \overline{y}_{X_{ki}}\right)^2$ shows that it is the total squared residual error (error of prediction) for each value of the dependent variable within its own group.

10. Bouacha and colleagues (2010) present data on the impact of cutting speed of a Cubic Boron Nitride (CBN) Tool, on the roughness of extremely hard steel used to manufacture bearings.

Cutting Speed (m/min)	Average Absolute Roughness (mm)	Cutting Speed (m/min)	Average Absolute Roughness (mm)
125	0.37	246	0.19
125	0.37	246	0.2
125	0.35	246	0.21
125	0.63	246	0.36
125	0.64	246	0.38
125	0.66	246	0.39
125	0.77	246	0.45
125	0.75	246	0.47
125	0.73	246	0.46
176	0.24		
176	0.24		
176	0.25		
176	0.41		
176	0.44		
176	0.43		
176	0.55		
176	0.55		
176	0.56		

(a) Perform an Analysis of Variance on this Data. Go through the whole process.

Given I know quite a bit about manufacturing parts of this nature, my Alternative Hypothesis is that faster feed rates will end up with smoother surfacing, as opposed to slower feed rates. So I will make a one-tailed, left-tailed hypothesis:

H_0 : *Groups with slower cutting speeds will result in bearings with surfaces equal to or less than groups with higher cutting speeds.*

H_1 : *Groups with slower cutting speeds will result in bearings with surfaces rougher than groups with higher cutting speeds.*

The consequences of making a Type I error are not too dire, so I will use $\alpha = 0.05$.

A scatterplot of the data reveals some apparent curvilinearity in the pattern. This looks like, as speeds get faster, the roughness gets asymptotically less. I may have to transcribe this data to be able to fit the linear model for ANOVA (remember *Its just regression!*).

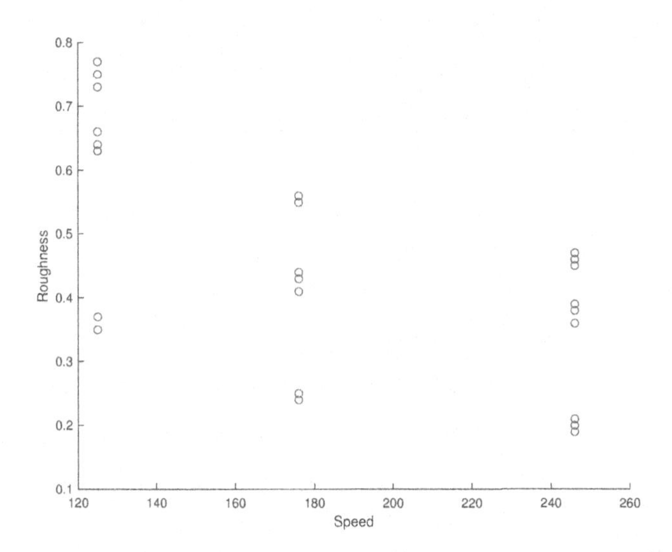

For now, for illustrative purposes, I am going to go ahead as if I didn't know this fact...

After entering the data into two column vectors, Matlab return the following ANOVA table:

Analysis of Variance

Source	Sum Sq.	d.f.	Mean Sq.	F	Prob>F
X1	0.27923	2	0.13961	6.79	0.0046
Error	0.4938	24	0.02058		
Total	0.77303	26			

Hierarchical (Type II) sums of squares.

This tells me that at least one of my speed groups is significantly different from at least one other. I now have to perform a post hoc test to see which one(s) are significantly different from the other(s). I chose the Dunn-Bonferroni correction as my method here, because I don't need to be too conservative, given the consequences of making a Type I error, and I want the extra power it gives over Scheffé

$Group_1$	$Group_2$	95% CI Lower Bound	Mean Difference	95% CI Upper Bound	*p*
1	2	0.004	0.178	0.352	0.044
1	3	0.066	0.240	0.414	0.005
2	3	−0.112	0.062	0.236	1

This tells me that the group of specimens manufactured at 125 m/min (Group 1) show significantly higher roughness than those manufactured at 176 (Group 2) or 246 (Group 3) m/min. There is a non-significant difference between 176 and 246 m/min conditions. I know this because in the first two rows of the results table, the 95% confidence intervals do not span zero, the hypothesized difference in means if cutting speed made no effect. The probabilities for these first two comparisons are within the rejection region given $\alpha = 0.05$. The final row, comparing 176 and 246 m/min conditions shows non-significant differences because the 95% confidence interval does span zero, and the probability that the difference could be due to random chance is very close to certainty.

(b) **There is at least one violation of the assumptions of Multiple Linear Regression in this analysis. Work through what it might be and make appropriate corrections.**

I already have a feeling that the assumption of linearity of the model has been violated by looking at the scatterplot of the raw data. I am going to double-check the residuals to see if we have violated the assumption of normality:

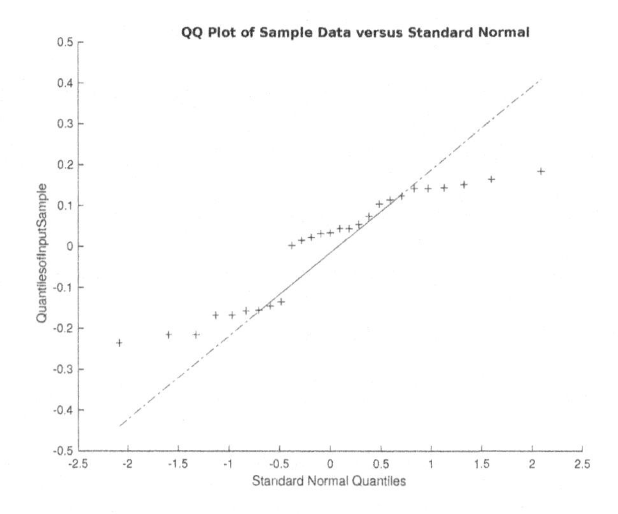

This looks pretty wonky. I think we don't have a good case for a normally distributed dependent variable across all levels of the independent variable. So, I think we might have to transform the relationship. Since this looks like an inverse proportion,–a power function, I will use what is called a "Box-Cox" transformation for power functions. I provided the basic form for this transformation back in Chapter 10: Simple Linear Regression. The way to approach this is to assume the function of the raw data has the form $y = x^a$, where a is a Real, exponent ranging from -5 to $+5$. These exponents are usually restricted to the following: $(3, -2, -1, -0.5, +0.5, 1, 2, 3)$. Then, the inverse of $y = x^a$, $y' = \frac{1}{y^a}$, can be used as the transformation. A procedure for linearizing a power function is to start with the appropriately signed exponent closest to zero, and then working outwards towards exponents with the largest magnitude until the data is sufficiently straightened. I did this with our data and found $y' = \frac{1}{y^2}$ to straighten out the data enough, given we only have three values in the domain to work with.

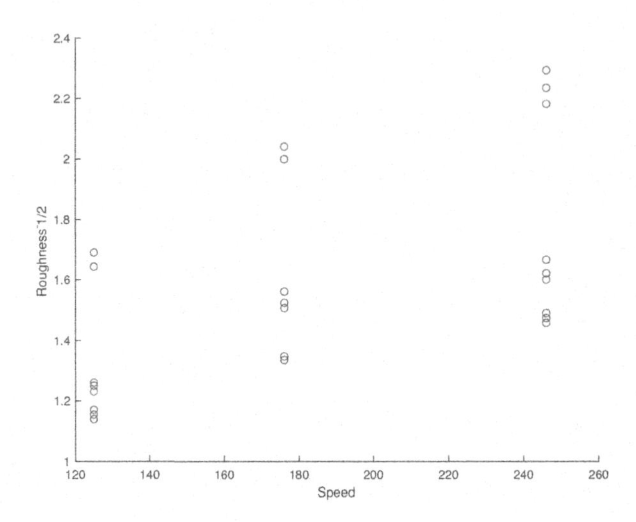

(c) Is there any relationship between cutting speed and absolute roughness?

After transforming the data, I ran the ANOVA again with y' as our new dependent variable:

Analysis of Variance

Source	Sum Sq.	d.f.	Mean Sq.	F	Prob>F
X1	0.84738	2	0.42369	4.69	0.0191
Error	2.17011	24	0.09042		
Total	3.0175	26			

Hierarchical (Type II) sums of squares.

(d) What is the goodness of fit of your analysis?

$R^2 = \frac{0.84738}{3.0175} = 0.28$. Only about 30% of the variation in *roughness*$^{-\frac{1}{2}}$ of the specimens can be attributed to cutting speed. On the other hand, only varying one thing in the manufacturing process, cutting speed, can dramatically improve the smoothness of the finished material up to a point...

(e) Do you need to compute post hoc comparisons for levels of cutting speed?. Use the method that is appropriate for this study, given its exploratory nature, and the fact that there are some violations of assumptions.

i. Given these results, is there any cutting speed that results in significantly different roughness from at least one of the others?

I will again use the Dunn-Bonferroni procedure.

$Group_1$	$Group_2$	95% CI Lower Bound	Mean Difference	95% CI Upper Bound	*p*
1	2	−0.645	−0.281	0.084	0.178
1	3	−0.792	−0.427	−0.062	0.018
2	3	−0.511	−0.146	0.219	0.937

Now, it doesn't look like there is a significant difference between the two lowest cutting speeds, but there is a significant difference in $roughness^{-\frac{1}{2}}$ between the lowest (125 m/min) and highest (246 m/min) speeds. If I were to use these results, I would definitely want to increase my cutting speed to at least 246 m/min to take advantage of this difference in quality of the manufactured specimens.

13

The General Linear Model: Randomized Block Factorial ANOVA

13.1 It is All Just Regression

The General Linear Model is powerful, because it is flexible, describing the relationship between a dependent (outcome, response) variable, and any number of independent (experimental, predictor) variables as a linear combination. This simplicity and flexibility make it an ideal conceptual model for you to put in the dusty file cabinets of your brain, so that (God forbid!) if you forget something you read or learned in this book you can easily reconstruct it from a few first principles when you need to use it to make some data-based decisions on the job.

The *t*-test and the Analysis of Variance are good models to start with, but sometimes we have several independent variables in which we are interested in testing their impact on our outcome. But we are in luck! Because of the flexibility of the General Linear Model, more complicated experimental hypotheses are hardly more difficult than a *t*-test to analyze. As we saw in Multiple Linear Regression, when we add independent variables (factors), the linear algebra *doesn't change at all.* Post hoc tests are a bit different, but we *use the same logic* as we have seen before with Dunn-Bonferroni and Scheffé. We will start with the simplest **Factorial Design**, the **Randomized Block Design**, using a simple dataset to illustrate the logic of the model, and then progress to the more complex Full Factorial Design that includes interactions among its independent variables.

> A **Factor** is a categorical (nominal scale) independent variable in an Analysis of Variance. It may have several **levels**, each of which is a different category within a factor. For example, if bullet shape is a *factor*, types of bullet shapes (Spitzer, boat-tail, etc.) are its *levels*.

As we are getting to more advanced models, there is a convention for naming the type of design you are using (See Table 13.1). In this convention, a 2 x 3 factorial ANOVA is a model with two independent variables, the first has 2 levels (like type of material: Aluminum or Steel) and the second has 3 levels (like Shape: round, elliptical, eccentric). A 2 x 2 x 4 factorial, likewise, has three independent variables, with 2, 2, and 4 levels respectively.

DOI: 10.1201/9781003094227-13

Number of Factors	Number of Dependent Variables	Number of Levels in Factors	Modeling Interactions among Factors	Blocking?	Type of Analysis
1	1		No	N	Oneway ANOVA
2	1		No	N	2-way ANOVA
n	1		No	N	n-way Factorial ANOVA
2	1	2 in each	No	Yes	2x2 Randomized Block Factorial ANOVA
3	1	n in the first, m in the second, p in the third	No	N	n x m x p Factorial ANOVA
n	1	n in the first, m in the second, p in the third... etc	Yes	N	n x m x p x... Full Factorial ANOVA
n	1	"	Yes	Yes	n x m x p x... Randomized Block Factorial ANOVA with Interactions
n	n	"	Yes or No	Yes or No	Multivariate ANOVA (MANOVA)

TABLE 13.1: Naming conventions for ANOVA with multiple independent variables

13.2 Randomized Block ANOVA

If you remember back in Chapter 6, we described this as one of our basic 10 ways to design experiments. The basic idea behind the Randomized Block Design is to take existing strata in the population we are interested in studying and take separate random samples from levels of each stratum. For example, I have been a builder of custom longbows for many years. There is an important relationship between the bending moment of the arrows I use, and the spring constant of the bow I shoot. If the arrows are too stiff, they tend to fly off to the left. If the arrows are a bit too "noodly" there is lots of random behavior as they bend first one way, and then the other as they go around the bow and continue towards the target. So I am very interested in this relationship if I am to properly engineer arrows that are accurate and precise. But arrows can be made out of many different materials such as wood (and different varieties of wood), fiberglass, aluminum, or carbon fiber. These materials all have different resistance to vibration, different densities, and different elastic moduli, that may influence how they perform, in addition to the "noodliness" (called "spine") of the arrows made from them.

This difference in behavior might add some "nuisance" variation that could mask, or **confound** the effect of the spine of the arrows. So, I took a set of wooden arrow shafts at random from the population. Then I took a set of carbon arrow shafts at random from the population (my stack of arrows in my workshop). I then randomly assigned shafts from these different materials into experimental conditions: Spine = 45 lbs and Spine = 65 lbs. I trimmed the length of the shafts and sanded them, measuring their spines to the nearest pound to ensure that experimental conditions were internally consistent. Figure 13.1 presents is a model of the two Factors and how I hypothesized they would predict Arrow Accuracy: Each of the variables, the primary independent variable, Spine, and the Blocking variable, Material are expected to contribute to Arrow Accuracy with some significant β .

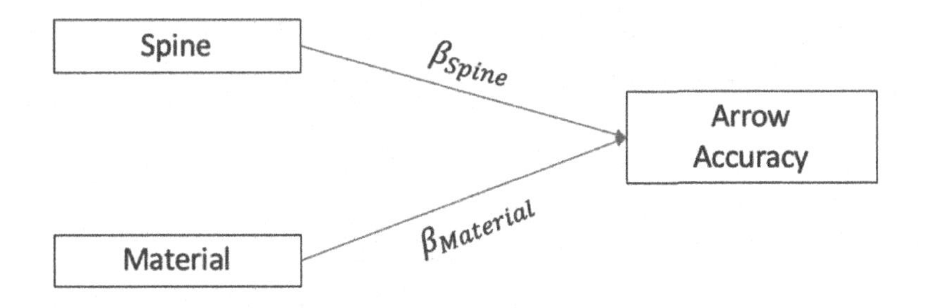

FIGURE 13.1: Hypothesized effects of spine and material on arrow accuracy

FIGURE 13.2: Testing the accuracy of arrows

THEN, I got to do the fun part! I shot all 20 arrows, using my best form, trying to aim exactly the same way, with the same stance, at the same spot on the target. I measured the distance each arrow ended up from the center of my target 20 yards away. My results are presented in Table 13.2:

Block (Material. Wood = 0, Carbon = 1)	**Factor 1** (Spine. 65 lbs = 0, 45 lbs = 1)	**Outcome** (Distance from Center (in))
0	0	2.16
0	0	2.86
0	0	2.90
0	0	1.37
0	0	1.58
0	1	2.60
0	1	2.53
0	1	2.93
0	1	3.26

0	1	3.37
1	0	2.36
1	0	3.37
1	0	3.20
1	0	3.37
1	0	1.32
1	1	2.51
1	1	4.20
1	1	3.84
1	1	5.00
1	1	4.83

TABLE 13.2: Data from an experiment testing the effect of arrow spine on accuracy, factoring in arrow material as a blocking variable

Figure 13.3 shows the randomized block design I used to test whether or not the Spine = 65 lbs arrows are more accurate than the Spine = 45 lbs arrows. The important thing to notice is that I took, essentially, *two* random samples, one from the population of wood arrow shafts, and another from the population of carbon arrow shafts. I then randomly assigned those different shafts into two experimental conditions: Spine = 45 and Spine = 65. Because of this, under the null hypothesis I would expect the performance of my arrows under each experimental condition, overall, to be equivalent. But because I have another factor measured, material, I can compute the variation it contributes to the accuracy of my arrows, and subtract this variation from the total, making my overall error of measurement less than if I did not take arrow material into account. Let's see how this works in practice:

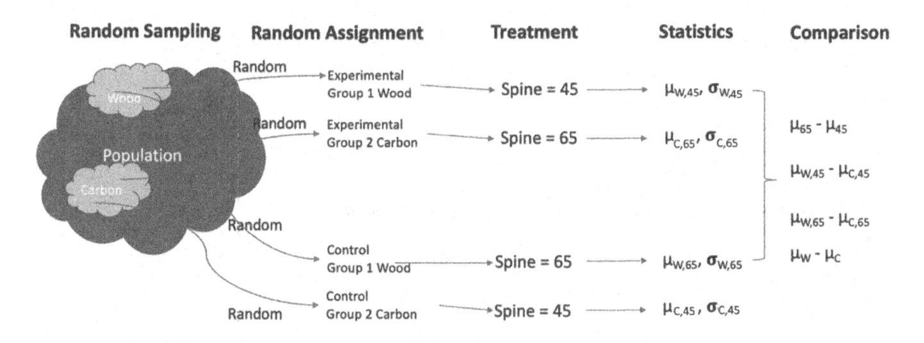

FIGURE 13.3: Randomized block design for arrow spine experiment

As we have established, Multiple Linear Regression allows us to determine the relationship between an outcome variable, y, and a set of independent predictors, $x_1, x_2, \ldots x_k$. This linear relationship doesn't matter if the independent variables are continuous or categorical in nature. When they are categorical, we call the regression analysis the Analysis of Variance, but it is *just regression.* We model the variation in this relationship as a simple linear combination:

$$SS_{Total} = SS_{regression} + SS_{\varepsilon}$$

$$SS_{Total} = \left(Y - \overline{Y}\right)^T \left(Y - \overline{Y}\right) = \left(X\beta - \overline{Y}\right)^T \left(X\beta - \overline{Y}\right) + (Y - X\beta)^T (Y - X\beta)$$

$$SS_{regression} = \left(X\beta - \overline{Y}\right)^T \left(X\beta - \overline{Y}\right) = \sum_{i=1}^{k} (\overline{y}_i - \overline{y}_.)^2$$

$$SS_{\varepsilon} = (Y - X\beta)^T (Y - X\beta) = \sum_{i=1}^{n} (y_i - \overline{y}_.)^2 - (\overline{y}_i - \overline{y}_.)^2$$

$$SS_{Total} = \left(Y - \overline{Y}\right)^T \left(Y - \overline{Y}\right) = \sum_{i=1}^{n} (y_i - \overline{y}_.)^2$$

If we added another independent variable to our matrix of predictors X, this just adds another term to the prediction equation, because as a linear combination, each variable added must be orthogonal to the others:

$$\hat{Y} = X\beta = \beta_0 + \beta_1 X_1 + \beta_2 X_2$$

This makes *two* variance components that constitute the $SS_{regression}$ terms: SS_{X_1} and SS_{X_2}.

$$SS_{regression} = SS_{X_1} + SS_{X_2}$$

In our example we would model the variation thusly:

$$\hat{y}_i = \beta_0 + \beta_1 Spine_i + \beta_2 Material_i$$

So, splitting the $SS_{regression}$ into the contribution (or *effect*) of Arrow Spine and Arrow Material, we can see that the two variables, because they are orthogonal, each add a separate component to the estimation of the variation in the dependent variable.

13.2.1 A Quick Note on Notation

In representing the matrix algebra in summation notation, we run into the problem of dimensions. When we have rows representing events, and columns representing variables, the summation for $X^T X$, the sum of the squares of our predictor matrix, is easy. The formula doesn't change as we add rows (new measured events) or columns (the measurements of those events). But summation, which is represented using conventional algebraic notation becomes very cumbersome. By convention, therefore, to keep the different columns of X in the matrices straight, we tend to show two or more variables, like arrow spine and material as 2-way tables when communicating the structure of the data (Table 13.3):

	Material	
Spine	**0 (carbon)**	**1 (wood)**
65 lbs	2.16	2.60
	2.86	2.53
	2.90	2.93
	1.37	3.26
	1.58	3.37
45 lbs	2.36	2.51
	3.37	4.20
	3.20	3.84
	3.37	5.00
	1.32	4.83

TABLE 13.3: Combinations of factors in a 2-way factorial ANOVA represented as a 2-way table Values in cells are measured distances from the point of aim

		Factor B					
	Levels	$j = 1$	$j = 2$	$\cdots$	$j = l$	Sum	Mean
Factor A	$i = 1$	$y_{111},\ y_{112} \cdots y_{11n}$	$y_{121},\ y_{122} \cdots y_{12n}$	$\cdots$	$y_{1b1},\ y_{1b2} \cdots y_{1bn}$	$y_{1..}$	$\overline{y}_{1..}$
	$i = 2$	$y_{211},\ y_{212} \cdots y_{21n}$	$y_{221},\ y_{222} \cdots y_{22n}$	$\cdots$	$y_{2b1},\ y_{2b2} \cdots y_{2bn}$	$y_{2..}$	$\overline{y}_{2..}$
	$\vdots$	$\vdots$	$\vdots$		$\vdots$	$\vdots$	$\vdots$
	$i = a$	$y_{a11},\ y_{a12} \cdots y_{a1n}$	$y_{a21},\ y_{a22} \cdots y_{a2n}$	$\cdots$	$y_{ab1},\ y_{ab2} \cdots y_{abn}$	$y_{a..}$	$\overline{y}_{a..}$
Sum		$y_{.1.}$	$y_{.2.}$	$\cdots$	$y_{.b.}$	$y_{...}$	
Mean		$\overline{y}_{.1.}$	$\overline{y}_{.2.}$	$\cdots$	$\overline{y}_{.b.}$		$\overline{y}_{...}$

TABLE 13.4: Convention for notation of factorial design, 2 x 2 example

Under this convention, if we added a third factor to the equation, the table would be 3d, with each combination of $Factor_i$, $Factor_j$, and $Factor_k$ represented by a cell *ijk* in the table (see Table 13.4). For our two-way example, the summations for Spine (*i*), and Material (*j*) follow standard conventions:

$$SS_{Total} = SS_{Spine} + SS_{material} + SS_{\varepsilon}$$

$$SS_{Spine} = b\sum_{i=1}^{a}(\bar{y}_{i..} - \bar{y}_{...})^2$$

The SS_{Spine} is the sum of the squared deviations from the grand mean for arrow spine, taken across each *column* in Table 13.4),

$$SS_{Material} = a\sum_{j=1}^{b}\left(\bar{y}_{.j} - \bar{y}_{..}\right)^2$$

The $SS_{Material}$ is the sum of the (sum of the squared deviations from the grand mean for arrow material, taken across each *row* in Table 13.4),

$$SS_{\varepsilon} = n\sum_{i=1}^{a}\sum_{j=1}^{b}\sum_{k=1}^{n}\left(\bar{y}_{ijk} - \bar{y}_{ij.}\right)^2$$

The SS_{ε} is just the sum of the (sum of the squared deviations of each value in a *cell* in the table and its *cell* mean), across each *cell* in Table 13.4, then subtracting out the grand mean which is counted twice, and finally:

$$SS_{Total} = \sum_{i=1}^{a}\sum_{j=1}^{b}\sum_{k=1}^{n}\left(\bar{y}_{ijk} - \bar{y}_{...}\right)^2$$

The SS_{Total} is just the sum of the squared deviations of each *measurement* from the grand mean across all the data—the numerator of the sample variance for the total sample.

This was all really confusing to me, notationally, when I first started studying statistics. What really helped was understanding that SS_{Spine}, for example *really is* just the **Sum of the Squared Deviations from the Mean** of the different spine conditions. The notational differences don't change the concept. The *i*s and *j*s are just ways of keeping track of calculations when you are doing it by hand, *i* and *j* (and *k*...) being conventions for denoting the different rows and columns of an *n-way* table. The matrix algebra is the same, and if you ever had way too much time on your hands, you would see that calculating the different sum of squares is exactly the same: *Compute the mean, subtract the mean from each measurement in a variable, square each of those differences, and then add them up.*

Wow, was I relieved 35 years ago when I learned that there was no real difference in the model when I added new factors to the prediction! It just adds another term to the linear equation, and the variation in the outcome variable the new variable(s) add just splits the $SS_{regression}$ into a set of individual effects. Computing the variance estimates for these components is just a matter of dividing each *SS* by its associated degrees of freedom.

$$MS_{Spine} = \frac{b \cdot \sum_{i=1}^{a} (\bar{y}_{i..} - \bar{y}_{...})^2}{a - 1}$$

$$MS_{Material} = \frac{a \sum_{j=1}^{b} \left(\bar{y}_{.j} - \bar{y}_{..}\right)^2}{b - 1}$$

$$MS_{\varepsilon} = \frac{n \sum_{i=1}^{a} \sum_{j=1}^{b} \sum_{k=1}^{n} \left(\bar{y}_{ijk} - \bar{y}_{ij.}\right)^2}{n - a - b}$$

13.2.2 Now Back to Analysis

Now let's do the ANOVA for this 2 x 2 Randomized Block Factorial Design experiment and see if 45 lbs spine or 65 lbs spine is better for my 65 lb draw weight bow. I am using the Matlab function anovan, the same one we did for the simpler Oneway ANOVA in Chapter 12. The output is in Figure 13.4.

```
» [p,table,stats]=anovan(distance,{spine,material},'sstype',2,'alpha',0.025);
```

Analysis of Variance

Source	Sum Sq.	d.f.	Mean Sq.	F	Prob>F
spine	5.5968	1	5.59682	9.41	0.007
material	3.5617	1	3.56168	5.99	0.0256
Error	10.111	17	0.59477		
Total	19.2695	19			

Hierarchical (Type II) sums of squares.

FIGURE 13.4: Randomized block factorial design ANOVA table

I ran this as a 2-tailed test, pretending I didn't know what the effect of spine really was. I listed 'alpha' in the command as 0.025 to account for the splitting of my 0.05 Type I error rate into the tails of the distribution. What I first notice is that spine really does seem to make a difference in the accuracy of my shots from the bow. The F is very high, and the probability that the mean difference represented by this F is about 7 in 1,000, meaning that, if I repeated this experiment 1000 times, I would only expect these results 7 times, if the difference in my accuracy with arrows of 45 lb spine and 65 lb spine was truly random.

The second thing I notice is that material predicts some of the variation in my accuracy. $SS_{material} = 3.5617$. This is about 20% of the total variation in the model, so accounting for it has made my model have overall less error of prediction.

$$SS_{Total} = SS_{Spine} + SS_{material} + SS_{\varepsilon}$$

To show how this works, I am going to re-run the analysis of spine, with the data for the block, material, eliminated (See Figure 13.5). This makes this new analysis just a Oneway ANOVA like we did in Chapter 12:

```
» [p,table,stats]=anovan(distance,{spine} ,'sstype',2,'alpha',0.025);
```

Analysis of Variance

Source	Sum Sq.	d.f.	Mean Sq.	F	Prob>F
Spine	5.5968	1	5.59682	7.37	0.0142
Error	13.6727	18	0.75959		
Total	19.2695	19			

Hierarchical (Type II) sums of squares.

FIGURE 13.5: Oneway ANOVA table for arrow data

$$SS_{Total} = SS_{Spine} + SS_{\varepsilon}$$

$$19.2695 = 5.5968 + 13.6727$$

Compare these results to those in the first analysis. The SS_{Total} is exactly the same! This shouldn't be too much of a surprise. SS_{Total} is the variation in each data point in the whole sample around the grand mean, $\bar{\bar{y}}$. We didn't do anything to the dependent variable, so this quantity has to remain the same. But additionally, our

SS_{Spine} is also exactly the same for the two analyses! Remember, all our independent variables are orthogonal to each other—they are linearly independent. So, the variation we can model with one factor is linearly independent from any other factor. What remains after we model SS_{Spine} in the Oneway ANOVA is SS_{ε}. *If we add another independent variable, like material, the variation it accounts for in the regression model has to come from SS_{ε}!* Adding a blocking variable really does reduce our error (see Figure 13.6)!

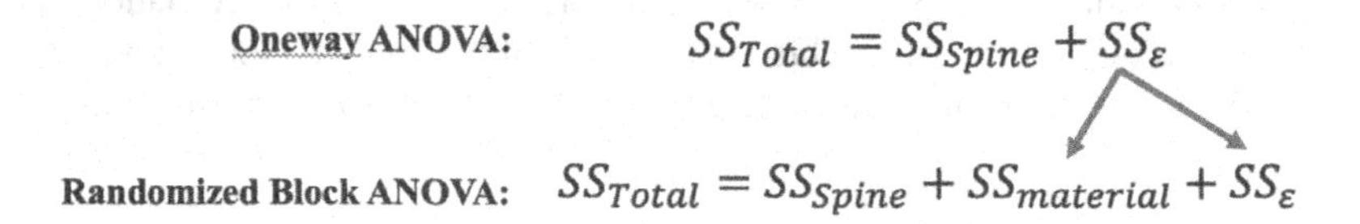

FIGURE 13.6: Adding a second independent variable partitions the SS_{ε} into two components: One describing the variation accounted for by the block, the second, the reduced SS_{ε}

You can check my math by adding $SS_{material} + SS_{\varepsilon}$ in the Randomized Block Factorial analysis and seeing that it equals SS_{ε} in the Oneway ANOVA.

$$19.2695 = 5.5968 + 3.5617 + 10.111$$

Recall, the reason I blocked my sample on material is that I suspected that there was some variation accounted for by the carbon shafts versus the wood shafts that I could use to reduce my overall error of prediction. If I calculate R^2 for this analysis, I find that for the Oneway ANOVA, it is $\frac{SS_{Spine}}{SS_{Total}} = \frac{5.5968}{19.2695} = 0.29$. For the Randomized Block ANOVA, however, $R^2 = \frac{SS_{Spine}+SS_{material}}{SS_{Total}} = \frac{5.5968+3.5617}{19.2695} = 0.48$! The addition of the Blocking variable really added to our goodness of fit.

Randomized Block Factorial ANOVA is an application of Multiple Linear Regression that accounts for the potential presence of a nuisance variable—one that might contribute variation to the performance of the outcome variable—that might mask or confound the effect being studied.

The process of blocking involves taking a stratified random sample from levels of the "nuisance variable" in the population, and then randomly assigning items from each stratum to experimental conditions.

The ANOVA calculations are exactly the same as in Oneway ANOVA; but by modeling the variation in the dependent variable due to the blocking variable, *you reduce the error of prediction in the model*, thus giving yourself more power to detect a real difference in the independent variable you are truly interested in.

The **efficiency of the Block**–the proportion it improves prediction–over a Oneway ANOVA or other completely randomized design, is:

$\frac{SS_{block}}{SS_{block}+SS_{\varepsilon}}$

Source	SS	df	MS	F	p
Between Groups	$b\sum_{i=1}^{a}(\bar{y}_{i..}-\bar{y}_{...})^2$	$a-1$	$\frac{b\cdot\sum_{i=1}^{a}(\bar{y}_{i..}-\bar{y}_{..})^2}{a-1}$	$\frac{MS_{Between\ Groups}}{MS_{Residual}}$	$\int_{-\infty}^{F_{Between\ Groups}} p(F)dF$
Block	$a\sum_{j=1}^{b}(\bar{y}_{.j}-\bar{y}_{..})^2$	$b-1$	$\frac{a\sum_{j=1}^{b}(\bar{y}_{.j}-\bar{y}_{..})^2}{b-1}$	$\frac{MS_{Blocks}}{MS_{Residual}}$	$\int_{-\infty}^{F_{Block}} p(F)dF$
Residual	$n\sum_{i=1}^{a}\sum_{j=1}^{b}\sum_{k=1}^{n}(\bar{y}_{ijk}-\bar{y}_{ij.})^2$	$n-a-b$	$\frac{n\sum_{i=1}^{a}\sum_{j=1}^{b}\sum_{k=1}^{n}(\bar{y}_{ijk}-\bar{y}_{ij.})^2}{n-a-b}$		
Total	$\sum_{i=1}^{a}\sum_{j=1}^{b}\sum_{k=1}^{n}(\bar{y}_{ijk}-\bar{y}_{..})^2$	n			

FIGURE 13.7: Conceptual formulas for randomized block factorial ANOVA. Each factor (between groups and block) is treated as a separate oneway ANOVA, with the exception that the within groups error (i.e., residual) is reduced as a result of adding additional predictive factors to the model

13.2.3 Partial R^2

Because each of the components in our linear model are independent and additive, we can estimate the individual contributions of each independent variable to the total variation in the model by computing their partial R^2 values.

Because $SS_{Total} = SS_{\text{Between Groups}} + SS_{Block} + SS_{\varepsilon}$, the proportion of the total variation that our Experimental factor causes is just $\frac{SS_{\text{Between Groups}}}{SS_{Total}}$, and the proportion accounted for by our Block is $\frac{SS_{Block}}{SS_{Total}}$.

This leads us to another nice feature of the General Linear Model: The proportion of the total variation accounted for by each factor, plus the proportion due to error adds up to the total variation:

Sum of squares total.

$$R^2_{Total} = R^2_{Factor_1} + R^2_{Factor_2} + \ldots + R^2_{\varepsilon}$$

The R^2_{Total} for my analysis of arrow spine is $\frac{SS_{Spine}+SS_{material}}{SS_{Total}} = 0.48$. The partial R^2_{Spine} is $\frac{SS_{Spine}}{SS_{Total}} = 0.29$, the partial $R^2_{Material}$ is $\frac{SS_{material}}{SS_{Total}} = 0.18$, and the partial R^2_{ε} is

$\frac{SS_{\varepsilon}}{SS_{Total}} = 0.52$. In general, the partial R^2_{Factor} for any factor in an analysis is a good indicator of its importance as a predictor in the model. Because arrow spine predicts more of the variability in accuracy than does material, if I had to choose between the two as a means of focusing my work to improve the accuracy of archery equipment, I would choose spine.

13.2.3.1 Efficiency of a Block

In my arrow experiment, my block was fairly efficient. It reduced the SS_{ε} by 26%! Doing so, greatly increased my power to detect a significant difference in the Factor I was really interested in, the spine of the arrows. The **efficiency of a block** is measured by the proportion it improves prediction over a Oneway ANOVA or other completely randomized design:

$$\frac{SS_{block}}{SS_{block} + SS_{\varepsilon}}$$

The denominator, $SS_{block} + SS_{\varepsilon}$, is the SS_{ε} we calculated with our Oneway Anova. Figure 13.8 will illustrate how this works:

Analysis of Variance

Source	Sum Sq.	d.f.	Mean Sq.	F	Prob>F
X1	1056.2	1	1056.25	1.25	0.2721
Error	28818.1	34	847.59		
Total	29874.3	35			

The Block Variation accounts for stuff we couldn't model in the Oneway!

Analysis of Variance

Source	Sum Sq.	d.f.	Mean Sq.	F	Prob>F
X1	1056.3	1	1056.3	12.22	0.0014
X2	26051.4	2	13025.7	150.66	0
Error	2766.7	32	86.5		
Total	29874.3	35			

FIGURE 13.8: Efficiency of a block (X2) is shown as a reduction in error of prediction over a oneway ANOVA

Notice how, in the lower ANOVA table, if X2 is our blocking variable, how much of the error variation it predicts!

$$\frac{SS_{block}}{SS_{block} + SS_{\varepsilon}} = \frac{26,051.4}{28,818.1} = 0.90$$

It is so efficient, that the $MS_{\text{Between Groups}}$, which is linearly independent of the block, now predicts enough of the variation in the model to be considered significant (say alpha = 0.01), whereas in the Oneway ANOVA, there was so much error variation in the model, the amount explained by X1 was non-significant. Choosing a good block, therefore, can be an important improvement to the sensitivity of your analysis. Of course, like any application of the GLM, one must also be careful about multicollinearity of the block with the factors of interest, and *overfitting*, adding a bunch of variables that really don't contribute significant variation to the linear model, just to reduce the residual error variation.

13.3 Summary

The Randomized Block Design is one of the most efficient of all analytic frameworks. We have studied it here in this chapter using analysis of variance, but the design can be used in almost any other test of experimental effects: χ^2, nonparametric analogs to the ANOVA, and even computational approaches like bootstrapping (Chapter 15). Understanding its logic and procedures is a cornerstone of any engineer's statistical knowledge.

The logic of blocking is to account for pre-existing variation in a population by splitting the population into strata and taking random samples from each strata. Using the block as an independent variable in the GLM reduces the error variation. Not all blocks contribute significant variation to an analysis, however. The efficiency of a block can be determined by comparing the variation it accounts for to the total residual error that would exist had the block not been included in the analysis.

Randomized Block designs, like all Factorial designs, have the same assumptions as ANOVA and other regression analyses. Examining the results for violations of these assumptions is critical for establishing the trustworthiness of the decisions made from the analysis.

13.4 References

Jahanbakhshi, A., Ghamari, B., & Heidarbeigi, K. (2016). Effect of engine rotation speed and gear ratio on the acoustic emission of John Deere 1055I combine harvester. *Agricultural Engineering International: CIGR Journal, 18*(3), 106-112.

Oguntunde, P. E., Adejumo, O. A., Odetunmibi, O. A., Okagbue, H. I., & Adejumo, A. O. (2018). Data analysis on physical and mechanical properties of cassava pellets. *Data in brief, 16*, 286-302.

Prosser, R. A., Cohen, S. H., & Segars, R. A. (2000). Heat as a factor in the penetration of cloth ballistic panels by 0.22 caliber projectiles. *Textile Research Journal, 70*(8), 709-722.

13.5 Study Problems for Chapter 13

1. Explain the logic of Blocking. Why would anyone ever use blocking in an experiment?

   Because the independent variables in the GLM are orthogonal, they each contribute unique variation to the total variation in the dependent variable. If random samples can be taken from each of the pertinent levels of an independent variable (the blocking variable), and then those samples randomly assigned to experimental conditions, the blocking variable will contribute unique variation to the total, thus reducing the error of prediction. This gives the experiment more power to detect differences caused by the experimental conditions. Blocking is most often used to account for the impact of pre-existing variation in a population that might mask or confound the expression of the dependent variable.

2. The noise produced by vehicles is an often cited problem in both urban and rural communities. Industrial and agricultural equipment, in particular can generate high-decibel noise pollution. In 2016, Jahanbakhshi, Ghamari, and Heidarbeigi studied the effect of engine RPMs and Gear Ratio on the level of sound produced by John Deere harvesters. They blocked their experiment on the distance at which the microphone they used to record the sound pressure, was placed in relation to the driver's ear (10 cm, 7.5 m, 20 m). The ANOVA table below shows the results of their experiment (some calculations are left empty):

Source	SS	df	MS	F	p
Microphone Distance	3865.768	2			
RPM	78.457	1			
Gear Ratio		2			
Residual	15.794				
Total	4561.710	53			

   (a) Complete the ANOVA table.

Source	SS	df	MS	F	p
Microphone Distance	3865.768	2	1932.884	5875.027	Nearly 0
RPM	78.457	1	78.457	238.47	Nearly 0
Gear Ratio	601.691	2	300.845	914.422	Nearly 0
Residual	15.794	48	0.3290		
Total	4561.710	53			

(b) **How well did the block do in reducing the error variation? Explain**
The block has an efficiency of $\frac{3865.768}{15.794+3865.768} = 0.996$. 99.6% of the error variation that would have existed without the block is accounted for by including Microphone Distance in the analysis.

(c) **Interpret the analysis. What is(are) the most important variables to consider in understanding the volume of sound produced by the harvesters?**

The block itself contributes the majority of variation to the model. The next most important factor in predicting the noise of John Deer harvesters is the Gear Ratio. Its SS is second largest of the independent variables, followed by RPM. All predictors are significant, so both RPM and Gear Ratio contribute to the sound pressure given off by harvesters, with Microphone Distance controlled for. We would need the means and standard deviations of the sub-samples to determine the groups that are significantly different from the others, and in the case of Gear Ratio, we need to follow up with post-hoc analyses.

3. **Prosser et al., (2000) studied bullet penetration in fabric panels for bullet-resistant body armor. They recorded the shape of the bullets, the thickness of the fabric panels, in number of layers, and the velocity at which the bullets would penetrate 50% of the panels. Their primary question was, what was the effect of panel thickness on the penetration of the bullets. They knew that Bullet Shape would affect penetration, independently of the panel thickness. So, they blocked their sample on bullet shape: 1=rounded, 2= sharp, 3=flat soft point. Their data is included in the file: Ballistic_cloth_RB.xlsx**

(a) **Perform a Randomized Block ANOVA on this data.**

Analysis of Variance

Source	Sum Sq.	d.f.	Mean Sq.	F	Prob>F
shape	6997.8	2	3498.9	17.03	0.0004
thickness	450585	11	40962.3	199.32	0
Error	2260.6	11	205.5		
Total	459555.2	24			

Hierarchical (Type II) sums of squares.

(b) **Is there a significant effect of panel thickness? If so, which panel thicknesses are most effective? Is there a drop-off in effectiveness at some point?**

Yes, the omnibus test yields significant differences ($F_{(11,11)} = 199.32, p \sim 0$). It is highly unlikely that, if we repeated this experiment many times, we would get mean differences as high as our samples. I need to do a post-hoc analysis to test these differences. For this one, I will use Scheffé, since we have 12 different groups, and the consequences, given the context is pretty dire—I want to be conservative in my claims.

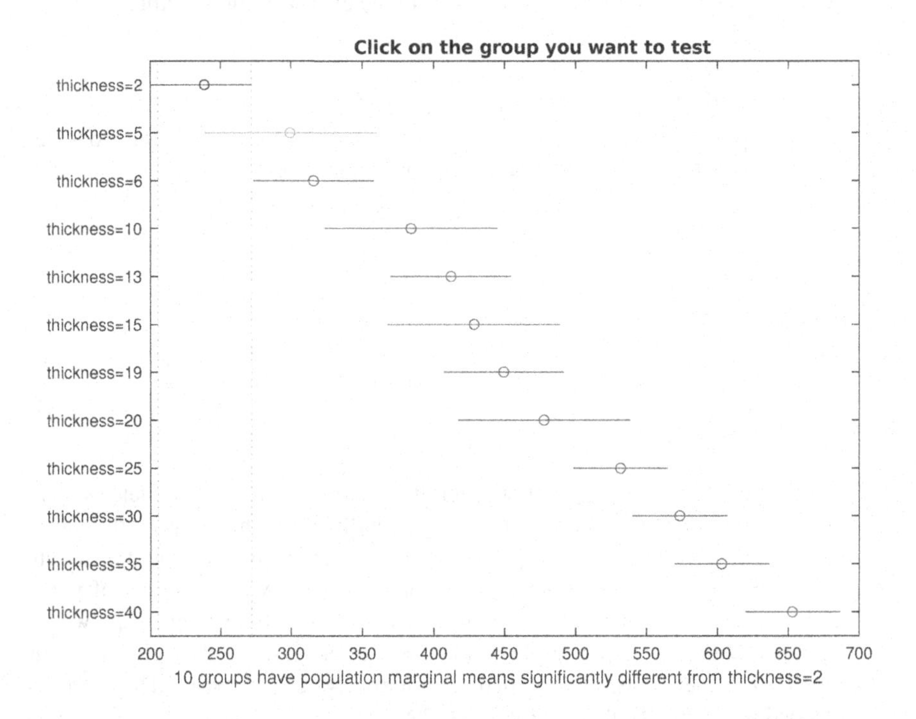

The plot of group means shows a monotonically increasing effect of thickness. The 2 layer thickness material shows significantly less penetration velocity than all the other groups except the 5 layer thickness. Checking the results table from Matlab anovan, we can see that Thickness=40 layers is significantly different from all other groups except Thickness = 35. There are some non-significant differences in the middle thicknesses, but overall, it looks like at the lower and upper ends of the thicknesses, the effect seems more pronounced. Because these are measured in millimeters, I would probably want to fit a curve using multiple regression (where the independent variable, thickness, is continuous, so I could model the effect of any number of different layers.

(c) **What was the efficiency of the Block?**

The block, shape, accounts for $\frac{6997.8}{6997.8+2260.6} = 75.59$. So about 75% of the error variance, if shape was ignored, is accounted for by including shape in the model.

(d) **What is the overall goodness of fit of the data?**
$R^2 = \frac{459555.2-2260.6}{459555.2} = 0.995$. This is an extremely good fit.

(e) **Are there any assumptions of ANOVA that were violated in your analysis? If so, how would you propose to address them?**

The Q-Q plot of the residual values looks very good. I have no problem with the assumption of normality:

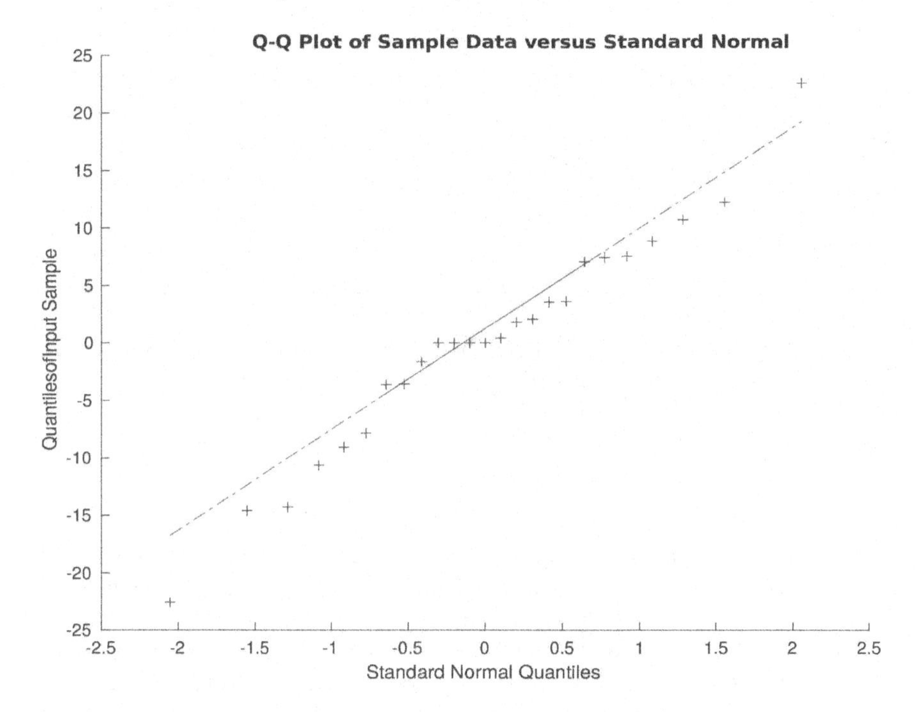

There are minor differences in sample variances within each of the two independent variables, and the residual plots look appropriately random, so I am happy that we are not significantly violating homogeneity of variance.

4. Cassava (manioc, tapioca) is a staple food across much of the equatorial world. Storing it in pellet form can be used to ward off starvation in time of famine in equatorial communities. Oguntunde, et al., (2018) studied the effect of the size of manufacturing dies and speed of manufacture of cassava pellets on their hardness. They knew that the moisture content of the pellets would have an effect on their hardness, so they used it as a blocking factor. I have reconstructed their Randomized Block analyses to the following ANOVA table (they included more terms in their analyses):

Source	SS	df	MS	F	p
Moisture Content	3068.848	3	1022.949	1.605	
Speed	293.327	3	97.776	0.153	
Die Diameter	255.827	3	85.276	0.134	
Error	28036.198	44	637.186		
Total	31654.250	53			

(a) How many groups were in the block, Moisture Content?

The degrees of freedom are 3, therefore there were 4 levels of Moisture Content.

(b) Did Moisture Content serve as a good blocking factor in this analysis?

The block, Moisture Content, accounts for $\frac{3068.848}{3068.848+28036.198} = 0.099$. So about 10% of the error variance, if Moisture Content was ignored, is accounted for by including Moisture Content in the model. Moisture Content accounts for the most variation of any of the independent variables in the model, but it still only accounts for a very small proportion. Further, neither Die Diameter, nor Speed were found to be significant predictors of cassava pellet hardness, so I think the Block did not do a great job, overall.

(c) Did the two independent variables make a significant impact on the hardness of Cassava pellets? Explain.

NO! Neither independent variable had an F greater in magnitude than $F_{critical} = F_{(3,44)} = 2.82$, so their probability is too close to chance to be considered a significant difference.

5. A company developing new sustainable pressboard for construction is investigating the effects of the *pressure* at which the mixture is cooked, and the *amount of time* at which the mixture is cooked on the modulus of elasticity of the finished product. They have two different formulas they currently use, and they know that the modulus of elasticity is impacted by the percentage of the hardwood concentration in these formulations. The data for this experiment is included in the file: Pressboard_RB.xlsx.

(a) Which of the three factors in this experiment would you use as a blocking factor? Explain.

Because the concentration of the hardwood in the formula is a pre-existing condition, it makes the most sense as a block.

(b) Run the Randomized Block analysis using the Block you chose in a. Do the independent variables make a significant difference in the pressboard's elasticity?

Analysis of Variance

Source	Sum Sq.	d.f.	Mean Sq.	F	Prob>F
conc	7.2839	2	3.6419	6.53	0.0044
press	16.8439	2	8.4219	15.11	0
time	20.8544	1	20.8544	37.4	0
Error	16.7267	30	0.5576		
Total	61.7089	35			

Hierarchical (Type II) sums of squares.

From this analysis, it appears that both pressure and time are significant predictors of modulus of elasticity. I chose $\alpha = 0.05$. The probability of $F_{pressure} = 15.11, p \approx 0$, is so small that it is highly unlikely that we could have gotten differences this big should we re-run this experiment. This probability is less than $\alpha = 0.05$, my Type I error rate, so the effects are significant.

Time appears to have more of an impact, given its SS is greater than any of the other predictors $R^2_{time} = \frac{20.8544}{61.7089} = 0.338$. Time alone accounts for about 1/3 of the total variation in elastic modulus, with greater time associated with higher elasticity.

Pressure has 3 levels, so I needed to perform a post-hoc analysis. Since this is exploratory, I will use the Dunn-Bonferroni procedure:

$Group_1$	$Group_2$	95% CI Lower Bound	Mean Difference	95% CI Upper Bound	p
1	2	−0.631	0.142	0.915	1
1	3	−2.148	−1.375	−0.602	0.0003
2	3	−2.290	−1.517	−0.744	7.493×10^{-5}

As you can see, the mean difference between the two lower pressures, 400 and 500 bar, are non-significant. But the difference between 600 bar and the two lower pressures are both significant. If we are interested in more elastic pressboard, then the 600 bar pressure setting would be the one we would choose, at 4 hours of cooking time.

(c) **What is the efficiency of the Block? Did it do a good job reducing the error variation?**

The efficiency of the Block, concentration of hardwood, is about 0.303. It accounted for just over 30% of the variation in elastic modulus of the measured specimens. In the ANOVA table below you an see the analysis performed without concentration as a block. The overall decision hasn't changed, but with concentration, there is less overall error variance, causing the F-ratio to be higher in the Randomized Block analysis. This is the sign of a good block.

Analysis of Variance

Source	Sum Sq.	d.f.	Mean Sq.	F	Prob>F
press	16.8439	2	8.4219	11.22	0.0002
time	20.8544	1	20.8544	27.79	0
Error	24.0106	32	0.7503		
Total	61.7089	35			

Hierarchical (Type II) sums of squares.

14

Factorial Analysis of Variance

There are many times when we can't block, even when there is a nuisance variable present in our population. It may be impractical or too expensive to sift through a set of carbon fiber structural members of an aircraft fuselage, for example, to develop a stratified random sample based on the geometry of the fibers layup.

But we can account for the variation of any number of factors in exactly the same manner *as if* they were randomized blocks. This is a form of statistical control. Adding good predictors (uncorrelated with the independent variables of interest, of course) will *always* reduce the error of prediction in the regression model, so if you have good predictors that are theoretically relevant to the relationship you are wanting to model, and if these predictors won't suppress the effects of other variables, it is often good practice to include them.

One of the real powers of Factorial Analysis of Variance is the ability to assess **interactions** among independent variables. For example, Nepal, Monty & Kay (2013) evaluated the yield strength per unit ultimate tensile strength of titanium alloy medical wire. The medical industry is subject to strict regulations regarding the consistency and quality of device components, so understanding these parameters and the factors that impact quality manufacturing is critical for the industry to create consistent-performing prosthetics and other high-cycle devices. Among the factors identified as potential causes of non-uniformity of the wire, the authors studied the diameter of the stock wire, the length of the bearing used to draw the wire to finer diameter, and the reduction angle at which the bearing was adjusted for the drawing process. The first few rows of their data are presented below in Table 14.1. The entire dataset can be downloaded in the following file: Medical Wire Tensile Strength Factorial ANOVA.xlsx.

DOI: 10.1201/9781003094227-14

Reduction Angle	**Bearing Length**	**Supply Diameter**	**yield strength/ultimate tensile strength**
0	0	0	93.3
0	0	1	92.5
1	0	1	92.9
1	0	0	93.5
0	1	1	93.2
0	1	0	92.4
1	1	0	93.7
1	1	1	92.9
⋮	⋮	⋮	⋮

TABLE 14.1: Multiple potential determinants of tensile strength of medical wire

One of the key hypotheses they tested was whether or not a combination of wire diameter, bearing length and reduction angle would increase the yield strength/ultimate tensile strength (YS/UTS) ratio. So, they made 120 runs, counterbalancing the order of the variables for the different machines to reduce any random effects caused by different machines, and measured the outcome for each run. Figure 14.1 shows some of the potential interactions in their data they had to contend with.

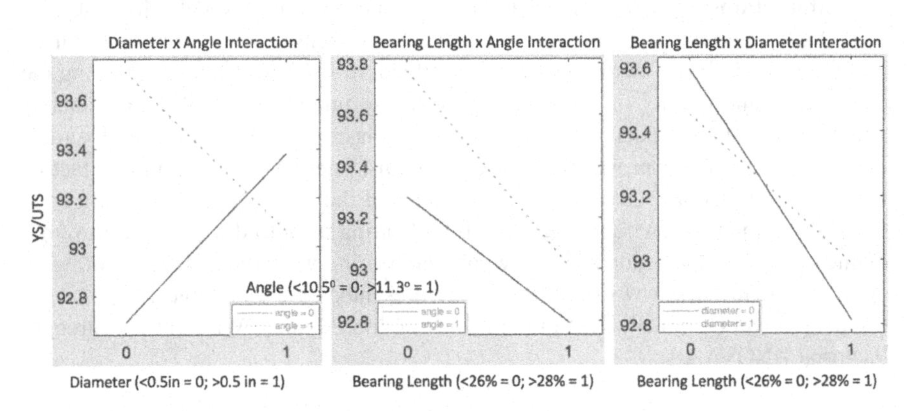

FIGURE 14.1: Potential interactions among diameter of stock, angle, and length of bearing. Mean values of YS/UTS for each level of the independent variables represent the endpoints of each segment

Each of these three graphs show something interesting about the relationships among the three factors. In all of the panels, the important feature to key in on is the different slopes of the lines representing the effects of the variables. In the leftmost panel, the slopes even have different signs. This shows that at higher bearing angles, the *larger* diameter wires show *less* YS/UTS, while at lower bearing angles, the *smaller* diameter wires show *greater* YS/UTS. We call this kind of relationship **non-monotonic**. The rightmost panel shows a similar overall pattern: larger bearing length appears to be negatively associated with YS/UTS for both diameter wires. But the larger diameter wires appear to be affected slightly *less* by bearing length, than the smaller diameter wires. Even though the slopes cross, we call this kind of a relationship **monotonic**—meaning that their slopes have the same sign, but different magnitudes. The middle panel in Figure 14.1 is also monotonic, you can see that the overall effect of Bearing Length is negative, with larger bearings resulting in less YS/UTS (consistent with what we saw with its interaction with wire diameter). But, when interacting with angle, the effect of Bearing Length appears to be less for lower angles, than higher ones. Figure 14.2 shows the differences in both monotonic and non-monotonic interactions.

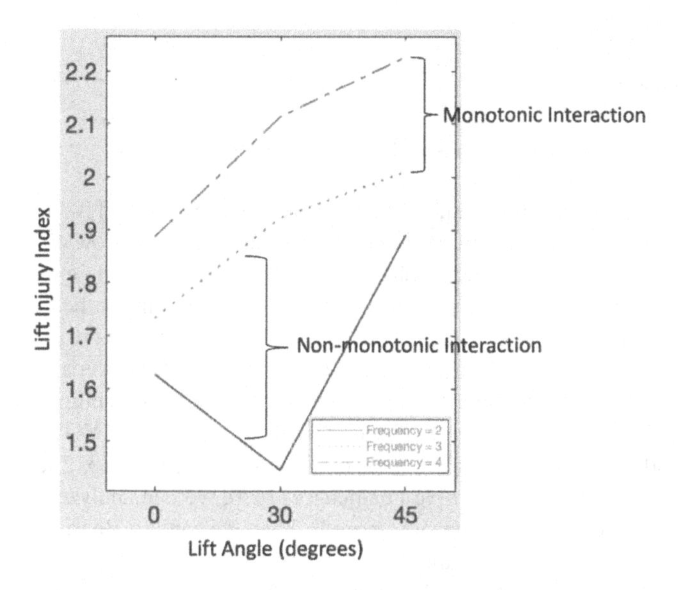

FIGURE 14.2: Monotonic interactions have different slopes with the same sign. Non-monotonic interactions have slopes with opposite sign

What, in the context of trying to figure out which combinations of these factors contribute to the relative tensile strength of medical grade titanium wire, do these patterns mean? Interpreting these graphs, I would hypothesize that fatigue due to greater work hardening, is impacted by greater bearing angle more for larger wires more than smaller wires. That is a reasonable interpretation of the non-monotonicity of the interaction between Angle and Diameter. Work hardening is also a big culprit for the relationship between Angle and Bearing Length. The longer the stock stays in the bearing, at higher angles, the more likely it is to experience work hardening and thus display less tensile strength. Lastly, in the rightmost panel, we can see that the smaller diameter wires, when worked less, tend to be stronger, and when worked more, tend to be weaker, relatively speaking, than thicker wires.

An **Interaction** is the situation when the effect a predictor variable has on an outcome depends on the state of a second predictor. Analytically, an interaction is a variable constructed by dividing up one factor into levels of the second, thus creating, in an i x j Factorial Design, where the i x j group means can be compared just like any other set of groups in Analysis of Variance. Interactions may be **monotonic**, where the direction of the relationship for one variable across levels of the second are the same, they just have different slopes, or **non-monotonic**, where the direction of the relationship has opposite slopes.

14.1 Interactions as Additional Factors

Now, from examining the data, we believe that there may be a few interactions impacting the tensile strength of medical wire in manufacturing, how do we determine if they are systematic, or if the graphs just show variation we might expect due to random chance? This is especially important, because just looking at the graphs, some of which have different scales, it can be difficult to tell if the slopes are *different enough* to be considered a real difference. If you look at the data from Nepal, Monty & Kay (2013), you will see that there are three factors, each with levels dummy coded 0 and 1. If we fix one level of a Factor, and then perform a Oneway ANOVA across levels of a second factor, we get six different interactions (3 variables x 2 levels each = 6 interactions). Fixing, for example Diameter to be 0, we can analyze whether or not Angle makes a difference in tensile strength. Then we can test Bearing Length to see if it makes a difference. We could then switch to Diameter = 1 and do the same thing across Angle and Bearing Length. After that, we could switch to Angle and perform the same kind of analysis, fixing it to Angle = 0, and testing if Bearing Length makes a difference, then again at Angle = 1.

But, this procedure is pretty cumbersome to perform one after the other. A better way is to treat the interactions as if they were additional independent variables—as factors like the **main effects** factors. This is the real power of Factorial ANOVA. It creates additional columns in the matrix of predictors in the regression model, treating them just like it does any of the other column vectors. The resulting linear model becomes something like this:

$$Y = \beta_0 + \beta_1 X_1 + \beta_2 X_2 + \beta_3 X_1 X_2 + \varepsilon$$

and the prediction model is:

$$\hat{Y} = \beta_0 + \beta_1 X_1 + \beta_2 X_2 + \beta_3 X_1 X_2$$

Where $X_1 X_2$ is the interaction between the two factors constituting the main effects in the model.

Mathematically speaking, an interaction is just another independent variable, and is computed just like any other. The ANOVA table below (see Figure 14.3), shows the conceptual formulae for a two-factor model with interactions. The computation for each factor is exactly the same as a Oneway Analysis of Variance. You can see how the AxB interaction term, because it is formed by the combination of Factor A and B is just the squared distances between the cell means and each factor mean. But because this is done for both factors, the grand mean is subtracted twice and needs to be added back in yielding $SS_{A \times B} = m \sum_{j=1}^{b} \sum_{i=i}^{a} \left(\bar{y}_{ij.} - \bar{y}_{i..} - \bar{y}_{.j.} + \bar{y}_{...}\right)^2$.

	SS	df	MS	F	p
Factor A	$mb \sum_{i=1}^{a} (\bar{y}_{i..} - \bar{y}_{...})^2$	$a - 1$	$\frac{b \cdot \sum_{i=1}^{a} (\bar{y}_{i..} - \bar{y}_{...})^2}{a - 1}$	$\frac{MS_A}{MS_\varepsilon}$	$\int_{-\infty}^{F_A} p(F)dF$
Factor B	$ma \sum_{j=1}^{b} (\bar{y}_{.j.} - \bar{y}_{...})^2$	$b - 1$	$\frac{a \sum_{j=1}^{b} (\bar{y}_{.j} - \bar{y}_{..})^2}{b - 1}$	$\frac{MS_B}{MS_\varepsilon}$	$\int_{-\infty}^{F_B} p(F)dF$
A x B	$m \sum_{j=1}^{b} \sum_{i=i}^{a} (\bar{y}_{ij.} - \bar{y}_{i..} - \bar{y}_{.j.} + \bar{y}_{...})^2$	$(a - 1)(b - 1)$	$\frac{m \sum_{j=1}^{b} \sum_{i=i}^{a} (\bar{y}_{ij.} - \bar{y}_{i..} - \bar{y}_{.j.} + \bar{y}_{...})^2}{(a - 1)(b - 1)}$	$\frac{MS_{A \times B}}{MS_\varepsilon}$	$\int_{-\infty}^{F_{A \times B}} p(F)dF$
Residual	$n \sum_{i=1}^{a} \sum_{j=1}^{b} \sum_{k=1}^{n} (\bar{y}_{ijk} - \bar{y}_{ij.})^2$	$n - ab$	$\frac{n \sum_{i=1}^{a} \sum_{j=1}^{b} \sum_{k=1}^{n} (\bar{y}_{ijk} - \bar{y}_{ij})^2}{n - ab}$		
Total	$\sum_{i=1}^{a} \sum_{j=1}^{b} \sum_{k=1}^{n} (\bar{y}_{ijk} - \bar{y}_{...})^2$	n			

FIGURE 14.3: ANOVA table showing conceptual formulae for sources of variation in an $a \times b$ factorial design

If we added a third factor to our analysis, the ANOVA table would include a third main effect, Factor C, 3 2-way interactions, AxB, AxC and BxC, and a 3-way interaction, AxBxC. *Any number of factors can be added to a factorial design*, given you have a large enough sample for each combination of factors.

Main effects in a Factorial Design are the estimated effects for each independent variable, without considering any interactions with the other independent variables. That is to say, primary Factors in a Factorial Design, constitute its Main Effects. Interactions, by contrast, constitute the effects of combinations of the primary Factors.

Because we now have both main effects and interactions as independent variables, we have two sets of null and alternative hypotheses:

For the Main Effects:

H_0 : Means for all levels of a factor are equal

H_1 : Means for at least one level of a factor is not equal to the others (may also be one-tailed)

There will be one set of hypotheses for each main effect in the analysis.

The null hypothesis of an interaction is that there is none–that the slopes across levels of a second factor should be parallel to each other (i.e., no interaction—no significant difference of one independent variable across levels of another):

H_0 : Means for all levels of a factor are equal, given a level of second factor is fixed

H_1 : Means for at least one level of a factor is not equal to the others, given a level of second factor is fixed (may also be one-tailed)

Like the Oneway ANOVA and Randomized Block ANOVA, you can run an omnibus test to see which factors show at least one significant difference in group means, then follow up with post hoc comparisons to pinpoint which specific groups show significant differences. My recommended procedure for Factorial ANOVA is as follows:

1. First, *run an omnibus test* as a Factorial ANOVA, including any interactions of interest.
2. *If main effects are significant*, perform post hoc comparisons on the group means;
3. *If an interaction shows a significant effect*, perform post hoc comparisons.

Let's walk through this procedure for the medical wire tensile strength data of Nepal, Monty & Kay (2013). I am running this as a two-tailed test. The overall Type I error rate I have chosen is 0.05, and I have partitioned it into $\alpha = 0.025$ in each tail of the distribution.

Analysis of Variance

Source	Sum Sq.	d.f.	Mean Sq.	F	Prob>F
Diameter	0.0101	1	0.0101	0.02	0.8934
Length	11.5941	1	11.5941	20.75	0
Angle	3.7101	1	3.7101	6.64	0.0113
DiaxLength	0.7521	1	0.7521	1.35	0.2484
DiaxAngle	13.4001	1	13.4001	23.99	0
LengthxAngle	0.5468	1	0.5468	0.98	0.3247
DiaxLengthxAngle	0.4687	1	0.4687	0.84	0.3616
Error	62.5693	112	0.5587		
Total	93.0513	119			

Hierarchical (Type II) sums of squares.

FIGURE 14.4: ANOVA table for a 2 x 2 x 2 factorial ANOVA with all interactions

The ANOVA table in Figure 14.4 shows a number of interesting things. First, as we discovered in Multiple Linear Regression, whenever we add predictors, the error of estimation drops. With all these factors and interactions, we have dropped the SS_ε considerably. But it is still pretty high! Second, there are a couple of effects that appear to be significant: Length of Bearing, Angle of Bearing, and Diameter x Angle interaction. Of these, Diameter x Angle appears to make the biggest contribution to the variation in the model. Its partial $R^2 = \frac{13.4001}{93.0513} = 0.144$. Length comes in second with partial $R^2 = \frac{11.5941}{93.0513} = 0.125$. And third we have poor little Angle, only accounting for $R^2 = \frac{3.7101}{93.0513} = 0.040$ of the variability in tensile strength of the medical wire.

You can see how performing the omnibus test first can eliminate lots of pairwise analyses across all the main effects and their interactions. We really don't have to worry much about the effect of Diameter. There really isn't much variation that this factor accounts for. Its *SS* is very small (0.01), and its partial $R^2 = \frac{0.0101}{93.0513} = 0.0001$. We also don't have to worry about the Diameter x Length interaction or the Length x Angle interaction. I am also happy to note that the 3-way interaction, Diameter x Length x Angle is non-significant. That many dimensions makes interpretation of the interactions very difficult (at least for me). But it can be done, and if we had found a significant result, I would go in and look at all those combinations to figure out which one(s) contributed most to tensile strength of our wire. Those combinations may be the optimal ones, of the ones studied, for manufacturing wire in the future.[1]

14.1.1 Post Hoc Tests

Post hoc tests for the Factorial ANOVA are hardly different than the methods we learned for Oneway ANOVA. After all, if each row in the ANOVA table is a separate

[1]There is a wonderful method called Response Surface Methodology that uses factorial design to efficiently find areas of local minimum or maximum on the hypersurface created by the regression equation. It can utilize a combination of categorical and continuous independent variables to optimize combinations for generating desired levels of the outcome variable. If you want to learn more about this extension of the Factorial ANOVA, refer to Myers, Montgomery, & Anderson-Cook, (2016).

Oneway Analysis of Variance, the post hoc tests have to be the same. You can use the Dunn-Bonferroni method, especially for the main effects analysis, or you can use Scheffé. But for the interactions, if they prove to be significant, it is often difficult to justify an a-priori hypothesis about which ones are important and which are less important by which to cut the number of post hoc comparisons to a manageable level. For that reason, Scheffé, because of its conservative nature, is the safest bet.

The significant main effects for this study only have two levels each. Because they returned a significant difference, there is no need for a post hoc test. We know that small bearing length corresponds with significantly greater tensile strength. And for Angle, larger angles result in significantly greater tensile strength. Their summary statistics are shown in Table 14.2:

	Bearing Length		**Angle**	
	$\bar{y}$	s	$\bar{y}$	s
Small	93.52	0.81	93.04	0.97
Large	92.90	0.85	93.39	0.75

TABLE 14.2: Summary statistics for the main effects: Bearing length and angle. Each has only two levels: Small and large

For the Diameter x Angle interaction, if we put the means of the different combinations of levels into a table (see Table 14.3), we get a slightly different look at how an interaction works.

		Angle		
Diameter of Wire		Small	Large	**Marginal Mean**
	Small	92.69	93.71	93.20
	Large	93.38	93.06	93.22
	Marginal Mean	93.04	93.39	93.21

TABLE 14.3: Interaction between wire diameter and angle

From the marginal means, we can see that Diameter makes a much smaller impact on YS/UTS than Angle (93.22 × 93.20 is only 0.02, compared to 93.39 × 93.04 = 0.35). But the cell means paint a more complicated picture.

If we fix the Angle of the wire to be Small, then Diameter makes a difference of 93.38−92.69 = *0.69*, whereas if we fix the diameter to be Large, then angle makes a difference of 93.06−93.71 = *−0.65*. In other words, for large values of Angle, Diameter has a *negative effect*! This is born out in Figure 14.1, in the leftmost panel showing as a nonmonotonic interaction. Going the other way, if we fix Diameter to be Small, then Angle makes a difference of 1.02, while when we fix Diameter to Large, the Angle makes a difference of 0.32. Again, this nonmonotonic interaction shows that the effect of one variable is dependent on the levels of the second. A similar analysis can be performed for the interactions of Length x Angle and Length x Diameter interactions.

But even though we can see that these interactions are significant, we still don't know which one(s) are different enough to reject the null hypothesis. To do this, we can run a Scheffé test, making pairwise comparisons of each of the 4 combinations in the table (See Table 14.4). This results in $\binom{4}{2} = 6$ comparisons (Table 14.4):

Group 1	**Group 2**	**Lower CI Bound**	**Mean Difference**	**Upper CI Bound**	$p\left(F_{Scheffé}\right)$
Dia=0, Angle=0	Dia=1, Angle=0	−1.2344	−0.6867	−0.1389	0.00729
Dia=0, Angle=0	Dia=0, Angle=1	−1.5678	−1.02	−0.4722	1.51E−05
Dia=0, Angle=0	Dia=1, Angle=1	−0.9178	−0.37	0.1778	0.3039
Dia=1, Angle=0	Dia=0, Angle=1	−0.8811	−0.3333	0.2145	0.3982
Dia=1, Angle=0	Dia=1, Angle=1	−0.2311	0.3167	0.8645	0.4449
Dia=0, Angle=1	Dia=1, Angle=1	0.1022	0.65	1.1978	0.01255

TABLE 14.4: Scheffé post hoc contrasts for diameter x angle interaction

Walking through the table of contrasts, we can see that comparing **Small Diameter, Small Angle** with **Large Diameter, Small Angle** (Row 1), we get a significant effect. So when Angle is Small, Diameter has a significant effect! In Row 2, **Small Diameter, Small Angle**, compared with **Small Diameter, Large Angle** is also significant. So, when diameter is Small, Angle has an effect! Rows 3, 4, and 5 show non-significant effects. Row 6, shows that **Small Diameter, Large Angle** compared with **Large Diameter, Large Angle** is significant. This means that for Large Angle, Diameter shows an effect.

Put together, this confirms the nonmonotonic interaction we suspected existed by examining the interaction plots early on. Diameter of our wire has a significant effect on tensile strength of our medical wire across both levels of Angle, while Angle has a significant effect only for small Diameter wire.

More importantly, this simple analysis shows that, even when the main effects factors show small effects (e.g., Diameter) by itself, the effect of their interactions can be significant.

14.1.2 Fixed vs. Random Effects

Up to now in this book, we have modeled effects (levels of an independent variable) as if they were the only possible way to categorize events. This makes the math much easier, because any factor we measure will have levels that are mutually exclusive and collectively exhaustive. But in a couple of examples, I have snuck in some factors that don't necessarily fall into this definition. For example, in the example we just did, the authors chose two levels of Angle, small and large. These categories are mutually exclusive, but by their definition, they are not collectively exhaustive: recall that angles less than 10.5 degrees were categorized as small, and those greater than 11.3 degrees were considered large. What about the bearing angles in-between? The intervals of 0 – 10.5 degrees is not the same as 11.3 – 90 degrees. Because of this, if the authors are trying to model the effect of angle on tensile strength across the entire range of angle measure, they have some sampling error to deal with—all those (infinitely many) angles between 10.5 and 11.3 degrees that will never appear in their analyses (yes, it is probably small, but it is still error).

This kind of error is attributed to the error of sampling a set of levels from all possible levels that could be sampled from a population. For this reason we call the effect of *sampled levels* a **random effect**. In other words, a random effect is a measured effect of an independent variable when the levels of that variable are sampled from a population of values.

As a better example, we could assess the effect of temperature in curing epoxy resin by selecting several temperatures at which to bake the curing samples. Let's say we layup a set of epoxy wafers and bake them at one of three temperatures 23 °C, 30°C, and 40 °C, while they cure. Following the curing time, we can assess their compressive strength in Mpa. Our factor is temperature, and it has three levels: 23 °C, 30°C, and 40 °C. We could just as easily have chosen 25 °C, 32°C, and 41 °C, or fewer, or more levels of different values. Our levels, therefore, represent a *sample* of the possible levels of the independent variable, temperature.

Like all samples, there is some inherent error in its sample statistics from those of the population from which the sample was taken. So, not only do we have error as a result of measurement, and as a result of sampling from the dependent variable, compressive strength, we also have error in our model as a result of sampling levels of our *independent* variable (see Table 14.5).

The majority of our analyses won't involve these random effects, but some will, so it is important to keep in mind. So I have made a simple table (Table 14.5) that helps you recognize Random Effects from **Fixed Effects**, those situations when we are able to account for all possible levels of the independent variable (see Table 14.5):

	Definition	**If the Experiment were to be Repeated**	**What is the Desired Inference to be Made?**	**Example**
Fixed Effect	A statistical model where levels of the independent variable(s) are unchanging	You would use the same levels because those are the only ones there are, or because those are the most convenient levels, ***AND***...	You are only generalizing your results to the actual levels used in your analysis	You are looking at the effect of temperature on the curing of Epoxy. You choose levels of Low, Medium and High, where Low = 0K – 296K, 296K<Med <335K, High>335K. You want to generalize the results to only these three temperature settings.
Random Effect	A statistical model where levels of the independent variable(s) random sampled from a population of levels	You might choose different levels, sampling them from the population of levels, ***AND***...	You are generalizing your results to ALL levels in the population	You are looking at the effect of temperature on the curing of Epoxy. You choose levels of Low, Medium and High, where 273K<Low< 296K, 310<Med <335K, 345K<High<365K. You want to generalize your findings across all temperature settings.

TABLE 14.5: Examples of criteria for recognizing fixed versus random effects.

So, professor, why does it matter? What's that you say? You want to know *why* knowing whether or not your variables represent fixed or random effects is important? Well, here goes...

1. When you have Random Effects, you will have more than one error term in your general linear model: one associated with the error of estimation from your data, and a second associated with the error of sampling your levels:

$$\text{Fixed Effect:} \quad y_i = \beta_0 + \beta_1 X_i + \varepsilon$$

$$\text{Random Effect:} \quad y_i = \beta_0 + \beta_1 X_i + \varepsilon_{random} + \varepsilon$$

2. We estimate the random error by the covariance of *X*, the degree to which any two levels of *X* are dependent upon each other. The fixed error variation, like we have used throughout this book, is estimated by the variance of *X*: the average SS_e.

   (a) These dual error terms can be somewhat confusing, particularly for factorial designs that contain both types of effects (we call this a particularly unclever term: a **mixed effects** model); and

   (b) The addition of the second error term means that, *if you analyzed a random effect using a fixed effect model, you will undoubtedly underestimate the standard error and increase your Type I error rate.*

3. If you are clever, you can define many variables as Fixed, if:

   (a) You can choose levels of your variables judiciously (like the temperature example above) that span the domain of the independent variable.

   (b) This is most easily done by splitting the domain into a number of equal, mutually exclusive and collectively exhaustive categories.

   (c) In this way, a Fixed effect can still be generalized to the entire population of levels, because you defined the population to be only the categories you chose.

Suppose Bearing Angle really *was* a random effect that we wanted to generalize our findings across all possible angles at which we can adjust the drawing bearing. Let's re-do our ANOVA to see what happens:

Fixed effects model

Analysis of Variance

Source	Sum Sq.	d.f.	Mean Sq.	F	Prob>F
Diameter	0.0101	1	0.0101	0.02	0.8934
Length	11.5941	1	11.5941	20.75	0
Angle	3.7101	1	3.7101	6.64	0.0113
DiaxLength	0.7521	1	0.7521	1.35	0.2484
DiaxAngle	13.4001	1	13.4001	23.99	0
LengthxAngle	0.5468	1	0.5468	0.98	0.3247
DiaxLengthxAngle	0.4687	1	0.4687	0.84	0.3616
Error	62.5693	112	0.5587		
Total	93.0513	119			

Random effects model

Analysis of Variance

Source	Sum Sq.	d.f.	Mean Sq.	F	Prob>F
Diameter	0.0101	1	0.0101	0	0.9825
Length	11.5941	1	11.5941	21.21	0.1361
Angle	3.7101	1	3.7101	0.28	0.6918
DiaxLength	0.7521	1	0.7521	1.6	0.4254
DiaxAngle	13.4001	1	13.4001	28.59	0.1177
LengthxAngle	0.5468	1	0.5468	1.17	0.4755
DiaxLengthxAngle	0.4687	1	0.4687	0.84	0.3616
Error	62.5693	112	0.5587		
Total	93.0513	119			

error term for random effects model

df_{error} term for random effects model

FIGURE 14.5: Fixed effects versus random effects models when angle is modeled as a random effect. The random effects model changes the error term for the F-ratio, and subsequently changes the degrees of freedom for evaluating the significance of F

If you examine the results of the two analyses shown in Figure 14.5, you can see that, because the error term changes from the overall error of estimation to the *SS* of the 3-way interaction, the MS_e, and its associated degrees of freedom change both the critical value of F and the calculated value in such a way that it must take a much larger F-value—a much larger effect—to conclude that, at a particular α, the effect is improbable as a function of random chance. None of the significant results in our fixed-effects model turned out to be significant when the random effect of angle was taken into account!

Determining the appropriate error term for F in a random effects model is beyond the scope of this course. It is not particularly difficult, but there are more important issues you need to understand about factorial ANOVA at this stage in your learning process. If you find that you encounter a random effect in the future, there are a number of helpful documents that can assist you in calculating the appropriate MS_e (e.g., Nelder 1977; Snee, 1974). Right now, you need to be able to recognize a random effect from a fixed effect so that you can go learn how to account for it in your analysis if it is necessary.

14.1.3 Assumptions of Factorial ANOVA

Like all analyses, factorial ANOVA depends on a set of assumptions about the characteristics the population from which your samples are being drawn. It is robust to mild violations of most assumptions, but like Oneway ANOVA, you will lose power to detect a difference that may exist.

Assumptions of Factorial ANOVA are the same as that of any Multiple Linear Regression analysis:

1. Normality of the parent populations from which samples are drawn;
2. Independence of observations; and
3. Homogeneity of variance: The levels of each factor must display (approximately) equal variances. In addition, each level of an interaction must display (approximately) equal variance.

Again, this is just a generalization of the *t*-test to multiple variables. Each set of means you test against each other must display approximately normal distribution, and homogeneity of variance. Independence of observations, as we have learned earlier, is an assumption of all hypothesis tests—it is the basis for the Null Hypothesis.

Checking for violations of these assumptions is exactly the same as in Oneway ANOVA: Performing a Residual Analysis including a Q-Q plot of the residuals. There are just a lot more variables to plot! We will go through these tests in a comprehensive example at the end of this chapter. In many software packages, the programmers will include a statistic called Levignes Test of Homogeneity. This is an analog to the residual analysis. I will not go into this test here, but you may come across it as default output if you are using a statistical package like Minitab, SPSS, or SAS.

14.2 Nested Factors in ANOVA

Sanclemente, & Hess (2007) studied the different factors that influence the loosening of bolts. Among the variables they studied were the elastic modulus of the bolts (Aluminum bolts vs steel), diameter of the bolt ($\frac{1}{4}$" versus $\frac{1}{2}$"), pitch (coarse vs fine), preload (32% Yield vs 64% Yield), hole fit (loose vs tight), and lubrication (dry vs SAE 30 oil). They took a stratified random sample of 64 bolts across modulus and diameter and pitch, and then randomly assigned bolts to experimental conditions: Preload, Hole fit, and Lubrication. The issue they faced was, because of the different sizes of bolt, thread pitch, preload and hole fit were dependent on the levels of the bolts, diameter. For example, the measure of coarse and fine thread pitch is different for smaller bolts than larger bolts (28 and 20 threads per inch for $\frac{1}{4}$" versus 13 and 20 for $\frac{1}{2}$").

Likewise, preload, which is a percentage of the yield strength of the bolt, is also dependent upon diameter. Lastly the fit of the bolt, because the holes drilled are of different diameters for $\frac{1}{4}$" versus $\frac{1}{2}$" bolts, is also dependent upon the diameter of the bolt. Tight fit hole sizes were 6.53 mm for 6.4 mm ($\frac{1}{2}$") bolts and 13.11 for 12.7 mm ($\frac{1}{2}$") bolts, while loose fit hole sizes were 6.75 for 6.4 mm ($\frac{1}{4}$") bolts and 13.49 for 12.7 mm ($\frac{1}{2}$") bolts. Lubrication as a factor could be crossed with all of the other factors. Their experimental design is summarized in this table (Figure 14.6):

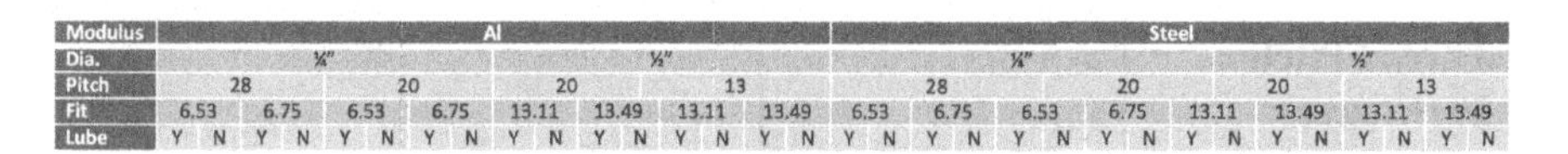
<table>
<tr><td>Modulus</td><td colspan="16">Al</td><td colspan="16">Steel</td></tr>
<tr><td>Dia.</td><td colspan="8">¼"</td><td colspan="8">½"</td><td colspan="8">¼"</td><td colspan="8">½"</td></tr>
<tr><td>Pitch</td><td colspan="4">28</td><td colspan="4">20</td><td colspan="4">20</td><td colspan="4">13</td><td colspan="4">28</td><td colspan="4">20</td><td colspan="4">20</td><td colspan="4">13</td></tr>
<tr><td>Fit</td><td colspan="2">6.53</td><td colspan="2">6.75</td><td colspan="2">6.53</td><td colspan="2">6.75</td><td colspan="2">13.11</td><td colspan="2">13.49</td><td colspan="2">13.11</td><td colspan="2">13.49</td><td colspan="2">6.53</td><td colspan="2">6.75</td><td colspan="2">6.53</td><td colspan="2">6.75</td><td colspan="2">13.11</td><td colspan="2">13.49</td><td colspan="2">13.11</td><td colspan="2">13.49</td></tr>
<tr><td>Lube</td><td>Y</td><td>N</td><td>Y</td><td>N</td><td>Y</td><td>N</td><td>Y</td><td>N</td><td>Y</td><td>N</td><td>Y</td><td>N</td><td>Y</td><td>N</td><td>Y</td><td>N</td><td>Y</td><td>N</td><td>Y</td><td>N</td><td>Y</td><td>N</td><td>Y</td><td>N</td><td>Y</td><td>N</td><td>Y</td><td>N</td><td>Y</td><td>N</td><td>Y</td><td>N</td></tr>
</table>

FIGURE 14.6: Data structure for Sanclemente & Hess (2007) bolt loosening study. Pitch, fit, and lube are nested within diameter. Diameter is crossed with modulus. Preload is not shown for simplicity. It was nested within diameter in the original study

Once establishing the experimental groups, they then subjected the bolts to the preload condition, then each bolt was tested for 1,750 cycles of tightening and loosening by +/−1mm. Following this cycling, they measured how much the bolt had loosened as a percentage of the preload. The first few rows of their data is presented below in Table 14.6 (0 represents the low level of each variable, 1 represents the higher level).

Fastener material elastic modulus	**Nominal diameter**	**Thread pitch**	**Preload**	**Hole fit**	**Lubrication**	**Loosening**
0	0	0	0	0	0	0.73
1	0	0	0	0	0	1
0	1	0	0	0	0	0.87
1	1	0	0	0	0	1
0	0	1	0	0	0	0.37
1	0	1	0	0	0	1
0	1	1	0	0	0	0.26
1	1	1	0	0	0	0.36
0	0	0	1	0	0	0.05
1	0	0	1	0	0	1
⋮	⋮	⋮	⋮	⋮	⋮	⋮

TABLE 14.6: First 10 rows of Sanclemente & Hess's data, showing coding

When we find that some levels of one variable do not occur across some levels of another variable, we call these variables **Nested**. In Sanclemente & Hess's study, pitch is considered nested within diameter.

	Diameter = $\frac{1}{4}$"	Diameter = $\frac{1}{2}$"	Marginal Mean
Thread Pitch = 28 (Fine0	$y_{111} y_{112}$ y_{113} $\vdots$ y_{11n}		$\bar{y}_{1..}$
Thread Pitch = 20 (Coarse)	$y_{211} y_{212}$ y_{213} $\vdots$ y_{21n}		$\bar{y}_{2..}$
Thread Pitch = 20 (Fine)		$y_{321} y_{322}$ y_{323} $\vdots$ y_{32n}	$\bar{y}_{3..}$
Thread Pitch = 13 (Coarse)		$y_{421} y_{422}$ y_{423} $\vdots$ y_{32n}	$\bar{y}_{4..}$
Marginal Mean	$\bar{y}_{.1.}$	$\bar{y}_{.2.}$	$\bar{y}_{...}$

TABLE 14.7: Pitch is nested within levels of diameter

If we are interested in the extent to which pitch of a bolt matters in its tendency to loosen, we worry a bit about the differences the diameter of the bolts—we expect there to be some random difference associated with bolt diameter. But there is also some *systematic* difference—given the coefficient of static friction is related to the surface area in contact between bolt and nut. So, there is variation associated with $\frac{1}{4}$" diameter that is *unique* to the 28 thread pitch (see Table 14.7 for the nested design). Likewise, there is variation associated with 13 pitch that is unique to $\frac{1}{2}$" bolts because there are no $\frac{1}{4}$" bolts with 13 pitch. The 20 pitch condition does seem to overlap, but this condition is defined as coarse for $\frac{1}{4}$" bolts and fine for $\frac{1}{2}$" bolts. It is expected to behave differently for the two diameters.

For this reason, any analysis we do of the pitch is dependent upon which bolt they came from. In other words, the effect of pitch is *dependent* upon the bolt diameter. Therefore, we have to take this dependency into account—there will be no interaction term for Diameter x Pitch. Or for Preload x Pitch, nor for Fit x Pitch. These variables are all nested within levels of the bolt diameter.

Suppose we were to, somehow equate pitches for both $\frac{1}{4}$" and $\frac{1}{2}$" bolts. Contrast the nested design above with a fully crossed Factorial Design presented in Table 14.8):

	Diameter = $\frac{1}{4}$"	**Diameter = $\frac{1}{2}$"**	**Marginal Mean**
Thread Pitch = Fine	$y_{111} y_{112}$ y_{113} $\vdots$	$y_{121} y_{122}$ y_{123} $\vdots$	$\bar{y}_{1..}$
Thread Pitch = Coarse	$y_{211} y_{212}$ y_{213} $\vdots$	$y_{221} y_{222}$ y_{223} $\vdots$	$\bar{y}_{2..}$
Marginal Mean	$\bar{y}_{.1.}$	$\bar{y}_{.2.}$	$\bar{y}_{...}$

TABLE 14.8: Fully crossed factorial design. Each level of diameter contains both levels of Pitch

In *this* analysis, because each pitch is found across each level of diameter, there is no variation unique to a single pitch condition. For this reason, we can compare the levels of pitch across all levels of thread in any significant interactions we might find.

Nested designs occur when, levels of one factor (say, thread pitch is Factor B) are hierarchically subsumed under (or **nested** within) levels of another factor (say diameter is Factor A). As a result, assessing the all combinations of A and B levels is not possible. Often called, *hierarchical* models, the variation occurring in the higher levels in the analysis (Factor A) must be assessed relative to the variation occurring in the lower levels (Factor B).

Nested, or *hierarchical* designs are common in environmental effects monitoring studies. They are also common in the study of manufacturing processes because it is impractical (or impossible) to swap out innovations for all models of a product. For example, fuel injectors may be nested within automobile model since they can't typically be swapped out across different models.

Many demographics are also nested variables. For example, students are nested within classrooms If you are testing the effect of a learning system, you have to take into account that there are different teachers in different classrooms, and that students are not randomly assigned to classes, thus there is variability unique to any class that you need to account for if you wish to compare across classes.

Thankfully, the GLM addresses nested variables in *exactly the same manner* as any other factor:

$$Y = \beta_0 + \beta_1 X_1 + \beta_2 X_2 (X_1) + \varepsilon$$

$$\hat{Y} = \beta_0 + \beta_1 X_1 + \beta_2 X_2 (X_1)$$

In the regression equation, the nesting of X_2 in X_1 is denoted $X_2\,(X_1)$ (say, " X_2 *within* X_1 ").

Calculations for sum of squares are also the same: The $SS_{B(A)}$ (B *within* A), is just the sum of the squared deviations of the cell means for B, with the mean of their respective levels of A. There can be no interaction term between Factor A and B(A) with only two factors, one nested in the other. If we added additional factors, they could be nested in A, or crossed with either A or B, so keeping track of which factors are nested and which are crossed is important.

$$SS_{Total} = SS_A + SS_{B(A)} + SS_{\varepsilon}$$

$$\sum_{i=1}^{a}\sum_{j=1}^{b}\sum_{k=1}^{n}\left(\bar{y}_{ijk} - \bar{y}_{...}\right)^2 = b\sum_{i=1}^{a}(\bar{y}_{i..} - \bar{y}_{...})^2 + n\sum_{i=1}^{a}\sum_{j=1}^{b}\left(\bar{y}_{ij.} - \bar{y}_{i..}\right)^2 + \sum_{i=1}^{a}\sum_{j=1}^{b}\sum_{k=1}^{n}\left(\bar{y}_{ijk} - \bar{y}_{ij.}\right)^2$$

Entering the variables into column vectors and then running a factorial analysis of variance with pitch, fit, lube, and preload nested within diameter yields the following ANOVA table (Figure 14.7).

The appropriate response to this table is, "What the holy living hell is this?" This full factorial model with 3 factors nested in diameter results in 6 main effects, 12 2-way interactions, 11 3-way interactions, 5 4-way interactions and one 5-way interaction. The 5-way interaction, down at the very bottom, has a *SS* value of zero with df = 0. This means that, due to the overall low sample size, there is only one data point in this cell, so there is zero variability. We can't even assess that. Moreover, the SS_{ε} has been reduced to a small fraction, but its degrees of freedom is so small that there is no power to detect a significant difference for each of the factors and their interactions. The lesson to be learned here is: When you have such a big model—meaning lots of factors and interactions—you run the risk of two thing: Overfitting and low-power.

A better approach is to identify a smaller set of interactions to analyze, perhaps limiting the analysis to only 2-way interactions. Because Pitch, Preload, and Fit are nested in bolt size, we are really only concerned with their interactions with lube and material, so limiting the analysis to only 2-way interactions is appropriate. This makes the overall analysis more powerful, and perhaps, just as importantly, more *interpretable* (see Figure 14.8).

This is still a prodigious output of information, but it is much easier to walk through than the full model. Plus, the degrees of freedom of the Error term is much higher, resulting in a smaller MS_{ε}. Partial R^2 for this analysis is $1 - \frac{SS_{\varepsilon}}{SS_{Total}} = 1 - \frac{1.5738}{11.0169} = 0.86$. This doesn't mean that the model isn't overfitted, but we can see that certainly reducing the number of terms has given more power to the analysis while not significantly decreasing our goodness of fit.

Analysis of Variance

Source	Sum Sq.	d.f.	Mean Sq.	F	Prob>F
X1	1.8282	1	1.82818	13.24	0.0358
X2	0.0081	1	0.0081	0.06	0.8242
X3(X2)	0.234	2	0.11702	0.85	0.5108
X4(X2)	4.1751	2	2.08753	15.12	0.0271
X5(X2)	0.366	2	0.18299	1.33	0.3869
X6	0.0232	1	0.02321	0.17	0.7094
X1*X2	0.4993	1	0.49927	3.62	0.1534
X1*X3(X2)	0.0622	2	0.03112	0.23	0.8106
X1*X4(X2)	0.0272	2	0.01359	0.1	0.9091
X1*X5(X2)	0.7601	2	0.38006	2.75	0.2095
X1*X6	0.0214	1	0.02137	0.15	0.7203
X2*X6	0.0513	1	0.05135	0.37	0.5851
X3(X2)*X4(X2)	0.1636	2	0.08178	0.59	0.6071
X3(X2)*X5(X2)	0.0575	2	0.02873	0.21	0.823
X3(X2)*X6	0.1437	2	0.07183	0.52	0.6398
X4(X2)*X5(X2)	0.0979	2	0.04893	0.35	0.7276
X4(X2)*X6	0.4155	2	0.20777	1.5	0.3528
X5(X2)*X6	0.0738	2	0.03688	0.27	0.7821
X1*X2*X6	0.2651	1	0.26505	1.92	0.26
X1*X3(X2)*X4(X2)	0.0656	2	0.0328	0.24	0.8021
X1*X3(X2)*X5(X2)	0.0282	2	0.01411	0.1	0.9059
X1*X3(X2)*X6	0.0011	2	0.00056	0	0.996
X1*X4(X2)*X5(X2)	0.0368	2	0.0184	0.13	0.8801
X1*X4(X2)*X6	0.117	2	0.05852	0.42	0.6885
X1*X5(X2)*X6	0.0679	2	0.03396	0.25	0.7964
X3(X2)*X4(X2)*X5(X2)	0.0434	2	0.0217	0.16	0.8612
X3(X2)*X4(X2)*X6	0.0528	2	0.02638	0.19	0.8354
X3(X2)*X5(X2)*X6	0.0786	2	0.03932	0.28	0.7705
X4(X2)*X5(X2)*X6	0.2618	2	0.13088	0.95	0.4797
X1*X3(X2)*X4(X2)*X5(X2)	0.0487	2	0.02435	0.18	0.8464
X1*X3(X2)*X4(X2)*X6	0.0131	2	0.00656	0.05	0.9543
X1*X3(X2)*X5(X2)*X6	0.0077	2	0.00383	0.03	0.9729
X1*X4(X2)*X5(X2)*X6	0.0106	1	0.01058	0.08	0.7999
X3(X2)*X4(X2)*X5(X2)*X6	0.0315	1	0.03151	0.23	0.6655
X1*X3(X2)*X4(X2)*X5(X2)*X6	0	0	0	0	NaN
Error	0.4143	3	0.1381		
Total	11.0169	63			

Hierarchical (Type II) sums of squares.

FIGURE 14.7: ANOVA table for the full factorial, nested design for the Sanclemente & Hess (2007) bolt loosening data

We can now see 5 significant effects: Two main effects of Modulus, Preload, and Fit; and three 2-way interactions, Modulus x Diameter, Modulus x Fit, and Preload x Lube. What the heck does all this mean? Let's start with a few graphs of the group means.

The omnibus test returned significant results for two of the main effects in our analysis. Looking at Figure 14.9, you can see on the left that the high modulus material (steel bolts vs aluminum) resulted in looser bolts than the lower modulus. The partial R^2 for modulus is about $\frac{1.955}{11.0169} = 0.18$. Not a great predictor, but a significant one. On the right, the main effects show that a low preload, on average, resulted in

Analysis of Variance

Source	Sum Sq.	d.f.	Mean Sq.	F	Prob>F
Modulus	1.955	1	1.95504	41	0
Dia	0.0081	1	0.0081	0.17	0.6829
Pitch(Dia)	0.287	2	0.14348	3.01	0.0631
Preload(dia)	4.0795	2	2.03976	42.77	0
Fit(dia)	0.3214	2	0.16068	3.37	0.0466
Lube	0.0171	1	0.01709	0.36	0.5535
Modulus*Dia	0.5008	1	0.50078	10.5	0.0027
Modulus*Pitch(Dia)	0.0446	2	0.02231	0.47	0.6304
Modulus*Preload(Dia)	0.0098	2	0.0049	0.1	0.9026
Modulus*Fit(Dia)	0.6323	2	0.31617	6.63	0.0038
Modulus*Lube	0.0634	1	0.06341	1.33	0.2572
Dia*Lube	0.0327	1	0.03268	0.69	0.4137
Pitch(Dia)*Preload(Dia)	0.2051	2	0.10255	2.15	0.1325
Pitch(Dia)*Fit(Dia)	0.0937	2	0.04684	0.98	0.3852
Pitch(Dia)*Lube	0.1192	2	0.05958	1.25	0.2999
Preload(Dia)*Fit(Dia)	0.1391	2	0.06957	1.46	0.2471
Preload(Dia)*Lube	0.5737	2	0.28685	6.01	0.0059
Fig(Dia)*Lube	0.0608	2	0.0304	0.64	0.535
Error	1.5738	33	0.04769		
Total	11.0169	63			

Hierarchical (Type II) sums of squares.

FIGURE 14.8: ANOVA table for the nested design for the Sanclemente & Hess (2007) bolt loosening data, only accounting for 2-way interactions

looser bolts. Its partial R^2 is about $\frac{4.0795}{11.0169} = 0.37$. This is a pretty good predictor, explaining over twice that of the next best predictor, modulus. The 95% confidence intervals indicated in Figure 14.9, show almost no overlap in the predicted distributions of the low versus high conditions for both of these variables.

Interpreting these results makes pretty obvious sense. As a bolt is being tightened, it is actually being stretched. This causes tension, increasing both the friction between the threaded surfaces, and some deformation of the threads. Aluminum bolts, made with the softer, lower-modulus material, will have both greater friction and greater deformation, resulting in less tendency to loosen. Preload is a measure of how much torque is put on a bolt when it is tightened. Tighter bolts, naturally tend to loosen less because of added tensile force.

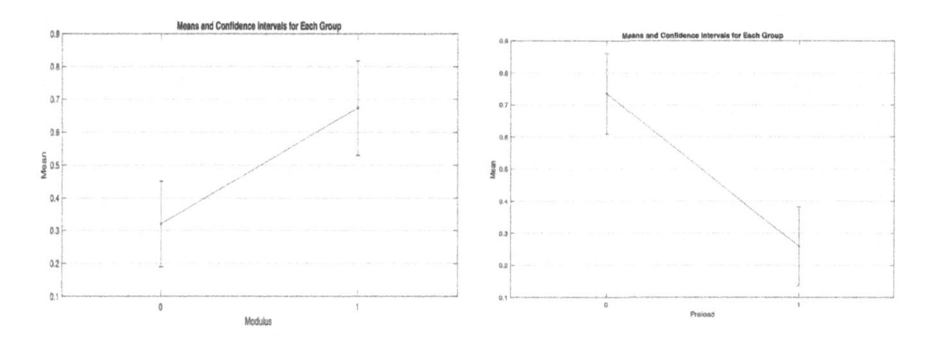

FIGURE 14.9: Main effects showing the mean difference in each factor. The left most graph is for modulus. The right graph is for preload. Bars are 95% confidence intervals around the group mean

But the picture is not that simple. There are *interactions* between these two factors and other important factors in the analysis, namely diameter, fit, and the presence or absence of lubrication (see Figure 14.10).

For Modulus x Diameter, we see a monotonic interaction that shows that modulus matters more for small diameter bolts than large diameter. Small diameter bolts made from aluminum loosened much less than large diameter aluminum bolts. This effect was much more pronounced than for steel bolts (Partial R^2 = 0.05). Modulus x Fit shows that modulus matters much more for tight bolts than loose bolts. Aluminum bolts with loose fit showed *looser* bolts after testing, than steel bolts with loose fit (Partial R^2 = 0.06). Finally, preload and lubrication interact significantly. For higher preload, lubrication lessens the tendency for bolts to loosen, and under lower preload lubrication tends to increase the tendency for bolts to loosen (Partial R^2 = 0.05).

Putting all this information together suggests that choosing larger diameter bolts with lower modulus of elasticity, and then placing them under higher preload, in tighter nuts with a bit of oil results in the best combination of factors, on average, for reducing the tendency for the fasteners to loosen under repeated stress. But if I were to have to choose only one factor, the one that makes the most difference of all these, I would choose preload—duh! Tightening a bolt will tend to make it resistant to loosening. Another drab discovery of the obvious by your engineering professor.

Just to double-check my assumptions, I run a Q-Q plot of the residuals (see Figure 14.11):

There is a little "snakiness" to the data, especially at the upper end of the distribution, we might consider those two points out there in space to be outliers and remove them from the analysis, but overall I don't see a big violation of the assumption of normality.

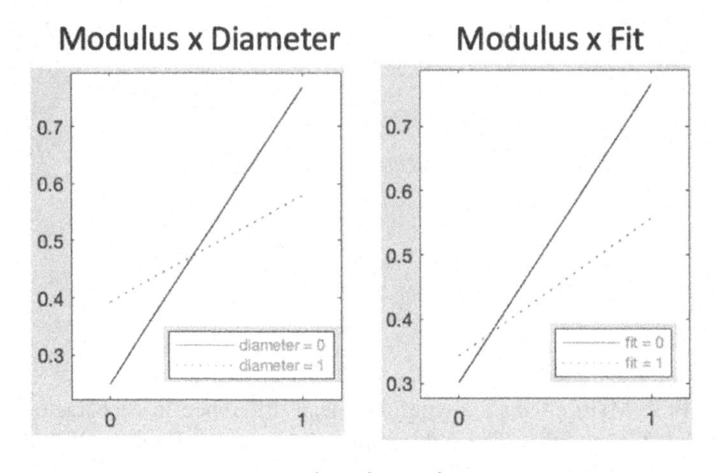

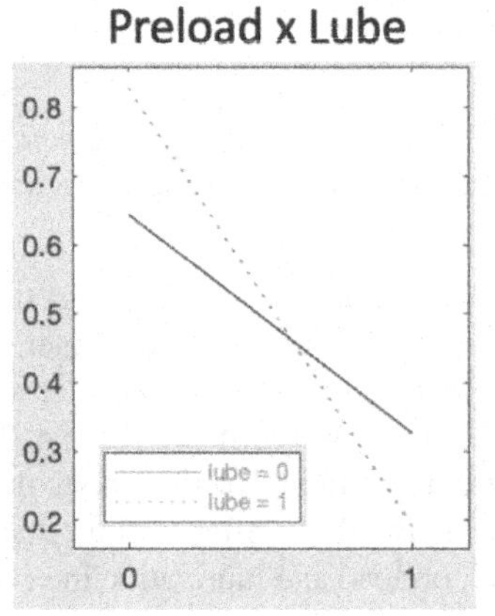

FIGURE 14.10: Significant 2-way interactions

14.3 Summary

Factorial Analysis of Variance is one of the most flexible and useful models for testing group means, whatever the experimental design. We have worked through three powerful methods in this chapter and in Chapter 13: The Randomized Block Design, Factorial Design with Interactions, and Nested Design. There are only a couple more experimental models that add to these three: Repeated Measures design, and Response Surface Methodology as the chief methods employed in engineering contexts. We will not cover these two designs in this book, but if you have a decent grounding in Factorial models, they are not difficult to understand and apply.

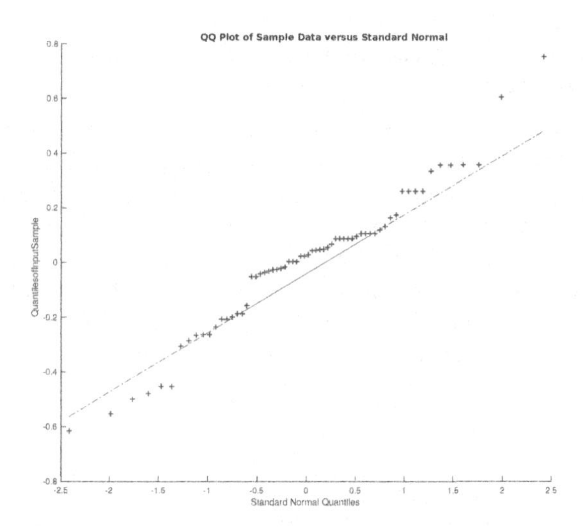

FIGURE 14.11: Q-Q plot of the residual error in the Sanclemente and Hess (2007) bolt loosening data

The most important concept in this chapter that you need to keep in your toolkit for later is the fact that all of these designs are just examples of the General Linear Model. Adding factors or interactions, or attending to nesting is just a matter of keeping your dummy coding clear, and adding another column to the matrix of predictors in the GLM. The formulas stay the same, and the method of computing variance components is the same. These designs all have the same assumptions. It is all just regression!

But another concept that I think you should keep in mind is the fact that, as you add more predictors, particularly interactions and nested factors that will always show some statistical dependency, you run the risk of suppressing real effects like we saw in our extended example of our nested factorial model, reducing your power to detect those real effects and thus increasing your Type II error rate. You can also, quite often, actually, detect effects that are difficult to impossible to interpret. If it is necessary to interpret such interactions, I would suggest Response Surface Methodology, that has a particular method for optimizing combinations of factors. For most factorial analyses, limiting interactions to 2-way or at most 3-way is generally a more robust and useful approach.

Finishing this chapter, you have now gone well beyond what is typically expected of a first course in engineering statistics. You will undoubtedly forget some (many) of the details of these analyses. But you should commit the logic of hypothesis testing, and the General Linear Model and its logic to memory. Armed with these concepts, you can reconstruct any of the analytic techniques we have studied thus far (including ones like the χ^2 test of Independence that operate on categorical dependent variables), and you are prepared for advanced statistical techniques that do not assume linear models, or normally distributed populations such as the.

The next chapter, we will explore one of the simplest, yet most powerful computational techniques for getting around some of the more restrictive assumptions that the parametric approach to inferential statistics have. Many times, we can't make the case for normally distributed populations, or homogeneity of variance. In such cases, our *t*-tests and ANOVA can be woefully underpowered. So, we have to pick ourselves up by our bootstraps—Bootstrap is the name of the general computational method for estimating population characteristics without worrying about parameters, continuous distribution functions, or shape of parent populations.

14.4 References

Box, G. E., Hunter, J. S., & Hunter, W. G. (2005). Statistics for experimenters. In *Wiley Series in Probability and Statistics*. Wiley Hoboken, NJ, USA.

Myers, R. H., Montgomery, D. C., & Anderson-Cook, C. M. (2016). *Response surface methodology: process and product optimization using designed experiments*. John Wiley & Sons.

Nelder, J. A. (1977). A reformulation of linear models. *Journal of the Royal Statistical Society: Series A (General)*, *140*(1), 48-63.

Nepal, B., Mohanty, S., & Kay, L. (2013). Quality improvement of medical wire manufacturing process. *Quality Engineering*, *25*(2), 151-163.

Sachdeva, A., Sharma, V., Bhardwaj, A., Singh, S., & Kumar, S. (2012). Factorial analysis of lifting task to determine the effect of different parameters and interactions. *Journal of Manufacturing Technology Management.*

Sanclemente, J. A., & Hess, D. P. (2007). Parametric study of threaded fastener loosening due to cyclic transverse loads. Engineering failure analysis, 14(1), 239-249.

Snee, R. D. (1974). Computation and use of expected mean squares in Analysis of Variance. *Journal of Quality Technology*, *6*(3), 128-137.

14.5 Study Problems for Chapter 14

1. In 2012, Sachdeva, and colleagues studied the impact of a number of factors related to lifting weights on the potential for back injury. The primary factors they were interested in were the weight of the object being lifted (10, 15, and 20 kg), the twisting angle of the body as the weight was lifted (0, 30, or 45 degrees), and the frequency at which weights were lifted again and again (2, 3, or 4 times per minute). The outcome measure was a "lifting index," a set of biometric measurements that predict injury potential (higher Lifting index means higher potential for back injury), at the origin of the lift, and at the destination. Their data is available in the file "Lifting_Factorial_ANOVA_Interactions.xlsx"

   (a) Are there any significant main effects of these factors on potential for injury at destination of the lift? If so, what levels of those effects seem to make the most impact on potential for injury?

   Including all interactions in the analyses for each of the dependent variables yields NaN (Not a Number) errors, because the model is overfitted given the sample size (i.e., the residual error is zero). Here is the ANOVA table for lifting index at the origin of the lift using the Matalb code:

   »[p,table,stats]=anovan(origin,{ weight,angle,freq} ,'alpha', 0.05, 'model', 'full' ĂŹ, 'sstype', 2, 'varnames', ["weight","angle", "freq"]);

Analysis of Variance

Source	Sum Sq.	d.f.	Mean Sq.	F	Prob>F
weight	8.8001	2	4.40003	Inf	NaN
angle	0.4148	2	0.20738	Inf	NaN
freq	0.8014	2	0.40071	Inf	NaN
weight*angle	0.232	4	0.058	Inf	NaN
weight*freq	0.1061	4	0.02652	Inf	NaN
angle*freq	0.1835	4	0.04588	Inf	NaN
weight*angle*freq	0.3461	8	0.04326	Inf	NaN
Error	0	0	0		
Total	10.884	26			

Hierarchical (Type II) sums of squares.

   Looking at the SS for each of the predictors, including the interactions, we can see that none contributes a lot of variation to the overall model compared to the main effects. The 3-way interaction contributes about 3% to the total. In this case, I would re-run the analysis using just the main effects:

   » [p,table,stats]=anovan(origin,{ weight,angle,freq} ,'alpha',0.05,'model',2, 'sstype',2,'varnames',["weight","angle","freq"]);

Analysis of Variance

Source	Sum Sq.	d.f.	Mean Sq.	F	Prob>F
weight	8.8001	2	4.40003	101.42	0
angle	0.4148	2	0.20738	4.78	0.0201
freq	0.8014	2	0.40071	9.24	0.0014
Error	0.8677	20	0.04339		
Total	10.884	26			

Hierarchical (Type II) sums of squares.

We can see that each of the main effects shows a significant impact upon lifting index at the origin of the lift.

```
» [results,means]=multcompare(stats,'ctype','bonferroni','dimension',1);
» [results,means]=multcompare(stats,'ctype','bonferroni','dimension',2);
» [results,means]=multcompare(stats,'ctype','bonferroni','dimension',3);
```

Bonferroni post hoc analyses show significant differences for each successively larger weight.

Angle shows significant differences between the highest lift angle, 45 degrees, compared to 0 degrees, but non-significant differences between 0 and 30 degrees or between 30 and 45 degrees.

Frequency also shows significant differences between the highest level, 4 times per minute and 2 time per minute, but nonsignificant differences between 2 vs 3 times per minute nor 3 vs 4 times per minute.

For the lift index at destination, we see the similar overfitting when we run a full factorial with all interactions. So, we will suppress the 3-way interaction and focus on the 2-way interactions:

```
» [p,table,stats]=anovan(destination,{ weight,angle,freq} ,'alpha',0.05,'model',2,
'sstype',2,'varnames',["weight","angle","freq"]);
```

Analysis of Variance

Source	Sum Sq.	d.f.	Mean Sq.	F	Prob>F
weight	5.95156	2	2.97578	6180.47	0
angle	0.28725	2	0.14363	298.3	0
freq	0.31614	2	0.15807	328.3	0
weight*angle	0.01717	4	0.00429	8.92	0.0048
weight*freq	0.02281	4	0.0057	11.85	0.0019
angle*freq	0.00606	4	0.00151	3.15	0.0786
Error	0.00385	8	0.00048		
Total	6.60485	26			

Hierarchical (Type II) sums of squares.

» [results,means]=multcompare(stats,'ctype','bonferroni','dimension',1);

» [results,means]=multcompare(stats,'ctype','bonferroni','dimension',2);

» [results,means]=multcompare(stats,'ctype','bonferroni','dimension',3);

Again, we see that weight (not surprisingly) shows the greatest impact on lift index at the destination of the lift, followed by frequency and then the Weight x Frequency interaction.

Weight, again, shows that each successively higher level is significantly different than the lower levels. Angle and Frequency also shows that each successively higher level contributes significantly more to lifting index than the lower levels.

(b) **Are there any significant interactions? If so, which ones, and what combinations seem to impact potential for injury the most?**

From the ANOVA table, we can see that all 2-way interactions are significant, with the SS for Weight x Frequency accounting for the most variation. An interaction plot for Weight x Frequency using lifting index at destination as the dependent variable shows that

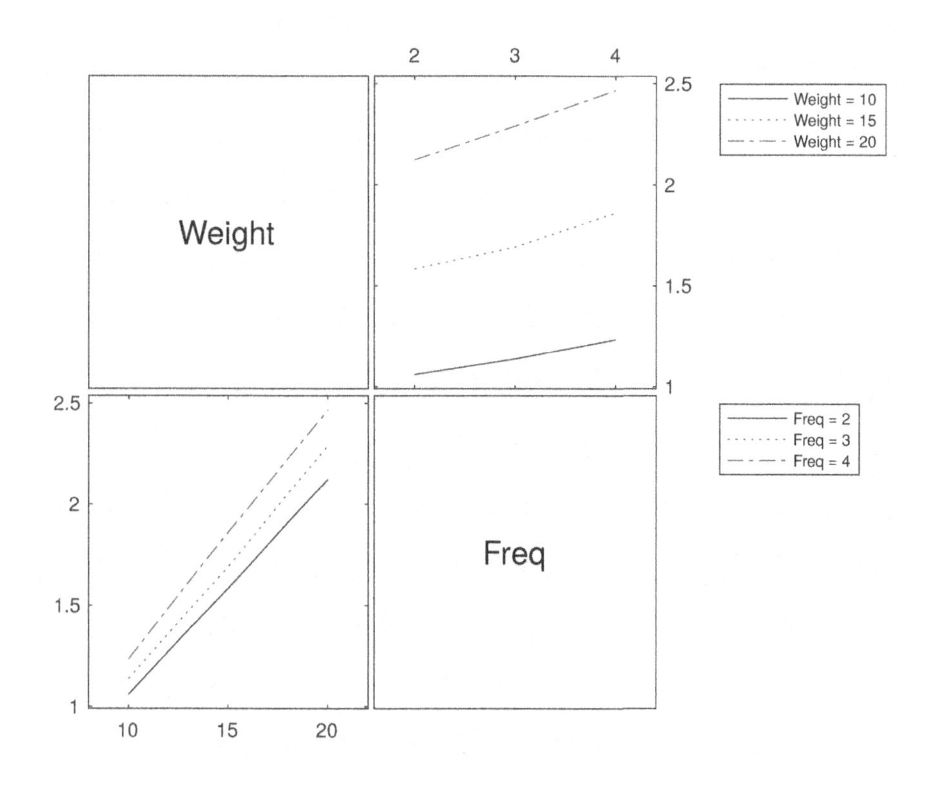

There doesn't appear to be a big difference in slopes across the different groups. In the upper right panel, we can see that the slope for Weight= 20 appears to be greater than the slope for either of the other two groups of weight. In the lower left panel, we can see that Frequency = 4 times per minute appears to have a steeper slope than either of the other two groups.

» [results,means]=multcompare(stats,'ctype','bonferroni','dimension',[1,3]);

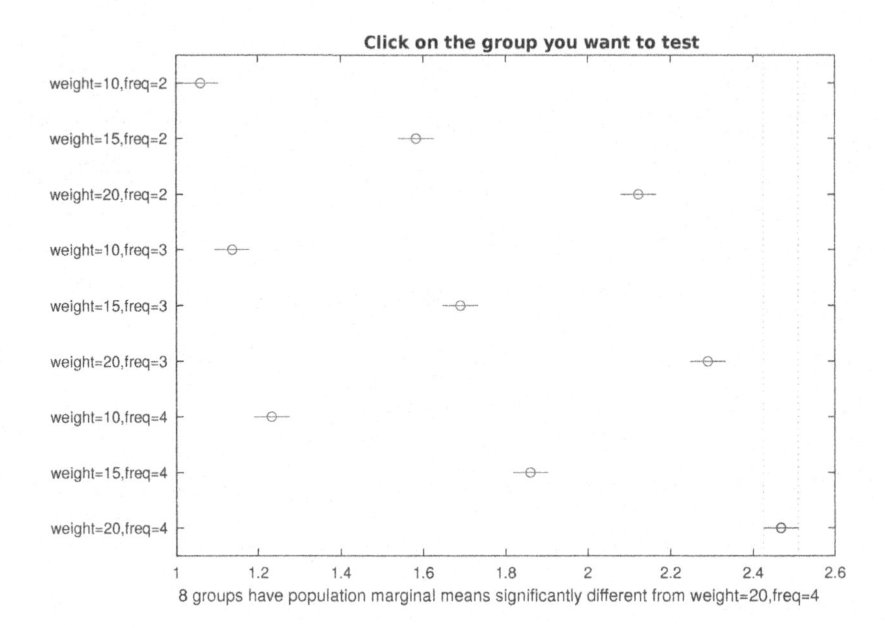

The plot of the group means and confidence intervals in this interaction shows clearly that the combination of weight and frequency is highest at Weight=20, Freq = 4 , and lowest at Weight = 10, Freq = 2. The only non-significant comparison is between Weight=10, Freq=3, and Weight=10, Freq=2. Interpreting these results, we can see that as weight and/or frequency of lifting gets to very low levels, the difference they make in the danger of back injury becomes insignificantly less, when lift index is measured at the destination of the lift.

(c) **Evaluate the practical significance of each significant effect? Are any so small that they could potentially be ignored?**

Here is the hard part of this analysis. Even though Weight x Frequency shows a significant interaction, and this interaction makes perfect sense from a biomechanics perspective, the amount of variation it accounts for in the GLM is only slight. Its $R^2_{WxF} = 0.003$ only 3 tenths of 1% of the variation is accounted for by the biggest interaction effect. I, personally, don't think this is practically something I am going to worry about if I am a physical trainer.

(d) **What is the overall goodness of fit of the data?**

For the analysis of variance on the lifting index at origin, $R^2 = 0.92$. This is a great fitting model, even when the interactions returned are non-significant.

For the ANOVA of lifting index at destination, $R^2 = 0.999$. This almost certainly is a result of overfitting. Considering the low impact of each of the interactions, I can remove them, making a more parsimonious model. Removing the interactions in this analysis gives $R^2 = 0.992$. I am satisfied that any significant interactions found are not of any practical value.

(e) **Are there any assumptions of Factorial ANOVA that were violated in your analysis? If so, how would you propose to address them?**

The analysis of lifting angle at origin shows a Q-Q plot of the residuals that looks "textbook" good.

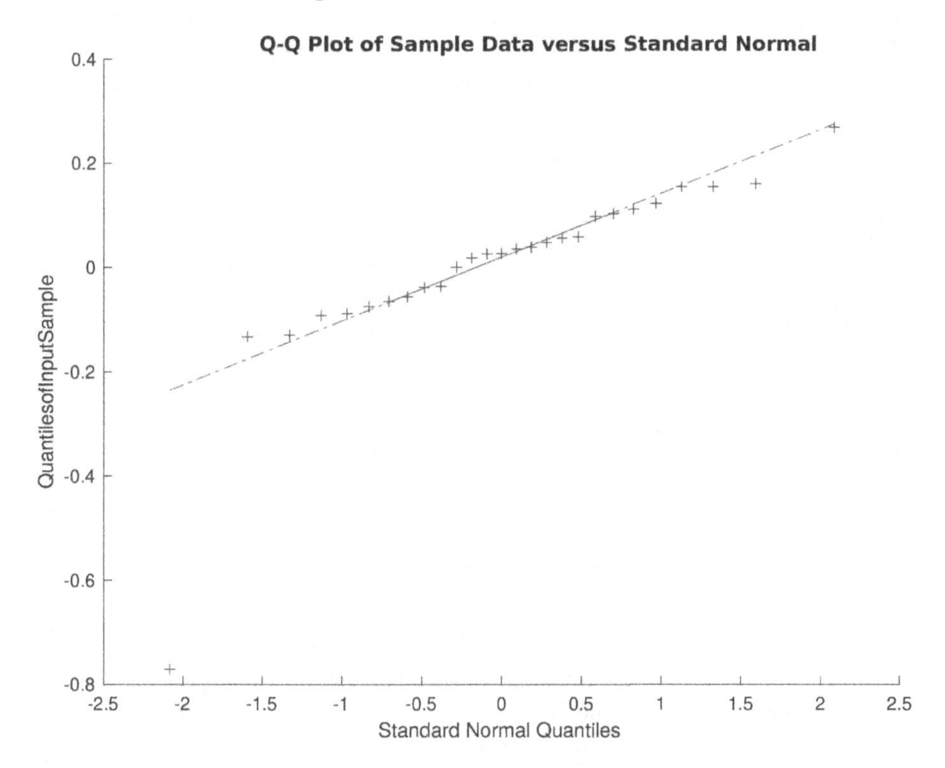

The Q-Q plot of the residuals for the analysis of lifting index at destination looks just right:

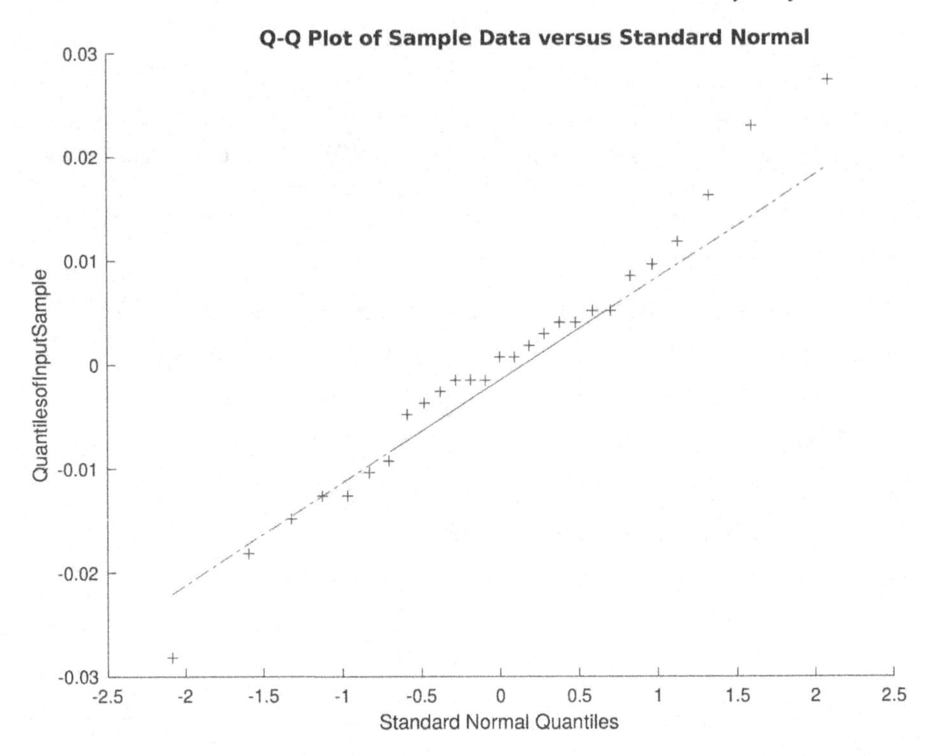

There is a telltale tail that deviates slightly from the y=x line, but this is very minor compared to the overall linearity of the model.

Standard deviations for all groups in the analyses appear to be fairly close to each other. I am not worried about non-homogeneity of variance.

(f) **Evaluate the impact of the factors and interactions on the potential for lifting at the origin of the lift. Is there any difference between these effects and those you found for the destination?**

Evaluating both origin and destination indices shows that each of the main effects appear to have more of an effect at the destination than at the origin. This is expected given the biomechanics of lifting. In particular, angle does not appear to have much of an effect at the origin, but has a greater effect at the destination. Perhaps we should follow up on this, given the moment put on someones limbs when muscles are extended may exacerbate the problems that may cause back injury.

Analysis of Variance

Source	Sum Sq.	d.f.	Mean Sq.	F	Prob>F
X1	8.8001	2	4.40003		
X2	0.4148		0.20738	0.415	
X3		2	0.40071	0.801	
X1*X2	0.232	4	0.058		
X1*X3		4	0.02652		
X2*X3	0.1835	4	0.04588		
X1*X2*X3	0.3461		0.04326		
Error	4.000	8	0.500		
Total	14.884	34			

Constrained (Type III) sums of squares.

2. Complete the following ANOVA table. Which of the main effects and/or interactions are significant?

   Here you can see that only X1, one of the main effects, shows a significant F, p=0.01. All the others are nonsignificant.

3. In the bolt loosening example by Sanclemente, & Hess (2007), they treated all the factors and subsequent interactions as fixed effects.

   (a) How did this choice constrain the kinds of conclusions they could make with their results?

   Because fixed effects are only generalizable to themselves and not to the (potentially) rest of potential values of an independent variable, it limits the usefulness of the study to the specific values chosen. For variables like pitch, this could be problematic, because there are many pitches depending on the size of the bolt and its application. Likewise for modulus, the results are limited to only moduli that are associated with (a specific set of) Al and Steel alloys.

   (b) In their study, modulus only has two levels: Al and Steel bolts. Likewise, levels of bolt size are just $\frac{1}{4}$" and $\frac{1}{2}$" . What would their assumptions have to be if they treated these variables as random effects?

   If the researchers wanted to generalize to the entire population of bolt moduli, including brass, high carbon steel, other Al alloys, etc., they have to assume that the Al and Steel bolts they selected are representative of the whole of moduli. That is a pretty tenuous assumption. If they limited their generalization to only Al and Steel bolts, but assume other alloys are similar to their current ones (perhaps okay), then the random effects model would be appropriate.

4. You are studying the heat transfer of the automobile cooling system you are designing. You test the vehicles under real conditions in a number of sites around

the world. You randomly assign 2 automobiles to Treatment and 2 to Control conditions in each site and run 30 trials for each car. You have 4 sites in each region of the world you are testing: North America, Europe, China, and Australia, that represent extreme thermal conditions.

Sites	**United States**	**Australia**	**China**	**Brazil**
Mountains	Car1 Treat Car2 Control Car3 Treat Car4 Control			
Low Desert		Car5 Treat Car6 Control Car7 Treat Car8 Control		
High Desert			Car9 Treat Car10 Control Car11 Treat Car12 Control	
Tropical				Car13 Treat Car14 Control Car15 Treat Car16 Control

(a) Discuss what (if any) factors are nested factors in your experiment. Justify your reason(s) why you can say that they are nested factors.

A nested factor is one in which some of its levels are not represented in the levels of another factor in the experiment. Here Car, the experimental unit, is nested within treatment (this is always the case with experimental units). Treatment/Control is crossed with both Site and Region. Each site and each region have cars in both the Treatment and Control conditions. Region is nested within Site because each geographical feature is only found within one of the nations in which the cars were tested.

5. The efficacy of fume-extractors was studied in an electronic parts manufacturing facility. Three different fume extractors were randomly assigned to 6 different workstations at the facility (see table below). The amount of Colophony (rosin) fumes (ppm) given off by soldering was measured for 5 air samples, taken randomly at different points in the day for each workstation. These data are presented in the table below:

	Fume Extractor		
Station	Ductless Fixture	Ducted Fixture	Ductless Portable
1	17 7 9 12 15		
2	13 23 14 18 22		
3		21 24 14 19 17	
4		25 29 26 24 21	
5			26 34 32 30 28
6			32 32 37 34 35

(a) Perform the appropriate analysis for this data.

Because there are two values of station for every value of fixture, we will model station nested within fixture.

```
»[p,table,stats]=anovan(fumes,{ fixture,station} ,'alpha',0.05,'sstype',
2,'nest',m,'varnames',["fixture","station"]);
```

(b) Is there evidence to conclude that one or more of the fume extractors is more efficacious than the others?

From the ANOVA table, it looks like the omnibus test returned a significant difference for both fixture and station(fixture). We should now do a post hoc analysis to see which levels of fixture appear to be most efficacious.

Like before, I will choose the Dunn-Bonferroni method. Matlab doesn't compute multiple comparisons for nested data. So I ran a 1-way ANOVA for fixture, and used the $\sqrt{MS_\varepsilon}$ as my denominator for the t-test. Below are the pairwise t-tests (familywise α =0.05).

Group 1	**Group 2**	**Lower CI Bound**	**Mean Difference**	**Upper CI Bound**	$p\left(F_{Scheffe}\right)$
1	2	−12.009	−7.000	−1.991	0.004
1	3	−22.009	−17	−11.991	8.44 x 10^{-9}
2	3	−15.009	−10	−4.991	7.07 x 10^{-5}

The confidence intervals all do not span zero, so for fixture, each successively larger value is significantly different from lower values.

(c) Is there any difference caused by the workstations in the facility?

For station(fixture), gain, I had to do some on-the fly computation. Remember that, because station is nested within fixture, each level of fixture estimates the population parameter to which each nested level of station can be compared. Here are the means and standard deviations of all the workstations:

	Fixture 1		**Fixture 2**		**Fixture 3**	
Station	$\overline{x}$	s	$\overline{x}$	s	$\overline{x}$	s
1	12	4.123				
2	18	4.528				
3			19	3.808		
4			25	2.915		
5					30	3.162
6					34	2.121

Now we can run 2-sample t-tests for each of these stations against the other, adjusting our Type I error rate by the number of contrasts $\binom{6}{2} = 15$! That would drop our Type I error rate per contrast to $\frac{0.05}{15} = 0.0033$. Very small! To make my job easier, I will start with the largest mean difference (Station 6 at 34 Station 1 at 12 = 22). Then I will systematically work to successively smaller differences until I reach non-significant results. At that time, I will stop the procedure. A table of all of the t-values for the Bonferroni contrasts is provided below:

			Station		
Station	6	5	4	3	2
1	−67.103	−48.992	−36.410	−17.638	−13.856
2	−45.254	−30.728	−18.383	−2.390	
3	−48.667	−31.429	−17.694		
4	−35.307	−16.442			
5	−14.857				

The critical value of t at $5 + 5 - 2 = 8$ degrees of freedom, with an α per comparison at 0.0033 is: −3.639 (low ppm of fumes is better). We can see from the table that all of our 15 contrasts returned significant differences in favor of the higher station, with the exception of the comparison between station 2 and station 3. The mean difference between the two stations was only 1 ppm and was non-significant.

(d) **Why isn't there an interaction between workstation and type of fume extractor?**

Because levels of station are nested within fume extractor, there can be no interaction. Multiple levels of station do not exist across multiple fume extractors!

(e) **Using their partial R^2, which of the two factors, or their interaction makes the most difference in the quality of air in the workplace?**

Definitely fixture. Its $R^2 = \frac{1{,}460}{1{,}980} = 0.74$, compared to station with a partial $R^2 = \frac{220}{1{,}980} = 0.11$

(f) **If you were to write a memo to the facility manager regarding the results of your study, what would you conclude?**

From this analysis, we can conclude that the ductless fixtures appear to work best, but that there is also variation among station, even within fixture type. This man necessitate the need to install more efficient fixtures in our manufacturing facilities, but also look at the layout and practices being used in each of the workstations to see why workstation number 1 appears to be so much better than the others.

Part IV

Introduction to Computational Methods and Machine Learning

15

The Bootstrap

Since about Chapter 6, I have continually emphasized the importance of making explicit assumptions about the nature of the population from which we are taking our samples. Each of the hypothesis tests we have explored has a basic assumptions that any two (or more) samples we are testing, come from the same population, and that any differences in the samples is only due to randomness. It is a big help if you know what the population's characteristics are! If you know that it is symmetric and bell shaped, then the mean and standard deviation are all you need to know to be able to plot the function and compute every value of the PDF from every value of x.

The Central Limit Theorem helps us in this regard, when the exact shape and location of a distribution are unknown. We are assured that, if we take infinitely many samples of a given size, their means will converge on the population mean, no matter what the shape of the parent distribution is. But with finite distributions, and with distributions that are decidedly non-normal in shape, the number of measurements we need to take to approximate the normal distribution can vary from very few (where the parent distribution is close to normal in shape), to very many (negative exponential, or other highly skewed phenomena). This makes our sampling expensive, requiring many observations to help reduce the standard error of the mean.

So, in instances where we violate many of the assumptions of what are called, *Parametric statistics*—those that estimate parameters of a smooth, continuous population function, from discrete, sample data, we often lose power. The usual remedies for such violations are to transform the data (as we have seen in Multiple Regression and the General Linear Model), or to resort to *nonparametric* analyses that do not assume the shape of the parent distribution. nonparametric tests do not treat statistics as estimates of parameters in the algebraic sense of the word. Instead, they are treated like descriptive characteristics of a population without making assumptions about the specific shape of the PDF.

It must be noted that there are nonparametric analogs to each of the tests we have studied in this course. The t-test has a nonparametric analog called the Mann-Whitney U test. It is essentially a t-test using ranked data. The Oneway ANOVA analog is called the Kruskal-Wallis test. The Wilcoxon and Friedman tests are nonparametric analogs to the Paired-sample t-test, and the Aligned Rank Transform can handle factorial models (Wobbrock, et al., 2011). These and other nonparametric

DOI: 10.1201/9781003094227-15

tests and their application are beyond the scope of this book, but some excellent explanations are found in Kvam & Vidakovic (2007).

15.0.1 What It Means to Be *Nonparametric* (and What It Does NOT Mean)

The most basic description of what nonparametric analysis entails concerns its assumptions, or lack thereof. In general, when we free ourselves of the assumption that the parent distribution of a sample, the population, has a specific form, then the mean and variance are no longer parameters, but are descriptors at both the sample and population levels. Some people call nonparametric analyses *distribution free* methods, but this is *not* true. Many times we assume that the characteristics of a population distribution have some general form, i.e., that it is symmetric, for example, or unimodal, but we don't assume that these characteristics describe a *particular* algebraic function. In other words, nonparametric are defined by what they are not. When we are not sure of the underlying probability distribution that may have generated a given sample, and for which the sample is serving as an estimate, we need to either ignore the warnings that our brain is telling us about the nature of the data and just use ANOVA like the proverbial hammer—when all you have is a hammer, everything looks like a nail—OR we need to relax our assumptions and utilize methods that are more robust to their violation (Kvam & Vidakovic, 2007).

Notice that I have continued to use the word, *assumptions*. nonparametric tests are just as dependent upon a set of assumptions as parametric analyses. There are just fewer of them. Violation of such assumptions as normal population and homogeneity of variance can cripple power in the t-test and its ANOVA extensions. But so can violation of assumptions of nonparametric tests. The Mann-Whitney U test, a nonparametric analog to the t-test, for example, still assumes that the two samples come from populations with identical distributions—even though we don't hypothesize what the shape of those distributions might be. If they are drawn from populations with different shape characteristics, Mann-Whitney loses power just like any other t-test.

Parametric Statistics involve estimating values of a population that follows a given probability distribution. Measures of the sample serve as parameters-variables that when fixed, make the probability known for all values of the independent variable.

Nonparametric Statistics involve estimating values of a population with no exact specification of the probability distribution of that population.

15.1 The Bootstrap Method

Relatively recently in the history of statistics, Bradley Efron accelerated the modern era of statistical analysis by introducing a concept so simple as to be almost absurd. He called the principle of the thing, *Bootstrapping*, and the method, *Bootstrap*. Since the late 1940s, statisticians have embraced the power of computational technology just to handle large sets of data at first, but increasingly to solve, computationally, problems that have no closed-form solution. Statistical thermodynamics and stochastic finite elements analysis are a few applications you will come across in engineering (cf., Stefanou, 2009). The basic idea behind these computational methods is to simulate a distribution of potential values of a variable. Because less likely values will appear less often, while more likely values will appear more often in the simulation, a probability distribution can be estimated.

Most of the time, researchers use Gaussian (e.g., Normal) distributions as the function that generates the simulated data. But in a lot of situations, we really don't know the shape of the parent distribution, and often we can see, using Q-Q plots and other tools, that the shape is decidedly *non-normal*. Subsequently, some ability to use a relatively small sample of data to generate a simulated population, is needed.

What Efron did, building on earlier work by Tukey and others, was to suggest that, *if a sample is unbiased*—that is, if a sample is an accurate representation of the population from which it was drawn—*we can use that sample as a proxy for the population*, and simulate the sampling distribution of a statistic to serve in lieu of a parametric distribution. There is a bunch of good math behind this, but basically his proposition, called the **Bootstrap Principle** was to suggest that the expected value of a statistic taken from an unbiased sample, is equal to the population value. *And* that taking many simple random samples from the unbiased sample, with replacement, can simulate the sampling distribution of that statistic with fidelity (Efron, 1979).

I learned about this idea within a decade of his proposition. And within another decade it was being used in engineering contexts to help with problems associated with non-normal data and heteroscedasticity. I have included it in this book because when I went to companies asking them what the most important ideas they wanted their employees to know and be able to use from statistics, Bootstrap rose high in the rankings. Traditionally, it has been an advanced topic, reserved for graduate school, but the concept is actually very simple, and the procedures so straightforward, it really will give you a leg up on other job applicants, on the kinds of problem

you will be able to solve. So learn it, and keep this method in your toolbox for use in your career.

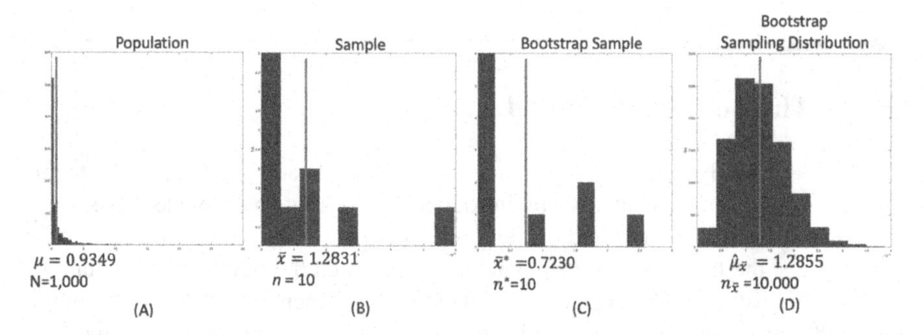

FIGURE 15.1: (A) a definitely non-normal population, (B) a sample (n=10) taken randomly from that population, (C), a bootstrap resample, taken from B, of the same size as B, and (D) 10,000 means, computed from bootstrap samples, taken randomly with replacement from B. The means of the different distributions are marked as vertical lines

The **Bootstrap Principle** is a mathematical formalism that shows that, when a sample is a good representation of the population characteristics, the sample itself can be used as a proxy for the population. Repeated resampling from the original sample will create a sampling distribution of a statistic that is a good estimate of the variability of that statistic in the population.

In other words, *a bootstrap sample is to the original sample as the original sample is to the population.*

"I also wish to thank the many friends who suggested names more colorful than *Bootstrap*, including *Swiss Army Knife*, *Meat Axe*, *Swan-Dive*, *Jack-Rabbit*, and my personal favorite, the *Shotgun*, which, to paraphrase Tukey, "can blow the head off any problem if the statistician can stand the resulting mess." Bradley Efron, (1979), on perhaps the best name of any statistical concept ever (though, *Shotgun* would have been great).

15.1.1 Basic Logic Computing a Bootstrap Confidence Interval

If you really believe that the Bootstrap Principle applies to your sample, then you can use that sample as a proxy for your population. Then, if you take many random samples from the original sample, you can generate a sampling distribution that will resemble that of the population of sample statistics.

Consider the negative exponential population in Figure 15.1 (A) above. I have taken a simple random sample of 10 values from this population (B). Though it is a bit ragged, I really did take a simple random sample, so I have every right to believe that my sample reflects the characteristics of the population. So, I computed the mean and found it to be 1.2831. This is a bit off–a bit high compared to the actual mean of 0.9349. In normal statistical practice, I don't know what the actual population mean is, generally. I just have to assume that the sample mean is my best estimate of the population mean and so, I will use it as a proxy for the real thing.

I next took a simple random sample of size 10 (the same n as my original sample), *from B*. This is called a *Bootstrap Sample*. Here you can see the values of the original sample, and the bootstrap sample side-by-side (Figure 15.1). Because I used simple random sampling—*with* replacement, I actually picked three of my sample values twice. This is unlikely, but it can happen if your sample is truly random.[1]

	Sample	**Bootstrap Sample**
	0.03834814	0.90354108
	1.59890898	0.03834814
	5.94178909	2.56115727
	0.00616014	1.59890898
	0.90354108	0.03834814
	0.0026814	0.13963914
	0.20498174	0.13963914
	2.56115727	0.20498174
	0.13963914	0.00616014
	1.43366766	1.59890898
Mean	**1.28308747**	**0.72296328**

TABLE 15.1: Comparison of original sample to bootstrap resample

The mean of my bootstrap resample is a bit low, compared with the mean of the sample taken directly from the population. That is not too unexpected. I fully expect to have variability in any set of bootstrap samples I might take from the original.

I placed this sample mean. 0.72296328 in a column vector, and then I computed another bootstrap resample, n = 10, just like the first and computed its mean. I put this second bootstrap sample mean in the column vector with the first. I did this *10,000 times*. This generated a sampling distribution of the mean, with 10,000 means, taken from 10,000 resamples of my original sample. Histogram (D) in Figure

[1]iTunes ® had to change their random shuffle algorithm, because people would occasionally get the same song twice in a row, or songs from the same genre several times in a row. They thought, because of this, the shuffle was biased, when in fact the shuffle was not biased-it was completely fair and these things happen from time to time. So Apple changed the algorithm to prevent these kinds of repeats happening on the shuffle setting.

15.1 shows this distribution. Take a look at the following procedure, bulleted out to make it clear:

Procedure for Computing a Bootstrap Confidence Interval

1. Draw a really good sample! Use good random sampling, blocking, and other controls to ensure that your sample has little error of measurement and little bias;
2. Use this sample as a proxy for your population;
3. From this sample-as-proxy, draw a (re)sample *of the same size*, using random sampling with replacement;
4. This new sample is called a *Bootstrap Sample*
5. Sampling with replacement allows you, if you sample many times, to get all possible combinations of sample values, in proportions approximately equal to those that would appear in the population of values;
6. Calculate a sample statistic for the Bootstrap Sample
7. Often this is the mean, but it could also be the median or the variance, or for bivariate data, the correlation coefficient, etc.;
8. Repeat step 2 and 3 many times (I typically do 10,000 iterations);
9. This will create a *Bootstrap Sampling Distribution* of the sample statistic;
10. The mean of the Bootstrap Sampling Distribution will tend to the Expected Value of the sample statistic over the long haul;
11. The standard deviation of the Bootstrap Sampling Distribution will tend to the Standard Error of the statistic over the long haul.
12. Select your Type I error rate;
13. Sort your Bootstrap Sampling Distribution from least to greatest;
14. Determine the $(1 - \alpha)\,\%$ Confidence Interval
15. Count up from the least value of the sample statistic $\frac{\alpha}{2}\%$ of the values to find the lower bound of the $(1 - \alpha)\,\%$ confidence interval. If you used 10,000 bootstrap resamples, this would be the 250th value;
16. Count down from the greatest value of the sample statistic another $\frac{\alpha}{2}\%$ to find the upper bound. If you used 10,000 bootstrap samples, this would be the $9{,}750^{\text{th}}$ value;

To compute the 95% confidence interval (2-tailed), I just go to my column vector of bootstrap sample means, sort the vector of means in the sampling distribution, and pick the 250th and $9{,}750^{th}$ as the lower and upper bounds of my confidence interval[2]. These values are 0.3736, and 2.4940. This means that, if my sample truly reflects the population, then I would expect the mean values of any sample of size 10 taken at random from that population to be between the values of 0.3736 and 2.4940 95% of the time. Only 5% of the time would I find values outside this interval, just due to random chance.

Notice that I made no assumptions about the shape of the parent distribution (I purposefully chose a weird distribution to make this point). I merely simulated the sampling distribution of the mean using the bootstrap procedure and counted the lowest 2.5% and the highest 2.5% to estimate my 95% confidence interval. The mean of the population, 0.9340 is *well* within the confidence interval, as it should be given we sampled randomly from it. The actual probability of the population mean of 0.9349 turns out to be 0.29 in this bootstrap sampling distribution.

15.2 Empirical Distribution Function

So how can we trust this magic? What makes a bootstrap simulation of a probability distribution a reasonable model for the long-term behavior of samples taken from a population? This stems from the idea of the **Empirical Distribution Function**.

If we have a set of real data, we can express its distribution —its frequencies, as a function of values of the independent variable:

$$f(x) = I(x_i = x)$$

Where I is the *indicator function* (meaning it is just counting the number each of x_i). This frequency distribution is finite and discrete: There are particular values of x, and each has some frequency of occurrence. A finite population can be fully defined, without parameterization, in this manner (although if it has large n, it might take a while, counting all the instances). If we have a population, we can just sum the relative frequencies from the least value to each successive value of x to create a cumulative distribution function (CDF).

[2]I have chosen to use the so-called percentile method of estimating a confidence interval for this chapter. There are a number of methods people use, including some confusing combinations of parametric distributions with nonparametric assumptions (???!). The method used here is the simplest, and easiest to understand, and serves as a good conceptual base for learning any of the other methods. Plus, for tests of the median, it appears to be pretty robust.

$$P(x) = \frac{1}{n}\sum_{i=1}^{n} I(x_i \leq x)$$

So P(x) is a discrete function that assigns a cumulative probability of x_i for each value of x. It has the same meaning as any other CDF, except that it is discrete. It has a lower bound of zero (for any x less than $x_{minimum}$), and an upper bound of 1 at $x_{maximum}$.

As an example, the following is a set of water hardness measures for a watershed in West Virginia that goes through different limestone layers (taken from Grand, 2005). Creating water treatment technology for homes, farms, and areas impacted by mining requires knowing how hard the water is. These measures are relative hardness measures, expressed as $\frac{s_g}{\bar{x}_g} \times 100\%$. This is a population of measures, all springs in the watershed were measured multiple times and their average hardness is listed:

$\frac{s_g}{\bar{x}_g}\%$
39
19
13
12
12
10
10
10
9
7
5
3
3

The empirical distribution for this population is illustrated in Figure 15.2. It is represented as a step function, where the steps are the probabilities of each value. You can see that the cumulative probability of each value in the distribution is the sum of the previous values. The lower bound is zero, like all probabilities, and the upper bound is 1. If we want to find the likelihood that any randomly drawn stream in the watershed has a relative hardness greater than 12, for example, we can find the sum of the probabilities of all $x_i \leq 12$ and subtract this cumulative probability from 1–just like the binomial distribution examples we studied back in Chapter 7.

When you have an actual population, like in our example, these values of *P(x)* are truly probabilities. We don't need to estimate them. But when we have a *sample*, taken from a population with unknown probability distribution, we have to estimate them in some way. Parametric statistics use the known features of theoretical

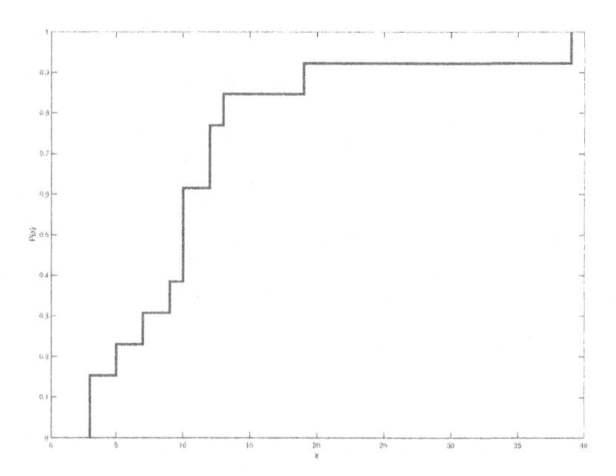

FIGURE 15.2: Empirical distribution of water hardness data

distributions to make these estimations. But when we can't make the case that the parent distribution approaches any of these theoretical models, the empirical probabilities can be estimated through bootstrap resampling.

To estimate the Empirical Distribution of the Water Hardness Data, supposing we did not know the population, we would take a random sample of water readings in the watershed, and, using the bootstrap method, generate a simulation. Taking a simple random sample of 6 measurements, I get the following hardness reading: 3, 13, 5, 9, 7, and 10. If I plot of this sample over that of the population I get the following (see Figure 15.3).

This is a fairly close estimate for one sample. If I take a 2nd sample and overlay its empirical distribution, I get the following data: 10, 10, 7, 5, 39, 12.

This 2nd sample is also a pretty good estimate (See Figure 15.4). I wonder what my estimate might be if I took many, many (say 1,000) such samples and plotted them:

Wow! I can estimate the empirical distribution of this water hardness data very precisely using simple computational methods (Figure 15.5). The simulated data nearly overlays that of the original population.

Bootstrap works on this tendency. Good samples, taken randomly, will tend to the characteristics of the population. If we take a single sample that fairly resembles the population, we can use that sample in lieu of a theoretical distribution to make probability estimates of that population.

The key point about bootstrapping is that we simulate a population of sample statistics—a *sampling distribution*—that represents the population of statistics all taken from the population from which the sample was drawn. The empirical

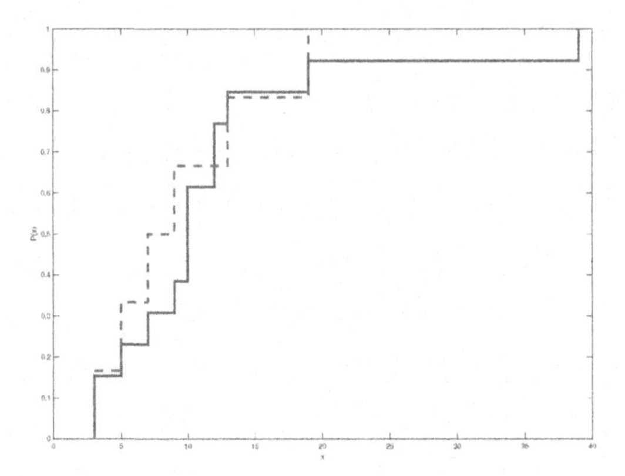

FIGURE 15.3: Empirical distribution of sample taken from water hardness population overlaid on the empirical distribution of the population

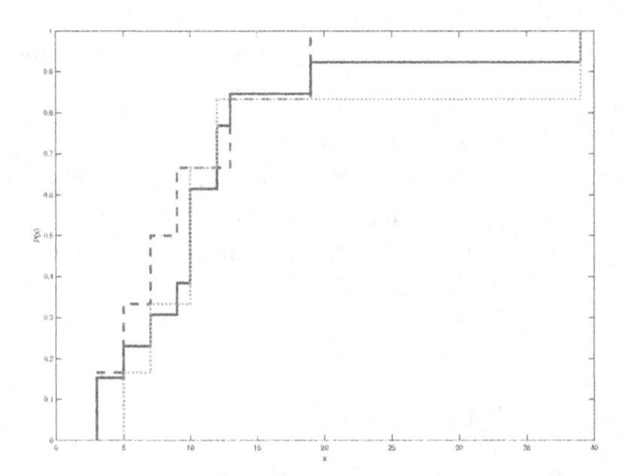

FIGURE 15.4: Empirical distribution of second sample (yellow) taken from water hardness population overlaid on the empirical distribution of the population

distribution of the sample statistic, then, serves as our means of determining the probability that any sample could have been taken from a population with the same characteristics as the original sample.

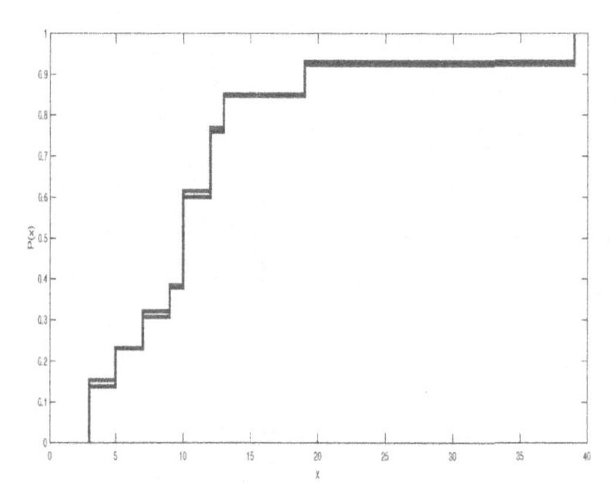

FIGURE 15.5: Empirical distribution of 1,000 samples taken from water hardness population overlaid on the empirical distribution of the population

15.3 Bootstrap Sampling Distribution of the Median

One of the things we looked at early on in the course is the problem of skewed data. I have used two skewed distributions in this Chapter to help illustrate this issue. The first was generated by a negative exponential distribution, and the second, Rand's water hardness data, is a skewed population with no underlying generating function. In cases where the data is skewed, often the mean is a poorer measure of the *location* of the dataset—the place in the histogram of the distribution where the highest probability of median is found. Because the mean is weighted, extreme values affect it more than the median. The median, therefore, can be more representative of the population because it is not affected by extreme values. Change any of the values in the data set other than the median, and it is not affected. It is always the value at the 50th percentile.

When data is skewed, or ordinal in scale, the median, in fact, is often more powerful as a statistic with which to compare the center of distributions than the mean (Fraser; 1957). Let's take a look at our original, negative exponential population and its resulting bootstrap confidence interval, when we replace the mean with the median. If you look back, you will find that, using the bootstrap method, with the mean as the statistic, we obtained a 95% confidence interval (2-tailed) of:

$$0.3736 \leq \mu_{\bar{x}} \leq 2.4940$$

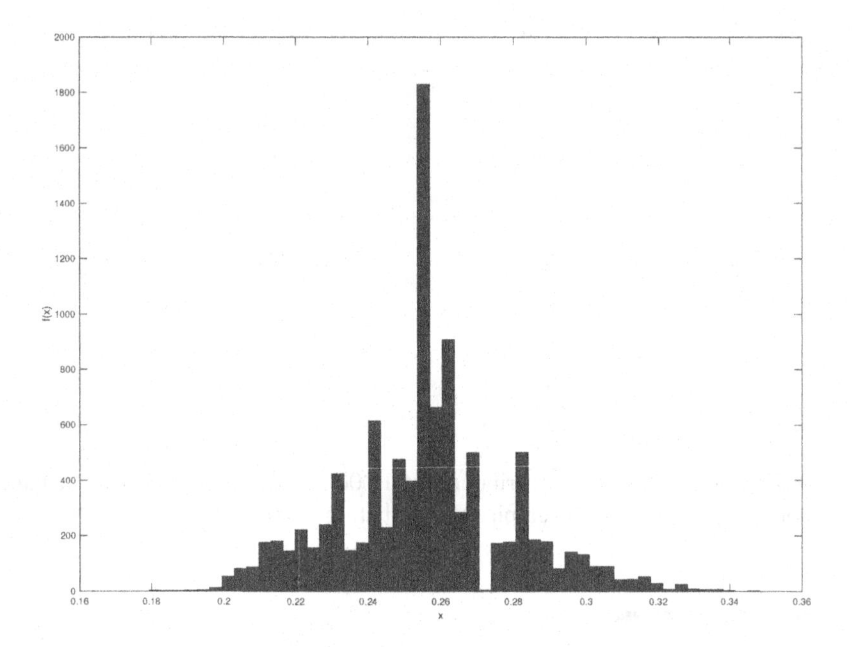

FIGURE 15.6: Bootstrap sampling distribution of the median (10,000 bootstrap resamples), n = 10, taken from negative exponential function in Figure 15.1(A)

The population mean, μ is 0.9340 is well within the interval, and I was pretty proud of myself for making such a good estimation.

Substituting the *median* as our sample statistic, I re-ran the 10,000 bootstrap resamples, taking the *median* each time. The resulting sampling distribution looks like this (Figure 15.6):

and its empirical distribution function looks like this (Figure 15.7):

Using these simulated distributions, the median of the sampling distribution is 0.2553, and the 95% confidence interval of the median (2-tailed) is:

$$0.2095 \leq med_{\overline{x}} \leq .3059$$

The population median is 0.2553. From this analysis we can see a critical difference from our original analysis using the mean. Namely, the confidence interval of the median is *much* narrower than the one we computed for the mean. Using the same sample as our proxy for the population, the median produces a much more precise interval than the mean. This makes utilization of the median as our measure of

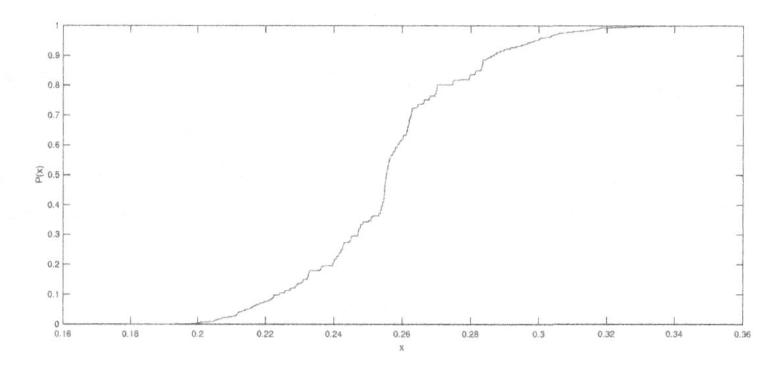

FIGURE 15.7: Empirical distribution of sampling distribution of the median from Figure 15.4

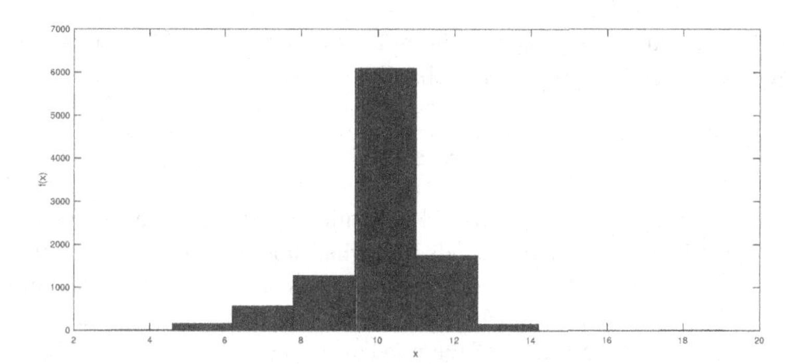

FIGURE 15.8: Histogram of bootstrap sampling distribution of the mean (10,000 bootstrap resamples), (n = 6) taken from water hardness original sample

the location of the distribution—its center—more powerful in this instance, than the mean.

Let's try the same analysis for the water hardness data (see figures 15.8 and 15.9).

The 95% confidence interval for the bootstrap sampling distribution of the median is:

$$7 \leq med \leq 12$$

The real median of the population is 10, fairly symmetrically located between the two confidence interval extremes. Comparing this result to that of the mean, I

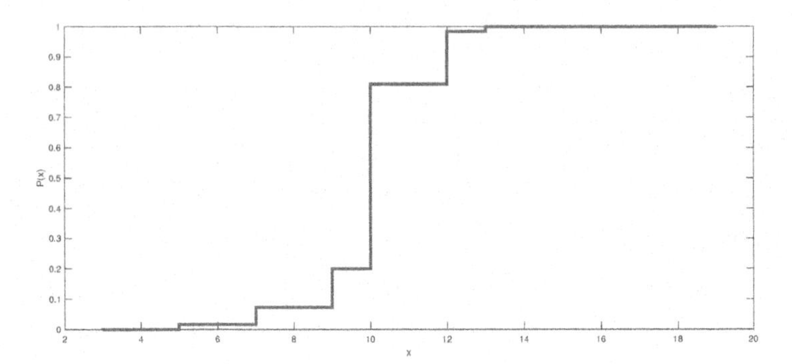

FIGURE 15.9: Empirical distribution of the medians of 10,000 samples taken from water hardness original sample

take 10,000 bootstrap samples from the *same* original sample. The 95% confidence interval of the mean for this sampling distribution is:

$$5.33 \leq \mu_{\bar{x}} \leq 14$$

The real mean of the population is 11.69, pulled a bit to the right in the confidence interval, but still well within the bounds. But what should be evident is that the width of the CI for the *mean*, in this skewed distribution, is greater than the width of the CI for the *median*. Again, this illustrates that, for a skewed distribution, the median can provide a more precise estimate of the population center (i.e., a smaller confidence interval), than the mean.

In general, when a population is skewed, and when we are not concerned about the specific shape of the parent distribution, tests of the median are, on average, **locally more powerful** than tests of the mean (Li, 1999). This applies to bootstrap methods as well as other nonparametric methods such as the Wilcoxon and Aligned Ranks Transform (don't worry about these right now. They are just names of other tests that use the median as opposed to the mean). In addition, this practice of just counting the percentiles to get a confidence interval of the median is often better than so-called "corrected" methods (Efron, 1979). I really love the simplicity of the method, and its general applicability. We will now examine bootstrap analogs to the *t*-test and regression so that you can witness its utility.

A test that is **locally most powerful** is the test, among alternatives, that has the greatest probability of correctly rejecting the null hypothesis at a given Type I error rate.

For parametric tests, when data violate their assumptions, often nonparametric alternatives like tests of the median, or of rank-ordered data, are locally most powerful.

15.3.1 2-sample Confidence Interval

So now that we have the basic logic and procedure down for bootstrap, let's do some real analyses! Going back to the basic Random Sample design, we listed the assumptions of the 2-sample *t*-test to include independence of observations, normally distributed parent population, and homogeneity of variance in the samples being compared. Violation of the last two assumptions, as we saw in our modeling of the sampling distribution for skewed distributions, can result in loss of power for tests that utilize the mean, and that reference the normal distribution like the *t*-test and ANOVA. When you suspect these assumptions are not met by your data, bootstrap is a method that can increase your power to detect a difference that really exists.

The basic procedure for estimating the confidence interval of the difference between 2 samples is essentially the same as one sample: 1) Take your samples from the population. 2) Using these two samples as a proxy for your population(s), create a bootstrap sampling distribution for each. 3) Because each statistic in the sampling distributions is random, you now can subtract the distribution drawn from Sample 2 from the distribution drawn from Sample 1 to create a sampling distribution of difference statistics. 4) Sort the vector of differences from least to greatest; and 5) Under the null hypothesis, the center of this distribution should be zero, indicating no real difference in location of the populations from which the two samples were drawn. So if we compute the $(1 - \alpha)\,\%$ confidence interval of the differences, if the null hypothesis is true, zero should be located within its bounds. If the null hypothesis is not true, zero should be located outside the bounds of the confidence interval.

This is exactly the same logic as the 2-sample *t*-test, with the exception that we do not assume the population the samples were drawn from is normally distributed, nor do we need to assume that the samples have similar variances. Additionally, we can make this hypothesis test a test of the difference in means, like the *t*-test, or the differences in medians. The process and interpretation are the same. I will perform two analyses on the same data, like I did for the water hardness data. I will assume that you will use the *t*-test or Oneway ANOVA for situations where their assumptions hold—they are the locally most powerful tests for those conditions. So let's choose a dataset that has definite problems with skew, in particular.

The following data are taken from Schechter et al., (1973, see Table 15.2). The researchers measured the sodium intake of patients with hypertension, and of patients with no hypertension. Their data is presented below. I have made the two samples equal in number of observations, so our results will be a bit different than the authors', but the conclusion will be the same.

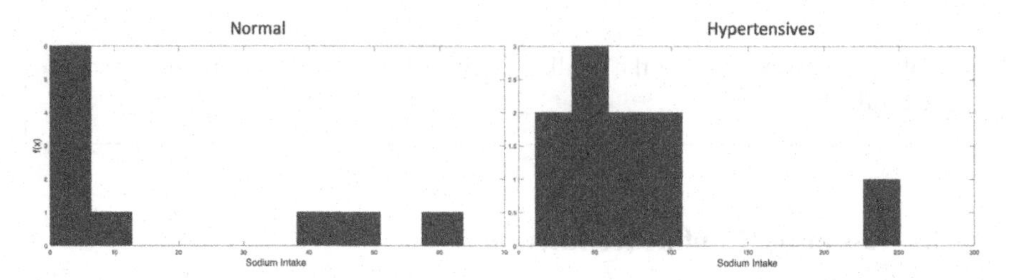

FIGURE 15.10: Histograms of the non-hypertensive and hypertensive patients' sodium intake

Normal	Hypertensive
10.2	92.8
2.2	54.8
2.6	51.6
43.1	61.7
45.8	250.8
63.6	84.5
1.8	34.7
0	62.2
3.7	11
0	39.1

TABLE 15.2: Sodium intake of normal patients versus hypertensive patients

First, let's run a regular old 2-sample *t*-test on the data. This is an epidemiology study, and the results might be used to create treatments for hypertensives, modifying a necessary nutrient in their diet, so I want to be a bit conservative. I am going to choose a Type I error rate of 1% (one-tailed). My working hypothesis is that hypertensives, on average, exhibit greater sodium intake than patients with normal blood pressure.

The 99% confidence interval of the mean for this test is:

$$-121.20 \leq \mu_{\bar{x}_1} - \mu_{\bar{x}_2} \leq 7.17$$

Zero, the mean difference predicted under the null hypothesis is found in this interval. Therefore, I have to conclude that the difference in sample means is too small to be considered significant with a probability of Type I error less than 0.01. Why is this. The data in the table look very different to me. The *t*-test should return a significant result! Let's take a look at the histograms of the two samples (See Figure 15.10):

Two features of these graphs leap immediately to mind: The first is that neither distribution appears anywhere close to normal. They are both highly skewed in the positive direction. Second, the variability in the Hypertensives looks much higher than the Normal patients. Both of these give me pause to think that perhaps my t-test is underpowered, despite the data hinting at a real difference.

Knowing what I know about the median and how it won't be pulled quite so far out into the right tail of the distributions as the mean, I might get a better estimate of the bulk of the data and perhaps boost my power. So I compute bootstrap distributions for both the normal and hypertensive samples. Then I take the differences of these two distributions just by subtracting across rows.

$$diff = \bar{x}_{\text{boot hypertensive}} - \bar{x}_{\text{boot normal}}.$$

And finally, I sort them and find the value of the first percentile by counting up to the 100th value in the sorted distribution. The bootstrap confidence interval of the median is found to be:

$$7.4 \leq Median \leq \infty$$

Zero, the value of the difference in medians under the null hypothesis, is not found in this interval, so I can conclude that hypertensives DO, in fact, show elevated sodium intake, compared to their non-hypertensive counterparts, with $p < 0.01$. The actual value of zero in this confidence interval is 32/10,000, or 0.0032. I found this by locating zero in the sorted difference vector.

This discussion was meant to draw your attention to two things: First, that non-parametric—methods that do not assume a particular probability distribution, can be useful methods when the assumptions of parametric tests are violated, giving you more power to detect a real difference when it exists. The second point I wanted to make is that the median, as a **robust** statistic, is not easily swayed one way or the other by extreme values. As such, it is just as viable as the mean for estimating the location of your distributions. There is a slight difference in interpretation of the median, in that you have to make clear that you are estimating the sampling distribution of the median: the estimate of the 50th percentile of the population, not the center of mass. All values of the distribution are weighted equally when computing the median, as opposed to the weighting that is inherent in the mean, so for asymmetric distributions, the median typically provides a better estimate of the location of the bulk of the data.

Robust statistics are those that can estimate the characteristics of data drawn from populations that display variety of shapes. They are especially useful for non-normal distributions.

The median is a robust estimator of the center of a distribution whereas the mean is not, being subject to "pull" by the weight of extreme values.

For variability, the mean absolute deviation and the interquartile range are robust, whereas the standard deviation is not.

In situations where the population of interest is non-normal, robust statistics can give more stable estimates of location and spread than non-robust statistics. But for distributions that are normal, the mean and standard deviation are the most powerful unbiased estimators.

Alternatives to using the median for situations where violation of normality and homogeneity of variance exist include the Box-Cox transformation, which is a type of $\log(x)$ transformation of the dependent variable to make the distribution more normal. It can be thought of like any other transformation in a regression analysis, you have to transform the data back into the original units to interpret the results. I, personally, prefer the simpler test of the median, and bootstrapping, as I think it makes interpretation of the results simpler than a Box-Cox transformation. Any hypothesis you wish to test using ANOVA and other regression methods can be bootstrapped. But like all decisions in statistics, the best one is the one that best helps you understand the true nature of the data and thereby make a decision about the meaning of the data so that you can design and build better engineering marvels to make the world a better place.

Note that bootstrap methods *do not change the ways in which statistics are computed.* They are used to describe the distribution of those statistics so that probability estimates can be made regarding their likelihood. In the previous examples, we computed the difference in means or medians, which is analogous to the t-test. Nothing changed there.

Bootstrap was used to estimate the *probabilities* of different values of the difference in means (or difference in medians) between two samples by generating an *empirical* distribution. Student's t-test estimates those probabilities by scaling the difference in sample statistics and comparing those scaled values to a continuous *theoretical* distribution. The criteria for judging the likelihood of that difference is the chief distinction between the two methods.

15.4 Regression Coefficients Bootstrapped!

One of the problems we have with regression concerns too little data to effectively model the relationship to where we have both goodness of fit and significance of our prediction equation. For example, I asked my graduate students to

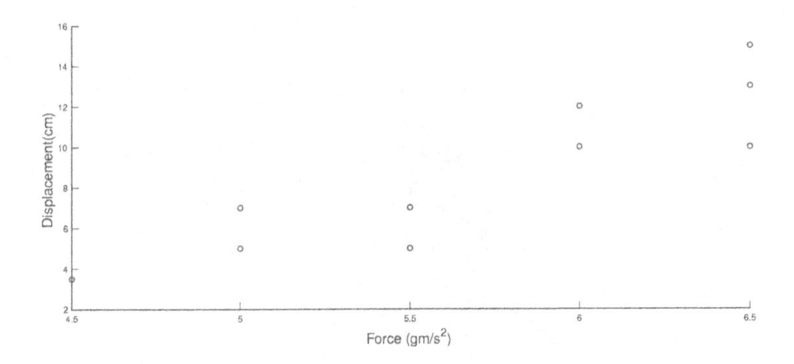

FIGURE 15.11: Scatterplot of spaghetti bending data

measure the force versus displacement of uncooked spaghetti (yes I really do this). The reason for this is that none of my students know what the bending moment for spaghetti is, so this analysis is a genuine exploration—I really don't know the answer. They only collected 10 data points before rebelling. Here is the data in Table 15.3:

Force	Vertical Displacement
5	5
6	12
5	5
5.5	7
6	10
6.5	13
4.5	3.5
5.5	5
6.5	10
5	7
6.5	15

TABLE 15.3: Force versus displacement of strands of uncooked spaghetti

I know from Hooke's law that, at least for the first portion of the curve, the force vs. displacement graph should be a line, but after that, it might go into a non-elastic phase and the graph will tend to curve dramatically. Like all regression analyses, I first made a scatterplot to see if a linear model is appropriate (see Figure 15.11):

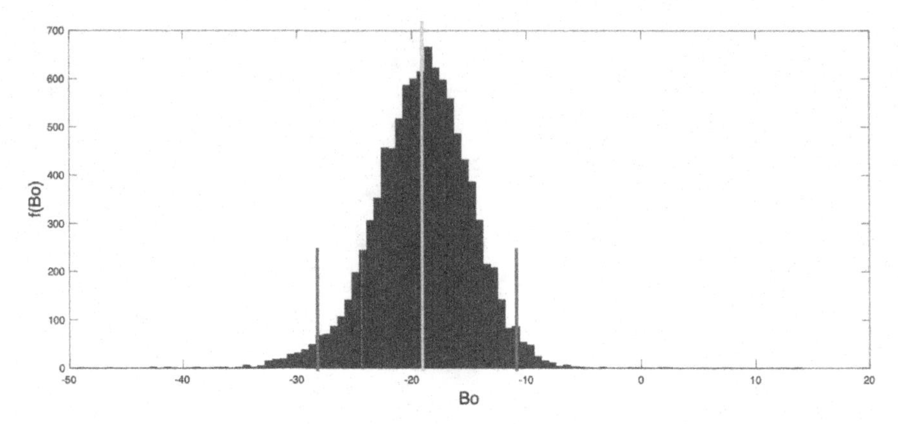

FIGURE 15.12: Frequency distribution of bootstrapped values of β_0 for spaghetti data. The mean is the predicted value of β_0 from the empirical distribution: -19.11. The grey lines indicate the lower and upper bounds of the 95% confidence interval: $-29.29 \leq \beta_0 \leq -11.21$

This looks kinda linear, so I think I am going to be okay within the contextual domain of the independent variable. I wish I had more data than just the small sample my lazy-assed students obtained for me...the sample size is so small... Lightbulb! Maybe I can bootstrap a sampling distribution to estimate the confidence intervals of the regression parameters, β_0 and β_1!

The process for this is basically the same as any regression problem we have studied thus far (we are still projecting a vector of the values of a dependent variable on the values of a matrix of predictor values). The measure of the sum of squared deviations from the regression model is still the criteria for the "best" fit line, and the system is still assumed to be linear. But now we don't assume normality of either independent or dependent variable. This frees us from some of the problems associated with this small sample size and its lack of ability to generate a narrow confidence interval around the regression line.

The bootstrap process will look familiar to you: 1) Take a sample of bivariate data. 2) generate k bootrstrap re-samples of the *bivariate* samples and calculate the regression coefficients of each. This yields a k x 2 matrix of estimated parameters: β_{0i} and β_{1i} . 3) Now, using your Type I error rate, count up to the lower bound, and down to the upper bound to determine your (2-tailed) confidence interval for each parameter.

For β_0 the mean of the bootstrapped distribution is -19.11, and the 95% confidence interval is $-29.29 \leq \beta_0 \leq -11.21$. This interval tells me that 95% of the time, with samples taken from a population represented by our original sample, we would expect to obtain an intercept between -29 and about -11 (see Figure 15.12).

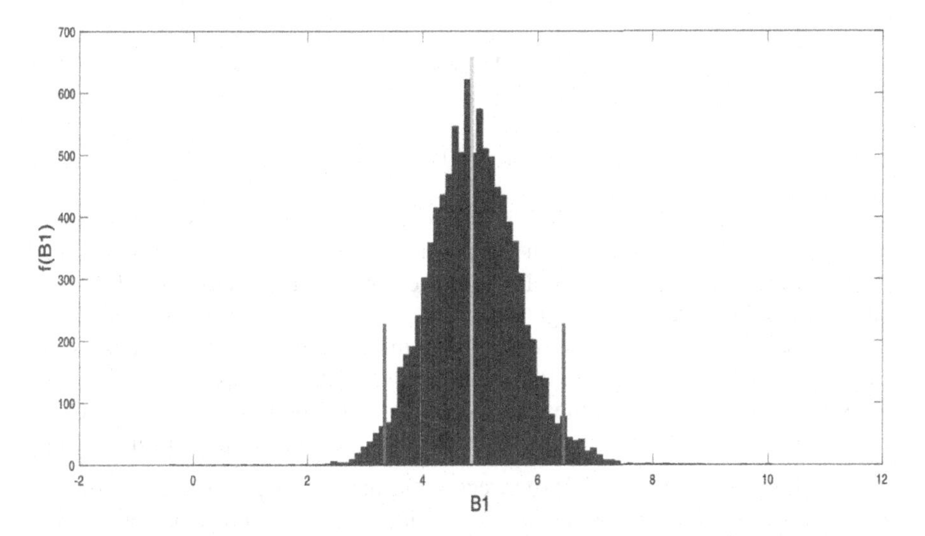

FIGURE 15.13: Frequency distribution of bootstrapped values of β_1 for spaghetti data. The mean is the predicted value of β_1 from the empirical distribution: 4.87. The grey lines indicate the lower and upper bounds of the 95% confidence interval: 3.39 $\leq \beta_0 \leq$ 6.46

For β_1 the mean of the bootstrapped distribution is 4.87 and the 95% confidence interval is $3.39 \leq \beta_0 \leq 6.46$ (see Figure 15.13). This tells me that the slope of this linear relationship is significantly different from zero—this is a nonparametric test analogous to the t- or F-tests of the slope.[3] The Sum of Squares and R^2 can be calculated by comparing the residuals of the fitted model to the original data:

$$\hat{y}_i = \beta_0 + \beta_1 x_i$$

$$\hat{y}_i = -19.11 + 4.87x_i$$

$$R^2 = 0.81$$

Compare these to the values estimated by Least Squares Regression, assuming Normality and homoscedasticity:

$$\hat{y}_i = \beta_0 + \beta_1 x_i$$

[3]This example is a random effects model, so we can bootstrap from the regressors directly (i.e., Case resampling). A fixed model would be bootstrapped from the vector of residuals of a fitted model of the original data. I chose this as the canonical example because it is simpler to program. The principle is the same, however for either model.

$$\hat{y}_i = -18.83 + 4.83x_i$$

$$R^2 = 0.81$$

The two models are remarkably similar, returning strong fitting, significant models. The slight skew of the data in the negative direction makes the bootstrapped model estimates both parameters slightly lower than the parametric model, but because the original data is only slightly skewed, and the underlying physics predicts a linear model, we would expect the data to meet the assumptions of Least Squares Regression as we learned it earlier. But had the data distribution shown more skew, the bootstrapped model would be the one I would prefer to report as my best guess of the relationship between the force required to bend spaghetti to specified displacements.

Like other parametric tests, regression has its nonparametric analogs, of which bootstrap is an important set of techniques. In these other modeling techniques, bootstrap can also be used to augment sample sizes, or to estimate confidence intervals. Its flexibility across applications is one of its chief appeals to the engineering statistician. Again, like our discussion of analyses of the differences in distributions—the t-test analog—bootstrap methods for regression don't change the calculations, they just change the way we estimate the probabilities of the parameters occurring in their sampling distribution.

15.5 Summary

Lest you forge ahead, bootstrapping the hell out of all your statistical problems, do reflect a bit on what such methods buy you and what they do not. The pros of bootstrap lie primarily in the situations where you do not know the parent distribution generating your samples, or where your samples, regardless of hypothesized parent distribution, violate the assumptions of the parametric tests you propose to use. Because it is free of these specific assumptions bootstrapping can help you estimate important characteristics of these unknown parent distributions and solve problems that you just cannot with more traditional, parametric, methods.

But they aren't a panacea. You still must make assumptions about the data that, if they do not hold, render any significant effect you might detect (literally) null. Moreover, many people do not yet have much experience with bootstrap models—their entire statistical history has been with parametric models. If you do choose to utilize bootstrapping and other nonparametric techniques, you will have to do more work justifying their use, than the standard methods. So, this brings me to a nice rule-of-thumb for when to use bootstrap and when to not use it.

Deciding when to run a Bootstrap analysis

1. Run the traditional parametric analysis, and check to see if it returns a significant result. If it does, great. Report your results. The *t*-test, ANOVA and regression are robust to violations of their assumptions;

2. If the parametric analysis returns non-significant results, perform a residual analysis;

3. If the residual analysis returns evidence that there is obvious violation of normalcy or heterogeneity, run the bootstrap analysis using the appropriate statistic, such as the median for skewed data.

4. If the test utilizes an alternative, robust, statistic and returns a significant result, report this result. The median and mean, for example, are independent of each other, and so analyses using them are independent;

5. If you bootstrap with the same statistic as the one you used in your original model (typically the mean or regression coefficient), and the result is significant, report this as a *post hoc* finding, describing your analysis of the residuals and the rationale for following up with the bootstrap test. Because a follow-up analysis based on the same statistic is *not* independent of the original analysis, it carries with it a conditional probability that may inflate your Type I error rate. So report your confidence interval estimates as potential values for the relationship being modeled.

Lastly, it is important to note that bootstrapping is a set of techniques, not a statistical test. It, therefore, requires some craftsmanship in its application. The techniques I have introduced in this chapter are the base model upon which other, more elaborate or specific, bootstrapping techniques can be understood. Nevertheless, it is one of my favorite methods for its simplicity and applicability.

I also recommend that you continue to study nonparametric statistics, particularly if your career trajectory moves you in the direction of statistical analyses. These methods are gaining popularity, and more importantly, they are helping engineers solve problems where classical statistics fall short.

15.6 References

Efron, B. (1979). The 1977 RIETZ lecture. Bootstrap Methods: Another look at the Jacknife. *The Annals of Statistics*, *7*(1), 1-26.

Fraser, D. A. (1957). Most powerful rank-type tests. *The Annals of Mathematical Statistics*, *28*(4), 1040-1043.

Fox, J. (2002). Bootstrapping regression models. *An R and S-PLUS Companion to Applied Regression: A Web Appendix to the Book. Sage, Thousand Oaks, CA. URL http://cran. r-project. org/doc/contrib/Fox-Companion/appendix-bootstrapping. pdf.*

Grand, R V. (2005) *Controls, characterization and small scale chemical variations of a Karst system: A geochemical assessment of Tuscarora Creek watershed, West Virginia.* West Virginia University: Unpublished Masters' Thesis.https://researchrepository.wvu.edu/etd/2287

Li, Q. (1999). Nonparametric testing the similarity of two unknown density functions: local power and bootstrap analysis. *Journal of Nonparametric Statistics*, *11*(1-3), 189-213.

NIST (2003). https://www.itl.nist.gov/div898/education/datasets.htm

Kvam, P. H., & Vidakovic, B. (2007). *Nonparametric statistics with applications to science and engineering* (Vol. 653). John Wiley & Sons.

Schechter, P.J., Horwitz, D., & Henkin, R.I. (1973) Sodium chloride preference in essential hypertension, *Journal of the American Medical Association,* 225, 1311-1315.

Stefanou, G. (2009). The stochastic finite element method: past, present and future. *Computer methods in applied mechanics and engineering*, *198*(9-12), 1031-1051.

Wobbrock, J. O., Findlater, L., Gergle, D., & Higgins, J. J. (2011, May). The aligned rank transform for nonparametric factorial analyses using only anova procedures. In *Proceedings of the SIGCHI conference on human factors in computing systems* (pp. 143-146). ACM.

15.7 Study Problems for Chapter 15

1. **State the bootstrap principle in your own words. How is it logical that we can use a sample, taken from a population, as a proxy for the population itself?**

   The bootstrap principle essentially asserts that, if a sample is a reasonable facsimile of the population from which it was drawn, its characteristics will mirror the population, and therefore, the sample can serve as a proxy for the population in creating a sampling distribution by which to estimate probabilities of events occurring. Of course, if you didn't believe the sample was a reasonable facsimile of the population, you wouldn't use it anyway!

2. **What is the procedure for drawing a bootstrap sampling distribution? Try to describe this procedure in a way that you both understand it, and are able to carry out the procedure.**

   First, draw a random sample from the population of interest. Use any blocking or other controls to ensure that the sample is a good representation of the population. Then, from the sample, draw a bunch of *resamples* all with the same number of events as the original sample. Compute the appropriate sample statistic from each of these resamples and place them in a column vector. This is the bootstrap sampling distribution that will be used to estimate probabilities of samples randomly drawn from the population.

3. **Box, Hunter & Hunter (1978) present data of a bicycle commute, where they adjusted the seat height of the bicycle (0=low), (1=high) and recorded the time it took for the commuter to travel to and from fixed points.**

   (a) **Given the data in the table below, what is the value of the bootstrap confidence interval of the mean for Seat Height = 0?**

Time	Seat Height
69	0
52	1
60	0
83	1
71	0
50	1
59	0
88	1

   (b) **What is the value of the bootstrap confidence interval of the mean for Seat Height = 1?**

First I compute the bootstrap sampling distribution using Matlab:

» bootx=bootstrp(10000,@mean,Seat0);

Then I sort the sampling distribution:

» bootsort=sort(bootx);

Now I can choose any α I want and compute the ($1-\alpha$)% confidence interval. For this example, I will choose a 95% CI. Since there are 10,000 means in bootx, I just have to find the 250th for the lower bound of the CI, and the 9,750th for the upper bound. These values, in my analysis are: (51, 85.5). 95% of all the means in the bootstrapped sampling distribution fall between 51 and 85.5.

(c) **Did seat height determine the speed at which the commuter got to work?**

Here I will perform the bootstrap analog to the 2 sample t-test. After already getting a bootstrapped sampling distribution for Seat Height = 1, I can do the same procedure for Seat Height = 0:

» bootx=bootstrp(10000,@mean,Seat0);

Now I will get a sampling distribution of the random differences in means by subtracting the means in the bootstrapped sampling distributions:

» bootdiff=bootx-booty;

» bootsort=sort(bootdiff);

I now find the values in bootsort that correspond to the 2.5% and 97.5% percentiles in the distribution of mean differences. The estimate 95% CI is (-9.25,27). Since 0, the hypothesized difference in means under the null hypothesis is contained in this interval, we cannot conclude that the two seat heights are significantly different in their influence of the commute time.

(d) **Compare the results of your bootstrap analysis with that of the 2-sample *t*-test. How much did your bootstrap results differ from the *t*-test?**

This is nice! I get to go back to something I already know:

» [h,p,ci]=ttest2(y,x);

The resulting 95% CI is (-35.04, 17.54). This is very different than the CI estimated in the bootstrap method. This is because the variances of the two samples are widely different:

$$s_0^2 = 60.33,\ s_1^2 = 401.58$$

4. **Here is the lift-to-drag ratio of a number of micro-air vehicles (MAVs), all manufactured using the same plans, as fabricated and measured by my students.**

L/D Ratio
3.6
4.6
2.8
2
2
3.2
6.8
3.2
3.8
4.8
2.4
1.6
6.2
4

(a) Draw the empirical distribution function of this set of data.

Here is a small Matlab script to overlay this graph with a plot of the normal CDF within the domain of L/D ratio:

```
cdfplot(data)
hold on
x = linspace(min(data),max(data));
plot(x,evcdf(x,0,3))
legend('Empirical CDF','Theoretical CDF','Location','best')
```

The resulting graph looks like this:

(b) Compare this empirical distribution function to the CDF of a normal distribution with the same mean and standard deviation. Is the empirical distribution a reasonable depiction of the cumulative probabilities of the MAV L/D ratio distribution?

The data is not distributed normal. The normal CDF is a poor approximation compared to the empirical distribution. You can see this because there is no point of inflection at the mean of the data. We can see that the distribution of cumulative probabilities in the empirical distribution more closely matches the proportions of the values in the table.

(c) Now compare this distribution to the empirical distribution function of the bootstrap sampling distribution of the mean. How does THIS compare? What are the key differences in the smoother bootstrapped distribution, compared to the empirical distribution of the original sample?

```
»cdfplot(data)
```

Notice here that the bootstrap sampling distribution has a point of inflection right at the mean of the original sample. It is steeper, because it is a sampling distribution of samples of size n=14.

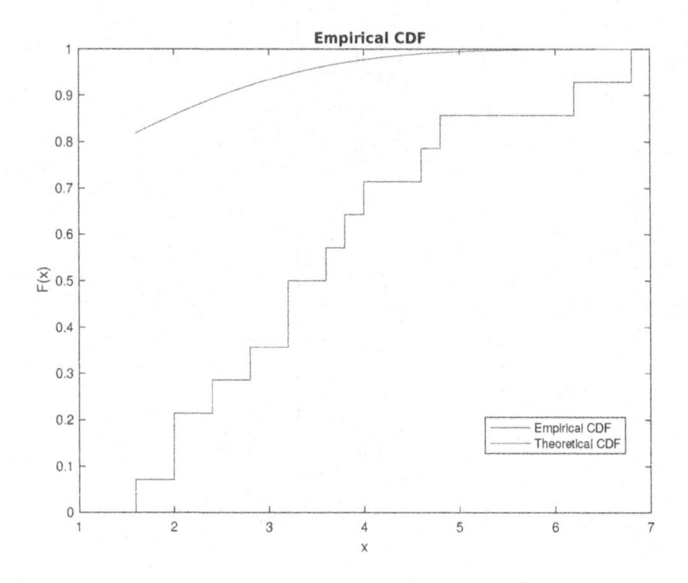

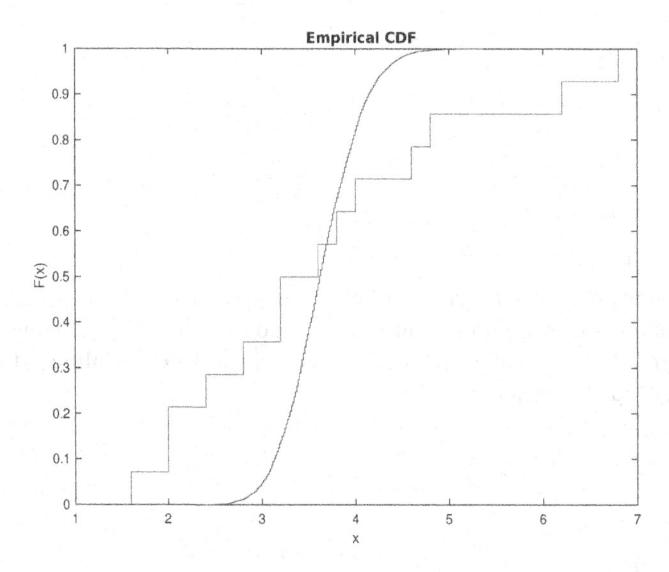

5. I went out to the rifle range the other day and, using my phone, recorded the sound pressure at randomly selected times during live firing. I first collected the data with my phone without a sound muffler. I then put a muffler over the microphone and collected another set of random sound pressure measurements. Here are my results:

Control (db)	**Muffled (db)**
73.4	57.2
53	63
63.6	69.2
79.1	56.2
60	67.8
80.3	50.9
58.8	55.3
84.2	83.4
78.7	76.9
100.3	66.4
80.9	63.3
99.5	67
98.2	58.3
91.5	72
85.8	60.8
61	57.5
116.9	64.6

(a) **Using the bootstrap test of the median, did muffling significantly reduce the sound pressure experienced by my phone?**

This is the bootstrap analog to the t-test, using the median as my estimate of the location of the two sampling distributions.

```
» x=data(:,1);
» y=data(:,2);
» bootx=bootstrp(10000,@median,x);
» booty=bootstrp(10000,@median,y);
» bootdiff=booty-bootx;
» bootsort=sort(bootdiff);
```

Because the muffling is meant to *reduce* the noise, I estimate a 1-sided confidence interval with $\alpha = 0.05$:

$$(-\infty, -7)$$

Zero, the hypothesized value of the median differences under the null hypothesis, is not in the interval, so we can conclude that our muffler worked to reduce the sound pressure measured by my phone.

6. The following data was collected by my students, where they varied the length of very long, thin, cantilevered polymer beams (spaghetti) and measured the force needed to displace them 5 cm.

Length (cm)	**Force (N)**
20	0.2
7.6	0.37
12.5	0.32
5	0.75
20	0.22
7.3	0.35
8.5	0.45
18	0.25
3	0.68
7.8	0.4
20	0.21
11.5	0.33
7	0.5
9.2	0.38
8.5	0.43
8.5	0.4
3.2	0.8
7	0.5

(a) Perform bootstrap regression of the data.

The first thing I need to do if I am going to use the regression function in Matlab is to put the data into appropriate form: Force in a column vector, and Length into an augmented matrix (adding a column of 1s). Then I can run a bootstrap regression analysis:

»bootx=bootstrp(10000,@regress,force,length);

This gives me a 10,000 x 2 matrix of slopes and intercepts—10,000 parameters defining lines of best fit for the bivariate relationship.

(b) How will you account for any non-linearity in the data?

Nonlinearity can be handled by transformation if necessary, just like least squares regression. I would do this prior to creating the bootstrap sampling distribution. Like any regression analysis, I take a look at the scatterplot of the untransformed data first:

Because this relationship is, theoretically, inverse cubic: $\frac{deflection}{Force} \sim length^3$,

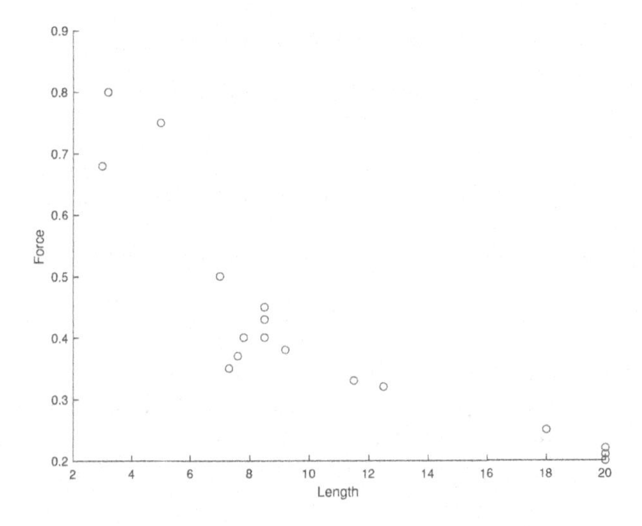

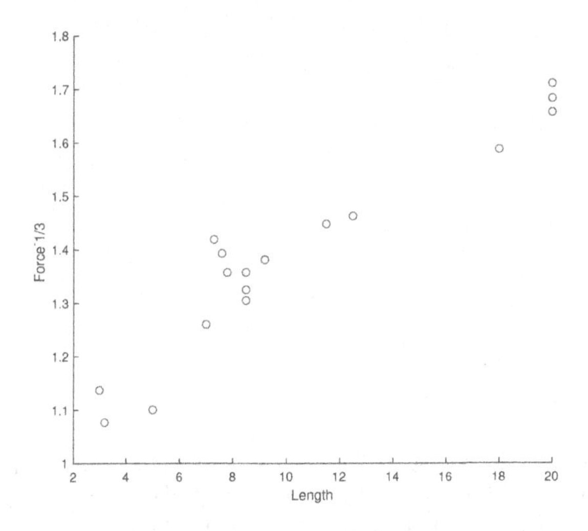

I would take the cube root of the values of Force, then invert them prior to running the bootstrap procedure.

Doing so straightens out the curve nicely so that we can fit a linear model:

(c) **What is the regression equation of the linear model, using the boostrapped estimates? What is the prediction equation, put in the original units?**

The mean of the transformed bootstrap sampling distribution for the slope, β_1 , is 0.0320. The mean of the intercept, β_0 , is 1.0574. So the prediction equation using these parameters is:

$$Force^{-\frac{1}{3}} = 1.0574 + 0.0320\,(Length)$$

(d) **Perform the regression using Least Squares regression.**

Regressing $Force^{-\frac{1}{3}}$ on length yields the following regression equation:

$$Force^{-\frac{1}{3}} = 1.0573 + 0.0319\,(Length)$$

(e) **What is the prediction equation, put in the original units?**

$$Force = (1.0574 + 0.0320\,(Length))^{-3}$$

(f) **Compare the bootstrap model to the Least Squares model. How close are they? Are there any violations of assumptions that should be accounted for that bootstrapping helps with?**

They are exceptionally close! A Q-Q plot of the residuals of the transformed data model shows that the distribution is not particularly normally distributed, violating one of the assumptions of least squares regression. So, the bootstrap model is more trustworthy in this example. Coincidentally, the residual distribution is very symmetric (look at both tails of the Q-Q plot below), so the overall difference between the bootstrap distribution and the Least Squares model

are not far off each other in value. Computing the SS_{resid} for each model shows small differences:

For the bootstrap analysis, $SS_{resid} = 2.7870$

For the standard Least Squares analysis, $SS_{resid} = 2.7924$

In this case, the bootstrap analysis has less total error than the traditional, parametric, approach.

7. **It is also possible to bootstrap sampling distributions of the proportion for χ^2, and to perform multiple linear regression and ANOVA using bootstrapped confidence intervals. I have not covered these methods in this text, but think about what hypotheses you are testing in each of these cases and describe how you would go about bootstrapping if their assumptions were violated.**

For χ^2, I would worry if my expected frequencies for a few cells were too small. This happens quite often, actually, in practice. If this is the case, we can treat the contingency table as a bivariate distribution. We could create a bootstrap distribution from the bivariate data, and then compute the probabilities of each cell difference appearing in the bivariate sampling distribution. Same logic as any bootstrap analysis!

For multiple regression and ANOVA, we have multivariate analyses, but the logic is the same. Since it is all regression, we can use the GLM as a guide, creating many many regression outcomes, using the mean if appropriate, or the median if the data are highly skewed. With confidence intervals found for each of the independent variables, we can determine the significance of the effects in just the same way as we did using parametric analyses!

16

Data Reduction: Principal Components Analysis

Thus far in the book we have dealt with models that require non-correlated predictors—ones—that show no collinearity. We have seen that, for the General Linear Model, when one predictor variable is a linear combination of another, or is very close, that the effects of significant predictors may be suppressed or distorted by this non-independence. This is true for traditional multiple regression–regressing a continuous outcome variable on a set of continuous predictors, as well as ANOVA models–regressing a continuous outcome variable on a set of categorical predictors (and if you take more stats, other models that are basically any combination of continuous and categorical variables).

We have also been confused by models when the number of predictors gets very high. Not only have we seen that the potential interactions among predictors becomes combinatorically large, but that these "large" models have a higher probability that one or more of the independent variables, including interactions, exhibit collinearity. That assumption of independence sure can be a pain! What if there were methods that embrace these interactions and co-dependencies? That would make data analysis richer and more useful for many situations. This chapter diverts from the General Linear Model and presents one of the fundamental ways of examining the structure of multiple interacting variables. It is one of the bases for machine learning and analytics. Plus, it is just great math!

16.1 Data Reduction

One of the plagues of engineering data analysis is the fact that we often collect (or are given) huge datasets including many variables, only a few of which are really good predictors. For this reason, it is often convenient to use some kind of disciplined approach to determining which, of the all the variables in a dataset, contribute significant variation to a model, and which contribute negligible variation. There are two primary approaches to this, that go by terms coined by the machine learning community: **Feature Elimination**, and **Feature Extraction**.

DOI: 10.1201/9781003094227-16

16.1.1 Feature Elimination

Feature elimination is a process of filtering irrelevant or redundant variables from the dataset. To do this, you need to employ some means of determining which variables are "true" and which constitute mainly residual variation that can be safely ignored. In multiple regression, we use the β -weights of our predictors to determine which most impact a prediction equation, their partial R^2 to determine if they contribute a significant proportion of the variation to the model, and an F -test to determine if their contribution is likely due to some systematic factor or just due to random chance. If a variable is found to have a small β -weight, and small partial R^2, relative to the other predictors in the model, and if its variation is not significantly different from zero for a given Type I error rate, we can eliminate it from the model with little change in our predictive power.

This is especially useful when we have lots of predictor variables that we assume are *not* independent. Reducing the number of predictors by eliminating some variables that appear to be poor predictors leaves us with a smaller set that has less potential to violate the assumptions of the General Linear Model. In general, this can be a good practice. This can make the overall prediction model easier to understand, with clearer, independent partial slopes for each independent variable. Reducing dimensionality in this fashion retains the things that matter: The variables that contribute the greatest amount of variation to the predictive model, and getting rid of variables that cause spurious associations, or that muddy interpretability without adding much in the way of predictive power. **Feature Elimination** is a method of lowering the complexity of a model by condensing the number of features (variables) in the model from an initial set to a final, reduced set that retains much of the variability of the original model. It is typically used to improve interpretability of a model by dropping extraneous variables that add unimportant levels of variation. For an $m\ (records) \times n\ (features)$ matrix, Feature Elimination reduces the column space of the model.

16.1.2 Feature Extraction

Sometimes, however, we need to retain all our predictors, but still reduce the size of the dataset. Sometimes, for example, the predictors are not independent because each contributes to a higher-order variable made up of their linear combination. Capturing this relationship among predictors, and using it to visualize these higher-order variables can be a valuable asset, when we cannot measure those higher-order variables directly, or when the number of "independent" predictors is high relative to the proportion each dimension adds to the overall predictive model. One of my favorite examples of the need to reduce dimensions in this fashion concerns facial recognition (or visual pattern recognition, in general). Faces are very complex, and trying to program a computer to sort through a database of faces, to pick out a face from a photograph can be very tricky. Facial features are difficult to break down into meaningful subunits, like nose length, distance between eyes, forehead length, skin tone, and other variables, and the sheer number of features that interact in the dynamics of facial expression, when combined with the size of a database of faces (say for criminal identification), make traditional classification and call-up retrieval systems impractical.

Feature extraction is a process of re-expressing the matrix of predictors by creating a new set of variables, each of which are combinations of *all* the predictors. Then, performing feature elimination on the new variables reduces the size of the dataset to some tractable dimension. This can be done iteratively to check whether the new, reduced models, indeed adequately predict the criterion variable without muddying your interpretation.

Feature Extraction, therefore, is a method of making the size of a dataset more manageable through the creation of new variables that consist of combinations of the original variables. These new variables are then subject to Feature Elimination to determine which ones are most important to keep in a reduced model. For an $m\ (records) \times n\ (features)$ matrix, Feature Extraction transforms the matrix to a new column space.

These two approaches are different in philosophy, but in practice we often employ both to get a reduced set of predictors, and to use those predictors to describe a latent variable (one that is made up of the values of several related variables) using a linear combination of that reduced set. The primary difference between feature elimination and feature extraction is that feature selection reduces the dataset to a subset of the original features while feature extraction creates brand new features and reduces the dataset by eliminating the non-significant latent variables.

If we could take a number of different predictive variables, and make a linear combination of them, we may be able to use their interrelationships to make a predictive model that doesn't have to take all of them into account separately. One of the simplest ways to do this is what is called Principal Components Analysis (PCA). Principal Components Analysis is a really clever method that helps us reduce the dimensionality of complex data from a high dimensional space, to a lower dimensional space. It has become one of the most common methods for facial recognition, among other uses in engineering (Yang et al., 2004).

Sanguansat (2012) lists a number of advantages of reducing dimensionality using Principal Components Analysis. 1) First of all, like examining the β -weight and partial R^2 of regression coefficients, PCA retains the information that adds predictive power, while reducing noise and other undesirable artifacts. 2) Secondly, the reduced model generated by PCA can be much more efficient, computationally, than one generated in the original high dimensional space. 3) Thirdly, like regression, PCA provides a way to understand and visualize the structure of complex data sets; and 4) lastly, because it generates a new set of predictor variables from linear combinations of measured predictors, PCA can help us identify higher-order variables that are more meaningful and explanatory in their linear combination, than were the original measured variables.

16.2 PCA as a Projection

We saw back in Chapter 11, that linear regression, geometrically, is projecting a vector of outcomes, Y onto the column space of a matrix of predictors.

$$\hat{Y} = X\left(X^T X\right)^{-1} X^T Y$$

$X\left(X^T X\right)^{-1} X^T$ is the matrix that orthogonally projects Y onto the span of the vectors comprising X. The "shadow" of Y in the column space of X is $\hat{Y}$, the vector of projected values. This is true for any dimension X, so long as X and Y have an equal number of rows. In data terms, each row of X and Y constitutes measurements of a single phenomenon, one dependent variable, Y, and a linear combination of predictors, X.

Like linear regression, Principal Components Analysis is also a projection into a subspace of X. Only there isn't necessarily any "dependent" variable to project—*all* the variables are predictors! Since I can't visualize much beyond 2 dimensions, a simple bivariate example will have to illustrate how this works:

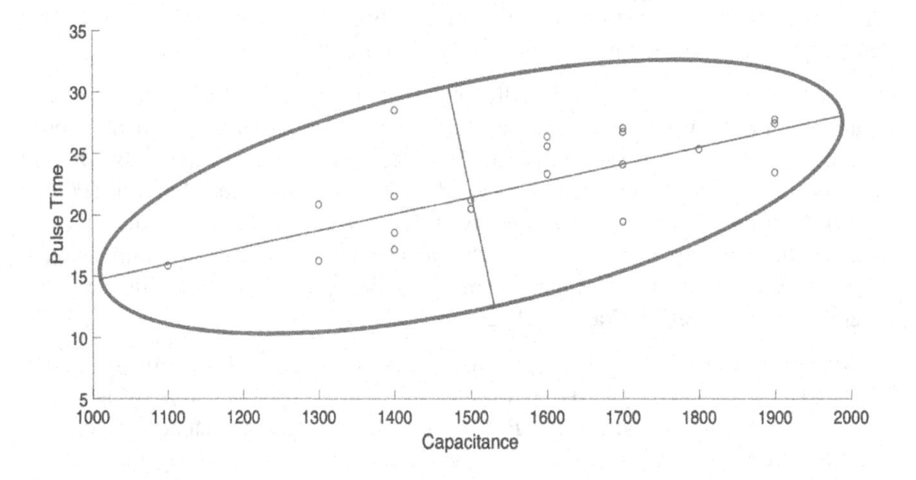

FIGURE 16.1: Bivariate relationship between capacitance and pulse time

Figure 16.1 shows the relationship between the measured capacitance of a capacitor and its pulse constant. Pulse time is not determined by capacitance, nor capacitance by pulse time. Yet the two are related.

A cloud of data in 2-space can be visualized as an ellipse. I have marked the major axis of the ellipse in Figure 16.1 with the Least Squares line of best fit. If the two variables are both normally distributed, the Minor Axis will intersect the major axis orthogonally at $(\overline{x}, \overline{y})$. We know from this figure that the two original variables, capacitance and pulse time, are *not* orthogonal to each other. They have a linear relationship!

But if we rotate the space around the major axis of the ellipse representing the data, we can re-express the data using these axes as dimensions (see Figure 16.2). The line of best fit represents the dimension in the plane that explains the majority of the variation between data, and the minor axis represents the dimension, orthogonal to the line of best fit, that explains the next most variation. The variation explained is a function of the lengths of the major and minor axes:

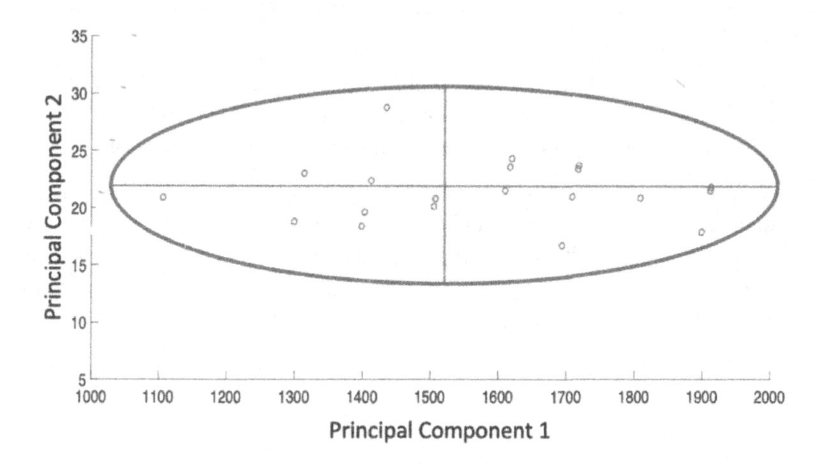

FIGURE 16.2: Pulse time and capacitance expressed as rotated data with the line of best fit as the horizontal axis. Notice that there is no relationship between PC2 and PC1: The scatter of PC2 about PC1 is random

We call these rotated axes, "Principal Components." **Principal Components Analysis** is a method for identifying orthogonal dimensions that underlie the data and that can define higher-order variables that constitute the major, semi-minor, and minor axes of an ellipsoid (a hyper-ellipsoid in multiple dimensions). The value of principal components is that, since these re-expressed dimensions are constrained to be orthogonal to each other, they have no multicollinearity. We can perform regression on them as if they were real measured variables. Additionally, because each successive principal component represents a minor axis to a hyper-ellipsoid, they descend in the amount of variability in the data each principal component is able to explain. A three-dimensional relationship, for example, will look like a slightly squashed football, with the major axis taking up most of the variability, and two minor axes of descending length taking up successively less.

In other words, we can re-express a matrix of data as a set of orthogonal components that have no collinearity. *AND* we can reduce the number of principal components—the number of variables in the space–to the few that add significant variation, eliminating those that add only negligible variation. This simplifies and clarifies many analyses of complex data. I want you to keep this geometric perspective in mind as we move through the mathematics that makes it possible.

Principle Components Analysis transforms a matrix of data into a set of linearly independent variables called principal components. As linearly independent, the set of principal components are orthogonal to each other and can be interpreted as underlying variables that are linear combinations of the original, measured variables.

The eigenvalues of these linear combinations constitute the amount of variation each principal component accounts for in the overall variability of the data. As such, they are analogous to partial R^2 in Least Squares Regression.

PCA, specifically, is Singular Value Decomposition of the covariance matrix of a data set. It transforms the data into a set of eigenvectors (the principal components), that describe the direction along which the new orthogonal dimensions lie, and their eigenvalues (the variance accounted for by each principal component).

One of the key uses of PCA is data reduction. The variables in a multivariate data set can be thought of as axes defining the set of coordinates made up by the data. By analyzing eigenvalues, components that account for significant variability can be retained in the model, while components that account for negligible variability can be eliminated, thus reducing the dimensionality of the transformed data.

16.3 PCA as Matrix Factorization

The Singular Value Theorem is one of the great theorems in Linear Algebra. This theorem proves that any matrix, X, can be factored into a product of three matrices:

1. U, a set of the orthogonal eigenvectors of XX^T ;
2. V^T, a set of the orthogonal eigenvectors of X^TX ; and
3. Σ, a diagonal matrix containing the eigenvalues of both X^TX and XX^T .

The implications of this theorem are vast:

1. If we have data that has a lot of collinearity, we can re-express it as a set of *orthogonal* (read *independent*) eigenvectors, which tell us the primary directions along which our data are intrinsically organized;
2. These eigenvectors can be thought of as axes that rotate our data into an orthogonal space; and
3. The set of eigenvalues tell us the *variability* of the data along those axes. Eigenvalues of lesser magnitude are associated with eigenvectors that explain less of the variation in the data.

As we have seen in using Linear Algebra for regression, if we have any $n \times m$ matrix, X, $X^T X$ is a square matrix. Moreover, if you can express a $n \times m$ data matrix as the product of a $n \times 1$ vector of unique eigenvalues and an $n \times m$ matrix of eigenvectors, the matrix of eigenvectors multiplied by itself is the Identity matrix $\left(X^T X = I\right)$. In other words, *if the square of the eigenvectors of a matrix form the Identity matrix, then the eigenvectors are mutually orthogonal and therefore form an orthonormal basis in m-dimensional space*. Because of this we can express the Singular Value Theorem as:

$$X_{n \times m} = U_{n \times n} \Sigma_{n \times m} V^T_{m \times m}$$

where

$$U^T U = I_{n \times n}$$

$$V^T V = I_{m \times m}$$

Doing a little algebra, if X is a square *symmetric* matrix, then U and V will be equal to each other.

Σ is a diagonal matrix of the eigenvalues (called the singular values) of X. As a reminder, a diagonal matrix has values in the diagonal, but contains zeros everywhere else. What's more, these diagonal entries of Σ are arranged in descending order of magnitude. In the examples we will use, U and V will be real matrices, but there may be instances in electrical systems particularly, where you might encounter complex matrices. Then U and V will be complex, but Σ will always be real. Lastly, the eigenvectors of $X^T X$ form the columns of V , and the eigenvectors of XX^T form the columns of U .

In somewhat plain English we can state the **Singular Value Theorem** in this manner:

The Singular Value Theorem

For any matrix X, we can rewrite it as the product of three matrices: $X = U\Sigma V^T$ where U and V have orthonormal columns and Σ is a diagonal matrix, the entries of which are known as singular values:

$$X_{n\times m} = U_{n\times n}\Sigma_{n\times m}V^T_{m\times m}$$

where

$$U^T U = I_{n\times n}$$

$$V^T V = I_{m\times m}$$

This theorem can be visualized, just like our example above, as a transformation of the vector space by matrix X. First, we rotate the space by V^T, then we scale the dimensions by the eigenvalues in Σ, and lastly, we rotate the scaled vectors by U to generate a new space wherein the vectors of our original matrix have been transformed into a set of linear combinations, *all of which are orthogonal to each other,* **and** *all of which are ordered in descending magnitude!*

Principle Components Analysis is an application of Singular Value Decomposition where the non-square data matrix is first reduced to a symmetric, square, matrix of covariances, and then U, V, and Σ are factored from the covariance matrix.

"Why might this be useful?" I hear you asking. Well, Sparky, remember when I was lamenting the effect of collinear predictors in our discussion of Multiple Regression? If you recall, when we have significant collinearity of predictors, the angle between predictors is not 90 degrees, therefore the β-weights in the regression equation will be distorted and some may appear significant when they are not, and some may be masked when they are significant predictors. If I could re-express the matrix of predictors, X in a way that eliminates the effect of collinear terms, and in fact create a new space within which those predictors *can* interact, but the linear combination of their interactions are orthogonal, I might be able to account for these misleading relationships among variables in my analysis.

Let's see how this works in a classic applicational context: Analysis of vibrations.

16.3.1 Extended Example: Acoustics

The field of acoustics is the study of vibrations. I know, *acoustics* is most often applied to the science of sound, but it actually deals with any kind of vibration (i.e., mechanical radiation) through almost any medium: gas, liquid, or solid. Mechanical and aerospace engineers are justifiably concerned with vibrations stemming from the engines they design, the structures they build, and the environments in which those products operate. So being able to model the vibrations in an airplane wing, or an engine, or a pump is a valuable tool for designers of such systems.

Gareth Forbes, from Curtin Australia (2012) put a set of 4 accelerometers on a centrifugal pump that is used to cool water (see Figure 16.3). The accelerometers were placed at 1) the pump drive end (vertical alignment), 2) the inlet flange (vertical alignment), 3) the outlet flange (horizontal alignment), and 4) the volute casing (horizontal alignment) (see Figure 16.3). The pump was run at 3,000 rpm, and the measurements were taken for 30 seconds at 25,600 Hz.

FIGURE 16.3: Accelerometer placement on the centrifugal pump. Acknowledgement is made for the measurements used in this work provided through, http://data-acoustics.comDatabase

The question we would like to ask, is: "What are the magnitudes and directions of the vibrations measured across the pump as a whole?" It is unlikely that vibrations are perfectly in line with the accelerometers' axes, so a linear combination of the readings will provide the precise directions, relative to the accelerometers' placement and orientation. Such information can help us determine if a shaft or bearing is not true and round, or if there is build-up of scale or other contaminants.

Forbes' dataset has 4 columns, one for each accelerometer placement, and 768,000 rows, the vibration measurements taken in the 30 second sampling period. Needless to say, I will not copy the dataset here! But, I have an Excel file labeled, Vibrations_PCA_dataset.xslx, if you would like to enter the data and follow along.

The first thing we need to do is reduce the dataset to the covariance matrix describing the interrelationships among the four accelerometer readings. If you remember, the covariance between any two variables, x and y is modeled as the average area each point in the bivariate distributions makes when it is compared with the overall expected value, $(\overline{x}, \overline{y},)$:

$$s_{xy} = \frac{\sum_{i=1}^{n} (x_i - \overline{x}_i)(y_i - \overline{y}_i)}{n - 1}$$

If, instead of comparing two vectors, we compare a set of k vectors in a matrix of variables, the covariance is modeled as:

$$S_{xx} = \left(X_1^T X_1 - 1\right)^{-1} \left(X - \overline{X}\right)^T \left(X - \overline{X}\right)$$

The traditional interpretation of a covariance is that it is a distance, the smaller s_{xy} is, the closer the actual point is to $(\overline{x}, \overline{y})$. In other words, the larger the average distance between (x_i, y_i) and $(\overline{x}, \overline{y})$, the more x and y hold variation in common. In Principal Components Analysis U (or V) represents the orthogonal directions (i.e., the axes) along which the majority of variance in the matrix *cov(X)* is aligned. The diagonals of Σ are the magnitudes of these axes, representing the amount of variation accounted for by each axis.

Put simply, *singular value decomposition on a covariance matrix provides the primary dimensions along which a matrix of variables vary, and the amount of variation accounted for by each dimension.*

The first issue we face is making the original 768,000 x 4 matrix into a square matrix of covariances. The covariance, as we learned earlier is just the average distance between each (x, y) pair between two variables, and $(\overline{x}, \overline{y})$.

$$S_{xx} = \left(X_1^T X_1 - 1\right)^{-1} \left(X - \overline{X}\right)^T \left(X - \overline{X}\right)$$

Applying this identity to the vibration data in Forbes' data, we get the following (see Table 16.1) square matrix of covariances in Table 16.1:

	Pump Drive End	Inlet Flange	Outlet Flange	Volute Casing
Pump Drive End	1.032	−0.034	−0.030	0.005
Inlet Flange	−0.034	2.121	−0.088	0.092
Outlet Flange	−0.030	−0.088	3.848	0.259
Volute Casing	0.005	0.092	0.259	1.358

TABLE 16.1: Covariance matrix of the location of the 4 vibration sensors in Forbes' (2012)

With this square matrix, we can now do the matrix factorization to find: 1) the diagonal matrix of eigenvalues, Σ, that estimate the variability that each principle component accounts for in the rotated space, and 2) the eigenvectors, U, V^T, that mark the directions—the principle axes along which the new set of orthogonal components lie.

The principal components of X are shown in the columns of U The structure of this matrix is shown in Table 16.2.

	PC1	PC2	PC3	PC4
Pump Drive	−0.0096	−0.0304	0.0424	0.9986
Inlet Flange	−0.0442	0.9908	−0.1230	0.0350
Outlet Flange	0.9939	0.0308	−0.1047	0.0149
Volute Casing	0.1004	0.1282	0.9860	−0.0370

TABLE 16.2: U, the matrix of principle component coefficients for each variable in the matrix of predictors

It also returns *sigma*, the values of the diagonal matrix, Σ (Table 16.3).

	Eigenvalue
PC1	3.88
PC2	2.13
PC3	1.32
PC4	1.03

TABLE 16.3: Σ, the vector of eigenvalues for each principle component

The eigenvalues are the sum of the squared deviations from the means of each Principle Component (see Table 16.3). Just like Multiple Regression, we can calculate the *SS* for each PC, and together they add up to the total variance in the data. The variance explained by each is the proportion that their *SS* is to the SS_{Total}.

$$VarianceExplained = \frac{SS_{PC_k}}{SS_{total}}$$

So the variance explained for PC1 is

$$VarianceExplained = \frac{3.88}{(3.88 + 2.13 + 1.32 + 1.03)}$$

$$= 46.40\%$$

	Variance Explained
PC1	46.40
PC2	25.49
PC3	15.78
PC4	12.32

TABLE 16.4: Variance explained by each principle component

Notice in Table 16.4 how the variance explained (see Table 16.4) gets less for each successive component. This is very helpful, as you can see that by the time we get to PC4, we are only explaining an additional 12 % of the variation. If this were negligible, I would consider reducing the dimensionality of the solution down to 3, or even 2 principal components. When there are dozens of variables, this ability to reduce the dimensionality of the space can both speed up processing, and help you focus on the variables that really matter. But right now, I don't think 12% of the variation is negligible. We will learn how to determine this a bit later in the chapter.

Lastly, we can use U and Σ to compute the value of each datum in the sample, calculated across the 4 principal components (see Table 16.6). The Principal Component Score is just the transformed value of each measurement in the matrix of predictors. For example, measurements at one time-point in the vibrations data have values shown in Table 16.5

Pump Drive End	**Inlet Flange**	**Outlet Flange**	**Volute Casing**
−0.7274	0.8125	2.0745	0.0668

TABLE 16.5: Vibration data for one point in time out of 780,000 measurements in Forbes (2012)

Using the identity:

$$X_{n \times m} = U_{n \times n} \Sigma_{n \times m} V^T_{m \times m}$$

We can compute all 780,000 values along the 4 principal components (see Table 16.6). The vibration data now is rotated and dilated to the following coordinates listed in Table 16.6:

PC1	**PC2**	**PC3**	**PC4**
0.6944	0.8165	−0.1728	0.6762

TABLE 16.6: Principal component scores across each of the principle components for the data in Forbes (2012)

There were 760,000 measured sets of vibrations, so these values will constitute a 760,000 x 4 matrix of these vectors across the 4 principal components.

Let's see how to interpret all this, hey? The first thing I want to look at is U the matrix of coefficients along each of the 4 rotated Principal Components. Here it is in Table 16.7..

	PC1	**PC2**	**PC3**	**PC4**
Pump Drive	−0.0096	−0.0304	0.0424	**0.9986**
Inlet Flange	−0.0442	**0.9908**	−0.1230	0.0350
Outlet Flange	**0.9939**	0.0308	−0.1047	0.0149
Volute Casing	0.1004	0.1282	**0.9860**	−0.0370

TABLE 16.7: U, the matrix of coefficients defining the eigenvectors for each of the 4 PCs

These Principal Components can be thought of as separate regression equations, each orthogonal to each other, and each pointing along the major and minor axes of a hyperellipse: Arranged in decreasing amount of variation in the data explained. One of the things we can see pretty immediately is that the PC with the greatest variation is oriented strongly towards the outlet flange. How do we know this? Well the coefficient for that location is the largest coefficient in PC1. Like β_1 , the partial slope of the regression equation, the PC coefficient of Outlet Flange is the one that contributes most variation to this variation, and Pump Drive contributes least.

Going from PC1 to successive PCs, we can see that each appears to be aligned with one of the accelerometer placements: Inlet Flange with PC2, Volute Casing with PC3, and Pump Drive with PC4. The coefficients of each of the other accelerometer locations within each PC is very enough to be negligible for most.

What does this tell us? Because, for example, in PC1, Outlet Flange has a huge coefficient relative to the other locations, we can conclude that, aside from some minor deviation (namely from Volute Casing), the direction of the vibrations on this part of the motor is aligned primarily along the horizontal axis of the accelerometer mounted there. Pump Drives, coefficient is near zero, as is Inlet Flange. Only Volute Casing contributes significantly to the deviation of the direction along the Flange.

Similarly, for PC2, the vibrations are oriented primarily along the vertical axis of the Inlet Flange, with some contribution of Volute Casing altering the direction of the vibrations slightly. PC3 has small contributions from both Inlet Flange and Outlet Flange, in the negative direction, but it is aligned primarily along the horizontal axis of the Volute Casing. PC4 is almost entirely aligned along the Pump drive end of the centrifugal pump.

Altogether, this tells me that the pump is vibrating pretty much as expected, along the primary inlet, outlet, near the motor, and along the volute casing where the impeller resides—nothing unexpected there! But now let's take a look at the eigenvalues that let us know how much variation each PC explains in the data (Table 16.8):

	Eigenvalue	Variance Explained
PC1	3.88	46.40
PC2	2.13	25.49
PC3	1.32	15.78
PC4	1.03	12.32

TABLE 16.8: Eigenvalues found in the diagonal of Σ, and the amount of variation each accounts for in the total variation across all 4 independent variables

These values, collapsed into a single table so that the relationship between eigenvalues and the variance each explains can be seen more easily, tell us that about 72% of the variation in the data is explained solely by PC1 and PC2, with only about 28% explained by PC3 and PC4. In other words, about $\frac{3}{4}$ of the variation in the direction of the vibrations is in the direction of PC1 (primarily due to the Outlet Flange), and PC2 (primarily due to the Inlet Flange). If we were concerned about vibrations in the pump, these are the places to focus our investigation more so than PC3 and PC4.

This is how PCA helps us reduce dimensionality: By focusing on the PCs that contribute the largest variation to the overall cloud of data in the column space of X (PC1 and PC2), we can focus on them, and less so (or even ignore) PCs that contribute significantly less variation (PC3 and PC4). In other words, Principal Components Analysis enables us to extract relevant information in the space of the predictor variables, encode it efficiently and compare the reduced, efficient encoding across the phenomena from which it was extracted (Turk & Pentland, 1991).

PCA is used extensively in the study of vibrations in ways quite similar to the previous example, vibrations in helicopter transmissions was studied by Tumer & Huff (2002). They put a triaxial accelerometer on the transmission of helicopters and analyzed the vibrations across the three dimensions measured by the instruments. Because X, Y, and Z directions are chosen arbitrarily, and because the vibration does not correspond perfectly to any specific direction, the accelerometer data was not perfectly orthogonal across the three axes—the vibrations were linear combinations across the three. The authors expressed the direction and magnitude of the vibrations across the three dimensions as vectors—as linear combinations of the accelerometer axes. Anytime you need to find the true directions of vibrations across a space, and are a bit clever figuring out accelerometer placement and orientation, PCA is a good analytic method.

16.4 Principal Components Regression

If we took the Principal Components Scores we compute (Table 16.5 shows one row of PC scores in the matrix), we have a fully transformed matrix where each vibration reading is now expressed in the space of the PCs. Because of this, these values can be used as a predictor matrix upon which we can regress a dependent variable such as outlet water pressure, or similarly important measure of the pump's performance. I have added a column of output pressures, in Mpa, generated from Forbes' pump data. Let's see how the fully orthogonal set of Principal Components predicts this output pressure. If we get a good model, we may be able to detect which of the components contributes to pressure fluctuations and then decide if we need to adjust or tune the pump to improve its performance.

The procedure for this just involves using the rescaled PCA scores in lieu of our original data, and regressing our outcome vector of outlet pressures on those scores:

1. Perform principal components analysis on the raw data;
2. Select a subset of the components based on the amount of variation they account for, if data reduction is a goal;
3. Regress the outcome variable on the (reduced) matrix of principal components using Least Squares multiple regression;
4. Interpret the results.

Using a regression function in my statistics package, I obtain the following results:

$$\beta = \begin{bmatrix} 30.046 \\ 1.922 \\ 3.251 \\ 1.172 \\ 5.902 \end{bmatrix}$$

$R^2 = 0.67\,3$, $F = 395,264$, $p = 0$, $MS_{resid} = 36.21$.

Not bad! Our rescaled variables, now called principal components, explain about 2/3 of the variation in our outlet pressure data, the model's partial slopes are significantly different from zero, and there is relatively low residual error. Now to the prediction equation:

$$\hat{y}_i = 30.046 + 1.922\,(PC1_i) + 3.251\,(PC2_i) + 1.172\,(PC3_i) + 5.902\,(PC4_i)$$

We would expect the typical outlet pressure of the pump, with zero vibrations along any of the PC directions to be about 30 MPa, on average. PC4, which is aligned primarily along the Pump Drive accelerometer has the greatest partial slope, and so contributes the most to the variation in outlet pressure. PC2 has the next biggest effect, indicating that vibrations along the Inlet Flange, contribute quite a bit to the variation in outlet pressure. PC1 and PC3 show that the Outlet Flange and Volute Casing contribute less to the overall variation in outlet pressure, their β weights are much lower than PC4 and PC2 respectively. All of the β weights are in the positive direction indicating that greater vibrations along those axes contribute to greater outlet pressure.

So, if I were interested in stabilizing the outlet pressure of this centrifugal pump, I would focus my attention on PC4 and PC2, adjusting the pump drive (makes sense!) and the inlet flange. These are the areas that appear to impact the pressure the most, and I interpret these as indicating fluctuation in the amount of water entering the pump contributes to the vibrations in these areas, and this contributes to the overall performance of the pump, more so than in the other two areas: Outlet Flange and Volute Casing.

Advantages and Disadvantages of PCA Regression. PCA as a method of pre-analyzing the data prior to regression has the advantage of insuring no collinearity of the independent variables. But it also has the disadvantage that the new Principal Components may be more difficult to interpret than the original variables. As an example take a look at PC3 in our matrix from the pump vibrations example (see Table 16.9):

	PC1	**PC2**	**PC3**	**PC4**
Pump Drive	−0.0096	−0.0304	0.0424	**0.9986**
Inlet Flange	−0.0442	**0.9908**	−0.1230	0.0350
Outlet Flange	**0.9939**	0.0308	−0.1047	0.0149
Volute Casing	0.1004	0.1282	**0.9860**	−0.0370

TABLE 16.9: PC3 is defined primarily by the vibrations at volute casing

We can easily see that Volute Casing makes a big contribution to the direction of this component, but Inlet Flange and Outlet Flange vibrations also contribute to the variation, albeit with modest rates of change in negative directions on their respective axes. What does this mean? I can only interpret this as a slight rotation of the Volute Casing vibrations away from the Inlet Flange (vertically), and slightly away from the Outlet Flange (horizontally). So this new, rotated axis is a combination of effects of the vibrations across all three of these areas. Luckily Volute Casing has such a high coefficient that it almost completely determines the axis, but in many cases two or more variables will contribute roughly equal effects on the PC. What then?

This is the primary drawback of PCA—interpretability of the new "variables" —the Principal Components. When the directions really are *directions* (like the accelerometer data), the interpretation is purely spatial—the new axes are just rotations in space of the pump vibrations. When the axes are not physical directions, but rotations of variables like Work, Time, and Volume, the linear combinations of these variables can be very difficult to interpret. In other words, we often sacrifice interpretability to establish linear independence of PCs. At other times, we sacrifice our assumption of linear independence in regression for the sake of interpretability.

16.5 Dimension Reduction: Feature Elimination

Reducing the number of variables in a model is super helpful, if you are like me and have difficulty visualizing in 3 dimensions, let alone 4 or 5 or 20 even. PCA, as I have said before, is a common method for dimension reduction. Up to this point, we have only talked about eliminating dimensions based on rough guesses based on the magnitudes of eigenvalues in the Σ matrix. We really need to get a principled (see what I did there?) method for determining what components contribute *significant* variation versus trivial amounts of variation.

It helps to have lots of dimensions to do this, so I think we might as well try something really fun, like pattern recognition using *eigenfaces* (Turk & Pentland, 1991). What this does is take a picture (we will only use static, black and white photos here, but the technique applies to color movies just as well) and assign each of its pixels a grayscale value—a number. It then arranges all of these values into a single vector representing the picture. So, if you had a picture that is 640×480, then it has $640 \times 480 = 307,200$ cells that are then arranged as a row vector containing 307,200 variables. If you had photos of 1,000 faces, the data matrix would be 1000 rows x 307,200 columns with each row representing a single face. Obviously, it would be helpful to reduce the number of variables from over 300,000 to a couple dozen or so: The computational time required to identify a face from the database would be reduced by several orders of magnitude. That is where PCA comes in.

A database of known faces (or photos of flowers, or text entries, or audio patterns) can be described as a row vector of values for each instance. By using PCA, the number of columns needed to adequately describe the database can be reduced to a manageable number to speed up recognition in future instances. For our facial recognition example, a set of known criminal mugshots can be reduced to a searchable database that can identify potential perpetrators in real time by only attending to a small number of dimensions.

These photos, when entered into a database, have many rows and many columns. A 1080p high definition image, for example, has 1,080 rows and 1,920 columns corresponding to the pixels of the image, each with a particular combination of red, green, or blue color values. So a 1080p image utilizes 1,080 x 1,1920 x 3 = 6,220,800 pieces of information to describe it! In our example below, the photo contains 770 rows x 576 columns x 3 different colors for a total of "only" 1.3 million pieces of information. A million pieces of information just to depict a face? What is the minimum number of dimensions we need to be able to recognize a face in a photo? If we knew this number, we could reduce the huge size of our photo database, and greatly increase the efficiency at which we could search the database and recognize faces in real time.

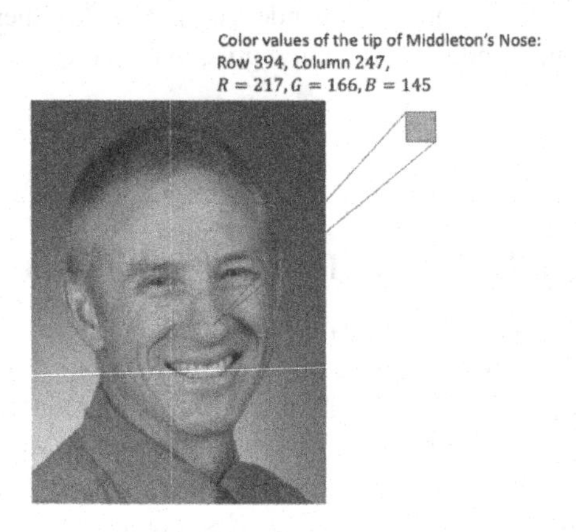

FIGURE 16.4: Definitely a "Person of interest"

For example, here is a handsome fellow in Figure 16.4. His mugshot has about a million pieces of information that specify the RGB color values for each little pixel in the frame. The tip of his nose is located at (394, 247) in the flat representation. It is represented in the vectors of RGB with values (217, 166, 145) respectively (see Figure 16.4). So in the 5 dimensions of the information space, the tip of his nose has a location (394, 247, 217, 166, 145). Only 770 x 576 more vectors to go and the photo is fully described!

But we really don't need a million pieces of information to recognize his mugshot. We could reduce the RGB values to grayscale and that alone would reduce the data from 1.3 million pieces of information to 443,000 (Figure 16.5).

Even this is too much information if we want to sort through the photos of a million bad guys. If I use PCA, I can reduce the column space of the original photo from 576 to only 10 dimensions (see Figure 16.6) and be able to recognize this character, even though he doesn't look quite as handsome anysmore (see Figure 16.6).

The following Matlab code shows how this was accomplished.

First, I read in the photo:

```
% Request filename of photo
        prompt = 'Type in filename of photo. Use single quotes';
        origphoto = input(prompt);
% Read in photo
        data=imread(origphoto);
```

FIGURE 16.5: Greyscale image

FIGURE 16.6: The photo reduced to only 10 column dimensions. Our "Perp" is recognizable!

```
% display original photo
        imshow(data);
        title('Input Image');
```

Then, I flattened the picture from a 3d data array to a 2d grayscale rendering.

```
% convert original to grayscale
        graydata=rgb2gray(data);
% display grayscale image
        figure,imshow(graydata);
% convert grayscale image to floating point format
        datamat=double(graydata);
```

When I get a 2d image, I can now run Principal Components Analysis to transform the data into a set of 576 orthogonal eigenvectors, and a vector of 576 eigenvalues. With this information, I have to determine how many, dimensions i.e., how many eigenvectors–to retain. I do this by making a scatterplot of the eigenvalues, the sigma vector, as a function of their principal component number (see Figure 16.7).

Because the eigenvectors represent the amount of variation each principal component accounts for in the data, and because they are arranged from largest to smallest, the scatterplot will show asymptotic behavior. We call this scatterplot a *Scree test*, because the shape represents a mountain on the left, where pieces of *scree*, little bits of broken stone lay scattered off to the right.

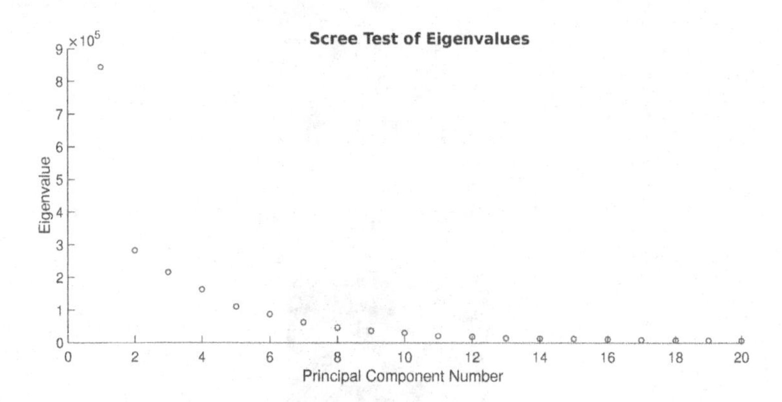

FIGURE 16.7: Scree test of the eigenvalues for each principal component. Variance explained gets negligible at or about 10 PCs

You can see in this graph that after we get past about 10 principal components, each successive eigenvalue becomes insignificantly different from each other. If we examine the variance explained by these first 10 principal components, we have accounted for 95% of all the variability in the data, the remaining 566 components account for only the remaining 5% (see Table 16.10). Heck, the first PC accounts for over 3/5 of the variability itself! But we need to get somewhere in the 90s to be able to tell one face from another—we are more alike, you and I, than not.

From examining the scree test, I tell the program to retain 10 principal components and eliminate the remainder. This is not a hard and fast rule. I will look at the results, and if they are acceptable, stop. If they are not acceptable, I will adjust the number of principal components higher for more resolution, or lower for less resolution until I hit a "sweet spot."

```
% Request number of PCs to retain
    prompt = 'What is the number of Principal Components you want
    to retain?';
    NumPCs = input(prompt);
% Truncate matrix to PCs
    NewU=U(1:length(U),1:NumPCs);
```

To produce the reduced image, I have to multiply the truncated matrix of PC scores and multiply them by their eigenvectors. This gives me a data matrix transformed back in the original units of my grayscale image.... and voila! A fuzzy, dark, but recognizable photo of yours truly represented by a fraction of the original 1.3

PC	Variance Explained
1	61.3007284
2	19.1291982
3	3.95066844
4	3.17848933
5	2.71006907
6	1.35941901
7	1.14807952
8	0.84854479
9	0.75474115
10	0.61030696
Sum	94.9902449

TABLE 16.10: Variance explained by each of the orthogonal dimensions (PCs) extracted in the PCA

million pieces of data. Only 7,700 grayscale values make up this photo! That is truly a reduction. PCA is now a standard machine learning algorithm for facial recognition specifically, and data reduction more broadly. I encourage you to choose your own photos (of yourself, your dog, flowers, etc.) and walk through this program, examining the process by which we can reduce the dimensionality of the photos while still retaining enough information to recognize their contents.

In particular, I think it is useful to really look at the scree test. You will probably have to play with the axes a bit to get an appropriate scale with which to view the "elbow" where the variation tapers off to insignificant levels. Try looking at the photos for several points around this elbow to determine the minimum number of PCs you need to retain to effectively identify the image.

16.5.1 The Scree Test

The scree test is one of the most effective means for selecting the principal components to retain and which to discard (Cattell, 1966). If you remember, because we first express our data as a matrix of covariances and then factor that matrix using singular value decomposition, the resulting eigenvalues are variance estimates. Like the sum of squares in a regression equation, they each add up because each is linearly independent of the other. In PCA this is guaranteed. Since the PCs are ordered by their eigenvalues, we can examine the PCs that account for the most variation, and then work down the list until we get PCs that only account for negligible variation and eliminate them from our prediction matrix, U. The scree test helps us do this visually.

FIGURE 16.8: Pett, M.A., Lackey, N. R., & Sullivan, H. (2003). Making Sense of Factor Analysis: The Use of Factor Analysis for Instrument Development in Health Care Research. Thousand Oaks, CA: SAGE Publications, Inc.

In the cartoon shown in Figure 16.8, we can see that the researchers are looking at the elbow in the scree test. The steepness of the line connecting the points is an indicator of the rate at which variation is accounted for in the column space of PCs. When the line is horizontal, successive PCs add no significant variation. When the line is steep, the PCs tend to add significant variation.

Combining this graph with the computed proportion of variation each PC adds, we can very quickly choose a value or two that would serve as the demarcation between significant variation and insignificant variation. This is done almost entirely qualitatively. There are no hard and fast rules, you just have to figure out at which point the variation added ceases to be worthwhile. In the case of our eigenface, this was about 10 PCs out of the 576 total.

Figure 16.9 shows how to interpret the scree test for our facial recognition example. I have eyeballed lines that fit three sets of eigenvalues: 1) the first set, where the slope is steepest represents the PCs that contribute the most variation to the model. I definitely want to keep those in the model for obvious reasons; 2) the second set, in the "elbow" of the plot represent PCs that are diminishing in the amount of variation they are contributing to the model. These are potential candidates for feature elimination; and 3) the third set, nearly parallel to the asymptote, represent the PCs that are contributing negligible variation. These I definitely want to eliminate. The transition area between the elbow and the PCs with negligible variation is the spot where I want to look to see where I get just enough variation to recognize the face, but not any extra so that my database is the smallest effective size it can be.

For contexts that don't involve facial recognition, we want to retain the minimum number of PCs that adequately explain the data, i.e., that gives us enough information to be able to make a decision with fidelity. If there is some question regarding whether to retain 10 PCs or 11 (in our example), then retain the higher of the two solutions. If there is a question about whether or not a PC contributes adequate variation, it is likely that it is not adequate for making a decision with high assurance that you are not making an error.

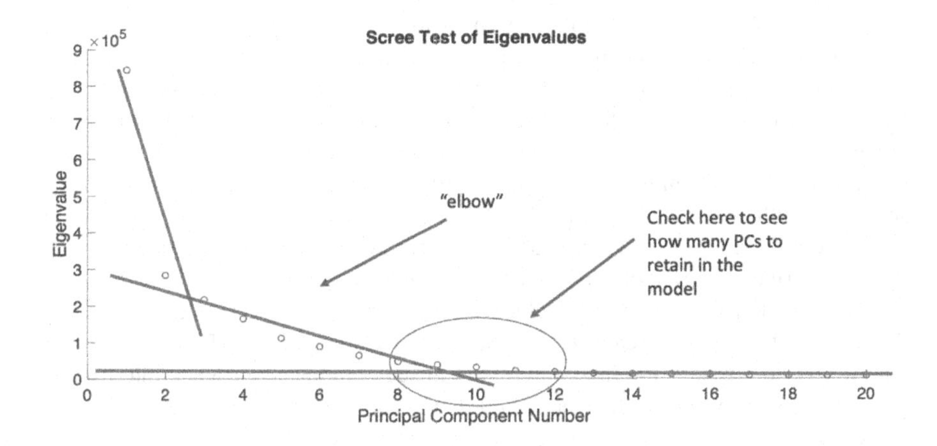

FIGURE 16.9: How to interpret a Scree test

In sum, it is important to note that PCA itself does not reduce the dimension of the data set. It only orders the data along the principal axes that define its maximum variance, while constraining the dimensions to be orthogonal to each other. YOU, the engineer, reduce the data by applying some rules of thumb for determining which dimensions contribute *significant* variation versus the others. But because this dimension reduction is done by using only the first few principal components as a basis set for the new space, the majority of the information value of the data can be retained, while the complexity and computational time is reduced significantly.

So now we know the process of Principal Components Analysis:

1. Factor the covariance matrix of a set of data using Singular Value Decomposition.
2. Interpret U and Σ, to understand the direction and magnitude of the variance components extracted from the data;
3. Potentially reduce the dimensionality of U, the components matrix, to retain those components that account for significant variance, eliminating those that add little variance;

    (a) Use feature extraction to identify the orthogonal dimensions along which the data are intrinsically aligned;

(b) Use feature elimination to reduce the dimensionality of the data; and then, *potentially*

4. Use the Principal Component Scores as predictors in Least Squares Regression to assess the performance of the phenomenon under study.

One of the problems many multivariate techniques have when assessing what predictors are important versus others is the issue of scale. If predictors are expressed in scales with different orders of magnitude, the predictors expressed in higher orders of magnitude will tend to overpower, or mask the effects of those expressed in lower orders of magnitude. This is true for Multiple Regression as well as PCA and other techniques that use Singular Value Decomposition. One way to get around this problem is to standardize all of the predictors prior to analysis. Standardization, if you recall, converts a continuous random variable of any scale to the standard normal distribution. Transforming the data in this manner places all predictors on the same scale so that the coefficients in the U matrix are all in the same units.

The disadvantages of standardization include having problems with interpretability of the Principal Components. It is hard enough to interpret a linear combination of several (if not hundreds) of variables without them being converted to units that don't retain the original scale. In general, it is my advice to NOT transform your data if you do not have to. Take differences in scale into account when interpreting your PCs, and only standardize the data if there is no other way to make sense of the individual contributions of your original variables to each PC.

16.6 Summary

That was so easy! PCA, and Singular Value Decomposition more generally, should be a tool that you keep in your toolbox for the rest of your career. It is a flexible, widely applicable technique that gets around the major problems of Multiple Regression, but it can be combined with Multiple Regression to make effective predictive models. In short, PCA is a means of dimension reduction that finds patterns in high dimensional data and expresses these patterns as orthogonal dimensions, ordered in the amount of variation each explains in the system. It can be used in machine learning for pattern matching, and image compression, in vibration analysis to express the direction and frequency of vibrations in 3d, and as a technique for reducing dimensionality of extremely complex data.

16.7 References

Cattell, R. B. (1966). The scree test for the number of factors. *Multivariate behavioral research*, *1*(2), 245-276.

https://data-acoustics.com/measurements/c_pump/

Pett, M.A., Lackey, N. R., & Sullivan, H. (2003). Making Sense of Factor Analysis: The Use of Factor Analysis for Instrument Development in Health Care Research. Thousand Oaks, CA: SAGE Publications, Inc.

Turk, M., & Pentland, A. (1991). Eigenfaces for recognition. *Journal of cognitive neuroscience*, *3*(1), 71-86.

Turner, I. Y., & Huff, E. M. (2002). Principal components analysis of triaxial vibration data from helicopter transmissions. In *Proceedings 56th Meeting of the Society for Machinery Failure Prevention Technology*.

Yang, J., Zhang, D., Frangi, A. F., & Yang, J. Y. (2004). Two-dimensional PCA: a new approach to appearance-based face representation and recognition. *IEEE transactions on pattern analysis and machine intelligence*, *26*(1), 131-137.

16.8 Study Problems for Chapter 16

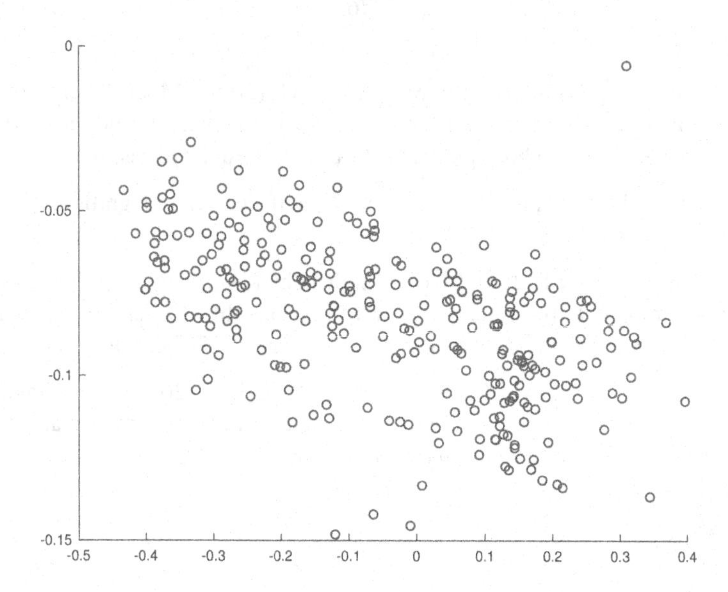

1. For this 2d scatterplot on (x,y), draw the two principal components using an eyeball estimate.

   Here is a reasonable Eyeball Estimate.

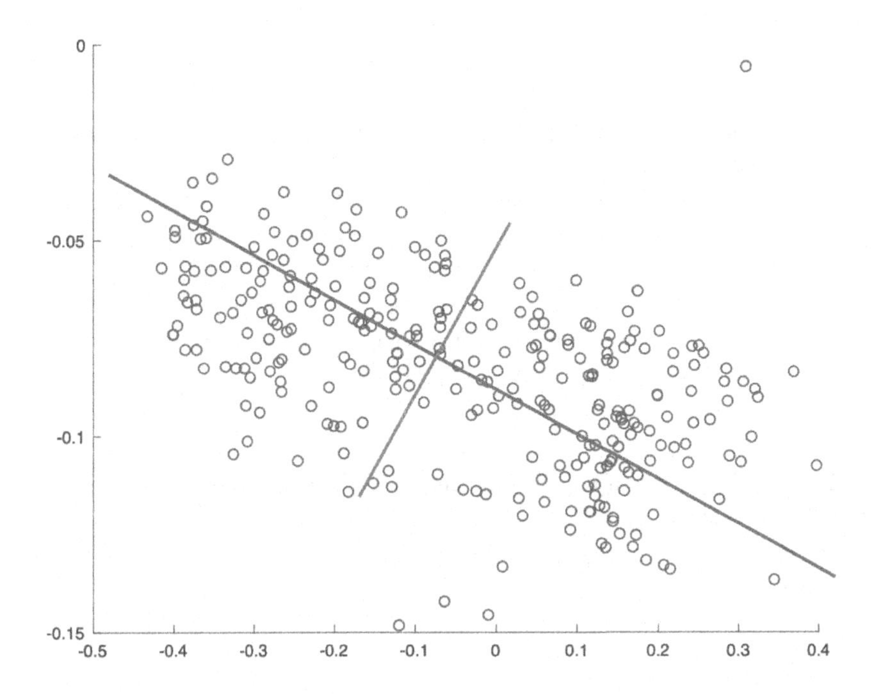

The relationship between the two variables looks pretty strong, with a clear difference in the width of the two axes. There is one outlier that may throw off the length of the short axis and pull the long axis a bit too high of where it should be. I might be tempted to eliminate this point...

2. **The principal components coefficients (eigenvectors) and eigenvalues for a 5-variable PCA analysis are presented below:**

PC1	PC2	PC3	PC4	PC5
−0.0512	0.0033	−0.0066	0.0472	0.0782
−0.0522	0.0047	−0.0030	0.0446	0.0889
−0.0529	0.0071	0.0011	0.0400	0.1018
−0.0533	0.0099	0.0039	0.0343	0.1115
−0.0531	0.0133	0.0068	0.0282	0.1197
λ = 844,480.00	λ = 283,174.21	λ = 216,855.57	λ = 164,540.65	λ = 111,906.89

   (a) **What percent of the variation is accounted for by each of the principal components?**
   The total variation accounted for is the sum of the λ. This sum is 1,620,957.32. So the percent variation accounted for by each PC is just its eigenvalue divided by the total.

PC1	PC2	PC3	PC4	PC5
0.52	0.17	0.13	0.10	0.07

(b) **Of the original variables, which contributes most to each PC? Which contributes the least?**

PC1, with 52% of the total variation explained, contributes the most, while PC5 contributes the least. Notice how the order of the eigenvalues corresponds to the order in which each principal component accounts for the variation in the model.

(c) **Determine which of the principal components above are necessary to explain at least 90% of the variability of the data.**

You need at least 4 PCs. This explains 93% of the variability.

3. **Explain the process of reducing dimensions for a high number of predictor variables, using PCA.**

Because the eigenvalues are a sum of squares, they each account for different amount of variation in the cloud of data. Because they are displayed in order, we can use the order of the variation explained to eliminate PCs with low eigenvalues those that don't contribute significant variation to the model. The Scree test is a good, easy, reliable method for doing this. If the eigenvalues are plotted against PC number, there tends to be an "elbow" where the variation tends to become asymptotically less following each successive PC. Examining the model right around this elbow is good practice for determining the final set of variables to keep in the analysis.

4. **Suppose you were conducting a facial recognition experiment. You run PCA on the matrix representing a photo and get the following scatterplot of the eigenvalues versus PC:**

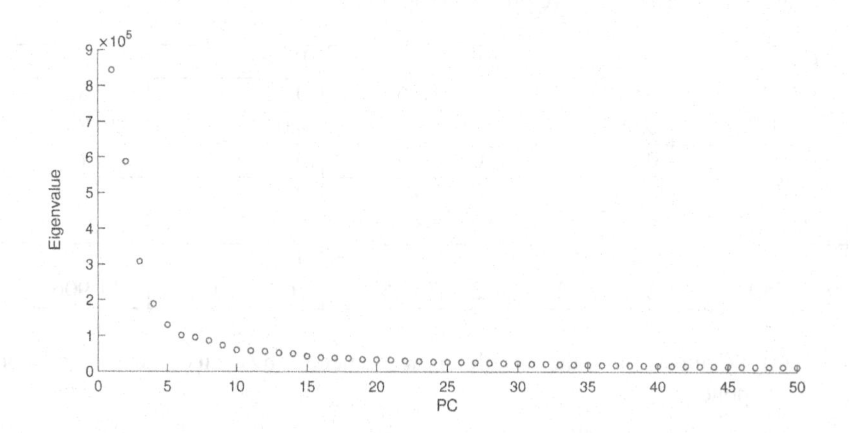

(a) About how many PCs would you select as the minimum required to recognize the face?

The elbow appears to hinge between 5 and 10 pcs. At 5, the amount of variability explained by each successive PC is very small, less than 5% of the total (eyeball estimate). So I would start with 5 PCs and work outward until the optimal number is found, constantly checking this against the ability of the program to produce a recognizable face.

(b) Describe the logic of how you would use the scree test to determine a first guess at the number of PCs to retain in the model.

See answer to a.

5. The following data was collected, measuring three predictors of automobile engine efficiency: Cylinder displacement, gross vehicular weight (converted to scale of 1 to 10), and rated number of passengers. It is clear that these variables are related to each other, so a standard regression analysis would find lots of collinearity. Use PCA to transform the matrix of predictors to orthogonal components:

Displacement (l)	**GVW (1 – 10)**	**Passengers**	**Fuel Efficiency Rating**
5.4	8	5	4
5.4	8	5	5
6.8	10	5	5
6.8	9	5	5
6.8	10	5	5
6.8	9	5	4
6.7	6	6	4
6.7	7	6	4
6.7	6	6	3
6.7	6	6	4

(a) What are the relative contributions of each variable to each PC extracted?

The coefficients matrix shows that, in PC1, the dimension that contributes the most variation to the model, GVW contributes the most in defining its direction. In PC2, Displacement contributes the most. In PC3, Fuel Efficiency contributes the most, and in PC4, Passengers contributes the most. But you can also see that Fuel efficiency contributes quite a bit to PC1, while Displacement contributes quite a bit to PC4. The 4 original variables are not orthogonal!

	PC1	PC2	PC3
Displacement.	0.0053	0.9468	−0.3218
GVW	0.9594	0.0856	0.2686
Passengers	−0.2820	0.3101	0.9079

	Eigenvalue	% Variation Explained
PC1	2.7602	87.98
PC2	0.3621	11.54
PC3	0.0151	0.48

(b) **Is there a PC that seems to contribute insignificant variation to the model compared to the others? If so, reduce the dimensionality of the model appropriately.**

PC1 contributes about 88% of the total variation! I might run with just PC1. But PC1 and PC2 contribute over 95% of the variation. PC3 contributes almost none. I will go with just PC1 and PC2.

(c) **Now run Least Squares Regression using the orthogonal components as predictors and Fuel Efficiency Rating as the outcome. How well does the reduced model predict fuel efficiency rating?**

	β
Intercept	4.300
PC1	0.307
PC2	−0.116

R^2	F	p	MS_{ε}
0.581	4.855	0.0476	0.245

This is ok. The two predictors account for about 60% of the variation in Fuel Efficiency Rating, and the relationship is significant. It looks like there are a lot more variables that might determine fuel efficiency other than these two dimensions.

(d) **Convert the predicted scores from the PCA regression back to the original units. How much error of prediction do you have?**

$\hat{y}$	Fuel Efficiency Rating
4.49	4
4.49	5
4.91	5
4.63	5
4.91	5
4.63	4
3.66	4
3.95	4
3.66	3
3.66	4

SSE	SST	R^2
1.718	4.100	0.58

The error is about 40% of the total variation in the data.

(e) **Run Least Squares Regression on the original data. Does PCA create a better model than the original, taking into account any violation of the assumptions of PCA?**

R^2	F	p	MS_{ε}
0.59	2.89	0.1248	0.2798

The model is about the same in terms of the variability it explains, but presumably because of the collinearity, the F-test returns a non-significant value. The Q-Q plot of the original data shows a big deviation from the y=x line. So, the power of the original data is not as great as that when the PCs are used. Of course, we sacrifice some of the interpretability of the data by using PCs as predictors.

6. **The following data was taken from a study of the use of a robotic exoskeleton to assist in the mobility of older patients undergoing orthopedic surgery on their hips. Three predictors of maximum blood oxygen volume (VOX) were recorded. Use PCA to answer the following questions:**

Systolic Blood Pressure	**Age**	**Weight**	**VOX (%)**
128	46	167	91
132	52	173	90
137	54	188	87
143	59	184	83
149	61	188	79
153	67	194	84
154	64	196	82
159	65	207	76
162	73	211	81
166	72	217	80
168	74	220	75

(a) What are the relative contributions of each predictor to each PC extracted?

	PC1	**PC2**	**PC3**
Blood Pressure	0.5725	−0.4805	−0.6643
Age	0.3763	−0.5658	**0.7336**
Weight	**0.7284**	**0.6700**	0.1431

	Eigenvalue	**% Variation Explained**
PC1	556.94	97.96
PC2	9.50	1.67
PC3	2.12	0.37

(b) Is there a PC that seems to contribute insignificant variation to the model compared to the others? If so, reduce the dimensionality of the model appropriately.

Here, it looks like there are two PCs that are insignificant. PC1 explains 98% of the variation in the model by itself! It is nice to reduce the dimensions to one predictor. In this predictor, it looks like all three variables contribute to the variation, with Weight being the biggest contributor, followed by Blood Pressure, and lastly, Age.

(c) Now run Least Squares Regression using the orthogonal components as predictors. How well does the reduced model predict blood oxygen volume?

	β
Intercept	82.55
PC1	−0.19

R^2	F	p	MS_{ε}
0.75	26.75	0.0006	7.57

This looks pretty good. PC1 accounts for 75% of the total variation in VOX, and it is significantly different than one would expect due to random chance.

(d) **Run Least Squares Regression on the original data. Does PCA create a better model than the original, taking into account any violation of the assumptions of** PCA?

	β
Intercept	155.909
Blood Pressure	−0.899
Age	0.657
Weight	0.106

R^2	F	p	MS_{ε}
0.85	13.05	0.003	5.867

The analysis using only one PC compares well with the regression analysis using the three original independent variables. Incidentally, there is tremendous amount of non-linearity in this data (due to all three predictors), and a huge amount of collinearity among the predictors. The covariance matrix of the original data looks like this:

	Blood Pressure	Age	Weight
Blood Pressure	185.69	121.55	229
Age	121.55	83.07	149.30
Weight	229	149.30	299.80

Compare these similarity measures to the ones obtained using PCs as predictors

	PC1	PC2	PC3
PC1	556.94	0	0
PC2	0	9.49948692259381	0
PC3	0	0	2.12477044578811

That is what orthogonality means! There is no relationship between the three predictors! But again, we do sacrifice some interpretability when we use them.

7. Back in Chapter 11, we did an extended example of multiple linear regression to predict the compressive strength of concrete from the relative amounts of a number of important ingredients in the mixture. The data for that analysis was a reduced set of variables. The full set is located in Concrete_data_full.xslx.

   (a) Run the regression again, but this time use PCA to extract orthogonal dimensions to get around the problem of collinearity;

	PC1	PC2	PC3	PC4	PC5	PC6	PC7	PC8
Cement	0.905	0.032	−0.154	−0.008	−0.151	0.306	0.194	0.007
Blast Furnace Slag	−0.262	0.786	−0.072	−0.199	−0.106	0.453	0.226	0.009
Fly Ash	−0.238	−0.303	0.051	0.687	−0.177	0.512	0.286	−0.005
Water	0.005	0.076	0.041	0.075	0.098	−0.482	0.824	0.253
Super-plasticizer	−0.001	−0.005	−0.024	0.02	−0.022	0.104	−0.233	0.965
Coarse Aggregate	−0.009	−0.274	0.76	−0.48	−0.076	0.27	0.185	0.041
Fine Aggregate	−0.21	−0.45	−0.61	−0.485	0.132	0.257	0.244	0.026
Age	0.098	0.069	0.118	0.126	0.948	0.234	0	−0.002

The coefficients matrix shows that PC1 is influenced primarily by the amount of Cement. PC2 is influenced primarily by Blast Furnace Slag with Fine Aggregate and Fly Ash contributing less. PC3 is influenced primarily by Coarse Aggregate, followed by Fine Aggregate. PC4 is influenced primarily by Fly Ash, with secondary impact of The two Aggregate sizes. PC5 is primarily influenced by Age. PC6 is about equally influenced by Fly Ash, Water, and Blast Furnace Slag, PC7 is primarily influenced by Water, and PC8 is primarily influenced by Superplasticizer.

	Eigenvalue	Percent of variation accounted for
PC1	12840.90	32.58
PC2	9809.96	24.89
PC3	7284.22	18.48
PC4	4243.65	10.77
PC5	3979.16	10.09
PC6	1176.42	2.98
PC7	71.66	0.18
PC8	11.34	0.029

The Eigenvalues drop of pretty precipitously towards the last two dimensions. I will check with a Scree Test:

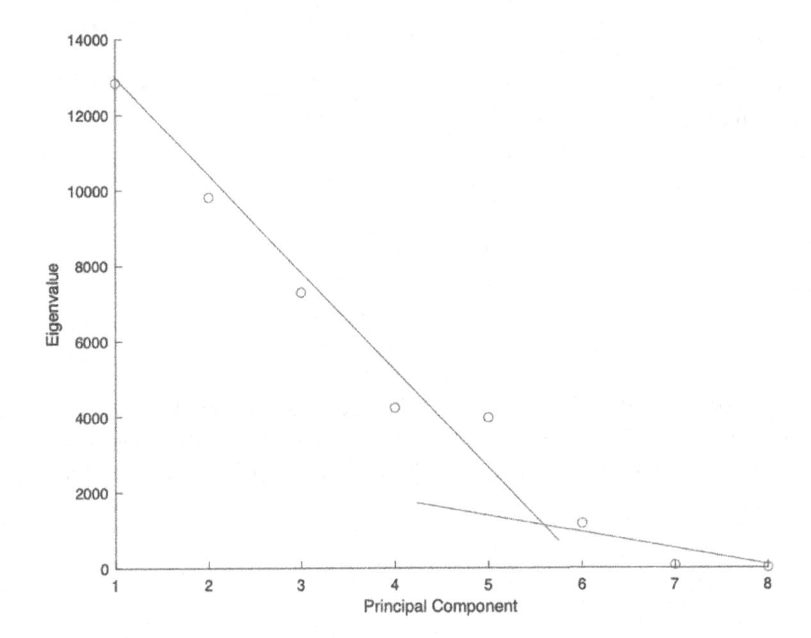

This looks like we should keep 5 or 6 PCs. I will keep 5, because PC6 only accounts for 3 percent of the total variation. I can now run the regression analysis with the reduced dimensionality.

1. Here are the results of the linear regression using PCs as predictors:

	β
Intercept	35.82
PC1	0.07
PC2	0.04
PC3	−0.02
PC4	0.03
PC5	0.04

R^2	F	p	MS_{ε}
0.31	90.22	1.02×10^{-78}	194.68

This is acceptable, though not as tight as the reduced model in Chapter 11 that only contained Water content, Super Plasctizer and Age. But, we don't have to deal with all of the transformations we had to do to linearize the relationship. Right now, Water Content, the most publicized contributor to concrete compressive strength, does not seem to have much impact.

2. **How interpretable are the PCs? Do they tend to have contributions from one of the mixture variables over the other?**

   Since there are one or two independent variables that contribute most to each PC, they are fairly interpretable. See response to a above.

3. **Is it reasonable to reduce the dimensionality?**

   I think in this case, it may help with figuring out which mixtures seem to make stronger concrete, but I would only reduce by 2 dimensions, down to 5, given the proportion of variation each PC accounts for.

4. **How good does the PCA regression compare to the original least squares regression in Chapter 11? Look at all the evidence you have available including R^2, F, and the interpretability of findings in light of the collinearity between the original variables.**

I don't think, in this instance, that keeping more variables is a good thing. The lack of interpretability makes this more difficult to understand than limiting the analysis like we did in Chapter 11 to 3 variables, transformed though they may be.

Index

Printed in the United States
by Baker & Taylor Publisher Services